CRANFIELD UNIVERSITY
LIBRARY

MOLECULAR NONLINEAR OPTICS

MATERIALS, PHYSICS, AND DEVICES

QUANTUM ELECTRONICS—PRINCIPLES AND APPLICATIONS

EDITED BY

PAUL F. LIAO

Bell Communications Research, Inc.
Red Bank, New Jersey

PAUL L. KELLEY

Lincoln Laboratory
Massachusetts Institute of Technology
Lexington, Massachusetts

A complete list of titles in this series appears at the end of this volume.

MOLECULAR NONLINEAR OPTICS

MATERIALS, PHYSICS, AND DEVICES

Joseph Zyss

Centre National d'Etudes des Telecommunications
Laboratoire de Bagneux, Bagneux, France

ACADEMIC PRESS, INC.
Harcourt Brace & Company, Publishers
Boston San Diego New York
London Sydney Tokyo Toronto

ACADEMIC PRESS, INC.
1250 Sixth Avenue, San Diego, CA 92101-4311

United Kingdom Edition published by
ACADEMIC PRESS LIMITED
24-28 Oval Road, London NW1 7DX

Library of Congress Cataloging-in-Publication Data:

Zyss, J.
Molecular nonlinear optics : materials, physics, and devices / Joseph Zyss.
p. cm. — (Quantum electronics—principles and applications)
Includes bibliographical references and index.
ISBN 0-12-784450-3
1. Nonlinear optics. 2. Quantum electronics. 3. Optoelectronics—Materials. 4. Polymers. I. Title. II. Series.
QC446.2.79 1993
621.36′9-dc20 92-42976 CIP

Printed in the United States of America

93 94 95 96 BC 9 8 7 6 5 4 3 2 1

Contents

Contributors

Numbers in parentheses indicate the pages on which the authors' contributions begin.

J. H. Andrews (245), *Case Western Reserve University, Department of Physics, Cleveland, OH 44106–7079*

Gregory L. Baker (433), *Michigan State University, East Lansing, MI 48824*

Shahab Etemad (433), *Bell Communications Research, Red Bank, NJ 07701*

Ryoichi Ito (201), *Department of Applied Physics, Faculty of Engineering, The University of Tokyo, Bunkyo-ku, Tokyo 113 Japan*

Takayoshi Kobayashi (47), *Department of Physics, University of Tokyo, Hongo, Bunkyo, Tokyo 113, Japan*

Takashi Kondo (201), *Department of Applied Physics, Faculty of Engineering, The University of Tokyo, Bunkyo-ku, Tokyo 113, Japan*

Mark G. Kuzyk (299), *Department of Physics, Washington State University, Pullman, WA 99164–2814*

Hilary S. Lackritz (339) (formerly Hilary L. Hampsch), *Department of Chemical Engineering, Purdue University, West Lafayette, IN 47907*

Pierre Le Barny (379), *Thomson-CSF, Laboratoire Central de Recherches, Domaine de Corbeville, 91404, Orsay Cedex, France*

Isabelle Ledoux (129), *Centre National d'Etudes des Télécommunications, Molecular Quantum Electronics Department, 196 avenue Henri Ravera, 92220, BP-107, Bagneux, France*

Vincent Lemoine (379), *Thomson-CSF, Laboratoire Central de Recherches, Domaine de Corbeville, 91404, Orsay Cedex, France*

Shaul Mukamel (1), *Department of Chemistry, University of Rochester, Rochester, NY 14627*

Jean-François Nicoud (129), *Groupe des Matériaux Organiques, Institut de Physique et Chimie des Matériaux de Strasbourg (ICPMS), 6, rue Boussingault, 67083, Strasbourg Cedex, France*

Jean Paul Pocholle (379), *Thomson-CSF, Laboratoire Central de Recherches, Domaine de Corbeville, 91404, Orsay Cedex, France*

Constantina Poga (299), *Department of Physics, Washington State University, Pullman, WA 99164–2814*

Philippe Robin (379), *Thomson-CSF, Laboratoire Central de Recherches, Domaine de Corbeville, 91404, Orsay Cedex, France*

Y. R. Shen (101), *Department of Physics, University of California and Materials Science Division, Lawrence Berkeley Laboratory, Berkeley, CA 94720*

Kenneth D. Singer (245), *Case Western Reserve University, Department of Physics, Cleveland, OH 44106–7079*

Zoltan G. Soos (433), *Department of Chemistry, Princeton University, Princeton, NJ 08544*

John M. Torkelson (339), *Department of Chemical Engineering, Department of Materials Science and Engineering, Northwestern University, Evanston, IL 60208*

Joseph Zyss (129), *Centre National d'Etudes des Télécommunications, Molecular Quantum Electronics Department, 196 avenue Henri Ravera, 92225, BP107-Bagneux, France*

Preface

Seven years after the publication, in that same series, of two volumes dedicated to the quadratic and cubic nonlinear properties of organic materials, time has come for an update to help the reader keep up with an active level of worldwide research effort in both fundamental and applied directions. While some of the scientific foundations of the field, as outlined in the two previous volumes, had already been firmly established at the time and have remained relatively untouched, many important pending issues were recognized at this early stage but left unsettled. Among them were the relative perspectives opened up by $\chi^{(2)}$ and $\chi^{(3)}$ related phenomena toward application endgoals; the comparative assets of crystalline versus polymeric materials; and the technological potential of organics, be they crystalline, polymeric, or others for integration in waveguiding passive and active optoelectronic devices, such as modulators and switches. As a result of intensive work combining the contributions of physicists, chemists, and device engineers, these basic questions have now been clarified to a significant extent with some important technological bottlenecks identified and addressed in the realm of both crystalline and polymeric materials. Owing to recent remarkable developments that encompass microlithographic patterning and organic molecular beam epitaxy, current attitudes towards the viability of an "organic technology" is much less biased by prejudice than a decade ago.

Electro-optic poled polymers, as reported mainly in Part III, have matured into patternable thin-film materials compatible with sophisticated multilayer architectures for integrated optics. Furthermore, their compatibility with semiconductor technology should ensure a smooth transition from present purely semiconductor-based systems to hybrid solutions whereby polymers and semiconductors are adequately combined in view of their respective advantages. Such developments have been based, and will further depend, on a deeper understanding of the physical properties of polymers, in particular as they relate to the quality of interfaces and the thermal relaxation dynamics of the induced polar order in view of its subsequent stabilization. Fundamental as well as application-related aspects have thus been closely associated in this book, with successive sections, such as 3-1, 3-3 and 3-4, jointly addressing closely related physical and technological issues from different perspectives.

As far as single crystals are concerned, various demonstrations in bulk formats have confirmed their potential for frequency doubling and optical parametric oscillation based on the availability of an increasing variety of molecular and lattice structures. Furthermore, the possibility of shaping crystalline structures in waveguides, as reported in Chapter 2-3, will further enhance the prospects for single crystals in quadratic nonlinear optics with near infrared laser frequency doubling to the blue as one of the main strategic goals.

Development and refinement of a molecular engineering approach toward quadratic nonlinear effects has helped point out new families of molecules more precisely targeted to satisfy precise transparency-efficiency requirements. While paranitroaniline has been serving as a dominant paradigm in the field since the mid-seventies, related dipolar intramolecular charge transfer systems may now be viewed as special cases of more general multipolar nonlinear systems. Organometallic and organomineral systems, self-assembling methods, and mesogenic nonlinear systems also appear as new and promising avenues with perspectives discussed in Chapter 2-2. Besides their intrinsic nonlinear behavior, molecules are very sensitive to environmental conditions that become dominant at interfaces; Chapter 2-1 reviews this increasingly important domain of surface science in the case of air–water interfaces.

Cubic nonlinear properties are more elusive than quadratic ones, and fundamental work, of both theoretical and experimental nature, is still needed to deepen our understanding of the underlying mechanisms so as to point out relevant structural and electronic features. This important issue is dealt with in Part I (Chapters 1-1 and 1-2) and again in Chapters 3-2 and 3-5. In this context, the magnitudes of cubic nonlinear susceptibilities do not suffice to account for subtle transient processes; the position and absorption cross-section of two-photon peaks as well as the dynamical features of induced absorption from photoexcited states play a dominant role. Advances in theoretical approaches (Chapter 1-1) and femtosecond probing techniques (Chapter 1-2) have significantly contributed to these issues.

While amorphous side-chain polymers qualify for quadratic nonlinear optics and conjugated ones for cubic effects, it would nevertheless seem erroneous to confine crystals to quadratic applications; results that have appeared in course of publishing this book tend to blur this barrier as they point out the relevance for $\chi^{(3)}$, by way of cascading mechanisms, of molecular crystals initially conceived for $\chi^{(2)}$.

While it may still be premature to predict the nature and extent of application breakthroughs in this field, the last decade has undoubtedly seen molecular nonlinear optics acquire full-fledged scientific status with recognition from both physicists and chemists. Few other fields have benefitted to such extent from a fruitful and convergent cooperation between the two communities, although motivations may be somewhat different: Nonlinear susceptibilities at both microscopic and macroscopic levels have now come to be widely adopted by chemists in view of their enhanced sensitivity to such features as conjugation, charge transfer, protonation, environment and local field effects, and crystalline or polymeric organization. As a consequence β and γ susceptibilities are now increasingly finding their way in chemical data bases, alongside absorption, dipole moment, dielectric constant, and other more traditional physico-chemical data. Conversely, physicists are attracted by the virtually unlimited possibilities to implement and manipulate properties at the ultimate molecular level. Advanced "guided" organic synthesis and molecular assembling techniques, combined with newly developed experimental tools, capable of probing and manipulating molecules at ultimate spatial, spectral, and time resolutions, are bound to open up a new frontier in molecular optical sciences.

This book will have fulfilled its dual goal when supplying state-of-the-art information to the benefit of currently committed scientists while helping renew the field by attracting needed future contributors.

JOSEPH ZYSS

PART I

FUNDAMENTALS

Chapter 1

MANY-BODY EFFECTS IN NONLINEAR SUSCEPTIBILITIES; BEYOND THE LOCAL-FIELD APPROXIMATION

Shaul Mukamel

Department of Chemistry, University of Rochester, Rochester, New York

ISBN 0-12-784450-3

1. INTRODUCTION

The systematic calculation of nonlinear susceptibilities of optical materials, and the precise relationship between individual molecular hyperpolarizablities and the macroscopic optical response, constitute a complex challenge that has drawn a considerable theoretical attention [1–5]. The design of new optical materials with specified characteristics (fast switching, large susceptibilities) and the interpretation of nonlinear optical measurements in terms of molecular properties and intermolecular forces require the development of a suitable theoretical framework.

The response of a medium to optical fields is most conveniently formulated in terms of wave vector and frequency-dependent optical susceptibilities, which are the expansion coefficients of the macroscopic polarization field in powers of the Maxwell electric field E [6–9]. The problem of incorporating intermolecular forces in the linear optical response (i.e., the dielectric function) has a long history [10–15]. The local-field approximation provides a simple way to relate the polarizabilities of isolated molecules to the macroscopic susceptibilities. In this approach, the effect of intermolecular forces is included in an effective local electric field. The problem of calculating the response of an interacting ensemble of molecules to the electromagnetic field is then reduced to the response of isolated molecules interacting with the local field $\mathbf{E}_L$, through an interaction Hamiltonian $-\boldsymbol{\mu}\cdot\mathbf{E}_L$, where $\boldsymbol{\mu}$ denotes the molecular dipole operator. The Lorentz relation between the local field and the Maxwell field $\mathbf{E}$ (Eq. 1.4.2) can then be used to calculate the dielectric function. This procedure, which reduces the complex many-body problem to a single-body problem, was subsequently generalized and applied also to the calculation of nonlinear susceptibilities [3,5,9,16–18]. The nonlinear susceptibilities at a given order are then given in terms of sums of products of molecular polarizabilities of that order and lower orders. This simple, back-of-the-envelope calculation of macroscopic susceptibilities is, however, not rigorous [5,19]. It fails to take properly into account the correlated dynamics of the interacting many-body system, i.e. correlations among the molecules, as well as correlations between the molecules and the radiation field. Short-range (e.g., exchange) forces are totally neglected in this procedure. Moreover, even the dipole–dipole forces are not fully taken into account. In addition, the resulting susceptibilities do not depend on the wavevectors, apart from the local-field contribution, but just on the frequencies. This indicates that processes such as exciton migration and energy transfer and transport (e.g., the Forster transfer) [20,21] are neglected. Such processes are often added phenomenologically in order to interpret transient grating spectroscopy [22,23], which is a four-wave mixing technique that measures transport processes by following the wave vector dependence of the susceptibilities. The common derivation of the local-field approximation [3] cannot be extended to include these processes, since it is intrinsically a mean-field single-molecule theory.

Moreover, the local-field approximation totally misses any collective effects resulting from cooperative interactions with the radiation field. Such effects are particularly important in the studies of optical nonlinearities of molecular [24–29] and semiconductor [4,30–36] systems with restricted geometries. Such studies are currently drawing considerable attention, due to the significant progress made in the fabrication of nanostructures such as microcrystallites, monolayers and quantum wires. Quantum well structures in semiconductors show sharp exciton resonances even at room temperature, and their nonlinear optical properties have been studied extensively [35]. One of the most fascinating open questions raised by these studies is the possibility of maintaining a large coherence size [28,29] which may give rise to enhanced nonlinear optical susceptibilities and ultrafast radiative decay rates [30,37,38], which result from collective (cooperative) interactions with the radiation field. These effects are totally absent in the local-field level of description since they involve explicitly intermolecular (non-local) coherences. Simple back-of-the-envelope argument [27] suggests that the transition dipole matrix element in an aggregate with N molecules should scale as $\sim\sqrt{N}$. $\chi^{(3)}$, which is proportional to the dipole moment to the fourth power, should therefore scale as $\sim N^2$, whereas if the molecules were independent it would scale as $\sim N$. We thus expect an extra N enhancement of $\chi^{(3)}$ upon aggregation. This argument was shown to be misleading [28,29]. Interference effects cancel this enhancement under off-resonant conditions. The resonant enhancement is expected to saturate with aggregate size [28]. The interaction with a thermal bath introduces fluctuations [39] that result in intermolecular dephasing and inhomogeneous broadening, which may destroy the coherence among different molecules. We expect that due to the dephasing processes, N should be replaced by N_{eff} where $N_{\text{eff}} < N$ is the effective number of coherently coupled molecules. N_{eff} is expected to attain the limiting value $N_{\text{eff}} = 1$ as the dephasing rate becomes sufficiently large to decouple the radiative dynamics of the various molecules. It is clear from the above arguments that the main problem in predicting the nonlinear optical response of optical materials is not how to better calculate the local field, but rather the fundamental failure of the local-field approximation to describe and predict some very important effects related to the many-body nature of the problem. A new theoretical approach is called for that goes beyond the local-field approximation.

In this review we present a microscopic theoretical framework for the calculation of the nonlinear optical response of molecular materials with localized electronic states that accounts for intermolecular interactions and dynamic correlations and overcomes the difficulties associated with the local-field approach. The present theory is based on the derivation of coupled reduced equations of motion for the material variables, which determine the optical response [5,19,28,39]. We find that in addition to the single-particle variables, we need consider also two-exciton and exciton-population variables which represent intermolecular coherences. The material system can thus be modelled as a collection of coupled anharmonic oscillators. An anharmonic oscillator picture of the nonlinear response has been suggested as a qualitative model [6,9,40] since the early days of nonlinear optics. The equation of motion approach [5,36] shows how such a picture can be rigorously established and applied toward the fully microscopic determination of the coherence size [28,36]. The effects of various factors (size, temperature, geometry, and exciton-phonon coupling) on the cooperative can be studied.

Many practical applications of optical material use laser frequencies that are far detuned from any molecular resonant frequency (electronic or vibrational). This is done to avoid absorption and other competing processes and to obtain faster switching timescales. Spectroscopic applications of nonlinear optics, on the other hand, often use resonant frequencies that provide a direct probe for specific energy levels and their dynamics [9,41,42]. We shall explore the nonlinear susceptibilities in both regimes. We show that for parametric (off-resonant) processes, cooperative effects will not be of great significance, but as some of the frequencies in the process are tuned near resonance, they may play an important role. Comparison with resonant spectroscopic measurements should establish the validity of the approximations made in the theory and allow us to predict with confidence the off-resonant optical response. For the sake of clarity we shall not consider here cooperative spontaneous emission (superradiance) [37,43–45] whose coherence size is closely related to that of optical nonlinearities. Polariton effects, which are important in materials with a large oscillator strength per unit volume at low temperatures [46–50] and require the incorporation of retarded interactions and a quantum description of the radiation field, were reviewed recently [5] and will not be discussed here.

In Sec. 2 we introduce the model system of a molecular crystal with Frenkel excitons. The basic equations of motion with intermolecular interactions are derived in Sec. 3. In Sec. 4 we present the single-particle level and in Sec. 5 we investigate the role of two-exciton variables in the enhancement of nonlinear susceptibilities in molecular aggregates. The role of exciton-population variables that are the source of intramolecular nonlinearities (known in the semiconductor literature as "phase-space filling") is investigated in Sec. 6. We further explore the effects of exciton transport using the Wigner phase-space distribution. In Sec. 7 we combine all bilinear exciton variables and present general equations of motion and a Green function expression for $\chi^{(3)}$. We analyze the nonlinear coherence size and discuss the limitations of the local-field approximation. In Sec. 8 we summarize our results.

2. MODEL HAMILTONIAN FOR MOLECULAR MATERIALS; FRENKEL-EXCITONS

In order to explore the role of many-body effects in the macroscopic nonlinear optical response, we consider a model system consisting of a lattice of polarizable (nonpolar) two-level molecules with transition frequency Ω and one molecule per unit cell [11,51,52]. In the dipole approximation, the Hamiltonian for our system is [5]:

$$H = H_{\text{mat}} - \int P(\mathbf{r}) \cdot E(\mathbf{r}, t) d\mathbf{r}. \tag{1.2.1}$$

Here, H_{mat} denotes the material Hamiltonian which can be partitioned as follows:

$$H_{\text{mat}} = H_{\text{ex}} + H_{\text{ph}} + H_{\text{ep}}. \tag{1.2.2a}$$

H_{ex} is the Frenkel-exciton Hamiltonian

$$H_{\text{ex}} = \hbar\Omega \sum_m B_m^\dagger B_m + \frac{\hbar}{2} \sum_{m,n}{}' J(\mathbf{r}_{mn})(B_m^\dagger + B_m)(B_n^\dagger + B_n), \tag{1.2.2b}$$

$B_m^\dagger$ (B_m) denotes the creation (annihilation) operator for an excitation on molecule m with transition frequency Ω. These operators commute for different molecules, whereas for any single molecule they obey the Pauli anticommutation relation

$$B_m^\dagger B_m + B_m B_m^\dagger = 1. \tag{1.2.3}$$

Equation (1.2.3) can alternatively be written in the form $[B_m, B_m^\dagger] = 1 - W_m$ with the exciton population operator $W_m \equiv 2B_m^\dagger B_m$. For harmonic oscillators (bosons), Eq. (1.2.3) does not hold and $[B_m, B_m^\dagger] = 1$. Therefore, neglecting the W_m operator for two-level molecules is usually denoted the Bose approximation. We note that in the present model the Pauli exclusion is the only source of nonlinearities. In systems with multilevel (and polar two level) molecules, other sources arise from intermolecular interaction terms that are cubic and quartic in the molecular exciton creation and annihilation operators [21,53]. Such terms give rise to nonlinearities even if a Bose approximation is applied. The second term in H_{ex} accounts for the instantaneous dipole–dipole interactions between the molecules in their equilibrium positions and orientations (the prime excludes terms with $m = n$ from the summation),

$$\hbar J(\mathbf{r}) = \vec{\mu}\cdot\left(\frac{1}{\mathbf{r}^3} - \frac{3\mathbf{rr}}{\mathbf{r}^5}\right)\cdot\vec{\mu}. \tag{1.2.4}$$

Here $\mathbf{r}_{mn} \equiv \mathbf{r}_m - \mathbf{r}_n$, where $\mathbf{r}_m$ denotes the equilibrium position of molecule m, with $\vec{\mu}$ denoting the molecular transition dipole in the equilibrium configuration (equal for all molecules on the lattice). In the present model the excitons have therefore two types of interaction: the dipole–dipole interaction and the Pauli exclusion which can effectively be viewed as a repulsive interaction.

The second contribution to H_{mat}, H_{ph}, represents the nuclear (phonon) motions and the last contribution to H_{mat} is the exciton–phonon interaction, which arises from the dependence of the intermolecular interactions on the displacements of the nuclear coordinates from their equilibrium values. The coupling with phonons plays an important role in determining the coherence size and the magnitude of optical nonlinearities. For the sake of clarity we shall use here simplified models for the exciton–phonon coupling, although a more rigorous treatment of this coupling may be carried out [5.51].

The second term in H represents the coupling to a classical transverse electric field $E(\mathbf{r})$. $P(\mathbf{r})$ is the optical polarization operator, which may be written as

$$P(\mathbf{r}) = \sum_m \vec{\mu}(B_m + B_m^\dagger)\delta(\mathbf{r} - \mathbf{r}_m), \tag{1.2.5}$$

where we assumed that all molecular transition dipole matrix elements are parallel and given by $\vec{\mu}$. We shall further introduce the material polarization field in momentum space

$$P(\mathbf{k}) \equiv \sqrt{N}\,\vec{\mu}(B_{\mathbf{k}} + B^\dagger_{-\mathbf{k}}), \tag{1.2.6}$$

where we have defined the exciton annihilation $B_{\mathbf{k}}$ and creation $B^\dagger_{-\mathbf{k}}$ operators in the momentum representation

$$B_{\mathbf{k}} \equiv \frac{1}{\sqrt{N}} \sum_m B_m \exp(i\mathbf{k}\cdot\mathbf{r}_m), \tag{1.2.7a}$$

$$B^\dagger_{-\mathbf{k}} \equiv \frac{1}{\sqrt{N}} \sum_m B_m^\dagger \exp(i\mathbf{k}\cdot\mathbf{r}_m). \tag{1.2.7b}$$

3. EQUATIONS OF MOTION: THE ANHARMONIC-OSCILLATORS PICTURE

Nonlinear susceptibilities of simple quantum systems are usually calculated using the time-dependent density matrix (the Schrödinger picture) [6–9]. The susceptibilities are then given in terms of multiple summations over eigenstates. In order to map the problem into coupled anharmonic oscillators, we adopt in the present work a different route, which is based on the Heisenberg picture. The polarization will be calculated by solving a set of coupled nonlinear equations of motion. The equations are obtained by identifying a set of relevant dynamical variables and deriving equations of motion for their expectation values [5,28,36]. Our starting point is the Heisenberg equation of motion for any material operator A

$$\dot{A} = \frac{i}{\hbar}[H, A]. \tag{1.3.1}$$

Since the polarization operator is expressed in terms of B_n and $B_n^\dagger$, it is natural to start with the equations of motion for these operators. Using the Hamiltonian Eq. (1.2.1) and the commutation relation (1.2.3), we obtain the following equations of motion (all operators taken at time t):

$$\frac{1}{i}\frac{d}{dt}B_n = -\Omega B_n - \sum_m{}'(J_{nm} - i\Gamma_{nm})B_m - \sum_m J_{nm}B_m^\dagger + 2\sum_m{}' J_{nm}(B_n^\dagger B_n B_m + B_n^\dagger B_n B_m^\dagger) + \frac{1}{\hbar}\tilde{\mu}\cdot E(\mathbf{r}_n, t)[1 - 2B_n^\dagger B_n]. \tag{1.3.2}$$

Equation (1.3.2) is the basis for all the equations of motion derived here. Γ_{nm} is an exciton relaxation term which is introduced here phenomenologically. The first three terms in the r.h.s. of Eq. (1.3.2) represent linear dynamics whereby B_n is coupled to B_m and $B_m^\dagger$. The other two terms are more complex. They represent nonlinear coupling to higher order operators $B_n^\dagger B_n$, $B_n^\dagger B_n B_m$ and $B_n^\dagger B_n B_m^\dagger$. Equation (1.3.2) is therefore not closed. We may proceed by taking $A = B_n^\dagger B_n$, $B_n^\dagger B_n B_m$, and $B_n^\dagger B_n B_m^\dagger$ and writing the Heisenberg equations for these operators. We shall then couple them to even more complex operators involving products of four B operators. In general, Eq. (1.3.1) will therefore result in an infinite hierarchy of coupled dynamical equations whereby single-body operators are successively coupled to more complex operators. Fortunately, the optical response to electromagnetic fields that are not too strong requires the explicit introduction of only few-particle states. This allows us to truncate the hierarchy at a very early stage. The main problem addressed in this review is the possible truncation schemes that yield a hierarchy of approximations for the nonlinear optical response. We shall demonstrate how a truncation at the two-particle level may be adequate for the calculation of a variety of optical measurements. This situation is formally very similar to zero-temperature many-body theory where a few quasiparticles dominate the dynamical behavior [54].

Let us first consider the linearized part of Eq. (1.3.2) by neglecting the last two terms, which represent nonlinear dynamics and the driving field. By considering an infinite periodic structure and switching to k-space, the equation reads

$$\frac{1}{i}\frac{d}{dt}B_{\mathbf{k}} = [-\Omega - J(\mathbf{k}) + i\Gamma(\mathbf{k}]B_{\mathbf{k}} - J(\mathbf{k})B^{\dagger}_{-\mathbf{k}}. \tag{1.3.3}$$

Here, $J(\mathbf{k})$ is the lattice Fourier transform of the intermolecular interaction

$$J(\mathbf{k}) = \sum_{m \neq 0} J(\mathbf{r}_m)\exp(-i\mathbf{k}\cdot\mathbf{r}_m). \tag{1.3.4}$$

For a centrosymmetric lattice, we have $J(\mathbf{k}) = J(-\mathbf{k})$. $\Gamma(\mathbf{k})$ is the damping rate of the exciton induced by the phonon bath. Equation (1.3.3), together with its Hermitian conjugate equation for $B^{\dagger}_{-\mathbf{k}}$, defines an eigenvalue problem whose solutions are the creation and annihilation operators for the Coulomb exciton at wave vector $\mathbf{k}$ in terms of $B_{\mathbf{k}}$ and $B^{\dagger}_{-\mathbf{k}}$. The Coulomb exciton frequency is determined by the secular equation of the problem and easily found to be $\Omega_{\mathbf{k}} - i\Gamma(\mathbf{k})$, with

$$\Omega_{\mathbf{k}} = [\Omega(\Omega + 2J(\mathbf{k}))]^{1/2}. \tag{1.3.5}$$

For $|J(\mathbf{k})| \ll \Omega$, which is by definition the case in molecular crystals [51], this yields $\Omega_{\mathbf{k}} \approx \Omega + J(\mathbf{k})$. This is known as the Heitler-London approximation whereby the Coulomb excitons are simply created (annihilated) by $B^{\dagger}_{-\mathbf{k}}$ ($B_{\mathbf{k}}$).

We now discuss the linear optical response, which is governed by the first-order susceptibility. To that end we define the discrete Fourier decomposition of the field,

$$\mathbf{E}(\mathbf{r}, t) = \sum_j [\mathbf{E}_j \exp(i\mathbf{k}_j\cdot\mathbf{r} - i\omega_j t) + c.c.], \tag{1.3.6}$$

where j labels a few modes that are relevant in the experiment. ω_j (>0) and $\mathbf{k}_j$ are related by the crystal's dispersion relation. Similar decomposition is possible for the polarization field $\mathbf{P}(\mathbf{r}, t)$. The susceptibilities are defined as the coefficients appearing in the expansion of the amplitudes $\mathbf{P}_j$ in terms of powers of the electric field amplitudes $\mathbf{E}_j$. The first order susceptibility is easily obtained from Eq. (1.3.2). The part of the polarization that is linear in the electric field is found by neglecting the nonlinearities (the $B_n^\dagger B_n B_m$, $B_n^\dagger B_n B_m^\dagger$ and the $B_n^\dagger B_m$ terms). After taking expectation values and substituting Eq. (1.3.6) and its analog for the polarization, we obtain

$$\mathbf{P}_j^{(1)} \equiv \chi^{(1)}(\mathbf{k}_j, \omega_j)\cdot\mathbf{E}_j, \tag{1.3.7a}$$

with the linear susceptibility tensor $\chi^{(1)}(\mathbf{k}, \omega) \equiv (\varepsilon(\mathbf{k}, \omega) - 1)/4\pi$, $\varepsilon(\mathbf{k}, \omega)$ being the frequency and wave vector dependent dielectric tensor. We finally get [11,46]

$$\varepsilon(\mathbf{k}, \omega) = 1 + \frac{8\pi\Omega\rho\hbar^{-1}\vec{\mu}\vec{\mu}}{-[\omega + i\Gamma(\mathbf{k})]^2 + \Omega_{\mathbf{k}}^2}, \tag{1.3.7b}$$

where $\rho \equiv N/V$ denotes the average molecular density in the crystal, and the superscripts in parentheses indicate the order in the electric field amplitudes.

We next turn to the calculation of optical nonlinearities. We note that $\chi^{(3)}$ is the lowest nonlinearity allowed by the present model, since $\chi^{(2)}$ vanishes for a centrosymmetric medium [6]. $\chi^{(3)}$ in general represents a broad class of processes known as four-wave mixing. In a frequency-domain four-wave mixing experiment we consider three fundamental fields ($j = 1, 2, 3$ in Eq. 1.3.6) and are interested in the signal at $(\mathbf{k}_s, \omega_s) \equiv (\mathbf{k}_1 - \mathbf{k}_2 + \mathbf{k}_3, \omega_1 - \omega_2 + \omega_3)$. To lowest order in the field amplitudes $\mathbf{E}_j$ this is determined by the third order susceptibility, which is defined through

$$\mathbf{P}_s^{(3)} \equiv \chi^{(3)}(-\mathbf{k}_s - \omega_s; \mathbf{k}_1\omega_1, -\mathbf{k}_2 - \omega_2, \mathbf{k}_3\omega_3)\mathbf{E}_1\mathbf{E}_2^*\mathbf{E}_3, \tag{1.3.8}$$

where $\mathbf{P}_s^{(3)}$ is the discrete Fourier coefficient of the polarization field with wave vector $\mathbf{k}_s \equiv \mathbf{k}_1 - \mathbf{k}_2 + \mathbf{k}_3$ and frequency $\omega_s \equiv \omega_1 - \omega_2 + \omega_3$ to third order in the electric field amplitudes. Other choices of $\mathbf{k}_s$ and ω_s can be obtained by changing the signs of one or more ω_j and $\mathbf{k}_j$ variables. When the expectation value of Eq. (1.3.2) is taken, we find that the single-particle variables $\langle B_n^\dagger \rangle$ are coupled to the two-operator variables $\langle B_n^\dagger B_n \rangle$ and to the three operator (two-particle) variables $\langle B_n^\dagger B_n B_m \rangle$ and $\langle B_n^\dagger B_n B_m^\dagger \rangle$. In most calculations presented in this article, we shall invoke the Heitler-London approximation, thereby neglecting certain off-resonant contributions to the nonlinear response. This amounts to neglecting the $B_n^\dagger$ and the $B_n^\dagger B_n B_m^\dagger$ terms in the r.h.s. of Eq. (1.3.2). In the discussion below we shall therefore not consider the $\langle B_n^\dagger B_n B_m^\dagger \rangle$ variables, although they may be treated very similarly to the $\langle B_n^\dagger B_n B_m \rangle$ variables (see Eq. (1.7.1).

The $\langle B_n^\dagger B_n \rangle$ variables represent exciton populations, and they are the source of nonlinearity in the single-particle formulation of nonlinear optics, which is based on the Bloch equations. In the semiconductor literature this contribution is known as "phase space filling" [35]. These single-particle variables pose no major theoretical problem; the story is very different for the $\langle B_n^\dagger B_n B_m \rangle$ terms. The rigorous way to proceed is to derive an equation of motion for $\langle B_n^\dagger B_n B_m \rangle$ which, when solved coupled with $\langle B_n^\dagger \rangle$ and $\langle B_n \rangle$, will yield the nonlinear response. It is important to recognize that all existing theories for the nonlinear optical response contain an (either implicit or explicit) approximation for these three-operator quantities. In this review we present a systematic approach and discuss the hierarchy of possible approximations. To gain some insight into the issue, it proves useful to switch for a moment to the Schrödinger picture and describe the system using the density matrix $\rho(t)$. It should be emphasized that eventually we are going to derive and solve equations of motion for the expectation values of dynamical variables; we are not going to calculate the density matrix. We use the density matrix here only in order to clarify the physical significance of the various factorizations. In the Schrödinger picture we have

$$\langle B_n^\dagger(t) B_n(t) B_m(t) \rangle \equiv Tr[B_n^\dagger(0) B_n(0) B_m(0) \rho(t)] \tag{1.3.9}$$

(where the time argument is explicitly indicated). Here

$$B_n(t) \equiv \exp(iH_{\text{mat}}t) B_n \exp(-iH_{\text{mat}}t)$$

are the operators in the Heisenberg picture and

$$\rho(t) \equiv \exp(-iH_{\text{mat}}t) \rho(0) \exp(iH_{\text{mat}}t)$$

is the density matrix in the Schrödinger picture. By introducing an ansatz for $\rho(t)$ it is possible to express Eq. (1.3.9) using lower order operators and thus close to hierarchy. In the coming sections we shall explore the following approximations for the three-operator quantities.

1. *Single-Particle Factorization.* The single-particle factorization is introduced in Sec. 4, where we set
$$\langle B_n^\dagger B_n B_m \rangle = \langle B_n^\dagger \rangle \langle B_n \rangle \langle B_m \rangle . \tag{1.3.10}$$
This is the simplest possible truncation of the hierarchy, since it retains only the exciton amplitude variables.
2. *Factorization into Two-Exciton Variables.* Neglecting phonons, the density matrix of the system represents a pure state $\rho(t) = |\psi(t)\rangle\langle\psi(t)|$, where $\psi(t)$ is the wave function. In this case we can rigorously factorize the creation and the annihilation operators in the calculation of $\chi^{(3)}$. [28]
$$\langle B_n^\dagger B_n B_m \rangle = \langle B_n^\dagger \rangle \langle B_n B_m \rangle . \tag{1.3.11}$$

The $\langle B_n B_m \rangle$ variables represent two-exciton dynamics. This level of the hierarchy will be discussed in Sec. 5.

3. *Factorization into Exciton-Population Variables.* Neglecting all correlations among different sites we set for $n \neq m$ [5,55]

$$\langle B_n^\dagger B_n B_m \rangle = \langle B_n^\dagger B_n \rangle \langle B_m \rangle. \tag{1.3.12}$$

The $\langle B_n^\dagger B_n \rangle$ variables represent exciton populations, and their dynamics may be related to standard transport equations (the Boltzmann and the diffusion equations). This level of the hierarchy will be discussed in Sec. 6.

4. *Inclusion of Two-Exciton and Exciton-Population Variables.* A more general factorization which includes the previous factorizations as special cases is

$$\langle B_n^\dagger B_n B_m \rangle = \langle B_n^\dagger B_n \rangle \langle B_m \rangle + \langle B_n^\dagger B_m \rangle \langle B_n \rangle + \langle B_n^\dagger \rangle \langle B_n B_m \rangle - 2\langle B_n^\dagger \rangle \langle B_n \rangle \langle B_m \rangle. \tag{1.3.13}$$

Equation (1.3.13) may be derived using an approximation maximum-entropy form for the density matrix $\rho(t)$ [39]. When Eq. (1.3.13) is used without introducing any further factorization, we incorporate all exciton populations ($\langle B_n^\dagger B_n \rangle$), exciton-coherence ($\langle B_n^\dagger B_m \rangle$), and two-exciton ($\langle B_n B_n \rangle$) variables. These variables, together with the single-exciton ($\langle B_n \rangle$) variables and their Hermitian conjugates, constitute the complete set of binary dynamical variables. Equation (1.3.13) will be discussed in Sec. 7. As indicated earlier, cases 1 to 3 are special cases of (4). Starting with Eq. (1.3.13), if we further factorize all operators, or creation and annihilation operators, or operators belonging to different sites, we recover Eqs. (1.3.10), (1.3.11), and (1.3.12) respectively.

4. THE SINGLE-PARTICLE LEVEL AND THE LOCAL-FIELD APPROXIMATION

The simplest approximation for the nonlinear response is obtained by adopting the single-particle factorization (Eq. 1.3.10). Taking the expectation value of Eq. (1.3.2) and its hermitian conjugate, substituting $\langle B_n^\dagger B_n B_m \rangle = \langle B_n^\dagger \rangle \langle B_n \rangle \langle B_m \rangle$ and $\langle B_n^\dagger B_n \rangle = \langle B_n^\dagger \rangle \langle B_n \rangle$, and using the definition of the polarization (Eq. 1.2.5), we finally get a closed equation for the polarization [5].

$$\ddot{P}(\mathbf{r}, t) + 2\Gamma \dot{P}(\mathbf{r}, t) + \Omega^2 P(\mathbf{r}, t) = 2\Omega\rho\hbar^{-1}|\mu|^2 \cdot \mathbf{E}_L(\mathbf{r}, t) - \frac{1}{\hbar\Omega\rho} \mathbf{E}_L; j(\mathbf{r}, t)|(\Omega + i\Gamma)P(\mathbf{r}, t) + i\dot{P}(\mathbf{r}, t)|^2. \tag{1.4.1}$$

In Eq. (1.4.1) all intermolecular interactions were formally eliminated by introducing the local electric field through

$$\mathbf{E}_L(\mathbf{r}_n, t) = \mathbf{E}(\mathbf{r}_n, t) - \frac{\hbar}{\mu} \sum_m{}' J_{nm}[\langle B_m \rangle + \langle B_m^\dagger \rangle] \tag{1.4.2a}$$

or in k space

$$\mathbf{E}_L(\mathbf{k}, t) \equiv \mathbf{E}(\mathbf{k}, t) - [\hbar J(\mathbf{k})/\mu^2 \rho] P(\mathbf{k}, t). \tag{1.4.2b}$$

The nonlocal character of the damping was neglected, and we have set $\Gamma_{nm} = \Gamma \delta_{nm}$. Otherwise, a local picture is impossible. The polarization at every site behaves as an oscillator driven by the local electric field and by anharmonic (nonlinear) forces. The linearized part of Eq. (1.4.1) obtained by neglecting the second term on the r.h.s. will result in our previous expression for the dielectric function (Eq. 1.3.7b).

For optical wave vectors and dipolar interactions in the ($\mathbf{k} \to 0$) limit, $J(\mathbf{k}) = -\frac{4\pi}{3\hbar} \rho \mu^2$ [56]. This gives the Lorentz local field [6,9,10,13]

$$\mathbf{E}_L(\mathbf{k}, t) \equiv \mathbf{E}(\mathbf{k}, t) + \frac{4\pi}{3} \mathbf{P}(\mathbf{k}, t). \tag{1.4.3}$$

In this limit, Eqs. (1.4.1) and (1.4.3) result in the Clausius-Mossotti expression for the dielectric function

$$\frac{\varepsilon(\omega) - 1}{\varepsilon(\omega) + 2} = \frac{4\pi}{3} \rho \alpha(\omega), \tag{1.4.4a}$$

with

$$\alpha(\omega) = \frac{2\Omega\hbar|\mu|^2}{-(\omega + i\Gamma)^2 + \Omega^2}. \tag{1.4.4b}$$

Here $\alpha(\omega)$ is the polarizability of an isolated molecule defined by Eq. (1.3.7a), with $\mathbf{E}_j$ representing the local field, and we have set $\mathbf{k} = 0$.

For subsequent manipulations we further introduce the local-field correction factor

$$S(\mathbf{k}, \omega) \equiv E_L(\mathbf{k}, \omega)/E(\mathbf{k}, \omega). \tag{1.4.5}$$

Combining this with Eq. (1.4.3), we get

$$S(\mathbf{k}, \omega) = \frac{\varepsilon(\mathbf{k}, \omega) + 2}{3}. \tag{1.4.6}$$

A common phenomenological procedure for calculating the nonlinear optical response is based on the cubic-oscillator model [6]. This amounts to adding a $\sim\mathbf{P}^3$ term in the potential which results in an equation similar to Eq. (1.4.1) but with the second (nonlinear) term in the r.h.s. replaced by $\sim P^2(\mathbf{r}, t)$. Equation (1.4.1) shows that this model is only of limited qualitative value. We find no $\sim P^2$ type of nonlinearity. Instead we have $E_L P^2$, $E_L \dot{P}^2$ and $E_L P\dot{P}$ types of terms. The anharmonicity is thus a function of both the "position" (P) and the "velocity" ($\dot{P}$) of the oscillator as well as the local electric field E_L (5.36].

The single-particle factorization used here leads to a local-field description. Consequently, it is possible to write the susceptibility Eq. (1.3.8) as the third order polarizability of a single molecule $\gamma\ (-\omega_s; \omega_1, -\omega_2, \omega_3)$ multiplied by local-field correction factors [5,8]

$$\chi^{(3)}(-\mathbf{k}_s - \omega_s; \mathbf{k}_1\omega_1, -\mathbf{k}_2 - \omega_2, \mathbf{k}_3\omega_3)$$
$$= \rho\gamma(-\omega_s; \omega_1, -\omega_2, \omega_3)S(\mathbf{k}_1\omega_1)S(-\mathbf{k}_2 - \omega_2)S(\mathbf{k}_3\omega_3)S(\mathbf{k}_s\omega_s) \quad (1.4.7a)$$

with

$$\gamma(-\omega_s; \omega_1, -\omega_2, \omega_3)$$
$$= 4\Omega \frac{\bar{\mu}\bar{\mu}\bar{\mu}\bar{\mu}}{\hbar^3} \frac{[-\Omega-\omega_2+i\Gamma(\mathbf{k}_2)][(\Omega+\omega_1+i\Gamma(\mathbf{k}_1)]}{[-(\omega_s+i\Gamma)^2+\Omega^2][-(\omega_1+i\Gamma)^2)+\Omega^2][-(-\omega_2+i\Gamma)^2)+\Omega^2]},$$
$$(+\text{permutations of } j = 1, 2, 3], \quad (1.4.7b)$$

and

$$S(\mathbf{k}, \omega) \equiv \frac{-(\omega + i\Gamma)^2 + \Omega^2}{-(\omega + i\Gamma(\mathbf{k}))^2 + \Omega_{\mathbf{k}}^2}. \quad (1.4.7c)$$

The numerators in the $S(\mathbf{k}, \omega)$ factors in Eq. (1.4.7a) cancel the three molecular (Ω) resonances in γ and introduce new exciton ($\Omega_{\mathbf{k}}$) resonances. The permutations of ($\mathbf{k}_1, \omega_1$), ($-\mathbf{k}_2, -\omega_2$) and ($\mathbf{k}_3, \omega_3$) account for the six different time orderings with which the electric fields can interact with the system.

5. THE ROLE OF TWO EXCITON VARIABLES; ENHANCED NONLINEAR SUSCEPTIBILITIES IN MOLECULAR AGGREGATES

In this section we explore the role of the two-exciton variables $\langle B_n B_m \rangle$ in the cooperative enhancement of optical nonlinearities [28]. These variables are particularly important in finite-size systems with restricted geometries such as molecular aggregates or monolayers [26,27,58]. We shall therefore analyze a specific model of confined excitons.

Consider a linear aggregate consisting of N coupled two-level molecules with transition frequency Ω and nearest neighbor dipole–dipole coupling V [28]. N is taken to be odd. The calculation for N even can be made along the same lines and will be omitted here for brevity. The nth molecule is located at $\mathbf{r}_n = n\mathbf{a}$ where $\mathbf{a}$ is the lattice vector, and $\mathbf{r}_{N+1} = \mathbf{r}_1$ (periodic boundary conditions). We further neglect the exciton-phonon coupling. $\chi^{(3)}$ for the present model can be expressed in terms of the single-exciton and two-exciton eigenstates. We shall first introduce these states, which can be exactly calculated for the present model in the Heitler-London approximation, and analyze the resulting $\chi^{(3)}$ obtained using the conventional sum-over-states expression. We then show how these results can be obtained using the present equation of motion approach by simply adopting the factorization (Eq. 1.3.11), which retains only two-exciton variables. We explore the roles of intra and intermolecular nonlinearities, show the limitations of the local-field approximation, and discuss the factors affecting cooperative enhancement. The relevant eigenstates of the aggregate are shown in Fig. 1. They are grouped in three levels. The lowest is the ground state $|g\rangle$. The next level includes N single-exciton states

$$|k\rangle \equiv B_k^\dagger|0\rangle, \tag{1.5.1a}$$

with energies $$\Omega_k \equiv \Omega + 2V\cos\left(\frac{2\pi k}{N}\right), \qquad \mathbf{k} = 0, 1, \ldots, N-1. \tag{1.5.1b}$$

Finally we have the two-exciton manifold

$$|k, q\rangle \equiv C_{k,q}^\dagger|0\rangle, \tag{1.5.2a}$$

with the two-exciton creation operators

$$C_{k,q}^\dagger \equiv \sum_R \sum_s \exp(i2\pi kR/N)\sin\frac{\pi qs}{N} B_{R-s/2}^\dagger B_{R+s/2}^\dagger, \qquad q = 1, 3, \ldots, N-2. \tag{1.5.2b}$$

These operators, when acting on the ground state, create two-exciton states with relative momentum $2\pi q/Na$, center of mass momentum $2\pi k/Na$, and energies

$$\omega_{k,q} = \Omega + 2V\cos\frac{\pi k}{N}\cos\frac{\pi(q-1)}{N}. \tag{1.5.2c}$$

By the Pauli exclusion we have N possibilities for the first excitation and only $N-1$ for the second. There are therefore $N(N-1)/2$ two-exciton states. (The factor 2 results from the permutation of these two excitations). Altogether we need consider $1 + N(N+1)/2$ eigenstates.

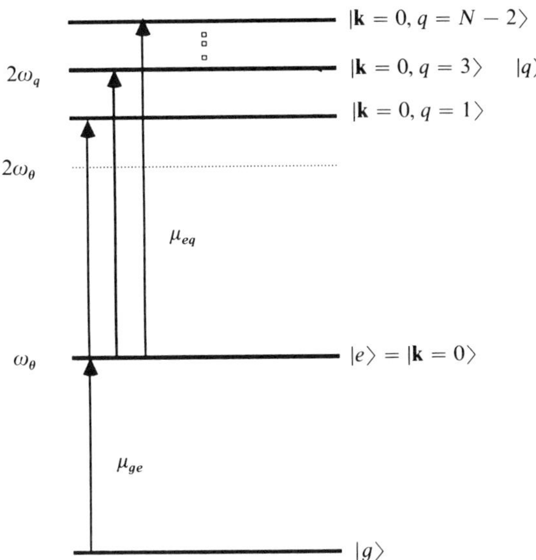

Fig. 1. Energy level diagram for a molecular aggregate, showing the single-photon allowed exciton states and the two-photon allowed two-exciton states. J is taken to be negative (attractive interaction , as in J aggregates) [37] and the exciton-two-exciton splitting $\omega_q - \omega_e$ is therefore positive. In this case strong two-photon absorption will occur to the blue of the exciton resonance.

For simplicity we hereafter specialize to a uniform excitation induced by an external electric field with a wave vector, oriented normal to the aggregate axis, which excites only the $k = 0$ states. We thus need consider only a single-exciton state $|e\rangle \equiv |k = 0\rangle$ and only $(N - 1)/2$ two-exciton states $|q\rangle \equiv C^{\dagger}_{o,q}|0\rangle$. Together with the ground state we have $1 + (N + 1)/2$ states. A general analysis for $k \neq 0$ excitons, excited when the aggregate axis not normal to the laser beam wavevector, is straightforward [28]; however, all of the essential physics is contained in the $k = 0$ exciton analysis. We further specialize to a particular $\chi^{(3)}$, which is responsible for nonlinear index of refraction and two-photon absorption of a single beam. Using the standard sum over states expression for $\chi^{(3)}$ [6] we then get

$$\chi^{(3)}(-\omega; \omega, -\omega, \omega)$$
$$= \rho \frac{1}{(\omega - \omega_e)^2 + (\gamma_e/2)^2} \left[\frac{|\mu_{ge}|^4}{\omega - \omega_e + i\gamma_e/2} - \sum_q \frac{|\mu_{ge}|^2 |\mu_{eq}|^2}{2\omega - 2\omega_q + i\gamma_q} \right], \quad (1.5.3)$$

with energies $\omega_e \equiv \Omega + 2V$ and $\omega_q \equiv \omega_e + 2V[\cos(\pi q/N) - 1]$; q = $1, 3, \ldots, N - 2$, γ_e and γ_q are phenomenological relaxation rates of the exciton and two-exciton states respectively. ρ is the number of aggregates per unit volume. The transition dipole moments are $\mu_{ge} = \sqrt{N}\mu$ and $\mu_{eq} = 2(2\mu/\sqrt{N}) \cot(\pi q/2N)$, μ being the transition dipole of a single molecule. They satisfy the sum rule

$$\sum_{q=1,3}^{N-2} |\mu_{eq}|^2 = 2|\mu_{ge}|^2 \frac{N-1}{N}. \tag{1.5.4}$$

$\chi^{(3)}$ is composed of two terms (Liouville-space pathways) [42]. The first term in the square brackets represents the contribution of the single-exciton level and scales as $|\mu_{eg}|^4 \sim N^2$. The second term consists of a series of two-exciton resonances. Using the sum rule (Eq. 1.5.4), the integrated area of these resonances scales as

$$(1/2)|\mu_{ge}|^2 \sum_q |\mu_{eq}|^2 = |\mu|^4 N(N-1).$$

When the laser beams are tuned far from an excitonic or two-excitonic resonance, the two terms intefere destructively, and the N^2 parts exactly cancel out, leaving an overall linear dependent of $\chi^{(3)}$ on size [29,29]

$$\chi^{(3)}(-\omega; \omega, -\omega, \omega) = \rho N |\mu|^4 \frac{1}{(\omega-\omega_e)^2 + (\gamma_e/2)^2} \frac{1}{\omega - \omega_e + i\gamma_e/2}. \tag{1.5.5}$$

In small aggregates the exciton and the two-exciton contributions may be well separated spectrally (i.e., $\omega_e - \omega_q$ is larger than the dephasing rates γ_e and γ_q). By carefully tuning the frequency near resonance, it may be possible spectrally to select either the first or the second term of Eq. (1.5.3), resulting in an $\sim N^2$ scaling of the nonlinear response due to cooperative enhancement. An increase in aggregate size N reduces the exciton-two-exciton splitting of the lowest q states, which contain most of the oscillator strength. When that splitting becomes comparable to the exciton linewidth γ, (where γ stands for γ_e or γ_q), it becomes impossible to spectrally select one of these terms. The interference will thus cancel the enhancement, resulting in an $\sim N$ scaling of $\chi^{(3)}$. The crossover size, whereby the magnitude of the aggregate response changes from $\sim N^2$ to $\sim N$, is a solution to $8 \sin^2(\pi/2N_c) = \gamma/V$, which gives [28]

$$N_c = \pi[2V/\gamma]^{1/2}. \tag{1.5.6}$$

In summary, the present model provides a simple expression for the nonlinear optical response of aggregates and demonstrates the origin of the crossover from the small aggregate to the bulk limit. Cooperatively enhanced optical nonlinearities ($\sim N^2$) are only possible in small aggregates with

$N < N_c$, where the two-level excitonic resonance is spectrally well separated from the two-photon resonances. In larger aggregates, the intermolecular nonlinearities due to exciton–exciton scattering are diminished, so that the hyperpolarizability is simply proportional to size. This reduction is caused by interference between excitonic nonlinearities and two-photon nonlinearities.

We shall now derive these results directly from our equations of motion, without making summations over eigenstates. This will provide an alternative interpretation for the aggregate size scaling, in terms of anharmonic-oscillators related to intermolecular coherences. Furthermore, the equation of motion result can be easily generalized by a systematic improvement of the factorization, taking into account other important many-body effects. This will be demonstrated in the coming sections. When phonons are neglected, we can make the pure-state ansatz (Eq. 1.3.11). We thus need to consider only the variables $\langle B_n \rangle$ and $\langle B_n B_m \rangle$ and their Hermitian conjugates $\langle B_n^\dagger \rangle$ and $\langle B_n^\dagger B_m^\dagger \rangle$. The equation of motion for $B_n^\dagger B_m^\dagger$ is obtained by substituting these operators in the Heisenberg equation (Eq. 1.3.1), resulting in the following closed equations in momentum space [28]:

$$\frac{d}{dt}\langle B_0^\dagger \rangle = (i\omega_e - \gamma_e/2)\langle B_0^\dagger \rangle + \sum_{q=1,3,\ldots}^{N-2} K(q)\langle C_q^\dagger \rangle \langle B_0 \rangle + i\frac{\mu}{\hbar} E(r,t)\left[N - \frac{2}{N}\langle B_0^\dagger \rangle \langle B_0 \rangle\right], \tag{1.5.7a}$$

$$\frac{d}{dt}\langle C_q^\dagger \rangle = (2i\omega_q - \gamma_q)\langle C_q^\dagger \rangle + 2i\frac{\mu}{\hbar} E(r,t) \cot\frac{\pi q}{2N}\langle B_0^\dagger \rangle \tag{1.5.7b}$$

with

$$K(q) = \frac{2i}{N^2}\sum_{q'=1,3} 4V \cot\left(\frac{\pi q'}{2N}\right)\left(\cos\frac{\pi q}{N} - 1\right). \tag{1.5.7c}$$

$B_0^\dagger \equiv \sum_0 B_n^\dagger$ represents delocalized single-photon coherences, which, to lowest order in the external electric field, oscillate at the exciton frequency, $\omega_e = \Omega + 2V$. The $(N-1)/2$ two-exciton operators $C_q^\dagger \equiv C_{k=0,q}^\dagger$ represent delocalized two-photon coherences. Equations (1.5.7) imply that calculating the optical nonlinearities requires solving the dynamics of a set of $(N+1)/2$ coupled anharmonic oscillators; a single-exciton oscillator that represents intramolecular coherence and $(N-1)/2$ two-exciton oscillators that represent intermolecular coherences. Equations (1.5.7) can be solved perturbatively in the field E, by expanding $\langle B_0^\dagger \rangle$ and $\langle C_q^\dagger \rangle$ in a Taylor series in E and determining the expansion coefficients order by order. Solving Eqs. (5.7) to third order in E and calculating the polarization yields Eq. (1.5.3). This shows that the two-exciton oscillators are responsible for the cooperative enhancement of the optical response.

The present anharmonic-oscillator picture offers an alternative interpretation of the cooperative $\sim N^2$ enhancement and the crossover to $\sim N$ scaling discussed earlier. In accordance with the Pauli exclusion principle, a single site cannot be doubly excited, and as a result excitons are not true bosons (or fermions). Linear optical properties can be exactly evaluated using the boson approximation for single excitons [13], but the approximation breaks down for the dynamics of two-excitons and higher excitons, which are vital for the nonlinear optical response, and can therefore be probed via second and higher order nonlinear optical techniques. Spatially confining the exciton enhances the nonbosonic nature and hence the nonlinear susceptibilities (which would vanish if Frenkel excitons were bosons). The combined influence of the Pauli exclusion and excitonic confinement on the third order nonlinear hyperpolarizability results in the cooperative $\sim N^2$ scaling for small sizes and the $\sim N$ scaling for larger sizes where the excitons become more Boson-like.

In order to illustrate the spectroscopic manifestations of the two contributions to Eq. (5.5), let us consider the third order nonlinear absorption coefficient, $W_{TPA} \equiv \text{Im}\,\chi^{(3)}(-\omega; \omega, -\omega, \omega)$. Since W_{TPA} may change sign, and we wish to display it on a logarithmic scale, we have plotted the following function

$$K(x) \equiv \frac{x}{|x|} \log(1 + |x|) \tag{1.5.8}$$

where $x = W_{TPA}$, for $|x| \gg 1$, $K = (\text{sign } x) \log|x|$ whereas for $|x| \sim 0$ we have $K \sim x$. The $K(x)$ function is thus simply equal to x, if x is small, and it switches over to a logarithmic scale if $|x|$ is sufficiently large. In Fig. 2 we show $K(W_{TPA})$ as a function of ω for several size aggregates (as indicated in each panel). In the small aggregate region, $\chi^{(3)}(-\omega; \omega, -\omega, \omega)$ reduces, near resonance, to that of a single excitonic two-level system with a transition dipole $N^{1/2}\mu$, with additional two-photon resonances at $2\omega = 2\omega_q$ on the blue side of the spectrum (we have assumed $V < 0$). Positive values of W_{TPA} correspond to two-photon absorption, whereas negative values represent a bleaching of the excitonic line (saturated absorption). As the aggregate size increases, the $q = 1$ resonance moves towards the exciton line and eventually interferes destructively with it. For larger sizes, the destructive interference is complete, resulting in the cancellation of the $\sim N^2$ prefactor in $\chi^{(3)}(-\omega; \omega, -\omega, \omega)$. At this point W_{TPA} scales linearly with N; the red side of the spectrum has converged completely, while the blue side still shows

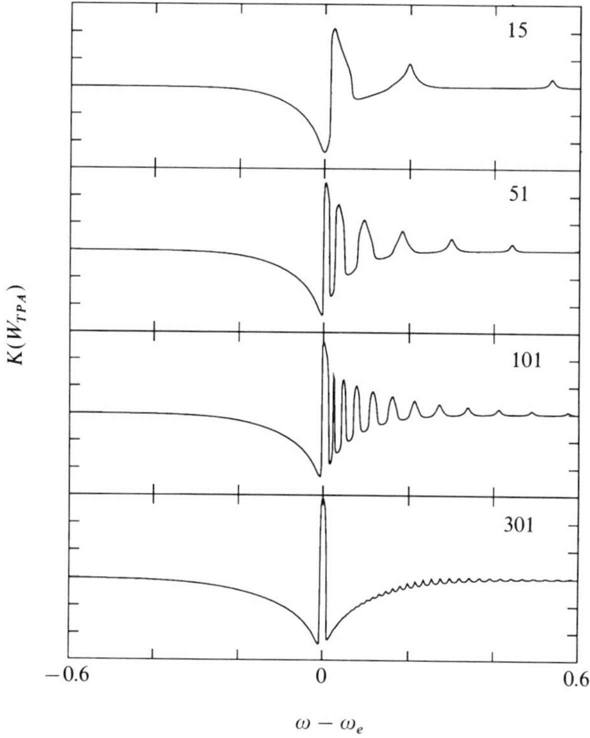

Fig. 2. The nonlinear two-photon absorption calculated using $\chi^{(3)}$ (Eq. 1.5.3). Shown is $K(W_{\text{TPA}})$ where $W_{\text{TPA}} = \text{Im}\,\chi^{(3)}(-\omega; \omega, -\omega, \omega)$ and the function K defined in Eq. (1.5.8). This function displays the signal on a logarithmic scale for large $|W_{\text{TPA}}|$ and on a linear scale for small $|W_{\text{TPA}}|$. Panels show the small aggregate and crossover to the bulk behavior. Note the reduction of the two-exciton blue-shift as the aggregate size is increased. $V = -1$, $\gamma_e = \gamma_q = 0.02$.

resolved two-photon resonances. These, however, becomes more congested as N increases, and eventually the finite experimental spectral resolution will exceed the line spacing, making the blue side a smooth function of $\Delta\omega$, with no N dependence.

In conclusion, we note that the analysis presented in this section and Eq. (1.5.3) constitute a special limiting case of the theory developed in [28b], in which the complete wave vector and frequency dependent susceptibility $\chi^{(3)}(-k_s - \omega_s; k_3\omega_3, k_2\omega_2, k_1\omega_1)$ was calculated. The temperature-dependent superradiant coherence size, which is related to the nonlinear coherence size N_c considered here, was also investigated [37]. It was also shown in [28b] that when $\langle B_n^\dagger B_m^\dagger \rangle$ is factorized into $\langle B_n^\dagger \rangle \langle B_m^\dagger \rangle$, we recover the single-particle expressions derived in Section 4. In general, when ω is close to the exciton resonance (ω_e), the local-field approximation is inadequate, completely missing the intermolecular nonlinearities; the local-field approximation does not reproduce the correct magnitude of $\chi^{(3)}$. In addition, within this approximation the small aggregate region is completely missed, the N^2 prefactor never appears, and the two-photon resonances are missed [28b].

6. EXCITON-POPULATION VARIABLES AND EXCITON TRANSPORT

In this section, we pursue further our study of nonlinear susceptibilities by adopting the factorization (Eq. 1.3.12) of the three-operator variables,

$$\langle B_n^\dagger B_n B_m \rangle = \langle B_n^\dagger B_n \rangle \langle B_m \rangle \equiv \langle W_n \rangle \langle B_m \rangle. \tag{1.6.1}$$

This factorization is expected to hold in the presence of sufficiently fast phonon-induced dephasing processes that destroy molecular coherences. Using the present factorization we therefore completely neglect intermolecular coherences and miss the cooperative enhancement associated with the two-excitonic variables as discussed in Sec. 5. However, the $\langle W_n \rangle$ variables allow the introduction of exciton transport into the nonlinear response. For the sake of clarity we focus our attention in this section on a specific four-wave mixing technique, degenerate four-wave mixing and its time-domain analogue, and the transient grating [55,22,23]. These techniques are particularly sensitive to the dynamics of the $\langle W_n \rangle$ variables.

6.1. Interaction-Induced Extra Resonances: Degenerate Four-Wave Mixing

Once the factorization (Eq. 1.6.1) is made, we need to derive an equation of motion for the exciton population variables $\langle W_n \rangle$. Instead of doing that, we consider the more general two-particle variables $\langle B_n^\dagger B_m \rangle$. In momentum space we define

$$W(\mathbf{k} - \mathbf{k}') \equiv \frac{1}{N} \sum_m W_m \exp[-i(\mathbf{k} - \mathbf{k}')\cdot \mathbf{r}_m] = \frac{2}{N} \sum_{\mathbf{k}''} B^\dagger_{\mathbf{k}'+\mathbf{k}''} B_{\mathbf{k}+\mathbf{k}''} \tag{1.6.2}$$

and

$$\begin{aligned} Q(\mathbf{k}, \mathbf{p}, t) &\equiv \langle B^\dagger_{\mathbf{p}-\mathbf{k}/2}(t) B_{\mathbf{p}+\mathbf{k}/2}(t) \rangle \\ &= \frac{1}{N} \sum_{m,n} \langle B_m^\dagger(t) B_n(t) \rangle \exp[-i\mathbf{k}\cdot(\mathbf{r}_m + \mathbf{r}_n)/2 + i\mathbf{p}\cdot(\mathbf{r}_m - \mathbf{r}_n)], \end{aligned} \tag{1.6.3}$$

which are the diagonal elements ($\mathbf{k} = \mathbf{0}$) and coherences ($\mathbf{k} \neq \mathbf{0}$) of the exciton density matrix in the momentum representation. From Eq. (1.6.3), it is clear that $\mathbf{k}$ is conjugate to the exciton center of mass, whereas $\mathbf{p}$ is related to the exciton momentum. The significance of the **k** and **p** variables may be clarified by switching to the Wigner representation for the exciton density matrix [59,60], which allows us to make the connection to common transport equations such as the Boltzmann equation [61]. This will be illustrated later in this section.

Using Eq. (1.6.1), we obtain the following equations of motion:

$$\frac{1}{i}\frac{d}{dt}\langle B_{\mathbf{k}}(t)\rangle = -(\Omega - i\Gamma(\mathbf{k}))\langle B_{\mathbf{k}}(t)\rangle + \frac{1}{\sqrt{N}}\rho\hbar^{-1}\sum_{\mathbf{k}'}\vec{\mu}\cdot\mathbf{E}_L(\mathbf{k}', t)[\delta_{\mathbf{k},\mathbf{k}'} - W(\mathbf{k}-\mathbf{k}', t)], \tag{1.6.4a}$$

$$\begin{aligned}\frac{1}{i}\frac{d}{dt}Q(\mathbf{k}, \mathbf{p}, t) = {} & [J(\mathbf{p}-\mathbf{k}/2) - J(\mathbf{p}+\mathbf{k}/2)]Q(\mathbf{k}, \mathbf{p}, t) - \sum_{\mathbf{p}'}\Sigma(\mathbf{k}; \mathbf{p}, \mathbf{p}')Q(\mathbf{k}, \mathbf{p}', t) \\ & - \frac{\sqrt{N}}{\hbar V}(\langle B_{\mathbf{p}+\mathbf{k}/2}(t)\rangle\vec{\mu}\cdot\mathbf{E}(-\mathbf{p}+\mathbf{k}/2, t) \\ & - \langle B^{\dagger}_{\mathbf{p}-\mathbf{k}/2}(t)\rangle\vec{\mu}\cdot\mathbf{E}(\mathbf{p}+\mathbf{k}/2, t)),\end{aligned} \tag{1.6.4b}$$

$$\mathrm{W}(\mathbf{k}, t) = \frac{2}{N}\sum_{\mathbf{p}} Q(\mathbf{k}, \mathbf{p}, t), \tag{1.6.4c}$$

with V being the quantization volume and the local-field E_L given by Eq. (1.4.2). The first r.h.s. term in Eq. (1.6.4b) describes the free-exciton motion, and the last term represents a source for two-particle coherences created by a combined effect of the electric fields and the polarization. The second r.h.s. term is due to the phonon bath, and $\Sigma(\mathbf{k}; \mathbf{p}, \mathbf{p}')$ is a complex self-energy matrix. Since the factorization Eq. (1.6.1) can be justified only in the presence of fast dephasing, it is essential to introduce phonon-induced relaxation into the present treatment (in contrast to the factorization used in Sec. 5, which is only valid in the absence of dephasing).

The limitations of the local-field approximation are apparent upon a close inspection of Eqs. (1.6.4). In Eqs. (1.6.4a) all intermolecular interactions are lumped into the local field. This is not the case however for Eq. (1.6.4b), where $J(\mathbf{k})$ enters not solely through the field and is responsible for transport processes, which the local-field approximation completely misses!

We shall now simplify the effects of phonons by adopting the following stochastic model for the exciton-phonon coupling (the Haken-Ströbl model) [60,62,63].

$$H_{\text{ex-phon}} = \hbar\sum_{n}\delta\Omega_n(t)B_n^{\dagger}B_n. \tag{1.6.5}$$

$\delta\Omega_n(t)$ is a stochastic Gaussian random variable with the following properties:

$$\langle \delta\Omega_n(t) \rangle = 0, \tag{1.6.6a}$$

$$\langle \delta\Omega_n(t)\delta\Omega_m(t) \rangle = \frac{\hat{\Gamma}}{2}\delta(\mathrm{t} - t')\delta_{m,n}. \tag{1.6.6b}$$

When the equations of motion are averaged over the stochastic part, we obtain the following relaxation equations [5,60]:

$$\left[\frac{d}{dt}\langle B_m(t)\rangle\right]_{ph} = -\tfrac{1}{2}(\hat{\Gamma} + \gamma)\langle B_m(t)\rangle, \tag{1.6.7a}$$

$$\left[\frac{d}{dt}\langle B_m^\dagger(t)B_n(t)\rangle\right]_{ph} = -\Big[\hat{\Gamma}(1 - \delta_{nm}) + \gamma]\langle B_m^\dagger(t)B_n(t)\rangle, \tag{1.6.7b}$$

where $[\cdots]_{ph}$ denotes the phonon contribution. These equations imply for the self-energies [5]

$$\Gamma(\mathbf{k}) = (\hat{\Gamma} + \gamma)/2, \tag{1.6.8a}$$

$$\Sigma(\mathbf{k}; \mathbf{p}, \mathbf{p}') = -i(\hat{\Gamma} + \gamma)\delta_{pp'} + i\hat{\Gamma}/N. \tag{1.6.8b}$$

The parameters $\hat{\Gamma}$ and γ are taken to be real. $\hat{\Gamma}$ is the pure dephasing rate, and γ represents the population relaxation rate. This model allows for analytical results while still preserving the essential physical aspects related to pure dephasing.

When Eqs. (1.6.4) are transformed to the frequency domain and solved iteratively, we obtain the nonlinear susceptibility $\chi^{(3)}$. For the sake of clarity we shall now restrict our attention to a specific technique: degenerate four-wave mixing (D4WM), which provides a direct probe for the exciton-population variables. This technique uses two incoming fields $\mathbf{k}_1$, $\mathbf{k}_2$, and the signal is observed at $\mathbf{k}_s = 2\mathbf{k}_1 - \mathbf{k}_2$ and $\omega_s = 2\omega_1 - \omega_2$. The degenerate four-wave mixing (D4WM) techniques studies the signal in the vicinity of $\omega_1 = \omega_2$, looking for a sharp resonance. The signal intensity $R(\mathbf{k}_s, \omega_s)$ is, within the slowly varying amplitude approximation, proportional to $|\chi^{(3)}(-\mathbf{k}_s - \omega_s; \mathbf{k}_1\omega_1, -\mathbf{k}_2 - \omega_2, \mathbf{k}_1\omega_1)|^2$. The solution of Eqs. (1.6.4) results in the following expression for $\chi^{(3)}$ relevant for D4WM [5,55]:

$$\begin{aligned}
&\vec{\chi}^{(3)}(-\mathbf{k}_s, -\omega_s; \mathbf{k}_1\omega_1, -\mathbf{k}_2 - \omega_2, \mathbf{k}_1\omega_1) \\
&= 2\rho\frac{\vec{\mu}\vec{\mu}\vec{\mu}\vec{\mu}}{\hbar^3}\frac{1}{\omega_s - \Omega_{\mathbf{k}_s} + i(\hat{\Gamma} + \gamma)/2}\frac{1}{\omega_1 - \Omega_{\mathbf{k}_1} + i(\hat{\Gamma} + \gamma)/2}\frac{1}{\omega_2 - \Omega_{\mathbf{k}_2} - i(\hat{\Gamma} + \gamma)/2} \\
&\times\left\{1 - \frac{\hat{\Gamma}}{N}\sum_{\mathbf{p}}\Big[-i\omega_{12} - iJ(\mathbf{p} - \mathbf{k}_g/2) + iJ(\mathbf{p} + \mathbf{k}_g/2) + \hat{\Gamma} + \gamma]^{-1}\right\}^{-1}.
\end{aligned} \tag{1.6.9}$$

The $\mathbf{k}_g \equiv \mathbf{k}_1 - \mathbf{k}_2$ is the grating wavevector. This result is the product of the rotating-wave (resonant) part of Eq. (1.4.7), obtained using the single-particle factorization, and a dephasing-induced correction factor given by the curly brackets, which is unity only if $\hat{\Gamma} = 0$. We note that the first three denominators in Eq. (1.6.9) depend on the frequencies ω_1, ω_2, and $\omega_s \equiv 2\omega_1 - \omega_2$ (single- and three-photon resonances) but not on $\omega_{12} \equiv \omega_1 - \omega_2$, which is a two-photon resonance. The only $\omega_1 - \omega_2$ frequency dependence in $R(\mathbf{k}_s, \omega_s)$ that may give a sharp resonance is then contained in the last (dephasing-induced) factor in Eq. (1.6.9), which is completely missed by the single-particle factorization. Three limiting cases are of special interest [5,55].

1. We first consider molecules with arbitrary interactions in the absence of dephasing. For $\hat{\Gamma} = 0$, the Haken-Ströbl model describes coherent exciton motion on the lattice, and from Eq. (1.6.9) it is clear that the D4WM signal exhibits no resonance as a function of ω_{12} in this limit.
2. For noninteracting molecules ($J(\mathbf{k}) = 0$) we find

$$R(\mathbf{k}_s, \omega_s) \propto 1 + \frac{\hat{\Gamma}(\hat{\Gamma} + 2\gamma)}{\omega_{12}^2 + \gamma^2}. \tag{1.6.10}$$

   This signal has a Lorentzian resonance at $\omega_{12} = 0$, whose width is the inverse of the excited state lifetime. The resonance disappears in the absence of dephasing. These dephasing-induced resonances have been observed by Bloembergen *et al.* [64] in the gase phase and have been denoted PIER4 (pressure induced extra resonances in four-wave mixing). In molecular crystals similar resonances were observed by Hochstrasser *et al.* [65] and have been denoted DICE (dephasing induced coherent emission). The unique and surprising aspect of these resonances is that usually dephasing results in line-broadening associated with the loss of coherence, whereas here it induces new sharp resonances as ω_{12} is varied. This is a consequence of a delicate interference of various terms contributing to $\chi^{(3)}$, which exactly cancel in the absence of dephasing. The addition of dephasing eliminates this cancellation and results in the new resonance [42].
3. We finally discuss the case of finite interactions in the strong dephasing (incoherent or diffusive) limit, defined by $\hat{\Gamma} \gg |J(\mathbf{p} - \mathbf{k}_g/2) - J(\mathbf{p} + \mathbf{k}_g/2)|$ (for all $\mathbf{p}$) and $\hat{\Gamma} \gg \gamma$. In this limit, the Haken-Ströbl model describes diffusive exciton motion, and we have

$$R(\mathbf{k}_s, \omega_s) \propto 1 + \frac{\hat{\Gamma}(\hat{\Gamma} + 2\gamma)}{\omega_{12}^2 + [\gamma + (\mathbf{k}_1 - \mathbf{k}_2)^2 D_e]^2}, \tag{1.6.11}$$

with the exciton diffusion constant given by

$$D_e \equiv \frac{1}{|\mathbf{k}_g|^2} \frac{1}{\hat{\Gamma} N} \sum_{\mathbf{p}} [J(\mathbf{p} - \mathbf{k}_g/2) - J(\mathbf{p} + \mathbf{k}_g/2)]^2$$

$$= \frac{4}{|\mathbf{k}_1 - \mathbf{k}_2|^2} \frac{1}{\hat{\Gamma}} \sum_m J^2(\mathbf{r}_m) \sin^2 \left[\frac{(\mathbf{k}_1 - \mathbf{k}_2) \cdot \mathbf{r}_m}{2} \right]. \tag{1.6.12}$$

In the incoherent limit the D4WM signal shows a Lorentzian resonance with a width that is the sum of the inverse excited state lifetime (γ) and a contribution from exciton diffusion ($D_e(\mathbf{k}_1 - \mathbf{k}_2)^2$). A similar result holds also for a disordered medium [19].

6.2. Exciton Transport in Real Space: The Wigner Representation

Further physical insight into the equation of motion (1.6.4b) for the two-particle coherences $Q(\mathbf{k}, \mathbf{p}, t)$ can be obtained by making the connection to macroscopic transport equations, such as the Boltzmann equation and the diffusion equation [61]. To this end, we define the Wigner phase-space distribution function $\phi(\mathbf{r}, \mathbf{p}, t)$ by [5,59,60]

$$\phi(\mathbf{r}, \mathbf{p}, t) \equiv \frac{1}{\sqrt{N}} \sum_{\mathbf{k}} Q(\mathbf{k}, \mathbf{p}, t) \exp[i\mathbf{k} \cdot \mathbf{r}]. \tag{1.6.13}$$

The equation of motion (1.6.4b) for the Wigner distribution reads

$$\frac{d}{dt} \phi(\mathbf{r}, \mathbf{p}, t) = 2 \sum_{\mathbf{a}} J(\mathbf{a}) \sin(\mathbf{p} \cdot \mathbf{a}) \phi(\mathbf{r} + \mathbf{a}/2, \mathbf{p}, t)$$

$$- \hat{\Gamma} \sum_{\mathbf{p}'} [g(\mathbf{p}')\phi(\mathbf{r}, \mathbf{p}, t) - g(\mathbf{p})\phi(\mathbf{r}, \mathbf{p}', t)] - \gamma\phi(\mathbf{r}, \mathbf{p}, t), \tag{1.6.14a}$$

with

$$\langle W_n(t) \rangle = \frac{2}{N} \sum_{P} \phi(\mathbf{r}_n, \mathbf{p}, t). \tag{1.6.14b}$$

The first term in Eq. (1.6.14a) describes coherent exciton motion on the lattice. When the interaction is symmetric ($J(\mathbf{r}) = J(-\mathbf{r})$) and has a short range, it reduces to $\mathbf{p}/m^* \cdot \vec{\nabla}_{\mathbf{r}} \phi(\mathbf{r}, \mathbf{p}, t)$ with the effective mass given by $(m^*)^{-1} = (1/d) \sum_n J(r_n) r_n^2$, d being the dimensionality of the system. Here, n runs over the lattice and $r_n \equiv |\mathbf{r}_n|$. The second term in this equation has the form of the BGK strong collision operator in the Boltzmann equation [61], in which collisions occur with rate $\hat{\Gamma}$, and the momentum after each collision is distributed according to the equilibrium momentum distribution $g(\mathbf{p})$. The strong collision operator conserves the number of particles (population); a population loss with rate γ is described by the last term. If we take $g(\mathbf{p}) = 1/N$, where N is the number of lattice sites, then Eq. (1.6.14a) is identical to the Haken-Ströbl model. The Haken-Ströbl model is therefore a high-temperature model, as the equilibrium distribution is then uniform over all momenta. Equation (1.6.14a) can also be connected to other standard transport equations (the diffusion and the master equation) [60].

6.3. Transient Grating: The Time-Domain Analogue of Degenerate Four-Wave Mixing

The exciton-population $\langle B_n^\dagger B_n \rangle$ variables can also be probed by an elegant time-domain technique: transient grating (TG) spectroscopy, which may be related to the Fourier transform of the degenerate four-wave mixing [5,22,23]. The following typical setup is considered. At time $t = 0$ two short excitation pulses, $(\mathbf{k}_1, \omega_1)$ and $(\mathbf{k}_2, \omega_2)$, crossed with an angle θ, interfere in the sample and create an excitonic grating. The decay of the grating as a result of dephasing and population relaxation is monitored by applying a probe pulse, $(\mathbf{k}_3, \omega_3)$, at $t = \tau$. The observable $R(\tau)$ is the time-integrated intensity of the nonlinear ("diffracted") signal with wave vector $\mathbf{k}_s = \mathbf{k}_1 - \mathbf{k}_2 + \mathbf{k}_3$ and frequency $\omega_s = \omega_1 - \omega_2 + \omega_3$ as a function of the pump-probe delay τ. We shall consider now the three cases discussed in section (6.1).

1. In the absence of dephasing, the TG signal never decays.
2. For noninteracting molecules we get

$$R(\tau) = \exp[-2(\gamma + \hat{\Gamma})\tau]. \tag{1.6.15}$$

   In the extreme case of $\hat{\Gamma} = 0$, no grating decay is observed, apart from the trivial population decay $\exp(-2\gamma t)$. This result is rigorous for all times, and is easily understood: for $\hat{\Gamma} = 0$ the initial state created by the incoming pulses is an exact eigenstate of the crystal. Dephasing destroys the grating and contributes to the decay of the signal.
3. We now turn to the opposite limit of strong scattering, $\hat{\Gamma} \gg \gamma$ and $\hat{\Gamma} \gg J$, where the exciton scattering length is much smaller than the grating length scale (diffusive or incoherent motion). The signal intensity is then given by

$$R(\tau) = \exp[-2\gamma\tau - 2D_e|\mathbf{k}_1 - \mathbf{k}_2|^2\tau]. \tag{1.6.16}$$

   The signal decay rate consists of a wavevector-independent contribution due to population relaxation (2γ) and a contribution from the exciton motion, which is proportional to $|\mathbf{k}_1 - \mathbf{k}_2|^2$. This is characteristic for diffusive motion [22,23] and leads for small cross angles θ between the two pump pulses, to a linear relation between the observed decay rate and θ^2. This relation provides a clear experimental distinction between diffusive and coherent exciton motion. In the incoherent limit, an interesting relation exists between the D4WM and TG signals. Namely, the amplitude of the D4WM signal (Eq. 1.6.11) can be obtained by evaluating the Fourier transform of the amplitude of the TG signal (Eq. 1.6.16) at the frequency ω_{12}.

Finally, we emphasize that we have only studied D4WM and the TG within the two-particle description. The single-particle description would lead to exponential decay independent of the magnitude of $\hat{\Gamma}$. Diffusive behavior cannot be found in the single-particle level of description, which is intimately related to the fact that in the single-particle factorization no two-photon ω_{12} resonance is found in the D4WM signal.

7. GREEN FUNCTION EXPRESSIONS FOR $\chi^{(3)}$ IN MOLECULAR NANOSTRUCTURES WITH ARBITRARY GEOMETRY

In the previous sections we have considered separately two important sources of optical nonlinearities. The $\langle B_n^\dagger B_n \rangle$ terms constitute local contributions whereas the two-excitonic variables $\langle B_n^\dagger B_m^\dagger \rangle$ represent nonlocal contributions. Applications were made to periodic assemblies (one-dimensional and three-dimensional). In this section we derive more general equations of motion that allow us to calculate the nonlinear susceptibility $\chi^{(3)}$ of assemblies of molecules with nonoverlapping charge distributions and an arbitrary geometry. The equations incorporate the contribution of all quadratic (bilinear) and cubic exciton variables and include the results of the previous sections as special cases. The equations can be solved using Green function techniques [39]. We provide an explicit solution in the absence of relaxation [66] and present some numerical calculations for one-, two-, and three-dimensional assemblies. We first invoke the Heitler-London approximation where the dipole dipole interaction in Eq. (1.2.2b) is replaced by

$$\frac{\hbar}{2} \sum_{m,n}{}' J(\mathbf{r}_{mn})(B_m^\dagger B_n + B_m B_n^\dagger). \tag{1.7.1}$$

This approximation implies that we work in first-order perturbation theory in the small parameter J/Ω, where J is the interaction energy between molecules and Ω is the molecular excited-state energy. It allows us to classify the variables by their nonlinear order with respect to the field. (It is possible to develop a perturbation theory and calculate corrections of higher order in J/Ω.) We further introduce the notion of *normal ordering* of operators, *i.e.*, in any product of operators, all $B^\dagger$ are to the left and all B to the right:

$$B_n^\dagger B_m^\dagger \ldots B_{n'} B_{m'} \ldots$$

By using the commutation rules we can bring any product to a normally ordered form. We next note that the expectation value of any normally ordered product of n operators (be it B or $B^\dagger$ must be at least to n'th order in the field. This is the case since initially the density matrix $\rho(0)$ is in the vacuum state. For the sake of evaluating the third-order response, it is thus sufficient to consider only products with up to three B or $B^\dagger$ operators.

Starting with Eq. (1.3.1), we now derive equations of motion for all relevant variables. In addition to the electronic degrees of freedom, we shall incorporate relaxation induced by coupling to nuclear degrees of freedom. We have outlined earlier a model for damping that results in a simple prescription of how to incorporate it in the equations of motion once expectation values are taken. The model depends on two real parameters $\hat{\Gamma}$ and γ. $\hat{\Gamma}$ is the pure dephasing rate, and γ represents the population relaxation rate. We use the notation $\Gamma \equiv \hat{\Gamma} + \gamma$ for the total dephasing rate in the sequel.

$$\left[\frac{d}{dt}\langle B_m^\dagger(t)B_n(t)\rangle\right]_{\text{ph}} = -[\hat{\Gamma}(1-\delta_{nm}) + \gamma]\langle B_m^\dagger(t)B_n(t)\rangle, \tag{1.7.2a}$$

and

$$\left[\frac{d}{dt}\langle B_n^\dagger(t)B_m(t)B_{n'}(t)\rangle\right]_{\text{ph}} = -[\hat{\Gamma}(\tfrac{3}{2}-\delta_{n,m}-\delta_{nn'}) + \tfrac{3}{2}\gamma]\langle B_n^\dagger(t)B_m(t)B_{n'}(t)\rangle. \tag{1.7.2b}$$

We further introduce the notation

$$\zeta_{nm} \equiv 1 - \delta_{nm}.$$

The equations of motion now read

$$\frac{1}{i}\frac{d}{dt}\langle B_n\rangle = \sum_m F_{nm}\langle B_m\rangle + 2\sum_m{}' J_{nm}\langle B_n^\dagger B_n B_m\rangle + \mu_n\frac{E_n(t)}{\hbar}(1 - 2\langle B_n^\dagger B_n\rangle), \tag{1.7.3a}$$

$$\frac{1}{i}\frac{d}{dt}\langle B_n B_m\rangle = \zeta_{nm}\sum_l (F_{n,l}\langle B_l B_m\rangle + F_{m,l}\langle B_n B_l\rangle) + \mu_n\zeta_{nm}\frac{E_n(t)}{\hbar}\langle B_m\rangle + \mu_m\zeta_{nm}\frac{E_m(t)}{\hbar}\langle B_n\rangle, \tag{1.7.3b}$$

$$\frac{1}{i}\frac{d}{dt}\langle B_n^\dagger B_m\rangle = \sum_l (F_{n,l}\langle B_n^\dagger B_l\rangle - F_{n,l}^*\langle B_l^\dagger B_m\rangle) - i\hat{\Gamma}\delta_{nm}\langle B_n^\dagger B_m\rangle$$

$$- \frac{\mu_n E_n^*(t)}{\hbar}\langle B_m\rangle + \frac{\mu_m E_m(t)}{\hbar}\langle B_n^\dagger\rangle, \tag{1.7.3c}$$

$$\frac{1}{i}\frac{d}{dt}\langle B_n^\dagger B_m B_l\rangle = \xi_{ml}\sum_i [F_{m,i}\langle B_n^\dagger B_i B_l\rangle + F_{l,i}\langle B_n^\dagger B_m B_i\rangle$$

$$- F_{n,i}^*\langle B_i^\dagger B_m B_l\rangle] - i\hat{\Gamma}\xi_{ml}(\delta_{nm} + \delta_{nl})\langle B_n^\dagger B_m B_l\rangle$$

$$- \frac{\mu_n E_n^*(t)}{\hbar}\xi_{ml}\langle B_m B_l\rangle + \frac{\mu_m E_m(t)}{\hbar}\zeta_{ml}\langle B_n^\dagger B_l\rangle$$

$$+ \frac{\mu_1 E_1(t)}{\hbar}\zeta_{ml}\langle B_n^\dagger B_m\rangle. \tag{1.7.3d}$$

with

$$F_{nm} \equiv -\Omega_n\delta_{nm} - J_{nm} + i\frac{\Gamma}{2}\delta_{nm}. \tag{1.7.4}$$

The ξ_{nm} factors in these equations result from the Pauli exclusion and guarantee that $\langle B_n B_n\rangle = 0$ at all times (only a single exciton is allowed at a given site). Their presence complicates the solution. Technically it is possible to use a different model in which excitons are bosonlike and omit the ξ factors [39,66]. To account for the Pauli exclusion we then add a repulsive exciton–exciton interaction and change the frequency of $\langle B_n B_m\rangle$ in Eq. (1.7.3b) from $\Omega_n + \Omega_m$ to $\Omega_n + \Omega_m + g$. This model is easier to solve. By sending $g \to \infty$ at the end, we exclude the two exciton states and obtain the solution of Eqs. (1.7.3).

We next introduce the polarization of the nth molecule

$$P_n \equiv \vec{\mu}(B_n + B_n^\dagger).$$

In order to calculate the linear response we neglect the nonlinear terms in Eq. (1.7.3a), which contribute only to third (and higher) orders in the field, and obtain

$$P_n^{(1)}(\omega) = \sum_m \alpha_{nm}(\omega)E_m(\omega),$$

with the linear polarizability

$$\alpha_{nm}(\omega) = |\mu|^2[G_{nm}(\omega) + G_{nm}^*(-\omega)]. \tag{1.7.5}$$

Here G_{nm} is the single-exciton Green function defined as follows:

$$G(\omega) = (\omega - F + i\eta)^{-1}. \tag{1.7.6}$$

The single-exciton Green function can be evaluated explicitly by introducing the one-exciton wavefunctions $\psi_\alpha(n)$:

$$\Psi_\alpha = \sum_n \psi_\alpha(n) B_n^\dagger |0\rangle, \tag{1.7.7a}$$

which satisfy the equation

$$H_{\text{ex}} \Psi_\alpha = \varepsilon_\alpha \Psi_\alpha. \tag{1.7.7b}$$

H_{ex} is given by Eq. (1.2.2b), using the Heitler-London approximation (Eq. (1.7.1)). In the frequency domain we then have

$$G_{nm}(\omega) = \sum_\alpha \frac{\psi_\alpha(n)\psi_\alpha^*(m)}{\omega - \varepsilon_\alpha + i\eta}, \tag{1.7.8}$$

where η is a small positive number that is set to zero at the end of the calculation.

We next turn to the calculation of the third-order nonlinear polarization $\mathbf{P}^{(3)}$. For weak fields, the polarization (and the expectation values of the B_n and $B_n^\dagger$ operators) is small. To find the nonlinear response we can therefore make a perturbative expansion in the number of B factors, provided they are normally ordered. We thus calculate $\langle B_n \rangle^{(1)}$ to first order, then solve for $\langle B_n^\dagger B_m \rangle^{(2)}$, $\langle B_n B_m \rangle^{(2)}$, and finally substitute these back in Eq. (1.7.3a) and solve for $\langle B_n \rangle^{(3)}$.

A Green function solution of Eqs. (1.7.4) was derived in [39] using the factorization (1.3.13). Here we consider the simpler case where relaxation is negligible $\hat{\Gamma} = \gamma = 0$, so that the system is in a pure state at all times, and the bras and the kets can be factorized. $\langle B_n^\dagger B_m \rangle = \langle B_n^\dagger \rangle \langle B_m \rangle$ and $\langle B_n^\dagger B_m B_l \rangle = \langle B_n^\dagger \rangle \langle B_m B_l \rangle$. We then need to solve only Eqs. (1.7.3a) and (1.7.3b). Their Green function solution is [66]:

$$\mathbf{p}_n^{(3)}(\omega_s) = \frac{1}{(2\pi)^2} \int d\omega_1\, d\omega_2\, d\omega_3\, \delta(\omega_s - \omega_1 - \omega_2 - \omega_3) \times \sum_{m_1, m_2, m_3} \gamma_{nm_1m_2m_3}(-\omega_s; \omega_1, \omega_2, \omega_3) E_{m_1}(\omega_1) E_{m_2}(\omega_2) E_{m_3}(\omega_3), \tag{1.7.9}$$

with the third order polarizability

$$\gamma_{nm_1m_2m_3}(-\omega_s;\omega_1,\omega_2,\omega_3)$$
$$= |\mu|^4 \frac{1}{6}\sum_p \sum_{n',n''} \{G_{nn'}(\omega_s)G^*_{n'm_3}(-\omega_3)G_{n''m_2}(\omega_2)G_{n''m_1}(\omega_1)\bar{\Gamma}_{n'n''}(\omega_1+\omega_2)$$
$$+ G^*_{nn'}(-\omega_s)G_{n'm_3}(\omega_3)G^*_{n''m_2}(-\omega_2)G^*_{n''m_1}(-\omega_1)\bar{\Gamma}^*_{n'n''}(-\omega_1-\omega_2)\}. \tag{1.7.10}$$

Here $\bar{\Gamma}$ is a two-exciton scattering matrix given by

$$\bar{\Gamma}_{n'n''}(\omega) = -2[\mathscr{F}^{-1}(\omega)]_{n'n''}, \tag{1.7.11a}$$

where the $\mathscr{F}$ matrix is

$$\mathscr{F}_{n'n''}(\omega) \equiv \int \frac{d\omega'}{2\pi i} G_{n''n'}(\omega')G_{n''n'}(\omega-\omega'),$$

or, in terms of the single exciton eigenstates,

$$\mathscr{F}_{n'n''}(\omega) = \sum_{\alpha',\alpha} \frac{\psi_{\alpha'}(n')\psi_{\alpha}(n')\psi^*_{\alpha'}(n'')\psi^*_{\alpha}(n'')}{\omega-\varepsilon_{\alpha'}-\varepsilon_{\alpha}+i\eta}, \tag{1.7.11b}$$

and the p sum is over all six permutations of three pairs (m_1,ω_1), (m_2,ω_2), (m_3,ω_3).

The above Green-function expression holds for a molecular assembly with arbitrary geometry. Nanostructures and superlattices often have a periodic geometry, where these expressions can be simplified further. Consider a periodic system in which the molecules occupy a d-dimensional lattice with lattice constant a. Each lattice site is occupied by a two-level molecule with a transition dipole moment μ.

To make use of the translational symmetry of the problem we recast the Green functions in momentum (**k**) space. We first perform a d-dimensional Fourier transform of the external field. We thus have

$$E(\mathbf{k},\omega) \equiv \sum_{\mathbf{n}} E(\mathbf{r_n},\omega)\exp(-i\mathbf{k}\cdot\mathbf{r_n}),$$

where the wavevector $\mathbf{k}$ [with components $k_j\, j=1,\ldots,d$] is defined in the first Brillouin Zone $0 \le k_j \le (2\pi)/a$. Similarly we define the Green function

$$G(\mathbf{k},\omega) \equiv \sum_{\mathbf{m}} G_{\mathbf{mn}}(\omega)\exp[-i\mathbf{k}\cdot(\mathbf{r_m}-\mathbf{r_n})].$$

In a periodic geometry with $\Omega_n = \Omega$ independent on n, the single exciton states have a well-defined momentum $\mathbf{k}$, *i.e.*, $\alpha = \mathbf{k}$ and energy $\Omega_\mathbf{k}$

$$\left.\begin{aligned} \psi_\alpha(\mathbf{n}) &= \exp(i\mathbf{k}\cdot\mathbf{r_n}), \\ \Omega_\mathbf{k} &= \Omega + \sum_\mathbf{n} \exp(-i\mathbf{k}\cdot\mathbf{r_n})J_{\mathbf{n}0}. \end{aligned}\right\} \tag{1.7.12b}$$

We then have:

$$G(\mathbf{k}, \omega) = \frac{1}{\omega - \Omega_k + i\eta}. \tag{1.7.13}$$

Eqs. (1.7.5) then yield

$$P^{(1)}(\mathbf{k}, \omega) = \chi^{(1)}(\mathbf{k}, \omega)E(\mathbf{k}, \omega), \tag{1.7.14}$$

with

$$\chi^{(1)}(\mathbf{k}, \omega) = -\frac{2\Omega_\mathbf{k}|\mu|^2}{(\omega + i\eta)^2 - \Omega_\mathbf{k}^2}. \tag{1.7.15}$$

Eqs. (1.7.9) and (1.7.10) then assume the forms

$$\begin{aligned} P^{(3)}(\mathbf{k}_s, \omega_s) = {} & \frac{a^{2d}}{(2\pi)^{2(d+1)}} \int d\omega_1\, d\omega_2\, d\omega_3\, d\mathbf{k}_1\, d\mathbf{k}_2\, d\mathbf{k}_3 \\ & \times \delta(\omega_s - \omega_1 - \omega_2 - \omega_3)\, \delta(\mathbf{k}_s - \mathbf{k}_1 - \mathbf{k}_2 - \mathbf{k}_3)E(\mathbf{k}_1, \omega_1) \\ & \times E(\mathbf{k}_2, \omega_2)E(\mathbf{k}_3, \omega_3)\chi^{(3)}(-\mathbf{k}_s, -\omega_s; \mathbf{k}_1\omega_1, \mathbf{k}_2\omega_2, \mathbf{k}_3\omega_3) \end{aligned} \tag{1.7.16}$$

$$\begin{aligned} \chi^{(3)}(-\mathbf{k}_s, -\omega_s; \mathbf{k}_1\omega_1, \mathbf{k}_2\omega_2, \mathbf{k}_3\omega_3) = {} & |\mu|^4 \frac{1}{6} \sum_p \\ & \times \{G(\mathbf{k}_s, \omega_s)\mathbf{G}^*(-\mathbf{k}_3, -\omega_3)G(\mathbf{k}_1, \omega_1)G(\mathbf{k}_2, \omega_2)\bar{\Gamma}(\mathbf{k}_1 + \mathbf{k}_2, \omega_1 + \omega_2) + \\ & \times G^*(-\mathbf{k}_s, -\omega_s)G(\mathbf{k}_3, \omega_3)G^*(-\mathbf{k}_1, -\omega_1)G^*(-\mathbf{k}_2, -\omega_2) \\ & \times \bar{\Gamma}^*(-\mathbf{k}_1 - \mathbf{k}_2, -\omega_1 - \omega_2)\}. \end{aligned} \tag{1.7.17}$$

Here the polarization field has a frequency ω_s and momentum $\mathbf{k}_s$, and

$$\bar{\Gamma}(\mathbf{k}, \omega) = -2\left(\frac{2\pi}{a}\right)^d \left[\int d\mathbf{k}_1(\omega - \Omega_{\mathbf{k}_1} - \Omega_{\mathbf{k}-\mathbf{k}_1} + i\eta)^{-1}\right]^{-1}. \tag{1.7.18}$$

In concluding this discussion, let us go back to the local field approximation, where we adopt the factorization (Eq. (1.3.10)) and we therefore need to solve only Eq. (1.7.3a). Consider a homogeneous excitation of an assembly of identical molecules so that $E_n(t) = E(t)$, $\mu_n = \mu$ and $\Omega_n = \Omega$. In this case we have

$$\psi(t) \equiv \langle B_n(t) \rangle,$$

independent on n.

Eq. (1.7.3a) then assumes the form

$$\frac{1}{i}\frac{d\psi}{dt} = \left(-\Omega - J(k=0) + i\frac{\Gamma}{2}\right)\psi + \mu\frac{E(t)}{\hbar}[1 - 2|\psi|^2] + 2J(k=0)|\psi|^2\psi.$$

This local field equation, which is equivalent to Eq. (1.4.1) [19], is closely related to the Landau-Ginzburg theory of phase transitions and has been used to interpret time-domain four-wave mixing measurements in semiconductor nanostructures [67,68,35].

We shall now present numerical calculations for periodic structures with dipolar interactions. We express energies in units of $J \equiv \mu^2/a^3$. We start with a three-dimensional lattice [39]. In the following calculations we aim for the infinite lattice, and our numerical results use finite lattices with periodic boundary conditions. For the three-dimensional calculations we have taken a $79 \times 79 \times 79$ simple cubic lattice, using the minimal-image convention [69]. We shall focus on two four-wave mixing techniques: The two-photon absorption signal is given by

$$W_{\mathrm{TPA}} = \mathrm{Im}\,\chi^{(3)}_{\mathrm{TPA}}, \tag{1.7.13a}$$

where

$$\chi^{(3)}_{\mathrm{TPA}} \equiv \chi^{(3)}(-\mathbf{k} - \omega; \mathbf{k}\omega, -\mathbf{k} - \omega, \mathbf{k}\omega). \tag{7.13b}$$

The two-exciton contribution to W_{TPA} was already displayed in Fig. 2.

The third harmonic signal is given by

$$W_{\mathrm{THG}} = |\chi^{(3)}_{\mathrm{THG}}|^2, \tag{7.27a}$$

where

$$\chi^{(3)}_{\mathrm{THG}} \equiv \chi^{(3)}(-3\mathbf{k} - 3\omega; \mathbf{k}\omega, \mathbf{k}\omega, \mathbf{k}\omega). \tag{7.28b}$$

The two-photon absorption as well as the third-harmonic generation depend in a trivial way on the isolated-molecule transition frequency Ω. We have taken $\Omega = 10^4/J$ in the following calculations.

We first consider two-photon absorption. There is one resonance for TPA. In Fig. 3 we present the TPA signal in three dimensions. We see that the cooperativity gives a large enhancement of the signal, compared with the local-field approximation even for the quite strong damping $\Gamma/J = 1$, and essentially no shift of the resonance.

We next consider third-harmonic generation (THG). In this case there are two resonances, a single-photon resonance at $\omega = \Omega$ and a three-photon

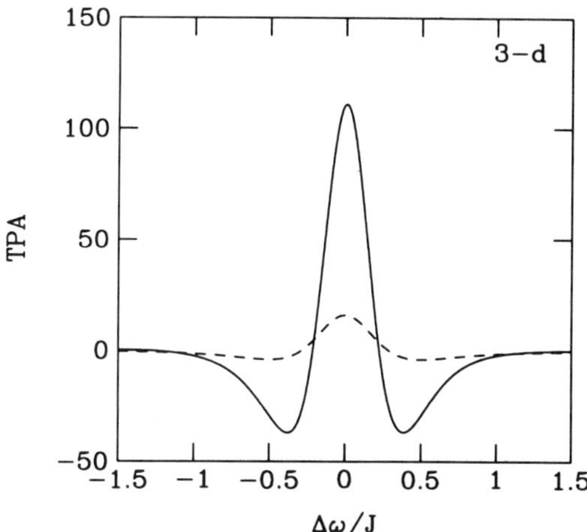

Fig. 3. Two-photon absorption W_{TPA} for a molecular assembly in three W_{TPA} dimensions, $\Delta\omega = \omega - \Omega_s$, at dephasing rate $\Gamma/J = 1, \Omega/J = 10^4$. $\Omega s = \Omega + J(k = 0)$ is the single exciton energy. Solid line: numerical results for a $79 \times 79 \times 79$ lattice. Dashed line: local-field approximation, note that $\bar{J} = 0$. The absorption is enhanced by a factor 7 compared with the local field approximation [39].

resonance at $\omega = \Omega/3$. In Fig. 4 we display the THG signal near these two resonances. Note the absence of cooperativity near $\omega = \Omega/3$, where the two-photon levels are nonresonant. In both Figs. 3 and 4, the local-field approximation misses the enhancement entirely.

We next turn to two-dimensional aggregates. We assume that the transition dipole is skewed to the plane at the angle $\theta = 35°$ with $3\cos^2\theta - 1 = 0$, where the one-exciton energy shift vanishes, and the local-field correction vanish. This is intermediate between the H aggregate configuration ($\theta = 90°$) and the J aggregate configuration ($\theta = 0$). The following calculations were made using a square 199×199 lattice. In order to observe directly the deviation from the local-field approximation (cooperative enhancement) we consider a two-color two-photon absorption experiment. We took frequencies such that single-photon transitions are off-resonant, so that we see only the two-photon resonances in the cooperative enhancement factor. The results are displayed in Fig. 5. Notice the complete absence of a resonance in the local-field approximation.

We have further solved Eqs. (1.7.3) approximately for a one-dimensional aggregate with N molecules, using a somewhat different relaxation matrix (this is not essential for the present discussion) [70]. The resulting expression

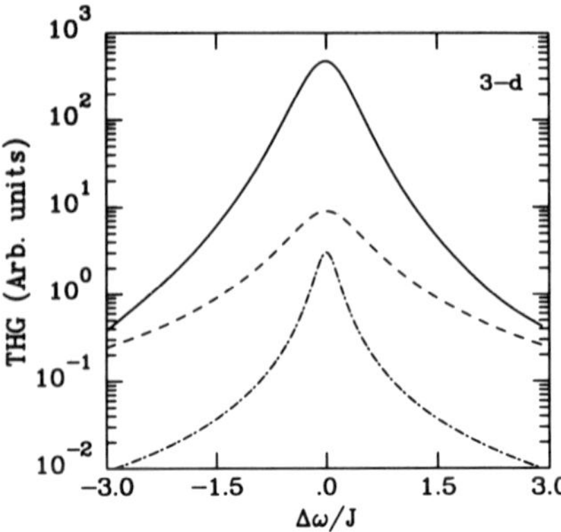

Fig. 4. The third-harmonic signal W_{THG} in three dimensions at dephasing $\Gamma/J = 1$. Solid line: Numerical results for the response near the single-photon resonance Ω_s, $\Delta\omega = \omega - \Omega_s$. Dashed line: local-field result at the same resonance. The enhancement now is 48, the square of the enhancement of the TPA (Fig. 3). The dot-dashed curve shows the three-photon resonance ($\Delta\omega = \omega - \Omega_s/3$). It is actually a superposition of two curves, one dashed, the other dotted corresponding to the numerical solution and the local-field results, respectively. Note the complete absence of enhancement for this resonance [39].

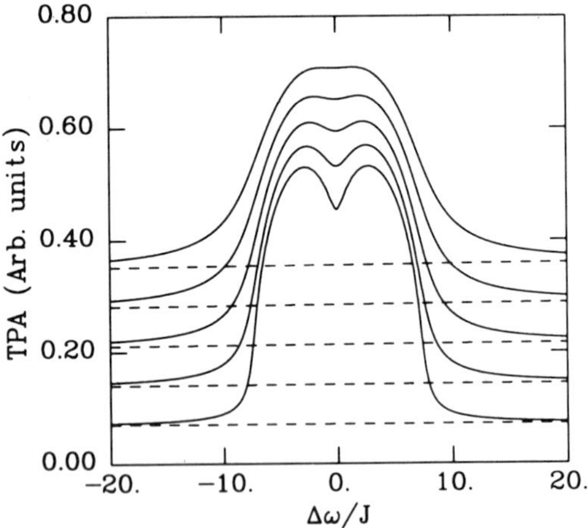

Fig. 5. Two-photon absorption W_{TPA}. The solid lines are numerical results for a two-dimensional assembly at the magic angle for a 199×199 lattice, the dashed line is the local-field approximation. Here we have taken $\omega_1 = 0.7\Omega_s + \Delta\omega$, and $\omega_2 = 1.3\Omega_s$. For these frequencies the one-photon frequencies are nonresonant, but the two-exciton term in the cooperativity-enhancement factor is resonant. We have plotted the signal for five dephasing rates, from bottom to top $\Gamma/J = 0.5$, 1, 1.5, 2, and 2.5. The TPA probes the two-exciton resonances which are missed completely by the local-field approximation [39].

for $\chi^{(3)}$ contains the contributions of local as well as nonlocal nonlinearities. The latter may be responsible for an enhanced (cooperative) nonlinear optical response. The enhancement can most conveniently be described in terms of an exciton coherence-size N_c, which represents the separation of two sites that can still respond coherently to the applied fields. In order to define the coherence size more precisely, gain a clear insight into its role, and make the connection with the optical response of aggregates discussed in Sec. 5, we proceed in the following way. We adopt the complete factorization into single operators (Eq. (1.3.10)) if sites n and m are separated by M bonds or more, and the more general factorization (Eq. (1.3.13)) if they are separated by fewer than M bonds. M is a cutoff size that can be varied at will. Nonlocal coherences are important only as long as $|n - m| < N_c$, N_c being the coherence size. We expect $\chi^{(3)}$ to vary with the cutoff M as long as M is smaller than thc coherence size N_c. As M exceeds N_c, the factorization (1.3.10) should hold, and $\chi^{(3)}$ should become independent on M. Observing the convergence of $\chi^{(3)}$ as M is varied should provide us with an operational definition of the nonlinear coherence-size N_c.

The present procedure maps the calculation of optical nonlinearities into solving the dynamics of coupled nonlinear oscillators. The oscillators correspond to the single-site variables $\langle B_n \rangle$ and nonlocal variables $\langle B_n B_m \rangle$ and $\langle B_n^\dagger B_m \rangle$, representing molecules separated by fewer than M bonds. In the present picture, the anharmonicities are short-range, and long-range interactions (among molecules separated by more than M bonds) are harmonic and enter only via the local field, which is the field at site n generated by the average field and the contributions of the long-range interactions. Eqs. (1.7.3) then provide a rigorous treatment of short-range dynamics and an approximate mean-field treatment of long-range interactions. The local-field approximation is obtained by taking $M = 1$. In this case, all intermolecular interactions enter via the local field, which, in the long-wavelength limit, it is given by Eq. (1.4.3). As M is increased, we treat the intermolecular interactions more rigorously, and the local field becomes closer to the average field. For $M = N$ we treat all intermolecular interactions explicitly. This interplay among local-field and intermolecular interactions has been discussed by Mukamel, Deng, and Grad[19].

Using the factorization Eq. (1.3.13), $\chi^{(3)}$ has three types of contributions originating from the $\langle B_n^\dagger \rangle \langle B_n \rangle \langle B_m \rangle$, $\langle B_n^\dagger B_m \rangle$, and $\langle B_n B_m \rangle$ nonlinearities. When expanded in powers of molecular density $\rho \equiv 1/a$, we find that the first contribution scales as $\sim \rho$, whereas the second and the third contributions scale as $\sim \rho^2$. This is to be expected, since the first represents the contribution of local nonlinearities whereas the other two are induced by nonlocal interactions. Nonlocal interactions enter therefore the nonlinear response in two ways; they modify the local term and induce additional terms. The appearance and the form of these new terms provide an excellent direct probe for nonlocal interactions, as was demonstrated in the previous sections.

The following calculations illustrate the relative role of the three types of nonlinearities [70]. In all the calculations we have used nearest neighbor interactions ($J_k = 2V \cos k$) and neglected any explicit dependence on light-wave vectors (the long wavelength approximation) setting $\mathbf{k}_1 = \mathbf{k}_2 = \mathbf{k}_3 = 0$. In Fig. 6 we show the contributions of the cubic and the exciton populations and the two-exciton nonlinearities to the TPA and to the THG signals. $\chi^{(3)}_{\mathrm{THG}}$ is shown in the vicinity of the single-photon resonance $\omega \sim \Omega_0$ and the three-photon resonance $\omega \sim \Omega_0/3$, where $\Omega_0 \equiv \Omega - 2|V|$ is the band-edge frequency. The convergence of $\chi^{(3)}$ with M is shown in Fig. 7.

In our model we have included only exciton repulsion through the Pauli exclusion. Attractive interactions may arise from a variety of sources either direct dipole–dipole interactions or phonon-mediated (very much like the Cooper pairs in superconductivity) [54]. We have calculated the TPA signal and its variation with the phonon-mediated attractive interaction [70]. We

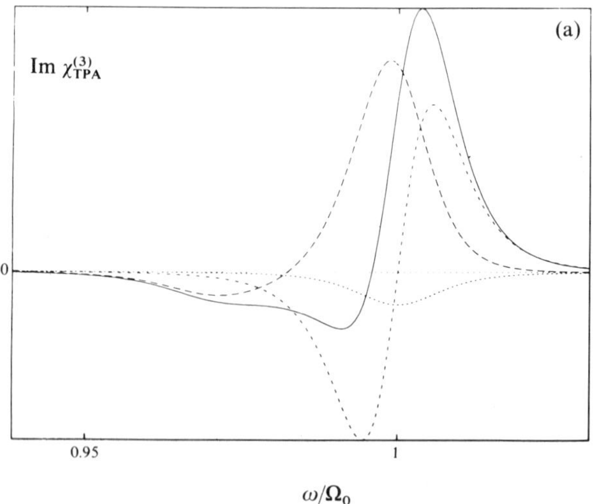

Fig. 6. The contribution of the cubic, exciton population, and two-exciton nonlinearities $\langle B_n^\dagger \rangle \langle B_n \rangle \langle B_m \rangle$, $\langle B_n^\dagger B_m \rangle$ and the $\langle B_n B_m \rangle$ terms in Eq. (1.3.13), to two-photon absorption and third harmonic generation. Solid line—total curve; long dash—two-excitonic contribution; medium dash—exciton population contribution; short dash—cubic nonlinearity contribution. (a) TPA Signal W_{TPA}; (*continued*)

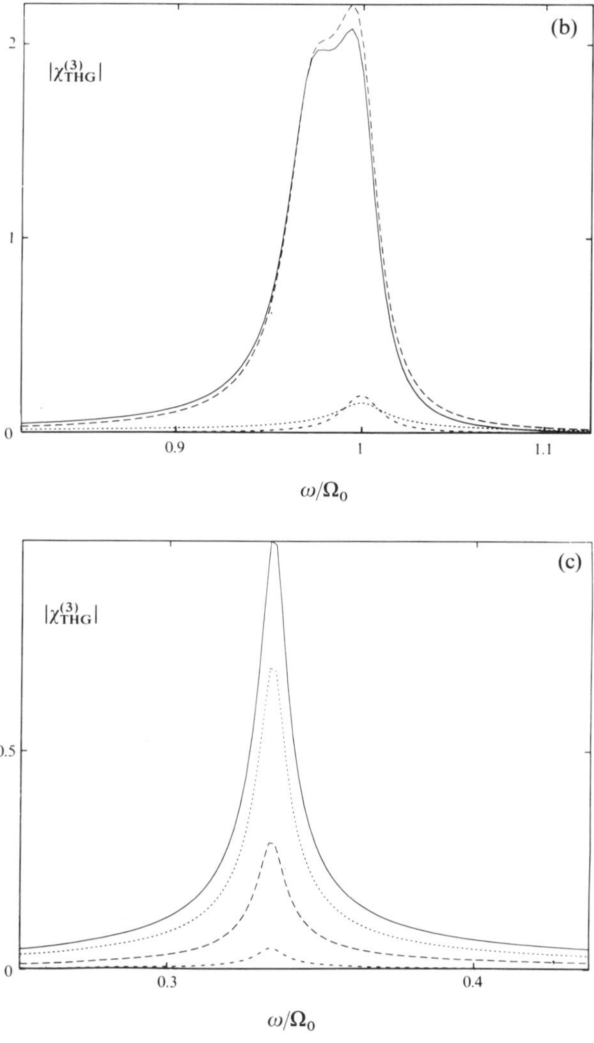

Fig. 6 (*continued*) (b) $|\chi^{(3)}_{THG}|$, square root of the THG signal in the vicinity of the band edge; (c) $|\chi^{(3)}_{THG}|$ in the vicinity of the three-photon resonances $\omega \sim \Omega_0/3$ [70].

found a progression of two-photon resonances in the two-exciton band. These resonances are blue-shifted compared with the band edge ($\omega > \Omega_0$). The blue shift results from the Pauli exclusion, which acts as an effective repulsion among excitons. These resonances are very similar to those displayed in Fig. 2. In addition, we noticed the appearance of strong resonances that are red-shifted with respect to the exciton frequency

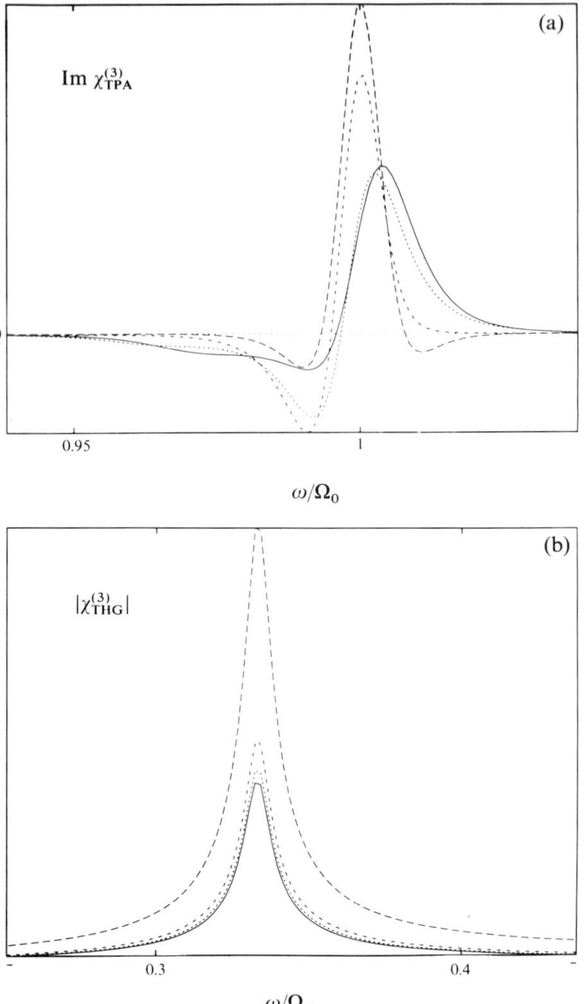

Fig. 7. Dependence of the nonlinear susceptibilities on the truncation size M, for an infinite chain $N \to \infty$ demonstrating the coherence size. (a) TPA signal Im $\chi^{(3)}_{TPA}$. Solid—$M = N$; long dash—$M = 1$; medium dash—$M = 3$; short dash— $M = 6$. (b) Square root of the THG signal $|\chi^{(3)}_{THG}|$ in the vicinity of the three-photon resonance. Solid line—$M = N$; long dash—$M = 1$; medium dash—$M = 2$; short dash—$M = 3$. (*Continued.*)

($\omega < \Omega_0$) and are induced by the attractive interactions. When the attractive part is sufficiently strong a new bound (biexciton) state can be formed, and the red shift is equal to the biexciton binding energy, and it naturally increases with the binding energy.

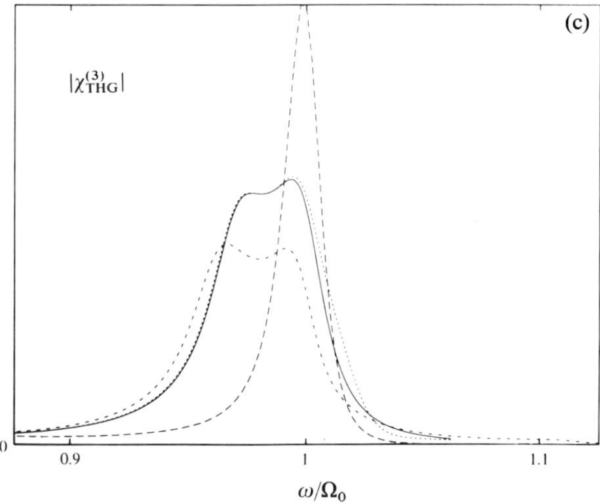

Fig. 7 (*continued*). (c) $|\chi^{(3)}_{THG}|$ in the vicinity of the band edge. Solid line—$M = N$; long dash—$M = 1$; medium dash—$M = 6$; short dash—$M = 12$ [70].

8. DISCUSSION

In this article we have presented a systematic unified methodology for calculating and interpreting the macroscopic nonlinear optical response of molecular materials. We have demonstrated a few manifestations of cooperativity that is totally missed by the local-field approximation. Cooperativity arises from the existence of non-local coherences and is the source of new resonances and possible enhancements (unusually large nonlinearities) as well as other phenomena. Enhanced radiative decay (superradiance) is another manifestation of cooperative radiative dynamics. Superradiance in molecular assemblies was calculated and observed recently [37,38]. The intermolecular nature of the nonlinear response is also reflected in exciton-transport whose signature is the wavevector dependence of $\chi^{(3)}$ and can be probed in the frequency domain (degenerate four-wave mixing) or in the time domain (transient grating spectroscopy).

When all frequencies are tuned far off resonance, the optical response is instantaneous (by the Heisenberg uncertainty, the process is completed in a time scale $\tau \sim \hbar/\Delta E$ with ΔE being an average detuning). Intermolecular interactions then do not have enough time to be effective and may be neglected. The optical response of a molecular material thus reduces to essentially that of a single molecule with minor local-field corrections. The local-field approximation fails completely to take into account any contribution of non-local nonlinearities. It therefore misses the cooperative effects, and exciton transport. In the design of optical materials it is important to optimize the detunings such that they are sufficiently large to avoid absorptive losses, but sufficiently small to make use of the cooperative effects. The present theory may be helpful in predicting the macroscopic nonlinearities based on individual molecular properties.

The approach developed here establishes an anharmonic-oscillator picture for the nonlinear medium. The linear optical properties in condensed phases can be calculated using a harmonic (Drude) oscillator picture. The optical polarization of a crystal of N two-level atoms is rigorously given by a sum of N coupled harmonic oscillators representing the individual molecular polarizations [10,11,13]. A natural extension of these ideas to nonlinear optics suggests the use of an anharmonic oscillator picture for the material polarization. This model proposed by Bloembergen [6] offers a simple qualitative physical picture, but was never rigorously established. The theory presented here is based upon a microscopic derivation of equations of motion for the optical polarization and other relevant dynamical variables, which are nonlocal in space and represent intermolecular coherences. These variables constitute a set of coupled anharmonic oscillators. The complex many-body problem thus reduces to the coupled nonlinear dynamics of relatively few oscillators. The common procedure, which is routinely used in the calculation of optical nonlinearities of atomic and molecular systems, is based on an expansion of the system's density matrix to third order in the applied fields. This results in a time-ordered expression for $\chi^{(3)}$ in which we keep track of the relative order in time of the interactions with the various electromagnetic fields [6,7,42] as given by the various Liouville-space pathways. When our results are compared with the density matrix time-ordered approach [28], we find the anharmonic-oscillator picture to yield a more compact expression that is much simpler to evaluate since it does not require calculating the molecular eigenstates.

The relevant anharmonic oscillators are identified as follows. We start with the optical polarization which is related to the exciton amplitudes $\langle B_m \rangle$. Its equation of motion couples it to additional variables, which in turn are coupled to more variables. In the present calculations the additional variables include the two-exciton $\langle B_n B_m \rangle$ and the exciton-population and coherence $\langle B_n^\dagger B_m \rangle$ variables. These variables together with their hermitian conjugates constitute the relevant set of anharmonic oscillators.

In this chapter we have applied this procedure for molecular crystals with Frenkel excitons and explored and compared various levels of description. The same procedure can be applied to materials other than molecular crystals, e.g., semiconductors [36,71], conjugated polyenes [72–75], and the like. The nature of the additional oscillators is going to be different for different materials. Semiconductors and molecular systems with extended (delocalized) electronic states constitute an important class of materials with interesting nonlinear optical properties. Extensive studies have been made on conjugated polymers such as polydiacetylene, polyacetylene, or polysilane. Resonant $\chi^{(3)}$ spectroscopies such as transient grating, optical stark, and coherent Raman reveal useful information regarding the nature of the excitons in these systems. Our reduced equations of motion developed for Frenkel (molecular) excitons may be extended to Wannier excitons by treating the electrons and the holes separately, using two-particle (electron-hole) variables [36,72].

It has been argued that as far as the magnitude of optical nonlinearities are concerned, there is no difference between different materials (molecular, semiconductor, or conjugated systems) [31]. This statement may hold some truth at the single-exciton level, but is clearly an oversimplification, since the nature of higher order variables is very different for various materials. The differences between materials can be very clearly identified and investigated by analyzing the nature of the nonlinear oscillators in each case. These physical differences are much less transparent when using the sum-over-states expression, which is formally the same for all materials, and whereby the differences enter through the complete set of eigenstates. The modelling of optical materials in terms of coupled anharmonic oscillators offers a simple and physically transparent picture that is a natural extension of the harmonic oscillator picture of linear optics. The incorporation of additional (*e.g.*, phonon) degrees of freedom in this picture may be carried out very easily. In the applications presented here, phonons simply result in relaxation and damping of the electronic degrees of freedom. Phonons can sometimes show up as explicit new resonances (rather than broadening). Raman and optical Stark effects are examples [36,71,76,77]. Such measurements can be described by adding more oscillator variables representing phonons in the equations of motion.

The present approach may be applied to various systems with restricted geometries (quantum wells, wires, and dots) where the non-Boson nature of the elementary excitations is amplified. Enhanced nonlinear susceptibilities have been reported in small semiconductor microcrystallites in glasses and polymers due to exciton confinement [34]. The systematic incorporation of disorder in nonlinear susceptibilities constitutes another important open problem in nonlinear optics. Many of the interesting optical materials are disordered. Dye-doped polymers [78] and glasses, mixed monolayers [79], and concentrated dye solutions are a few examples. There is extensive literature related to the effects of disorder in linear optics and the dielectric function $\varepsilon(\mathbf{k}, \omega)$ [14,15]. The equations of motion and Green function approach is ideally suited for incorporating the effects of disorder including static (substitutional) disorder as well as dynamical disorder (orientational, dielectric relaxation, and the like). An extension to second-order nonlinearities ($\chi^{(2)}$) in organized media such as Langmuir Blodgett films and multilayers is also possible [1]. Ultrafast time-domain spectroscopy with up to femtosecond resolution often provides invaluable information that is complementary to the frequency-domain susceptibilities [42]. The equation-of-motion approach naturally provides the necessary time-domain response functions, which are being probed by these techniques. The factors affecting the switching time scales can be analyzed. Finally, in highly ordered low-temperature materials with a large oscillator strength per unit volume, the radiation field combines with the material polarization to form new elementary excitations: polaritons. Polariton effects in optical nonlinearities, which were not covered here, include a significant change in damping and unusually fast transport rates [23,49]. The present methodology can be extended to incorporate polariton effects [5,50]. The anharmonic oscillators are then combined material and field degree of freedom (rather than purely material variables).

ACKNOWLEDGMENTS

The support of the Air Force Office of Scientific Research, the National Science Foundation, and the Center for Photoinduced Charge Transfer is gratefully acknowledged. I wish to thank my coworkers, in particular J. A. Leegwater, V. Chernyak, J. Knoester, F. C. Spano, O. Dubovsky, H. X. Wang, N. Wang and W. Bosma, who were instrumental in the developments covered in this review.

REFERENCES

1. D. S. Chemla and J. Zyss, *Nonlinear Optical Properties of Organic Molecules and Crystals*, Vols. I and II. Academic, New York, 1987.
2. D. J. Williams, *Nonlinear Optical Properties of Organic and Polymeric Materials*. ACS Symposium, Series 233, Washington, 1983.
3. D. Bedeaux and N. Bloembergen, *Physica* **69**, 67 (1973).
4. H. Haug, ed., *Optical Nonlinearities and Instabilities In Semiconductors*. Academic Press, New York, 1988; H. Haug and S. W. Koch, *Quantum Theory of the Optical and Electronic Properties of Semiconductors*. World Scientific, Singapore, 1990.
5. J. Knoester and S. Mukamel, *Physics, Reports* **205**, 1 (1991).

6. N. Bloembergen, *Nonlinear Optics*. Benjamin, New York, 1965.
7. P. N. Butcher, *Nonlinear Optical Phenomena*. Ohio University Press, Athens, Ohio, 1965.
8. C. Flytzanis, in *Quantum Electronics*. (H. Rabin and C. L. Tang, eds.). Vol. 1, p. 1, Academic Press, New York, 1975.
9. Y. R. Shen, *The Principles of Nonlinear Optics*. Wiley, New York, 1984.
10. H. A. Lorentz, *The Theory of Electrons*. Dover, New York, 1952.
11. M. Born and K. Huang, *Dynamical Theory of Crystal Lattices*. Oxford, London, 1954.
12. P. Mazur, *Adv. Chem. Phys.* **1**, 309 (1958); J. deGoede and P. Mazur, *Physica* **58**, 568 (1972); S. R. DeGroot, *The Maxwell Equations*, North Holland, Amsterdam, 1969.
13. U. Fano, *Phys. Rev.* **103**, 1202 (1956); S. I. Pekar, *Sov. Phys. JETP* **6**, 785 (1958); *ibid.* **11**, 1286 (1960).
14. C. J. F. Bottcher, *Theory of Electric Polarization*, Vol. I, Elsevier, Amsterdam, 1973; C. J. F. Bottcher and P. Bordewijk, *Theory of Electric Polarization*, Vol. II, Elsevier, Amsterdam, 1978.
15. J. Van Kranendonk and J. E. Sipe, in *Process in Optics* (E. Wolf, ed.). Vol. 15, p. 245. North-Holland, Amsterdam, 1977; P. Madden and D. Kivelson, *Adv. Chem. Phys.* **56**, 467 (1984).
16. G. R. Meredith, *J. Chem. Phys.* **75**, 4317 (1981); *ibid.* **77**, 5863 (1982); *Phys. Rev. B* **24**, 5522 (1981).
17. M. Hurst and R. W. Munn, *J. Mol. Elect* **3**, 75 (1987).
18. J. Knoester and S. Mukamel, *Phys. Rev. A* **39**, 1899 (1989); J. Knoester and S. Mukamel, *Phys. Rev. A* **41**, 3812 (1990).
19. S. Mukamel, Z. Deng, and J. Grad, *J. Opt. Soc. Am. B* **5**, 804 (1988).
20. Th. Förster, *Ann. Phys.* (Leipzig) **2**, 55 (1948); D. L. Dexter, *J. Chem. Phys.* **21**, 836 (1953).
21. V. M. Agranovich and M. D. Galanin, in *Electronic Excitation Energy Transfer in Condensed Matter* (V. M. Agranovich and A. A. Maradudin, eds.). North-Holland, Amsterdam, 1982.
22. H. J. Eichler, P. Günter, and D. W. Pohl, *Laser-Induced Dynamic Gratings*. Springer, Berlin, 1986.
23. J. R. Salcedo, A. E. Siegman, D. D. Dlott, and M. D. Fayer, *Phys. Rev. Lett.* **41**, 131 (1978); T. S. Rose, R. Righini, and M. D. Fayer, *Chem. Phys. Lett.* **106**, 13 (1984).
24. G. Roberts, ed. *Langmuir-Blodgett Films*. Plenum, New York, 1990.
25. D. Mobius and H. Kuhn, *Isr. J. Chem.* 18, 375 (1979); D. Mobius and H. Kuhn, *J. Appl. Phys.* **64**, 5138 (1988).
26. F. F. So, S. R. Forrest, Y. Q. Shi, and W. H. Steier, *Appl. Phys. Lett.* **56**, 674 (1990).
27. E. Hanamura, *Phys. Rev. B* **37**, 1273 (1988); *Phys. Rev. B* **38**, 1228 (1988).
28. (a) F. C. Spano and S. Mukamel, *Phys. Rev. A* **40**, 5783 (1989); (b) F. C. Spano and S. Mukamel, *Phys. Rev. Lett.* **66**, 1197 (1991); F. C. Spano and S. Mukamel, *J. Chem. Phys.* **95**, 7526 (1991).

29. H. Ishihara and K. Cho, *Phys. Rev. B* **42**, 1724 (1990); H. Ishihara and K. Cho, *Nonlin. Opt.* **1**, 287 (1992).
30. J. Feldmann, G. Peter, E. O. Gobel, P. Dawson, K. Moore, C. Foxton, and R. J. Elliot, *Phys. Rev. Lett.* **59**, 2337 (1987); F. C. Spano, J. R. Kuklinski, and S. Mukamel, *J. Chem. Phys.* **94**, 7534 (1991).
31. B. I. Greene, J. Orenstein, R. R. Millard, and L. R. Williams, *Phys. Rev. Lett.* **58**, 2750 (1987); B. I. Greene, J. Orenstein, and S. Schmitt-Rink, *Science* **247**, 679 (1990).
32. G. J. Blanchard and J. P. Heritage, *Chem. Phys. Lett.* **177**, 287 (1991); P. D. Townsend, W.-S. Fann, S. Etemad, G. L. Baker, Z. G. Soos, and P. C. M. McWilliams, *Chem. Phys. Lett.* **180**, 485 (1991); S. Etemad and Z. G. Soos, in *Spectroscopy of Advanced Materials* (R. J. H. Clark and R. E. Hester, eds.), p. 87, Wiley, New York, 1991.
33. G. J. Blanchard and J. P. Heritage, *Chem. Phys. Lett.* **177**, 287 (1991); W.-S. Fann, S. Benson, J. M. J. Madey, S. Etemad, G. L. Baker, and F. Kajzar, *Phys. Rev. Lett.* **62**, 1492 (1989).
34. M. G. Bawendi, M. L. Steigerwald, and L. E. Brus, *Ann. Rev. Phys. Chem.* **41**, 477 (1990).
35. S. Schmitt-Rink, D. S. Chemla, and D. A. B. Miller, *Phys. Rev. B* **32**, 6601 (1985); K. Leo, M. Wegener, J. Shah, D. S. Chemla, E. O. Göbel, T. C. Damen, S. Schmitt-Rink, and W. Shäfer, *Phys. Rev. Lett* **65**, 1340 (1990).
36. J. R. Kuklinski and S. Mukamel, *Phys. Rev. B* **42**, 2959 (1990); J. R. Kuklinski and S. Mukamel, *Phys. Rev. B* **42**, 295 (1990); *ibid.* **44**, 253 (1991).
37. J. Grad, G. Hernandez, and S. Mukamel, *Phys. Rev. A* **37**, 3835 (1988); F. C. Spano and S. Mukamel, *J. Chem. Phys.* **91**, 683 (1989); F. C. Spano, J. R. Kuklinski, and S. Mukamel, *Phys. Rev. Lett.* **65**, 211 (1990); F. C. Spano, J. R. Kuklinski, and S. Mukamel, *J. Chem. Phys.* **94**, 7534 (1991); F. C. Spano, J. R. Kuklinski, S. Mukamel, D. V. Brumbaugh, M. Burberry, and A. A. Muenter, *Molecular Crystals and Liquid Crystals* **194**, 331 (1991).
38. S. DeBoer, K. J. Vink, and D. A. Wiersma, *Chem. Phys. Lett.* **137**, 99 (1987); S. DeBoer and D. A. Wiersma, *Chem. Phys. Lett.* **165**, 45 (1990).
39. J. A. Leegwater and S. Mukamel, *Phys. Rev. A* **46**, 452 (1992).
40. P. N. Prasad, E. Perrin, and M. Samoc, *J. Chem. Phys.* **91**, 2360 (1989).
41. R. W. Hellwarth, *Prog. Quantum Electron.* **5**, 1 (1977).
42. S. Mukamel and R. F. Loring, *J. Opt. Soc. Am. B* **3**, 595 (1986); S. Mukamel, *Adv. Chem. Phys.* **70**, Part I, 165 (1988); S. Mukamel, *Ann. Rev. Phys. Chem.* **41**, 647 (1990).
43. M. Gross and S. Haroche, *Phys. Rep.* **93**, 301 (1982).
44. T. Itoh, T. Ikehara, and Y. Iwabuchi, *J. Lumin.* **45**, 29 (1990); T. Itoh and M. Furumiya, *J. Lumin.* **48** and **49**, 704 (1991); T. Itoh, *Nonlinear Optics* **1**, 61 (1991).
45. K. Kemnitz, K. Yoshihara, and T. Tani, *J. Phys. Chem.* **94**, 3099 (1990).
46. J. J. Hopfield and D. G. Thomas, *Phys. Rev.* **132**, 563 (1963); J. J. Hopfield, *Phys. Rev.* **112**, 1555 (1958); *ibid.* **182**, 945 (1969); V. M. Agranovich, *Zh. Eksp. Teor. Fiz.* **37**, 430 (1959) [Sov. Phys.-JETP **37**, 307 (1960)].

47. G. M. Gale, F. Vallée, and C. Flytzanis, *Phys. Rev. Lett.* **57**, 1867 (1986).
48. D. Frölich, S. Kirchhoff, P. Köhler, and W. Nieswand, *Phys. Rev. B* **40**, 1976 (1989).
49. C. K. Johnson and G. J. Small, in *Excited States* (E. C. Lim, ed.), Vol. 6, Academic Press, New York, 1982; S. H. Stevenson, M. A. Connolly, and G. J. Small, *Chem. Phys.* **128**, 157 (1988).
50. J. Knoester and S. Mukamel, *J. Chem. Phys.* **91**, 989 (1989); J. Knoester and S. Mukamel, *Phys. Rev. A* **40**, 7065 (1989).
51. A. S. Davydov, *Theory of Molecular Excitons*, Plenum, New York, 1971.
52. E. I. Rashba and M. D. Sturge, eds., *Excitons*. North-Holland, Amsterdam, 1982.
53. L. N. Ovander, *Usp. Fiz. Nauk* **86**, 3 (1965) [Sov. Phys.-Usp. **8**, 337 (1865)].
54. G. D. Mahan, *Many Particle Physics*. Plenum, New York, 1990.
55. R. F. Loring and S. Mukamel, *J. Chem. Phys.* **84**, 1228 (1986).
56. W. R. Heller and A. Marcus, *Phys. Rev.* **84**, 809 (1951).
57. M. Orrit and P. Kottis, *Adv. Chem. Phys.* **74**, 1 (1988).
58. F. C. Spano, V. M. Agranovich, and S. Mukamel, *J. Chem. Phys.* **95**, 1400 (1991).
59. E. Wigner, *Phys. Rev.* **40**, 749, (1932); M. Hillery, R. F. O'Connell, M. O. Scully, and E. P. Wigner, *Phys. Rept.* **106**, 121 (1984).
60. R. F. Loring, D. S. Franchi, and S. Mukamel, *Phys. Rev. B* **37**, 1874 (1988); S. Mukamel, D. S. Franchi, and R. F. Loring, *Chem. Phys.* **128**, 99 (1988).
61. H. Risken, *The Fokker-Planck Equation*. Springer, Berlin, 1984.
62. H. Haken and G. Strobl, *Z. Phys.* **262**, 135 (1973).
63. D. K. Garrity and J. L. Skinner, *J. Chem. Phys.* **82**, 260 (1985).
64. N. Bloembergen, H. Lotem, and R. T. Lynch, *Indian J. Pure Appl. Phys.* **16**, 151 (1978); Y. Prior, A. R. Bogdan, M. Dagenais, and N. Bloembergen, *Phys. Rev. Lett.* **46**, 111 (1981); L. Rothberg in *Prog. in Optics* (E. Wolf, ed.), Vol. 24, p. 38, North Holland, 1987.
65. J. R. Andrews, R. M. Hochstrasser, and H. R. Trommsdorff, *Chem. Phys.* **62**, 87 (1981); J. R. Andrews, and R. M. Hochstrasser, *Chem. Phys. Lett.* **82**, 381 (1981).
66. V. Chernyak and S. Mukamel, *Phys. Rev. B* (1993, in press).
67. W. Wegener, D. S. Chemla, S. Schmitt-Rink, and W. Schäfer, *Phys. Rev. A* **42**, 5675 (1990).
68. S. Schmitt-Rink and S. Mukamel; K. Leo and J. Shah, and D. S. Chemla, *Phys. Rev. A* **44**, 2124 (1991).
69. M. P. Allen and D. J. Tildesley, *Computer Simulations of Liquids*, Clarendon, Oxford 1987.
70. O. Dubovsky and S. Mukamel, *J. Chem. Phys.* **95**, 7828 (1991); O. Dubovsky and S. Mukamel, *J. Chem. Phys.* **96**, 9201 (1992).
71. S. Schmitt-Rink, D. S. Chemla, and D. A. B. Miller, *Adv. in Physics* **38**, 89 (1989).
72. A. Takahashi and S. Mukamel (to be published); M. Hartmann and S. Mukamel, *J. Chem. Phys.* **99** (in press).
73. A. F. Garito, J. R. Heflin, F. Y. Wong, and Q. Zamani-Khamiri, in *Organic Materials for Nonlinear Optics*, R. A. Hann and D. Bloor (eds.) (Royal Society

of Chemistry, London, 1988), Special Publication No. 69; J. R. Heflin, K. Y. Wong, O. Zamani-Khamiri, and A. F. Garito, *Phys. Rev. B* **38**, 1573 (1988).

74. S. Etemad and Z. G. Soos, in *Spectroscopy of Advanced Materials*, R. J. H. Clark and R. E. Hester (eds.), Wiley, New York (1991); S. N. Dixit, D. D. Guo, and S. Mazumdar, *Mol. Cryst. Liq. Cryst.* **194**, 33 (1991).
75. C. P. deMelo and R. Silbey, *Chem. Phys. Lett.* **140**, 537 (1987); D. Yaron and R. Silbey, *Phys. Rev. B* **45**, 11655 (1992).
76. W. B. Bosma, S. Mukamel, B. I. Greene, and S. Schmitt-Rink, *Phys. Rev. Lett.* **68**, 2456 (1992).
77. R. Zimmermann, *Phys. Stat. Sol.* (*b*) **159**, 317 (1990); R. Zimmermann and M. Hartmann, *J. Crystal Growth* **101**, 341 (1990).
78. S. Saikan, T. Kishda, Y. Kanematsu, H. Aota, A. Harada, and M. Kamachi, *Chem. Phys. Lett.* **166**, 358, (1990); S. Saikan, A. Imaoka, Y. Kanematsu, K. Sakoda, K. Kominami, and Iwamoto, *Phys. Rev. B* **41**, 3185 (1990).
79. D. Lupo, W. Prass, U. Scheunemann, A. Laschewsky, H. Ringsdorff, and I. Ledeoux, *J. Opt. Soc. Am. B* **5**, 300 (1988); J. S. Schildkraut, T. L. Penner, C. S. Willand, and A. Ulman, *Opt. Lett.* **13**, 134 (1988); L. M. Hayden, *Phys. Rev. B* **38**, 3718 (1988).

Chapter 2

TIME-RESOLVED NONLINEAR SPECTROSCOPY OF CONJUGATED POLYMERS

Takayoshi Kobayashi

Department of Physics, University of Tokyo, Hongo, Bunkyo, Tokyo, Japan

1. INTRODUCTION

Nonlinear optical processes are of essential importance in optoelectronics and/or photonics, which will be realized in the twenty-first century, in the

ISBN 0-12-784450-3

same way that nonlinear electronic processes such as modulation or frequency multiplication are key phenomena in well-developed electronics. Among various possibilities of materials to be used in devices in optoelectronics and photonics, conjugated polymers have many advantages, including large optical nonlinearity, ultrafast response, processability, high optical stability, and chemical stability. The present chapter describes first new measurement methods of the characterization of ultrafast nonlinear processes and then ultrafast relaxation in conjugated polymers with large third-order linearity.

2. VARIOUS METHODS OF TIME-RESOLVED MEASUREMENT OF ULTRAFAST OPTICAL NONLINEARITY

2.1. Application of Time-Incoherent Light to Ultrafast Nonlinear Optical Relaxation Measurement

The conventional methods using ultrashort pulses to study the ultrafast nonlinear phenomena have several difficulties and disadvantages:

1. Lasers for the generation of ultrafast pulses with enough intensity to be used in spectroscopies are complicated and expensive.
2. The wavelength of the ultrashort pulse shorter than a hundred femtoseconds is usually limited in the region around 615 to 625 nm from the colliding-pulse mode-locked ring dye lasers because of the limitation in the appropriate combination of gain medium and saturable absorber, and they have small tunable ranges. Recently mode-locked Ti sapphire lasers have a broader tunable range of 700 to 900 nm, but the most important wavelength region of 450 to 700 nm for organic or polymer materials is not available when the lasers are used even with the second harmonic.
3. An ultrashort light pulse is easily broadened due to linear and/or nonlinear dispersion in optical components because of its inevitably broad power spectrum due to Fourier transform relation.

A new spectroscopic technique with incoherent light utilizing transient coherent optical effects has been proposed to overcome the above difficulties. In the transient coherent spectroscopy for the studies of the ultrafast dynamics, the signal light generated or modulated by nonlinear optical effects is detected, using the correlation between the excitation and the probe light beams and the signal intensity expressed by the time integration of functions of the field amplitude (or intensity) and the response function of the matter.

Therefore, in such cases, the time resolution is expected to be determined by the correlation time instead of the pulse duration.

According to this principle, experiments with extremely high time resolution can be performed much more easily by using temporally incoherent light with a short enough correlation time. The applicability of this principle to studies of ultrafast processes has been verified for the electronic-dephasing-time measurement by degenerate four-wave mixing (DFWM) spectroscopy [1–7].

Though transient DFWM spectroscopy, either using coherent short pulses [8] or incoherent light sources as mentioned above, is powerful for the study of dynamic processes in condensed matters, it cannot be applied to optically forbidden transition, and the range of available wavelength is extremely limited. Dephasing of Raman active vibrational modes in molecules can be studied by so-called transient coherent Raman spectroscopy such as CARS (coherent anti-Stokes Raman scattering) or CSRS (coherent Stokes Raman scattering) [9–11]. We studied theoretically and demonstrated experimentally the applicability of the principle that the correlation time determines the time resolution of the vibrational-dephasing-time measurement by this method [12].

The techniques we mentioned above are concerned with the dephasing-time measurement. However, the population lifetime or energy decay time of electronically excited states can also be measured with the time resolution much shorter than the pulse width by using temporally incoherent light. For this purpose, two types of techniques have been reported. One is the pump-probe technique, in which the delay-time dependence of the change in the intensity of the transmitted probe beam is measured [13]. The other method is DFWM with three incident beams, in which we detect a fourth signal beam as varying the delay time of one of the three incident beams [14]. Further, as an extended case of the longitudinal relaxation measurement, we found that the relaxation of the nonresonant third-order susceptibility in optical Kerr media can also be studied by DFWM [15] and Kerr-shutter techniques with temporally incoherent light. The applicability of a Kerr shutter using incoherent light to the studies of dynamics of luminescence is also discussed.

2.2. Optical Heterodyne Detection of Induced Phase Modulation

Chesnoy and Mokhtari have investigated oscillatory features in femtosecond nonlinear optical responses of dye solutions that are resonant with the excitation light by means of detecting a pump-pulse-induced spectral shift of a probe light pulse [16,17]. This spectral shift is caused by the induced phase modulation (IPM) of the probe pulse and is expressed as the time derivative of the real part of the optical Kerr effect (OKE) response function in the time scale longer than the duration of the excitation light pulses. IPM technique, therefore, is suited for the study of the inertial molecular motions, being sensitive only to nonrelaxational responses. In a recently published paper [18] we apply optical-heterodyne-detection (OHD) technique to the IPM measurements resulting in highly enhanced sensitivity and high feasibility to apply the IPM technique to the study of the nonrelaxation response of nonresonant molecular liquids.

Extensive investigations have been performed on the molecular dynamics in a liquid phase by studying the dynamical light scattering (DLS) spectrum, which is proportional to the Fourier transform of the susceptibility-fluctuation correlation function (SCF) of the medium. The SCF is equated to the orientational correlation function (OCF) of the molecules if intermolecular interaction-induced effects are neglected [19–25]. Recently, however, several techniques employing nonlinear optical processes have been applied to the study of the molecular dynamics. These techniques include the time-domain measurements of the OKE response employing ultrashort optical pulses or incoherent light [18,23–36] and frequency-domain ones such as stimulated Raman gain [37] and tunable-laser-induced grating measurements [38,39]. The OKE response function is proportional to the time derivative of the SCF in the high-temperature limit [35]. What is directly obtained from the frequency-domain measurements is the Fourier transform of the OKE response function. These time-domain and frequency-domain nonlinear techniques, therefore, have high sensitivity in the high-frequency region of the SCF in proportion to the frequency, and can show oscillatory molecular motions more clearly than linear DLS measurements [35]. By employing the IPM technique we can further enhance the sensitivity in the high-frequency region of the response. Furthermore, the IPM response function, given by the time derivative of the nuclear contribution to the OKE response function, directly reflects the angular velocity correlation function (AVCF) of the molecules, which reveals that the inertial effects on the molecular motions more clearly than the OCF.

2.3. Time-Resolved FFT Interferometry with Reference Interferometer

We proposed a highly sensitive method that enables separation of the real and imaginary parts of the nonlinear susceptibility with femtosecond time resolution [40]. Our method is based on an interferometry to detect pump-induced transient change in both the absorption coefficient and

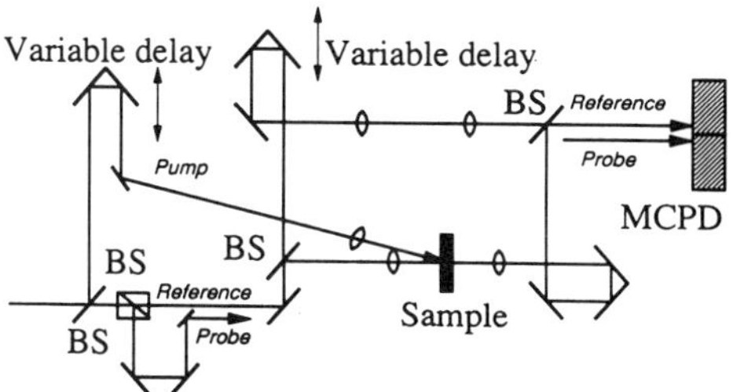

Fig. 1. Block diagrams of the femtosecond time-resolved Mach-Zender interferometer. BS's, beam-splitters; MCPD, multichannel photodiode.

refractive index of samples, with several improvements over the previous works related to time-resolved interferometry [41–45].

The block diagram of the apparatus is shown in Fig. 1 [40]. The light source is a colliding pulse mode-locked (CPM) ring dye laser amplified by a four-stage dye amplifier pumped by a Q-switched Nd-YAG laser at 10 Hz. The typical duration, energy, and peak wavelength of the amplified pulses are 100 fs, 200 μJ, and 630 nm, respectively. The beam was divided by a beam-splitter into two. One was led to a variable optical delay and used as a pump. The other was divided again by another beam-splitter into the probe and reference. The two pulses had an appropriate delay and were parallel with small spatial separation to each other, which enabled selective detection. Polarization difference has been used to select one that overlapped the other, but if the polarizations of the probe and reference beams rotated after passing through anisotropic materials, the two beams cannot be separated. Our method can be used even for anisotropic materials.

The interference fringe pattern was fast Fourier-transformed (FFT) to determine the complex nonlinear susceptibility. The real and imaginary parts were separately obtained by a single measurement, while in most previous studies two sets of experiments were needed [43,45] or only the materials of negligibly small absorption [40,41] were studied. The fast Fourier transform (FFT) has another advantage of improvement of accuracy by utilizing the whole fringe data over the point detection method instead of using the peak shift only [40].

In the first step, we will consider an ideal case that the intensity I of the interference fringe can be described by the following function of the position (ω) on the detection plane of the multichannel photodiode (MCPD), using a single wavenumber k_0 and a Gaussian envelope,

$$I(x) = [E_1^2 + E_2^2 + 2E_1E_2 \cos(k_0x - \phi_0)] \exp(-x^2/a^2) \quad (2.2.1)$$

Here E_1 and E_2 are the field amplitudes of the light in each arm of the interferometer, ϕ_0 is the has phase difference of the light in the two arms, and a is the width of the Gaussian envelope. We obtain Fourier transform, I, as follows,

$$I(k) = (a/2)\{(E_1^2 + E_2^2)\exp(-a^2k^2/4) + E_1E_2\exp[-a^2(k-k_0)^2/4]\exp(-i\phi_0) + E_1E_2\exp[-a^2(k+k_0)^2/4]\exp(i\phi_0)\}. \tag{2.2.2}$$

The spectrum of $I(k)$ has three Gaussian peaks at $k = 0, \pm k_0$, corresponding to the three terms in Eq. (2.2.2). When $k_0 \gg 4/a$, the peak at $k = k_0$ is expressed by only the second term of Eq. (2.2.2). In this case, we can obtain ϕ_0 and E_1E_2 directly from the phase and amplitude of $I(k_0)$, respectively.

In our measurement, both the probe and reference fringes give ϕ and E_1E_2. The shift and distortion of the fringe due to fluctuations were corrected for every shot and then averaged. Therefore this method is free from the reduction of the finesse, from which previous works without any correction suffered. The change in the phase and the transmittance induced by the sample excitation is given as

$$\Delta\phi = (\phi_{\rm pro}^{\rm ex} - \phi_{\rm ref}^{\rm ne}) - (\phi_{\rm pro}^{\rm ex} - \phi_{\rm ref}^{\rm ne}), \tag{2.2.3}$$

where $\Delta T/T$ is the relative change in the transmittance of the sample due to excitation. The subscripts of ex and ne correspond to with and without pump, respectively, and pro and ref represent the probe and reference interferometers, respectively.

The envelope of the actual fringe is not exactly a Gaussian, and the wave number k is not single-valued; therefore the spectrum of $I(k)$ cannot be expressed by three simple Gaussian peaks. However, we can obtain the spectrum with a sharp and intense peak at a specific value of k_0; thus the change in the phase and the transmittance due to the excitation of the sample can be determined in the same way.

The histogram of the observed phase change for 1000 shots is shown in Fig. 2. In this experiment, because the sample (CS_2) was not excited, the phase change must be zero. It shows that the standard deviation was reduced by a factor of 4, i.e., to 0.25 radian, by using the reference interferometer. High sensitivity is thus obtained in the fringe shift detection as small as 0.025 radian ($\lambda/250$) by averaging only over 100 shots, which can be easily attained even with the low-petition-rate laser using in the present study.

Next we show two examples that demonstrate our method. All the experiments were performed at 300 K. The first sample is CS_2, which has

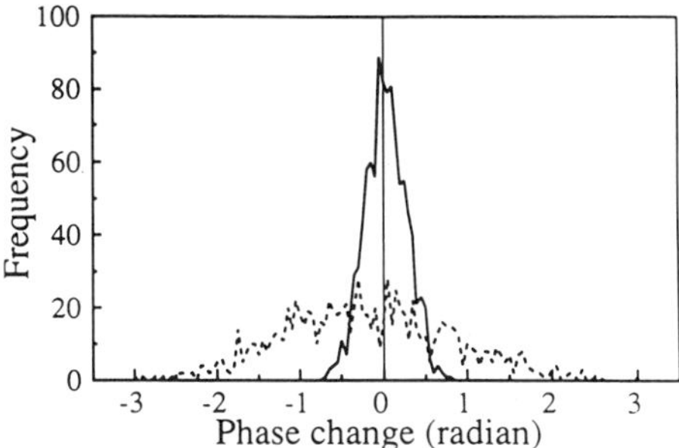

Fig. 2. Histogram of the observed phase change in CS_2 collected over 1000 shots. The solid and dashed lines correspond to the case with and without references, respectively.

essentially no absorption at 630 nm; thus only the real part of the nonlinear susceptibility is to be considered. The second one is CdS_xSe_{1-x} microcrystallites doped glass with average particle radius of 3.3 nm [46], which is the same as used in another work. In this case, both the real and the imaginary parts are to be considered.

The experimental results of the phase change of CS_2 induced by femtosecond pulse excitation are shown in Fig. 3. The polarization of the pump and probe light are parallel (a) and perpendicular (b) to each other. Interferometric method allows the several nonzero tensor components to be determined independently by changing the polarization of the pump and probe light. For example, we can obtain $\chi^{(3)}_{zzzz}$ and $\chi^{(3)}_{zzxx}$ for cases (a) and (b), respectively. The time-dependent phase change in CS_2 under parallel polarization was reproduced most simply by the convolution of the sum of instantaneous response and two exponential decay times of 0.30 ± 0.8 ps and 1.6 ± 0.9 ps to show the applicability of our method, even though it was

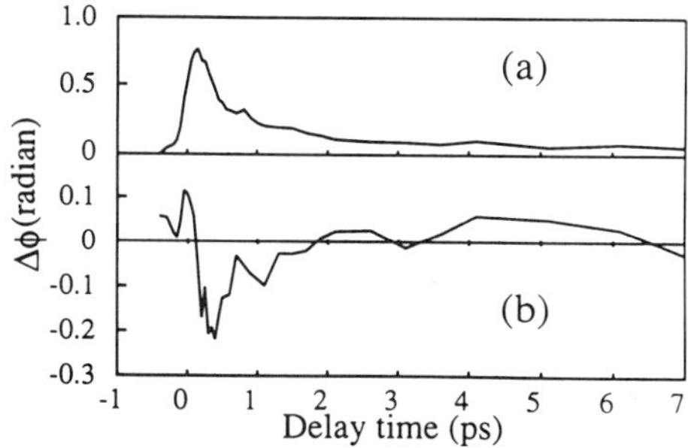

Fig. 3. Transient phase change ($\Delta\phi$) in CS_2. The polarizations of the pump and probe light were (a) parallel and (b) perpendicular to each other.

elucidated to be more complicated in a recent experiment of our group [18]. They agree well with 0.33 ps and 2 ps, a Mach–Zehnder interferometer [40]. The electronic response is instantaneous, and the decay times of 0.30 and 1.6 ps are both molecular origins.

The nonlinear refractive index n_2 was calculated as 3×10^{-12} esu at the peak. The values of n_2 determined by using much longer pulses than 1.6 ps are very much spread. They are between 11×10^{-12} and 20×10^{-12} esu [48]. Using our result of fitting parameters and assuming much longer pulse duration than 1.6 ps, we estimated time-dependent refractive index change expected in the literature. The nonlinear refractive index n_2 at the peaks near zero delay time were estimated to be four times larger than our value by femtosecond pulse. Therefore our result is consistent with the previously reported values, and furthermore, molecular contributions were time-resolved.

Figure 3 shows that the phase change in case (b) was positive just after excitation, then changed to be negative. On the other hand, in case (a) it was positive at all the delay times. In case (a), both the contributions of electronic and molecular origin are known to be positive. Therefore from the tensor analysis, in case (b), the former is considered to be positive and the latter is negative. Thus the obtained signal in case (b) is qualitatively expressed by the summation of these components.

Figure 4 shows the phase change (a) and the transmittance change (b—solid line) obtained by our method for the CdS_xSe_{1-x} doped glass. The transmittance change was also measured by blocking the beam in the arm without the sample, and it is shown by closed circles and was found to be in agreement with the result obtained by the interferometry as shown in Fig. 5. This verifies correct separation of the real and the imaginary parts by the

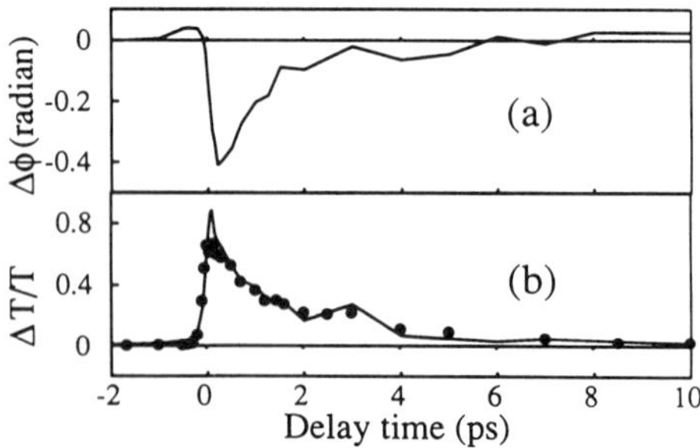

Fig. 4. Transient phase change, $\Delta\phi$ (a) and the normalized transmittance change, $\Delta T/T$ (b—solid line) obtained by the present method in CdS_xSe_{1-x} doped glass. Closed circles in (b) show $\Delta T/T$ by a conventional pump-probe absorbance change measurement only using the sample arm of the interferometer.

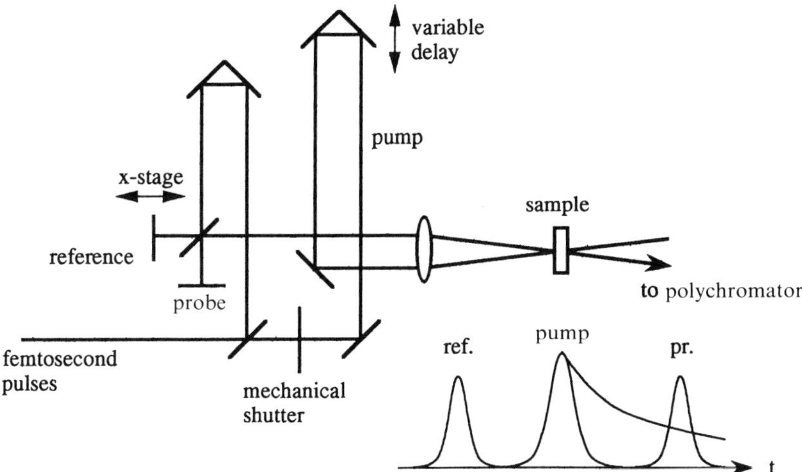

Fig. 5. The block-diagram of the experimental apparatus of time-domain interferometry. It is composed of the colliding pulse mode-locked ring-dye laser with six pass dye amplifier pumped with a copper vapor laser. The modified Michelon interferometer is utilized to obtain time-domain interference. The probe and reference spectra were measured with a polychromator with a multichannel detector.

present method proposed by us. The third-order nonlinear susceptibility of CdS_xSe_{1-x} doped glass was obtained as $(5 + 3i) \times 10^{-14}$ esu at the peak, which are several orders of magnitudes smaller than the results using much longer pulses such as nanosecond [47] which include the effect of accumulation over the pulse widths as mentioned above.

2.4. Femtosecond Frequency-Domain Interferometry

Here we propose a new interferometric technique, which utilizes interference fringes in frequency domain with femtosecond resolution for the first time [49–51].

Let us consider the case in which the same two optical pulses are displaced temporally by T. One is a probe (pr.) pulse and the other is a reference (ref.) pulse.

$$E_{pr}(t) = E(t)\exp(i\omega_0 t) \tag{2.2.4}$$

$$E_{ref}(t) = E(t - T)\exp(i\omega_0(t - T)) \tag{2.2.5}$$

$E(t)$ is assumed to be a real and symmetrical function in the case of a Fourier-transform-limited pulse for simplicity. It can be an arbitrary complex function. If the Fourier transform of the probe pulse is given by

$$F(E_{\mathrm{pr}}(t)) = E(\omega - \omega_0), \tag{2.2.6}$$

then the FT of the sum of the field of the two pulses is given by

$$F(E_{\mathrm{pr}}(t) + E_{\mathrm{ref}}(t)) = E(\omega - \omega_0)(1 + \exp(-i\omega T)). \tag{2.2.7}$$

Experimentally we can perform Fourier transformation of signals automatically with a grating in a spectrometer, and they are detected as power spectra. Therefore we must take the intensity of Eq. (2.2.7):

$$|F(E_{\mathrm{pr}}(t) + E_{\mathrm{ref}}(t)|^2 = |\mathbf{E}(\omega - \omega_0)|^2(2 + 2\cos\omega T). \tag{2.2.8}$$

This is the frequency domain interference we detect experimentally. This shows that we need not have an FT-limited pulse in order to obtain a regular oscillatory behavior as far as two pulses have the same spectral shape. This means that an incoherent white light source can usually be used for linear interferometry; but for time-resolved interferometry, we must use mode-locked laser pulses. A general expression of the electric field of a probe pulse propagating in a medium is

$$E_{\mathrm{pr}}(t) = d\omega E(\omega - \omega_0)\exp(i\omega(t - n(x)x/c + ik(x)x/c)), \tag{2.2.9}$$

where $n_c(\omega) = n(\omega) - ik(\omega)$ is a complex refractive index of a medium. The FT of Eq. (2.2.9) is given by

$$F(E_{\mathrm{pr}}(t)) = E(\omega - \omega_0)\exp((-in(\omega) - k(\omega))\omega x/c). \tag{2.2.10}$$

Suppose both the reference and probe pulses pass through the medium and the probe pulse alone undergoes the change in the complex refractive index, $\Delta n_c(\omega, \tau)$, induced by excitation, where

$$\Delta n_c = \Delta n(\omega, \tau) - i\Delta k(\omega, \tau)$$

$$F(E_{\mathrm{pr}}(t, \tau)) = F(E_{\mathrm{pr}}(t))\exp((-i\Delta n(\omega, \tau) - \Delta k(\omega, \tau))\omega x/c). \tag{2.2.11}$$

By the conventional pump-probe measurement we can measure difference transmission spectrum (DTS), which is expressed using the above formalism by

$$\begin{aligned}\Delta T/T(\omega, \tau) &= \{|F(E_{\mathrm{pr}}(t, \tau)|^2 - |F(E_{\mathrm{pr}}(t)|^2\}/|F(E_{\mathrm{pr}}(t)|^2 \\ &= \exp(-2\Delta k(\omega, \tau)\omega x/c) - 1.\end{aligned} \tag{2.2.12}$$

When the change is small, this becomes

$$\Delta T/T(\omega,\tau) = -2\Delta k(\omega,\tau)\omega x/c = -\Delta\alpha(\omega)x. \tag{2.2.13}$$

By the interferometric measurement, the intensity without excitation can be given by Eq. (2.2.8).

The intensity distribution in frequency domain, i.e., the frequency-domain interference signal with excitation, is given by

$$\begin{aligned}&|F(E_{\mathrm{pr}}(t,\tau) + E_{\mathrm{ref}}(t)|^2 \\ &\quad = E(\omega-\omega_0)^2(1 + e^{-2\Delta k(\omega,\tau)\omega x/c} + 2e^{-\Delta k(\omega,\tau)\omega x/c}\cos\omega(T - \Delta n(\omega,\tau)x/x)).\end{aligned} \tag{2.2.14}$$

Comparing Eqs. (2.2.8) and (2.2.14), we can obtain both $\Delta n(\omega,\tau)$ and $\Delta k(\omega,\tau)$ from the shifts and amplitude changes in the fringe peak in frequency domain, respectively. This means that by a single shot experiment we can in principle detect both the DTS and DPS.

The block diagram of the experimental apparatus is shown in Fig. 5 [49]. The output of a homemade colliding pulse mode-locked (CPM) ring-dye laser is amplified by a six-pass amplifier pumped by a copper-vapor laser. The wavelength, pulse duration, pulse energy, and repetition rate of the amplified pulses are 620 nm, 60 fs, 2 μJ, and 10 kHz, respectively. The reference and probe are focused into a sample, and the transmitted pulses are detected by a multichannel spectrometer. The pump passes through a variable time-delay line; it is focused on the sample and alternatively blocked and opened with a mechanical shutter to obtain a difference signal.

Examples of the interference fringes are shown in Figs. 6 and 7. A commercially available Toshiba R63 color glass filter containing CdS_xSe_{1-x} microcrystallites of a few weight percent was used. The pump and probe polarizations are parallel, and all data were taken at room temperature. The left part of Fig. 8 shows the interference spectra with pump on and off for several delay times, and the right part shows interference normalized by the probe spectrum (corresponding to Eqs. 2.2.8 and 2.2.14, respectively). The difference phase spectra (DPS) are given by open circles together with the average phase changes. The average phase shifts were obtained by Fourier-transforming the signal in the left of Fig. 8 into t-space. In Fig. 9 the time evolution of the average phase change is shown together with that of the average transmission change calculated from the DTS signals.

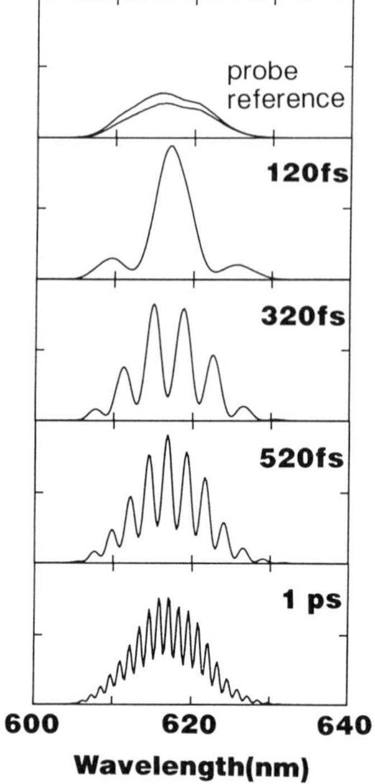

Fig. 6. Interference fringes of the reference and probe for several time displacements between them. The top shows the spectra of probe and reference with slightly different intensity because of the imbalance of the optical system. The fringe separation in the spectra is inversely proportional to the time separation, as expected.

3. ULTRAFAST RELAXATION OF PHOTOEXCITATIONS IN CONJUGATED POLYMERS WITH LARGE OPTICAL NONLINEARITY

3.1. Introduction

Recent progress in the field of high-speed optoelectronics has encouraged the search for new materials with large third-order nonlinear optical susceptibilities [52]. Conjugated polymers are promising candidate materials for future practical applications in nonlinear optical devices because of their varieties of main chains and side groups. Among conjugated polymers, polyacetylene (PA), polydiacetylenes (PDAs), and polythiophenes (PTs) are extensively investigated [53–63].

The results of measurements of photoinduced absorption and photoinduced

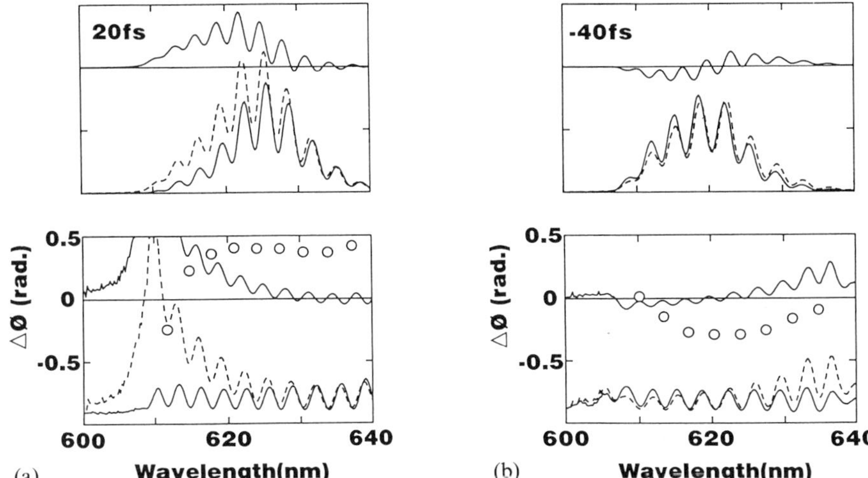

Fig. 7. The signals observed (a) for a Toshiba R63 color glass filter at 20 fs delay and (b) for CS_2 at -40 fs delay. The time displacements between probe and reference are 410 fs and 370 fs for (a) and (b), respectively. Each of the data is obtained by averaging 20 sweeps of wavelength.

bleaching (PB) on *trans*-PA under quasistatic conditions have confirmed the existence of solitons, and their relaxation dynamics have been also investigated so far with picosecond [60] and femtosecond [61–63] time resolution.

Vardeny et al. [64] observed the PB of interband transition in *trans*-polyacetylene with picosecond time resolution and that the bleaching decayed with time as $t^{-1/2}$ up to 50 ps. They also estimated the diffusion constant of charged excitations as 2×10^{-2} cm^2 sec^{-1} from the decay of polarization memory. The interchain excitation of polarons was observed in oriented PA films [60,64]. A polaron pair is generated from an electron-hole pair that is photoexcited on two different chains. After photoexcitation, polarons disappear by the interchain relaxation mechanism in nanoseconds to microseconds depending on the intensity of the interchain interaction.

PDAs possess the diacetylene-type configuration $\leftarrow\!\!(CR_1\text{—}C\equiv C\text{—}C)\!\!\rightarrow_n$ and have several remarkable characteristics. The properties of PDAs can be modified considerably by changing the side groups R_1 and R_2. The polymerization of a large number of diacetylenes with various side groups has been reported. PDAs can be obtained in the form of highly ordered single crystals, Langumuir-Blodgett films, vacuum deposited films, and solvent-case films [65]. Another characteristic of PDAs is dramatic color changes. Several PDA films and crystals have two phases, called blue-phase

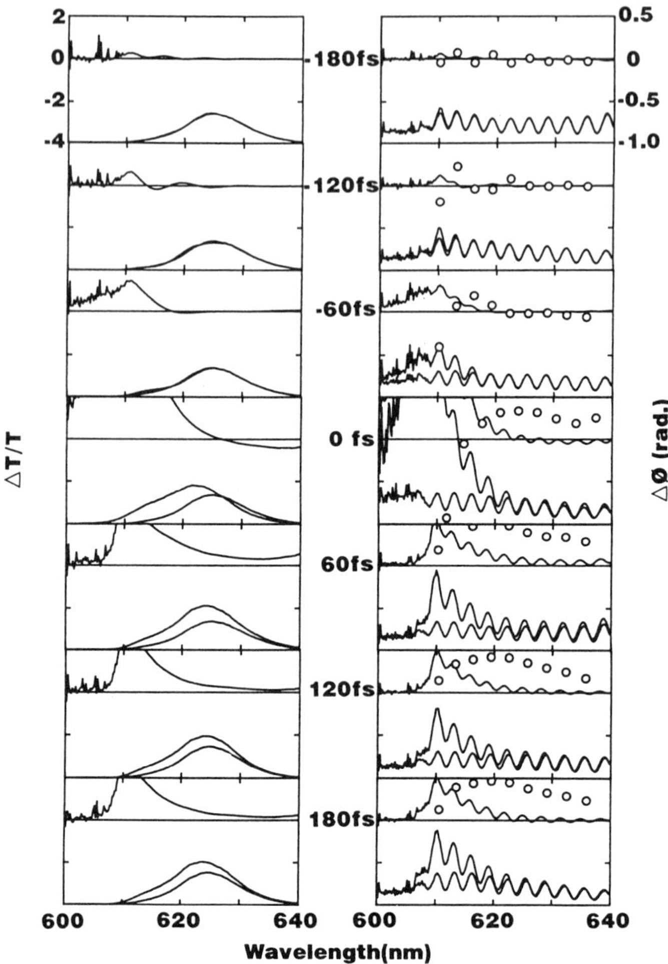

Fig. 8. The difference phase spectra of a Toshiba R63 filter are shown in the right-hand side, obtained from the transmitted probe and reference beams interfering with each other, detected by the polychromator and multichannel photodiode (MCPD) with and without excitation. The spectra of probe beam are shown also in the left-hand side.

and red-phase according to the color. The phase transition can be induced by photoirradiation, pressure, and heat [66]. Red-phase PDAs have weak fluorescence with the quantum yield of $<10^{-(3-4)}$ [67], while blue-phase PDAs have no fluorescence. Until several years ago it was supposed that the color changes were due to two possible chemical structures of

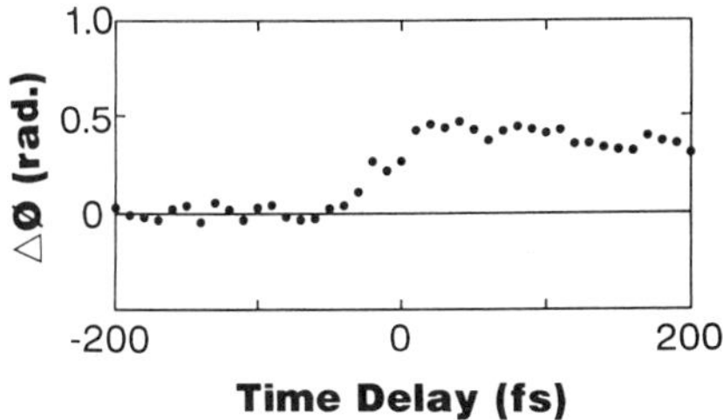

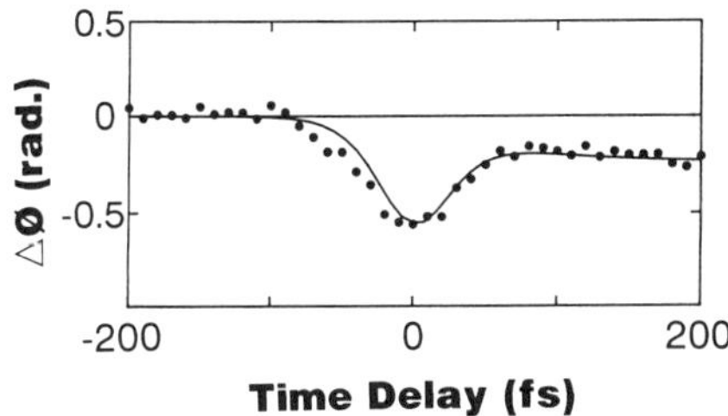

Fig. 9. The time evolution of the average phase change (a) in an R63 filter and (b) in CS_2 in 1 mm cell at 620 mm.

the main chain of PDAs, acetylenic $(CR{-}C{\equiv}C{-}CT)_n$ and butatrienic $(CR{=}C{=}C{=}CR)_n$. However, recent X-ray diffraction study has shown that both blue- and red-phase PDAs have the acetylenic main chain [68]. However, the electronic structures of these phases have not been clearly established.

Recent interest in PDAs has been focused on their large nonlinear optical susceptibility and ultrafast optical response. Various third-order nonlinear processes, such as the degenerate four-wave mixing [69,70], third-harmonic generation [71,72], coherent-Raman scattering [56–59,73], inverse Raman spectroscopy [74], optical Kerr effect [56–59,75,76], and absorption saturation [56–59,77] have been extensively studied. PDAs have ultrashort phase and energy relaxation times, T_2 [69,73] and T_1 [56–59,78,79], respectively. The time constants of 150 fs and 1.5 ps were observed in the decay kinetics of blue-phase PDA-3BCMU [56,57], and assigned respectively to the formation and relaxation processes of self-trapped excitons (STEs) [56–59].

PTs have a ring structure in their repeating unit. The backbone geometry resembles that of *cis*-PA and the simple *cis*-like structure is stabilized by the sulfur. The ultrafast dynamics have been also investigated in PTs [56–58, 80–82]. Ultrafast relaxation processes and several nonlinear optical effects, i.e., hole burning, Raman gain, and optical Kerr effect, were observed in poly(3-methylthiophene) (P3MT) films [58]. The obtained decay time was 800 ± 100 fs at 10 K, and the relaxation process was explained in terms of STEs. It is well established by stationary spectroscopies and photoinduced electron-spin resonance experiments that polarons and bipolarons are formed in PTs after photoexcitation or electrochemical doping [83–86]. The bipolaron formation is due to the nondegenerate ground state in PTs, which lead to the confinement of soliton-antisoliton pairs because of inequivalence of energy between phenyl and quinoid forms. The photoluminescence of PTs has been reported to be due to either free charge carriers or excitons [87]. It has been suggested that generation of photoluminescent excitons is independent of the formation of polarons or bipolarons.

In order to realize the application to actual devices the mechanism of the ultrafast optical nonlinearity must be clarified. For example, factors determine the size of the third-order susceptibility and ultrafast relaxation time must be solved to obtain guiding principles for materials search and synthesis.

3.2. Experimental Apparatus and Samples

The femtosecond absorption spectroscopy system consists of four parts, a colliding-pulse mode-locked (CPM) ring-dye laser, a four-stage dye amplifier pumped by the second harmonic pulses of a 10-Hz Q-switched Nd:YAG laser, a grating pair pulse compressor, and an optical system of pump-probe absorption spectroscopy [56,57]. The wavelength of intensity peak of the amplified pulses changed between 625 nm (1.98 eV) and 630 nm (1.97 eV) by the dye concentrations of the CPM laser and the amplifier. The duration and the energy are 80 fs and about 200 μJ, respectively. The preparation methods of samples used are described elsewhere [56–59,85,87–89].

3.3. Results and Discussion

Here we will describe the experimental results of the femtosecond time-resolved spectroscopy of several polydiacetylenes (PDAs), polythiophenes (PTs), polythienylene vinylene (PTV), and phenylpolyacetylene (PPA).

3.3.1. Blue-Phase PDA-3BCMU (cast film)

Figure 10 shows the transient photoinduced absorption spectra in an 11 μm cast film of PDA-3BCMU at 10 K after the femtosecond pulse excitation at the photon density of 9.5×10^{14} photons/cm^2 with parallel polarizations of the pump and probe [56,57]. Since the photon energy of the fundamental

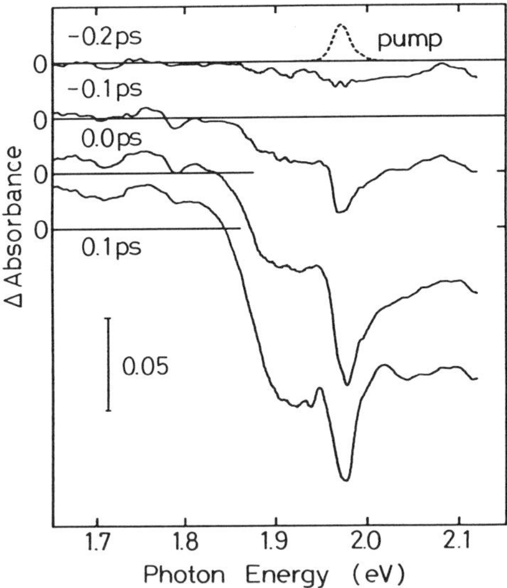

Fig. 10. The photoinduced absorption spectra in an 11 μm cast film of PDA-3BCMU at 10 K after 100 fs pulse excitation at the photon density of 9.5×10^{14} photons/cm^2 with parallel polarizations of pump and probe.

femtosecond pulse (1.97 eV) is very close to the absorption peak of the ^1B_u excitons (1.92 V at 10 K and about 1.95 eV at 290 K) in the blue phase PDA-3BMCU. An oscillatory structure around 1.97 eV is observed at negative delay times, due to the perturbed free induction decay observed also for several other materials [90–93]. The sharp bleaching peak at 1.97 eV is due to hole burning affected by the coherent pump polarization coupling, which sharpens the hole by the effect of interference.

Two small but reproducible downward peaks at 1.79 and 1.71 eV at zero delay time are due to the probe amplification by the stimulated Raman gain process [56,57,94] corresponding to the Stokes-Raman shifts of 1450 cm^{-1} (C=C stretching) and 2100 cm^{-1} (C≡C stretching), respectively [56,57,94]. The same vibrational modes as in the present study were found to be coupled strongly with the excitons, in the inverse Raman effect of PDA-TS (poly[2,4-hexadiyne-1,6-diol-bis(*p*-toluene sulfonate)]) [95,96].

The asymmetric bleaching near 1.97 eV in the transient spectrum of PDA-3BCMU cast film in blue phase at 0.0 ps, shown in Fig. 11, is induced by high excitation photon density of 3.8×10^{15} photons/cm^2 at 10 K. This is partially due to the optical (AC) Stark effect, which was recently discussed for several condensed-phase materials [90–93,96–104] and partially due to the induced phase modulation. The pump photon energy in the present study

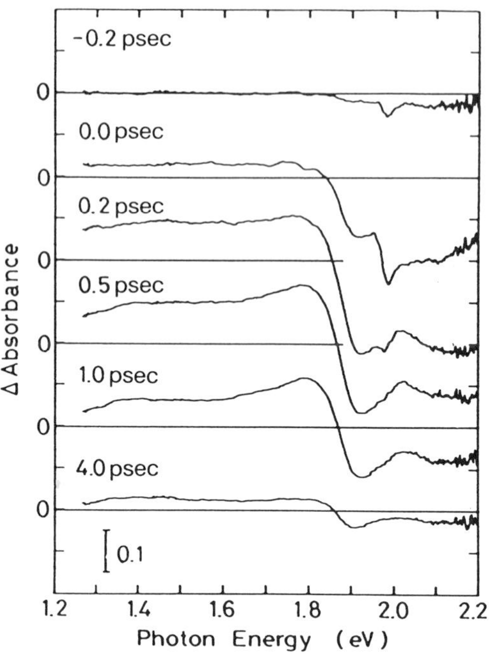

Fig. 11. The photoinduced absorption spectra of the PDA-3BCMU cast film in the blue phase at 0.0 ps showing the asymmetric bleaching near 1.97 eV at the high excitation photon density of 3.8×10^{15} photons/cm^2 at 10 K.

higher than the exciton transition energy in contrast to the previous reports. This leads to the red shift [56,57] as theoretically expected, and was recently reported for another system of a semiconductor by Hulin and others [103]. The effect is also partially due to the shift of the probe spectrum by the induced phase modulation, which is caused by the ultrafast modulation of phase by the optical Kerr effect [105]. This can be verified by the observed dispersion-type spectral change, which is not due to the AC stark effect for PDAs (red phase), which does not have an exciton peak near 2 eV, but due to the sharp peak of the probe spectrum [105].

3.3.2. Blue-phase PDA-3BCMU (evaporated film on a KCl single crystal)

Figure 12 shows the transient absorption spectrum of PDA-3BCMU/KCl. Because of the smaller inhomogeneity in the spectrum they exhibit a narrow homogeneous bleaching spectrum due to absorption saturation of excitonic transition. They show a very clear Raman gain spectrum at 0.0 ps delay. The bleaching peak at 1.97 eV of the laser photon energy is due to the interference

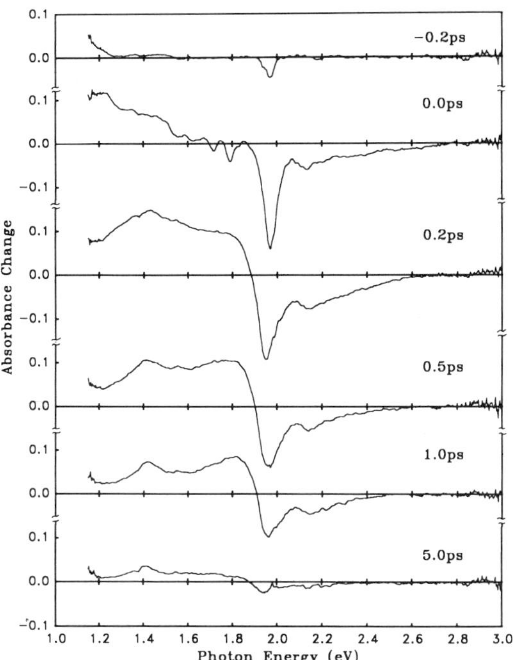

Fig. 12. The transient absorption spectra of the PDA-3BCMU sample evaporated on the (100) surface of KCl single crystal substrate at 290 K. The pump at 1.97 eV with the photon density of 2.2×10^{15} photons/cm^2. Pump and probe polarizations are parallel to each other.

effect of the pump polarization coupling. The peak at 200 fs is located at 1.95 eV, which is due to the hole burning not affected by the interference because of the long delay. These two peak positions at 10 K indicate that the homogeneous width (HWHM) is close to the separation of the two, i.e., 0.02 eV. This means that the excitonic dephasing time at 10 K is about 50 to 60 fs. This is consistent with the observation of the perturbed free induction decay at −200 fs, if we take into account the finite pulse width of both the pump and probe.

3.3.3. *Blue-Phase PDA-4BCMU (oriented film)*

Figure 13 shows the stationary and transient photoinduced absorption spectra of the blue-phase oriented PDA-4BCMU film at the delay times when the pump and probe pulses overlap in time in the sample. The

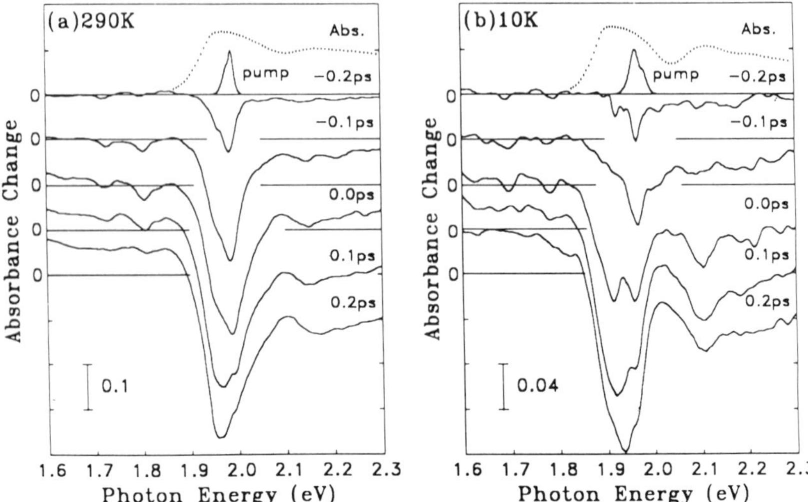

Fig. 13. Stationary and transient absorption spectra of the blue-phase oriented PDA-4BCMU film at (a) 290 K and (b) 10 K induced by the 1.98-eV pump pulse. The excitation photon density is 2.0×10^{15} photons/cm^2. The dotted curves are the stationary absorption spectra of the sample film. A pump laser spectrum is also depicted. Both the pump and probe pulses are polarized parallel to the main chain.

polarizations of the pump and probe pulses are parallel to the polymer main chain. The peak photon energy of the pump pulses was 1.98 eV when the experiment was done at 290 K, and it was 1.97 eV at 10 K as shown in Fig. 13. The photon energies are much lower than the absorption edge in the red phase (2.2 eV); the 1.97-eV pump pulse selectively excites the ^1B_u excitons in the blue phase even when the red-phase PDA-4BCMU is coexisting in the samples because of possible imperfect phase transition. The pump photon density was about 2.0×10^{15} photons/cm^2 at both temperatures.

The absorbance changes at -0.2 and -0.1 ps have a bleaching peak at the pump photon energy. The bleaching at 190 K becomes broader with time from -0.1 ps to 0.2 ps, and the peak shifts to 1.96 eV. At 10 K the bleaching peak at 1.97 eV disappears very rapidly and another bleaching peak appears at 91 eV where the excitons have the absorption peak. The 1.97 eV peak is observed when the pump and probe pulses overlap and is due to coherent coupling of the pump polarization. The 1.91-eV peak is due to the saturation of the excitonic absorption.

The absorbance changes below the pump photon energy have two minima at 1.80 and 1.72 eV at 290 K and 1.79 and 1.71 eV at 10 K. These minima were also observed in the PDA-3BCMU film and explained in terms of

Raman gain as discussed in Secs. 3.1.1 and 3.1.2 [56–59]. The corresponding Raman shifts are 1470 and 2110 cm^{-1}, which are assigned to the C=C and C=C stretching, respectively.

Absorbance changes due to the same photon modes are observed also at the photon energies above the pump. At the delay time of -0.1 ps two small maxima at 2.14 and 2.23 eV are clearly observed in the absorbance change at 10 K. They are due to the inverse Raman scattering. The inverse Raman scattering was observed and studied in detail in PDA-TS[74,95]. At positive delay times a bleaching peak appears at 2.10 eV. The energy difference from the 1.91-eV bleaching peak is 0.19 eV. The bleaching peak is due to the exciton phonon sideband, and the corresponding phonon mode is the C=C stretching vibration.

Two small maxima at 2.18 and 2.25 eV are observed also at 290 K. The signals are very small compared with the Raman gain signal, and the peak photon energies are different from the peak expected from the inverse Raman scattering signal. It is explained by the overlap of the maxima due to inverse Raman scattering and the minima due to the phonon sidebands, because the absorption peak of the excitons at 290 K is close to the pump photon energy.

Figure 14 shows the transient absorption spectra of the oriented PDA-4BCMU film in blue phase at 290 K up to 5 ps in the spectral range of 1.2 to 2.8 eV. At 0.0 ps, bleaching above 1.9 eV and absorption below 1.6 eV are observed in addition to the above-mentioned nonlinear optical responses. The bleaching is due to the absorption saturation of the ^1B_u exciton. The photoinduced absorption at 0.0 ps has a peak at the photon energy lower than 1.4 eV. Then the absorption spectrum shifts to higher energy with time from 0.0 ps to 0.5 ps. The absorption at 0.5 ps has a peak at 1.8 eV.

The absorption peaks at 1.8 and 1.4 eV are typical in the femtosecond transient absorption spectrum of polydiacetylenes at delay times of a few hundred to several hundred femtoseconds. This is due to the spectral change induced by the self-trapped (ST) excitons. The peak at 1.4 eV corresponds to the energy of $1.4 + 1.95 = 3.35$ eV from the ground state, if the relaxation energy due to self-trapping is neglected. The latter is estimated to be equal to or smaller than the observed Stokes shifts of 0.01 to 0.15 eV of the fluorescent PDAs. The energy of 1.8 eV peak measured from the ground state is then estimated as 3.3 to 3.2 eV. The peak at 1.4 eV is due to the transition from the ^1B_u self-trapped exciton to the m_1A_g exciton. The peak at 1.8 eV may be due to the biexciton (excitonic molecule). If the latter is the case, the binding energy of the biexciton is calculated to be $1.95 - 1.8 = 0.15$ eV.

The absorbance change at 5.0 ps has an absorption peak at 1.4 eV. This absorbance change remains for much longer than 100 ps. The absorption at

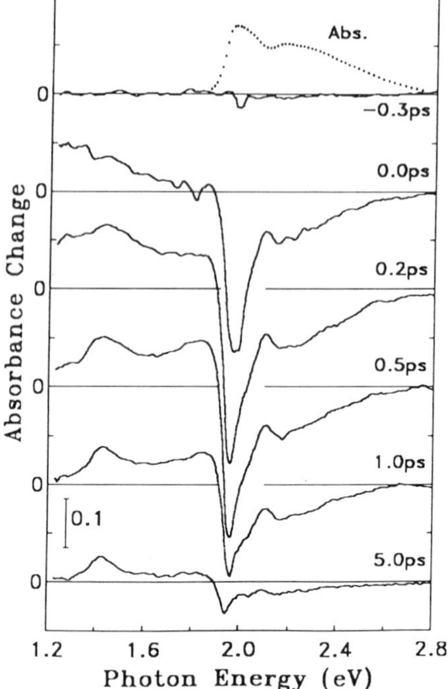

Fig. 14. The transient absorption spectra of the oriented PDA-4BCMU film in blue phase at 290 K up to 5 s in the spectral range of 1.2 to 2.8 eV. The excitation density is 2.0×10^{15} photons/cm^2. The dotted curve is the stationary absorption spectrum.

1.4 eV is assigned to triplet excitons [106,107,108]. The pump intensity dependence of the singlet excitons shows a linear response [108], while the long-lived component increases superlinearly with the pump intensity. The pump intensity dependence of the triplet excitons was studied in PDA-MADF in detail [108]. The result shows that the population of the triplet excitons is proportional to the square of the singlet exciton population even at higher-pump intensities where the singlet excitons exhibit saturation. The formation mechanism of the triplet excitons generated by the 1.97-eV pump pulse was concluded to be not due to the two-step two-photon absorption but due to the collision of the singlet excitons [108].

The decay curves of the absorbance changes in the blue-phase PDA-4BCMU can be fitted to single-exponential functions and long-lived component, $\Delta A \exp(-t/\tau) + \Delta A_L$. The obtained time constants are summarized in Table 1. The lifetime depends on the probe photon energy. The time constant around 1.9 decay kinetics below 1.7 eV is mainly the blue shift of

Table 1.

Decay Time Constants of Absorbance Changes in a Blue-Phase Oriented PDA-4BCMU Film.

	Decay time (ps)	
Probe photon energy (eV)	290 K	10 K
2.21	1.1 ± 0.1	1.2 ± 0.1
1.97	1.1 ± 0.1	1.5 ± 0.1
1.91	1.6 ± 0.3	2.1 ± 0.3
1.88	1.6 ± 0.1	2.0 ± 0.3
1.77	1.2 ± 0.1	2.1 ± 0.2
1.41	0.5 ± 0.1	1.1 ± 0.2
1.23	0.2 ± 0.1	0.3 ± 0.1

the photoinduced is shorter than 1 ps. The bleaching above 1.95 eV is because the free ground state is expected to be faster than that from the relaxed STEs. However, the decay kinetics of the free excitons and the relaxed STEs could not be distinguished in the blue-phase PDA-4BCMU. The observed lifetime of the bleaching is about 1 ps.

The transient absorbance changes of the oriented PDA-4BCMU film at 10 K are similar to the changes at 290 K. The bleaching due to the absorption saturation of the excitons, the broad absorption below the absorption edge, and the blue shift of the photoinduced absorption are observed. The formation time of the relaxed STEs is estimated as 120 ± 0.2 ps. It is longer than the main relaxation process, a tunneling in configuration space from the STEs to the ground state.

The transient absorbance changes of the oriented blue-phase PDA-4BCMU film at 10 K are similar to the changes at 290 K. The bleaching due to the absorption saturation of the excitons, the broad absorption below the absorption edge, and the blue shift of the photoinduced absorption are observed. The formation time of the relaxed STEs is estimated as 120 to 130 fs and is almost the same with that at 290 K. It is consistent with the barrierless potential of the ST exciton (STE) in one-dimensional systems. The lifetime of the relaxed STE is obtained as 2.1 ± 0.2 ps, which is longer than the lifetime (1.6 ps) at 290 K, but the difference is small. This suggests that the main relaxation process is a tunneling in configuration space from the STE state to the ground state.

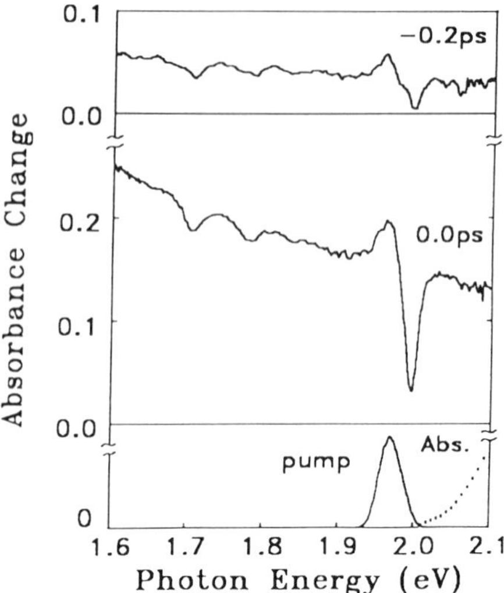

Fig. 15. Transient absorption spectra of a red-phase PDA-4BCMU cast film at 290 K induced by the 1.97-eV pump pulse, of which the spectrum is also shown. The spectra of the sample without excitation (Abs.) and of pump pulse is shown at the bottom. The excitation photon density is 4.0×10^{16} photons/cm^2. The dotted curve shows the edge region of the stationary absorption spectrum.

3.3.4. Red-Phase PDA-4BCMU (cast film)

Figure 15 shows the transient absorption spectra of the red-phase PDA-4BCMU cast film at 290 K excited by the 1.97-eV pump pulse. The observed absorbance change is positive in the whole photon energy region in the figure. Since the absorption edge of the cast film is 2.01 eV, the 1.97 eV pump pulse cannot directly create the exciton. The pump intensity dependence of the absorbance change is represented by I^a with $a = 1.9 \pm 0.1$. The observed transient absorption is induced by the two-photon absorption of the 1.97-eV pump pulse.

The absorbance change at -0.2 ps around 1.97 eV in the cast film has an antisymmetric oscillatory structure. The observed antisymmetric structure agrees with the differential spectra of the probe white continuum pulse generated from the 1.97-eV pulse by the cross-phase modulation. The amplitude of the oscillation is proportional to the pump intensity. The structure around 1.97 eV can be explained in terms of induced-frequency shift of the probe light [105,109]. The induced phase modulation (IPM) can be verified by changing the spectra of the probe pulse. The change in the probe pulse spectrum due to the induced frequency shift was clearly found in GaAs microcrystallites [105].

The absorbance change spectrum at 0.0 ps near 1.97 eV is asymmetric. This is due to both the IPM and the pump polarization coupling. Since the refractive index change causing IPM is due to the formation of the free exciton and self-trapped exciton, the effect is even larger at 0.0-ps delay time than -0.2 ps. If the refractive index change is the virtual effect as in the AC Stark effect, IPM disappears at 0.0 ps.

The formation time of STEs in the red-phase PDA-4BCMU is estimated from the spectral change around 2 eV as 120 ± 60 fs. After this initial change the spectrum of the red-phase PDA continues to change until about 5 ps. The time constant of the slow spectral change is obtained as 1.1 ± 0.1 ps. The decay kinetics of STEs in the red-phase PDA-4BCMU cannot be fitted to a single-exponential function, and the decay becomes slower at longer delay times. When the decay curve at 2.34 eV is empirically fitted to a biexponential function, the time constants of the short and long components are obtained as 660 ± 70 fs and 4.4 ± 0.5 ps. The time constants are listed in Table 2. The physical meanings of these time constants will be shown later.

3.3.5. PDA-MADF (single crystal)

The transient absorption spectra of PDA-DFMP single crystal in the spectral range of the Raman gain is shown in Fig. 16. Two peak minima in the

Table 2.

Decay Time Constants of Absorbance Changes in a Red-Phase Cast PDA-4BCMU Film at 290 K Excited by the 3.94 eV Pump Pulse

	Time constants (ps)	
Probe photon energy (eV)	τ_1	τ_2
2.58	0.4 ± 0.1	3.5 ± 0.6
2.34	0.7 ± 0.1	4.4 ± 0.5
2.00[a]	1.3 ± 0.3	—
1.77	0.8 ± 0.1	10.0 ± 0.6
1.70	0.9 ± 0.2	8.2 ± 1.4
1.55	1.3 ± 0.3	10.4 ± 2.2

[a] The decay curve at 2.00 eV can be fitted to a single-exponential function. The second time constant τ_2 could not be estimated.

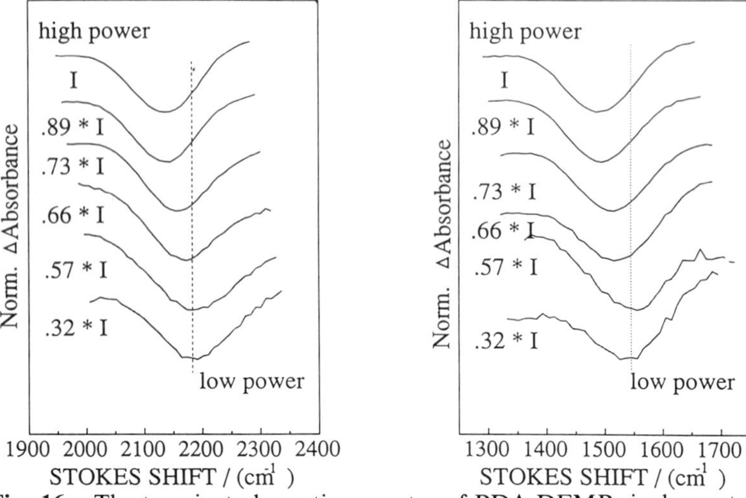

Fig. 16. The transient absorption spectra of PDA-DFMP single crystal.

photoinduced absorption due to the stimulated gain process are localized at approximately 1.80 and 1.72 eV, which are observed only when the pump and probe pulses overlap temporally. The Stokes shifts are 1550 and 2180 cm^{-1} corresponding to the ν_2 and ν_1 mode frequencies, in agreement with the spontaneous resonance Raman spectrum. The peak frequencies are shown in the figure to shift to higher photon energies by about 100 cm^{-1} with the factor of three increase in pump proton flux densities from 17 to 58 GW/cm^2.

To explain the behavior of the shift in terms of the dressed-state picture, the following four-level model was invoked. The energy levels $|a\rangle$, $|b\rangle$, and $|c\rangle$ are the ground level, $\nu_2(1500\ cm^{-1})$ and $\nu_1 = 2135\ cm^{-1}$) vibrational levels in the ground electronic state. The energy level $|d\rangle$ is the excitonic zero-phonon level. From the exciton transition energy of 2.27 eV and laser photon energy of 1.988 eV, the detuning is 0.282 eV. The laser is detuned to the virtual transition from the unpopulated vibrational excited level of ν_2 ($1500\ cm^{-1} = 0.186$ eV) and ν_1 ($2150\ cm^{-1} = 0.267$ eV) to the excitonic zero phonon level and 0.096 eV and 0.015 eV, respectively, by diagonalizing the super matrix, which is obtained by the unitary transformation and the rotating-wave approximation.

3.3.6. P3MT (cast film)

Figure 17 displays a series of absorption difference spectra of P3MT at 10 K measured at various delay times between the excitation and probe pulses.

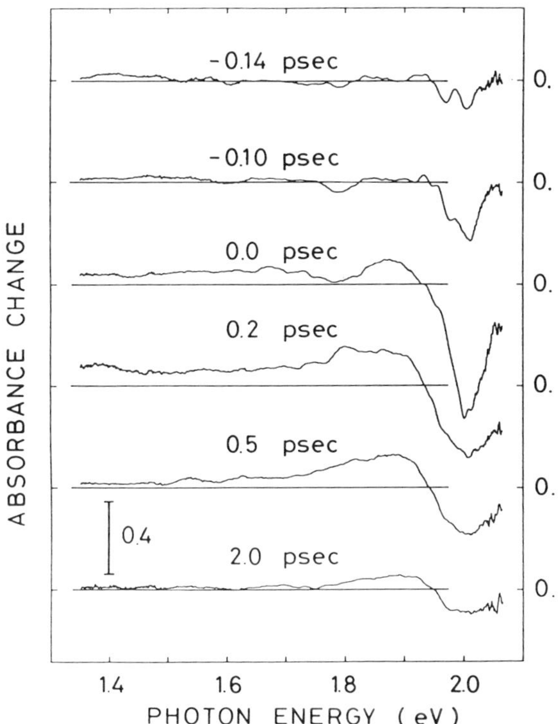

Fig. 17. The difference absorption spectrum of P3MT at various delay times at 10 K. The photon energy of the excitation pulse was 1.98 eV.

When the probe pulse arrives at the sample before the peak of the excitation pulse, clear oscillations in the differential absorption spectra are observed around the photon energy of the excitation pulse. These oscillations turn out to be very similar to those observed in femtosecond pump-probe experiments on semiconductors near the band edge. They result from the coherent interaction of the weak probe pulse with the polarization induced by the intense pump pulse in the P3MT film. As is expected from theory, the coherent oscillation vanishes when the maximum of the probe pulse arrives at the sample later than the pump pulse.

A striking feature of the photoinduced absorption spectra at early delay times (very pronounced at -0.1 and 0 ps) is the appearance of a strong minimum at 1.8 eV and a weaker one at 1.62 eV due to Raman gain. This assignment is also supported by the resonance Raman scattering spectrum of P3MT measured by Steigmeier *et al.* [109]. Their result shows that the phonon energy of the C=C stretching mode that has the strongest coupling with an exciton is 0.18 eV, the same photon energy shift which we observed. The one dominant mode in the Raman gain spectra suggests that only one photon mode with an energy of about 0.18 eV is much more strongly associated with the excitation of P3MT than any other modes, as was observed in the resonance Raman spectrum by Steigmeier *et al.* [109].

Just after excitation (0 ps), broad featureless absorption spectra were observed in the photon energy region of 1.35 to 1.9 eV in both P3MT and PDA. The common feature indicates that the absorption is due to the same transition, i.e., the free exciton to higher excited states. At a delay time of 0.2 to 0.5 ps, the spectrum of the self-trapped exciton can be seen with a peak in the spectral region 1.8 to 1.95 eV. This absorption is due to the transition from the self-trapped exciton to the excitons of higher energies or to the conduction band.

A second fast component is clearly apparent in the relaxation of the recovery of the bleached absorbance with a time constant of 800 ± 100 fs. The component of the same decay time was found in the broad band excited-state absorption. Thus this decay indicates the relaxation of self-trapped excitons to the ground state or to other long-lived species.

The time-resolved fluorescence measurement of PT has been reported by Wong *et al.* [67]. They estimated the fluorescence lifetime only to be shorter than 9 ps, which is the temporal resolution of their system. Since the excited-stated absorption has only one fast decay component shorter than 9 ps, this fast relaxation process is identical to that reported by Wong *et al.* [67] for PT. The Stokes shift of the transient fluorescence from PT is similar to that of the steady-state fluorescence from P3MT. Therefore, both fluorescence spectra are considered to be due to the self-trapped excitons. The steady-state fluorescence of P3MT has a peak around 1.9 eV. Assuming that the curvature of the potential curve of the self-trapped exciton is the same as that of the ground-state and that the free exciton energy is 2.1 eV, the deactivation energy due to the self-trapping is estimated to be 0.1 eV.

3.3.7. *P3DT (cast film)*

Figure 18 shows the stationary and transient absorption spectra of a P3DT film at 290 K and the pump pulse spectrum. The bleaching due to the depletion of the ground state and the photoinduced absorption below the absorption edge are observed also in P3DT. The absorbance change at 0.0 ps has a structure at the pump photon energy, probably due to the induced frequency shift of the probe pulse as observed in the red-phase PDA-4BCMU. A small minimum at 1.80 eV is due to the Raman gain. The

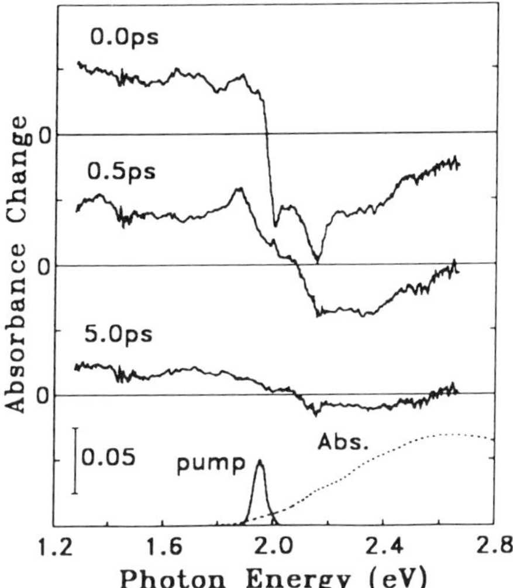

Fig. 18. The stationary (dotted curve) and transient absorption spectra (solid curves) of the P3DT film at 290 K and the pump pulse spectrum with peak at 1.97 eV. The excitation photon density is 7.5×10^{15} photons/cm^2.

corresponding Raman mode is assigned to the stretching vibration of the C=C bond. The bleaching peak at 2.14 eV is due to the 0 to 1 transition of the same phonon mode. At 0.5 ps the absorbance change around 2 eV becomes positive, and the photoinduced absorption has a peak at 1.88 eV. If excitons are assumed to be photoexcited in P3DT by the 1.97-eV pump pulse, this spectral change can be explained by the formation of the relaxed STEs in the same way as PDA-4BCMU. The time constant of the formation is estimated as 100 ± 50 fs.

P3DT has similar decay kinetics to the red-phase PDA-4BCMU. The decay kinetics has short- and long-lived components. The time constant of the long-lived component is much longer than 100 ps. The short-lived component can be fitted to biexponential functions. The sets of the time constants at 290 K are summarized in Table 3. The time constant τ_1 is shorter than 0.5 ps and is expected to correspond to the decay of the free excitons. The time constant τ_2 corresponds to the decay of the relaxed STEs and is estimated as 4.7 ± 1.2 ps.

If the decay curves are fitted to a power law decay and long-lived component, $\Delta A_1 t^{-\alpha} + \Delta A_L$, the power α at 290 K is obtained as 1.02 ± 0.06 at 1.63 eV. The power law decay was reported also in other PTs and the value of α was 0.37 in poly(3-hexylthiophene) at 1.7 eV, 0.22 in poly(3-octylthiophene) at 1.17 eV, and 0.9 in polythiophene at 2 eV [58,59]. The

Table 3.

Decay Time Constants of Absorbance Changes in a P3DT Film at 290 K

Probe photonenergy (eV)	Time constant (ps)	
	τ_1	τ_2
2.21	0.3 ± 0.1	4.7 ± 1.6
1.88	0.5 ± 0.1	4.7 ± 1.2
1.63	0.3 ± 0.1	7.9 ± 1.8
1.55	0.3 ± 0.1	7.9 ± 2.3

dependence of α on the probe photon energy and the spectral change at the delay time from 0.5 ps to 5 ps are difficult to explain by a simple recombination model.

Figure 19 shows the time dependence of the absorbance changes at 1.50 eV induced by the pump pulse polarized parallel (‖) and perpendicular (⊥) to the polymer chains. The short-lived component due to the STEs decreases rapidly and the long-lived component remains much longer than 100 ps. It is clearly seen that the long-lived component is induced more efficiently by the perpendicular pump pulse. This means the long-lived component is due to either polarons or bipolarons generated by the interchain photoexcitation.

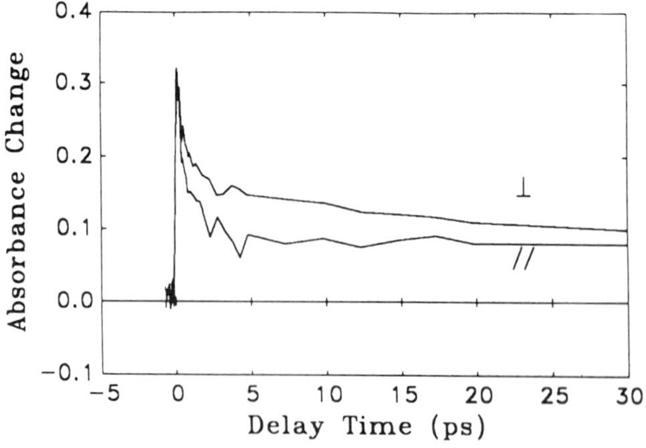

Fig. 19. The time dependence of the absorbance change at 1.50 eV in the P3DT film at 290 K induced by the pump pulse polarized parallel (‖) and perpendicular (⊥) to the polymer chain.

A polaron pair is formed from an electron-hole pair excited by a single photon, while a bipolaron is generated by a collision of two polarons with a charge at the same sign. The observed intensity dependence shows that the long-lived component in P3DT increases proportionally to the pump photon density up to 2×10^{16} photons/cm^2. Hence the long-lived component in P3DT is concluded to be due to polarons.

The absorbance change in poly(3-dodecylthiophene) has similar properties to the red-phase PDA-4BCMU. The bleaching due to absorption saturation and the broad absorption below 2-eV were observed. The decay kinetics of the absorbance change can be empirically fitted to the biexponential function, and the time constants are obtained as 0.3 ± 0.1 ps and 4.7 ± 1.6 ps at 2.2 eV.

3.3.8. PTV (dip-coated film)

The difference absorption spectra of PTV dip-coated film at 297 K and at 77 K shown in Figs. 20 and 21, respectively. The spectra show a dispersion-type structure near the spectral peak of the pump pulse at 1.97 eV. The spectral feature is accompanied by a maximum and minimum structure at 1.95 and 2.00 eV, respectively. Since at large negative delay times the pump pulse cannot populate the interband transition, the absorbance change single $t = -0.30$ ps has significant contributions from pump perturbed probe free-induced decay. The signal at $t = -0.05$ ps is due to the perturbed free induction decay pump polarization coupling and optical Stark effect. The gain signal at 1.80 eV can only be observed from $t = -0.15$ and 0.10 ps (during the temporal overlap of the pump and probe beams), so the signal is most likely due to Raman gain. The imaginary component of the third-order susceptibility $\chi^{(3)}(-\omega_2; +\omega_2, \omega_1, -\omega_1)$, where the subscripts p and pr represent the pump and probe, respectively, is $\text{Im}[\chi^{(3)}] = 1.8 \times 10^{-9}$ esu, for $\hbar\omega_1 = 1.97$ eV and $\hbar\omega_2 = 1.80$ eV.

In Fig. 22, the transient decay curve of the photoinduced absorption, recorded at six probe energies with the photon density of 2.4×10^{15} photons/cm^2 at 297 K, is displayed as a semilogarithmic plot. In this plot the long-time component $\Delta A(1)$ is subtracted. The slope of the semilogarithmic plot illustrates that there is significant deviation from linearity for the photon energies shown. Figure 23 shows that the transient decay curves can be numerically fitted to a biexponential decay function with a constant long-lived component, $\Delta A(t) \exp(-t/\tau_1) + \Delta A(t) \exp(-t/\tau_2) + \Delta A(1)$. (The fitted decay coefficients τ_1 and τ_2 are shown in Table 4.) Like the bleaching signals, the amplitudes of the photoinduced absorption signals are proportional to the pump laser intensity, which is varied from 5.9×10^{14} to

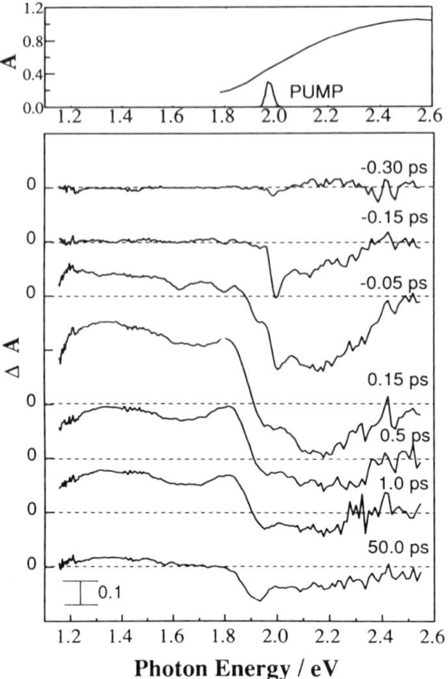

Fig. 20. Photoinduced absorbance change, ΔA, versus probe photon energy for seven pump-probe delay times for a PTV film after excitation with a 1.97 eV pump pulse at 297 K. The excitation photon density is 2.4×10^{15} photons/cm^2. The absorption spectrum and pump pulse spectral profile are shown for comparison.

Table 4.

The Two Decay Constants, τ_1 and τ_2, are obtained from the Biexponential Decay Fits of the Transient Absorbance Changes in the PTV Film recorded at Six Probe Photon Energies. The Uncertainties listed below represent One Standard Deviation Error.

Photon energy (eV)	Decay constant τ_1 (ps)	Decay constant τ_2 (ps)
1.82	0.86 ± 0.05	14.2 ± 1.4
1.80	0.73 ± 0.05	9.8 ± 1.0
1.70	0.56 ± 0.05	7.0 ± 1.0
1.59	0.36 ± 0.03	6.6 ± 0.9
1.48	0.21 ± 0.04	5.3 ± 0.8
1.44	0.16 ± 0.06	4.1 ± 0.9

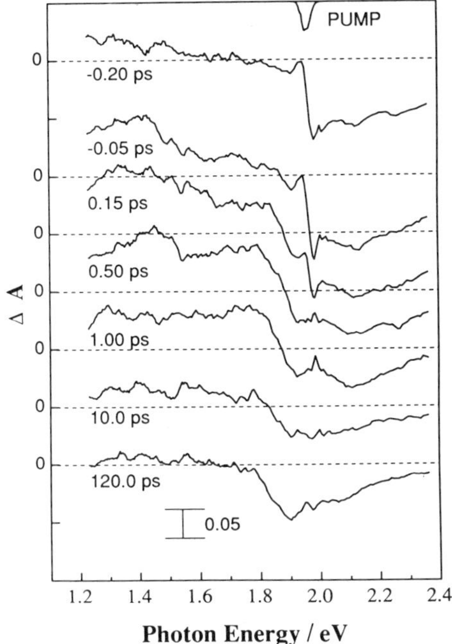

Fig. 21. Photoinduced absorbance change, ΔA, versus probe photon energy for seven pump-probe delay times for a PTV film after excitation with a 1.96 eV pump pulse at 77 K. The excitation-photon density is 5.9×10^{14} photons/cm^2. The absorption spectrum and pump pulse spectral profile are shown for comparison.

2.4×10^{15} photons/cm^2, and only a small decrease ($<20\%$) of the decay times is observed with increasing laser intensities. Thus, the kinetic behavior is not a bimolecular decay process. As seen in Table 4, both decay coefficients vary with probe photon energy. This is consistent with the self-trapped exciton relaxation model since the decay constant τ_1 decreases with lower photon energies. This suggests that there is a distribution of relaxation rate from the STE to the ground state that depends on the energy of the STE. Decay processes into the ground state potential energy surface are monitored at lower photon energies, corresponding to STE states with significant population in high-lying vibrational levels, and they are faster than STE states with population in low-lying vibrational levels.

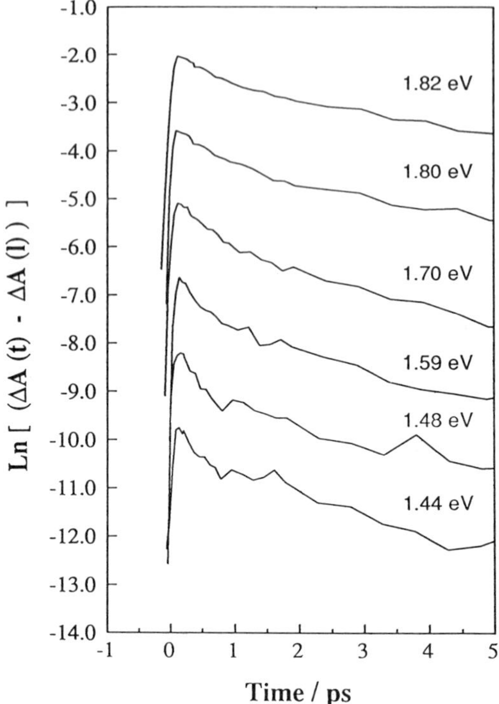

Fig. 22. The transient decay curves of the photoinduced absorption for six different photon energies at 297 K displayed as a semilogarithmic plot, where the long-time component has been subtracted.

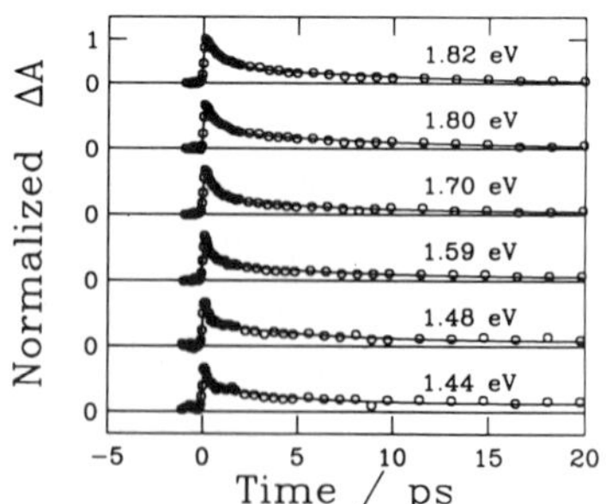

Fig. 23. The normalized transient decay curves of the photoinduced absorption for six different photon energies at 297 K fitted to the biexponential decay function, $\Delta A_1 \exp(-t/\tau_1) + \Delta A_2 \exp(-t/\tau_2)\Delta A(1)$. $\Delta A(1)$ is a constant amplitude absorbance change that does not decay within 150 ps.

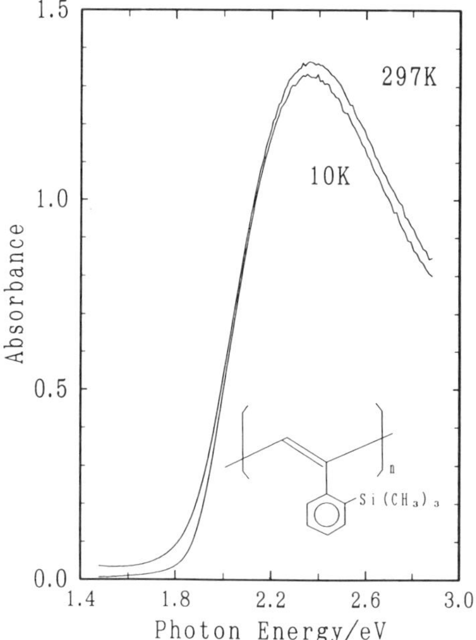

Fig. 24. The absorption spectra of PMSPA taken at 297 K and 10 K. Inset: The chemical structure of PMSPA.

3.3.9. PMSPA (cast film)

The stationary absorption spectrum of poly[*O*-(trimethylsilyl)phenylacetylene] (PMSPA) is shown in Fig. 24. Photoinduced difference absorption spectra taken at 10 K up to 1 ps after photoexcitation at $\hbar\omega = 1.98$ eV is shown in Fig. 25. The pump duration and excitation density are 100 fs and 2×10^{16} photons/cm^2, respectively. The following features can be seen from this figure. First, photoinduced absorption rises around 1.2 eV simultaneously with photoexcitation, and it shows a gradual blue shift to have a peak near 1.8 eV at delay time of 0.15 ps. Signal amplitude is kept constant during the shift. Secondly, bleaching is observed above 2.0 eV. The region is expanded to the higher photon energy region with increase in delay time. Since it has a peak at 2.3 eV, the same as that of stationary absorption spectra (Fig. 24), this bleaching is due to the saturation of interband transition. Thirdly, a peak minimum at 1.82 eV is observed at delay time of 0.0 ps, i.e., when the pump and probe pulses temporally overlap. This signal was also observed for several other polymers for the first time by our group. It is due to the Raman gain which appears at Stokes side from the pump frequency at 1410 cm^{-1}, corresponding to the C=C stretching mode of the

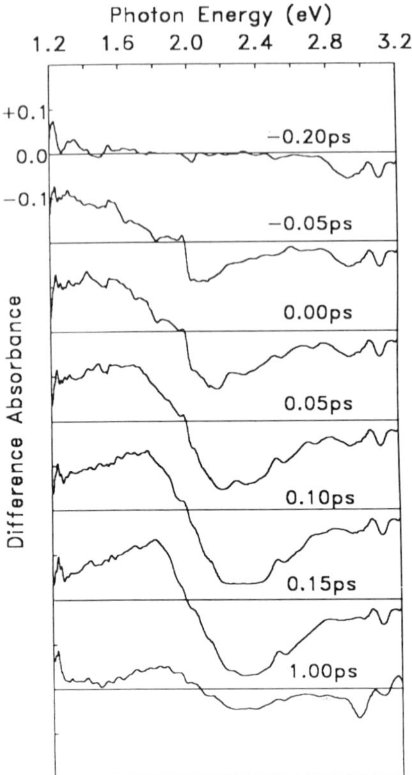

Fig. 25. Differential absorbance spectra of PMSPA taken at 10 K up to 1 ps after photoexcitation at $\hbar\omega = 1.98$ eV. Pump duration and excitation density are 100 fs, 2×10^{16} photons/cm^2, respectively. The positive change means photoinduced absorption, while the negative one means photoinduced bleaching.

ground state configuration. Finally, a sharp dispersion-type feature can be seen at small positive delay time, which is due to induced phase modulation, which was also observed for several polymers by our group for the first time. That is, at this delay time, a probe pulse experiences the frequency blue shift due to the modulation of refractive index of the sample caused by the trailing of the strong pump pulse. It should be noted that such an effect (change of refractive index) will be always superimposed more or less on the population-induced absorption change (change of extinction coefficient) in this time scale.

Figure 26 shows the temporal decay profiles at three different energy levels.

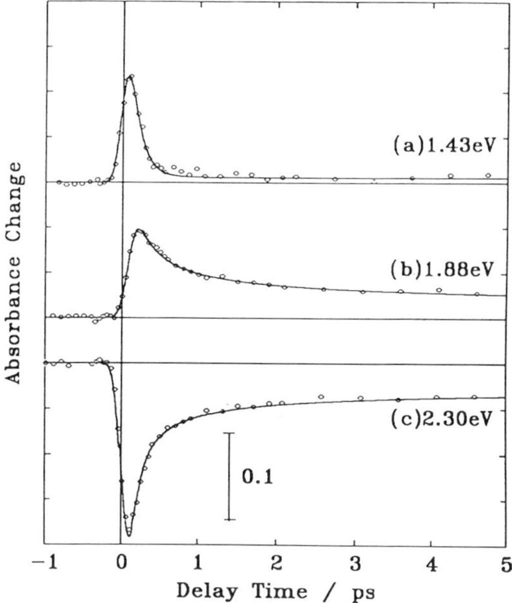

Fig. 26. Decay profiles at three different probe photon energy levels. Open circles represent the experimental data, and solid lines show the fitting curves with an exponential function for (a) and a power-law function for (b) and (c).

At a probe energy of 1.43 eV, the lower photon energy region, the photoinduction absorption almost disappears by 1 ps, and the ground state is recovered to some extent. In constant, the temporal behavior around the photoinduced absorption peak at 1.88 eV has some finite long-rising component that corresponds to the initial blue shift of photoinduced absorption peak mentioned above. The following decay rate is observed to become much slower than that at the lower energy region, which clearly can be seen by comparing (a) and (b) of Fig. 26. Furthermore, a long-lived component is also contained in the decay at the photoinduced absorption peak. At the bleaching peak (2.3 eV, (c)) again, a long-lived component is observed after the initial fast decay.

Since we failed in fitting of the decay profile (b) and (c) in Fig. 26 to the exponential functions and, in turn, successfully got the linear behaviors over the delay time of almost three orders of magnitude in the log-log plot as shown in Fig. 27, both photoinduced absorption and bleaching can be considered to follow power-law decay. So, we tried to decay profile (b) and

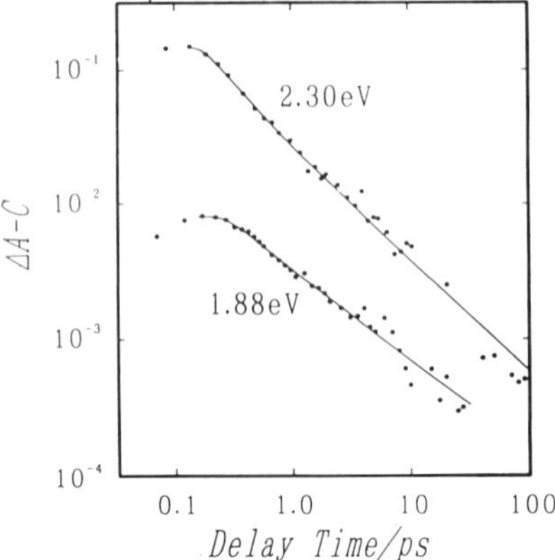

Fig. 27. Log-log plots of decay profiles at both PA (1.88 eV) and PB (2.30 eV) peaks after subtracting the long-lived components. Open circles and squares represent the experimental data, and solid curves show the results of fitting with Eq. (3.3.1). The data at PA peak is shifted by an order of magnitude.

(c) with the following function:

$$\Delta A(t) = A \, \mathrm{erf}[(\sigma, t]^{-n}] + C, \tag{3.3.1}$$

where erf is an error function. The first term represents the component that shows the approximately power-law decay, while the second term represents the long-lived component mentioned above. The fitting was performed by convoluting Eq. (3.3.1) with the pump pulse profile, which was assumed as gaussian type.

As a result of this fitting, we obtained a value of power n, which turned out to be 0.65 ± 0.05 for photoinduced absorption peak and 0.85 ± 0.05 for bleaching peak. The variation of n is less than 0.05 around the peak of both the absorption and bleaching. The fitting curves are also shown with solid lines in Fig. 27.

3.4. Relaxation of Photoexcitations in Conjugated Polymers

3.4.1. Mechanism of Exciton Self-Trapping in One-Dimensional System

The formation and decay time constants of ST excitons in several polymers are listed in Table 5. The mechanism of the photoinduced absorption and bleaching in polydiacetylene and the other conjugated polymers are explained using the potential curves of the free exciton. Self-trapped exciton and ground state are shown in Fig. 28.

Table 5.

The Formation Time (τ_f) and the Decay Time[a] (τ_d) of Self-Trapped (ST) Excitons and Fluorescence Property of Several Polymers

Polymer	τ_f(fs) (290 K)	τ_d (ps) (10 K)	τ_d(ps) (290 K)	Fluorescence property[b]
PDA-3BCMU (blue, oriented film[c])	150 ± 40	1.9 ± 0.2	1.6 ± 0.1	*n*
PDA-3BCMU (blue, cast film)	150 ± 50	2.0 ± 0.2	1.5 ± 0.2	*n*
PDA-DFMP (blue, single crystal)	100 ± 30	—	1.6 ± 0.1	*n*
PDA-4BCMU (red, oriented film)	120 ± 60	0.96 ± 0.09[d]	—	*f*
PDS-4BCMU (blue, oriented film)	140 ± 50	2.1 ± 0.1	1.6 ± 0.1	*n*
PDA-4BCMU (red, cast film)	<200	0.88 ± 0.08[d]	—	*f*
P3MT (electrochemical preparation)	70 ± 50	0.62 ± 0.07[d]	—	*f*
P3DT (electrochemical preparation)	100 ± 50	0.45 ± 0.06[d]	—	*f*

[a] The formation of the ST exciton corresponds to the emission process of the strongly coupled phonon (intramolecular vibration) to the excitonic transition.

[b] Polymers with higher and lower fluorescence quantum efficiency than 10^{-5} are indicated by *f* (fluorescent) and *n* (nonfluorescent), respectively.

[c] The sample is prepared by the polymerization of an evaporated monomer film of PDA-3BCMU on a KCl crystal. The sample is composed of small crystals and amorphous regions.

[d] This decay time is determined by the initial slope of the semilogarithmic plot in the early delay time.

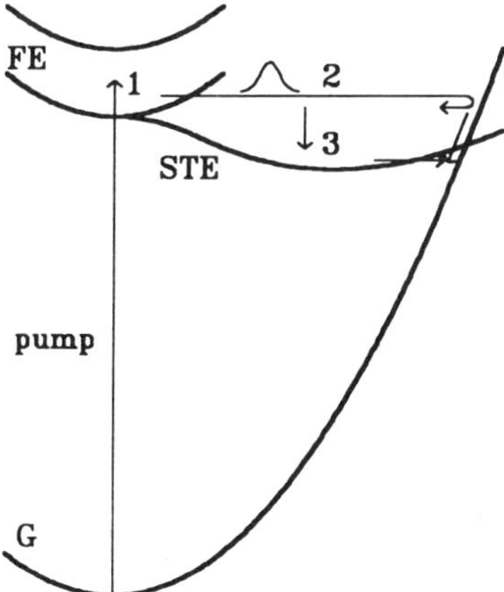

Fig. 28. Potential curves of the ground state (G), free exciton (FE), and self-trapped exciton (STE) of the nonfluorescent polydiacetylenes. There is no barrier between the free-exciton (FE) and self-trapped-exciton potential curves. Numbers 1, 2, and 3 indicate FE, STE before the emission of phonons with large frequency, and STE after the phonon emission but before thermal equilibration, respectively. A part of this figure is enlarged and shown in Figs. 30 and 31.

Since there is no barrier in the potential curve between free and self-trapped (ST) excitons in the one-dimensional system as shown in Fig. 28, the formation of the ST excitons is expected to take place within the period of the coupled phonon (intramolecular vibration) cycle $T_M = 2\pi/\omega_M$. The experimental results of the absorption spectra, the Raman gain, the resonance Raman scattering, and the phonon-mediated optical nonlinearity show that the singlet excitons inpolydiacetylenes are coupled with the C=C and C=C stretching modes. The coupled stretching mode frequencies are about 1500 cm^{-1} (C=C stretching) and 2100 cm^{-1} (C=C stretching) and correspond to the oscillation periods of 20 and 15 fs, respectively. Both of them are much shorter than the experi- mentally observed appearance times of ST excitons, which are 150 fs in PDA-3BCMU, 100 fs in PDA-4BCMU, 70 fs in P3MT, and 100 fs in P3DT.

In the following, the term "self-trapping" is used in two ways. One is the change in the adiabatic potential of excitons that have the same energy as the free excitons. The self-trapping time estimated above from the oscillation period is used in this meaning. The other is the emission of phonons of the most strongly coupled mode to the excitonic transition. After this process the exciton energy is lower than that of the free exciton, and the population of the ST excitons is distributed near the bottom of the "ST" exciton (STE) potential in Fig. 28.

According to Jortner [110], the rate of the self-trapping process of the second meaning in rare gas solids such as Ne and Ar at low temperatures is given by $k_{\mathrm{ST}} = \omega_{\mathrm{M}} \exp(-\omega_{\mathrm{M}}/\omega_{\mathrm{D}})$, ω_{M} and ω_{D} being the frequencies of the most strongly coupled mode and the Debye frequency, respectively. Using the values of ω_{D} in literature and ω_{M} given above, the calculated rate k_{ST} is several orders of magnitude smaller than the observed rate in the polymers. This is because of the too-simple phonon structure in the model, only suited to the rare gas solids. Much longer self-trapping time can be explained by the geometrical reorganization (relaxation) involving the lower frequency modes than by the oscillation period of the strongly coupled modes. The lower frequency modes may include those associated with the structural change in the bulky side chains, which are lacking in the rare gas solids.

There is difference in the formation time of ST excitons, namely "self-trapping" time of the second meaning, between PDAs and PTs. The formation times are 150 fs in PDA-3BCMU and 140 fs in PDA-4BCMU in the blue phase and 70 fs in P3MT and 100 fs in P3DT. This can be explained by the difference in the chemical structures of these two groups of polymers. PDA-3BCMU and PDA-4BCMU have side groups with hydrogen bonds connecting neighboring side chains, while P3MT and P3DT have rigid structures of thiophene rings conjugated with each other. The van der Waals interaction between the neighboring side chains of PTs results in the helical noncoplanar structure due to the steric hindrance among the side groups [111,112]. The van der Waals interaction between the side chains and that between main chain and side chains is more intense in P3DT than in P3MT because of the longer side group in the former. This explains the longer formation time in the former.

On the contrary, the PTs and both PDA-3BCMU and PDA-4BCMU have bulky substituent groups attached to the main PDA chain. They extend a few tens of angstroms from the backbone main chain and form a sheet in the backbone plane with the help of the hydrogen bonding between two neighboring side chains in the blue-phase PDA-3BCMU and PDA-4BCMU. After electronic excitation, the whole side-group structure must be reorganized in a way that it can fit the new most stable configuration in the ^1B_u exciton state. This change must be associated with the conformational reorganization of the whole substituent side groups in the polymer. It is expected to take longer time in PDA-3BCMU and PDA-4BCMU with the bulky side groups, which are connected with neighboring side chains by hydrogen bonding, than in P3MT and P3DT, which have small side groups and no strong interaction between neighboring side chains. From the experimental results, the reorganization of the whole configuration takes place after five to twenty periods of the oscillation induced by photogeneration of the excitons in PDA-3BCMU and PDA-4BCMU.

The self-trapping time constant of excitons in P3MT is 70 fs [58], which is different from 150 fs in PDA-3BCMU only by a factor of two even though the differences in the structure and size of the side chains are very large. This may be due to the following three possible mechanisms.

1. Finite time is needed for the self-trapped exciton to slide down along the adiabatic potential, because the potential curve of the self-trapped exciton deviates from the free exciton potential minimum, which is flat with null derivative [113].
2. According to Sumi's theory concerning the self-trapping process in reduced dimensionalities [114], the spontaneous geometrical relaxation, which takes place from infinitely extending free exciton to the self-trapped exciton, follows reaction pathway with very small slope along the reaction coordinate.
3. Because of the interchain interaction, the conjugated polymers may not be an ideal one-dimensional system. Then there may be a low barrier between two minima in the potential curves of the ground state and the ST exciton state.

These three mechanisms can also explain relatively small (0.05 to 0.15 eV) Stokes shift, observed for fluorescent polymers. Small Stokes shifts may also be explained as follows. There may be a very shallow minimum in the ground state potential curve of the polymers under the same configurations as the ST excitons. The geometrical configurations of the shallow minima are butatriene-like and quinoid-like forms in PDAs and PTs, respectively.

3.4.2. Decay Mechanism of Self-Trapped Excitons in Conjugated Polymers

In this subsection the following properties are addressed concerning the excited state dynamics in conjugated polymers, especially polydiacetylenes.

1. Many of conjugated polymers are nonfluorescent or only weakly fluorescent. The quantum efficiency of the nonfluorescent blue-phase polydiacetylenes is especially estimated to be lower than 10^{-5} and that of the fluorescent red-phase polydiacetylenes is only of the order of 10^{-4}.
2. There are common features in the transient absorption spectra of both PDAs (blue-phase PDA-3BCMU and PDA-4BCMU and red-phase PDA-4BCMU) and PTs either in random films or in oriented films. One feature is very broad absorption spectra in 1.35 to 1.9 eV region observed just at excitation (0 ps). The broad width indicates the transition to a continuum state, i.e., the free exciton to the conduction band.
3. Triplet excitons cannot be formed directly from the singlet exciton. They are formed only at high density excitation at exciton transition or by creation of electrons and holes by the interband transition.
4. The formation and decay times and photoluminescence properties of the ST excitons in several PDAs and PTs studied are summarized in Table 5.

Figure 29 shows the time dependence of the absorbance change due to the self-trapped excitons in several PDAs and PTs, after the component with very long lifetimes being subtracted. As can clearly be seen in the figure, the decay of nonfluorescent PDA-3BCMU in the blue phase and PDA-4BCMU in the blue phase are approximately given by an exponential function except the early stage within 500 fs. The other fluorescent polymers have nonexponential decay from very early delay time until 30 ps. The initial decay times defined by the slope of the semilogarithmic plot of the absorbance change against delay time just after excitation are determined as 890 ± 160 fs, 620 ± 60 fs, and 450 ± 50 fs in PDA-4BCMU in the red phase, P3MT, and

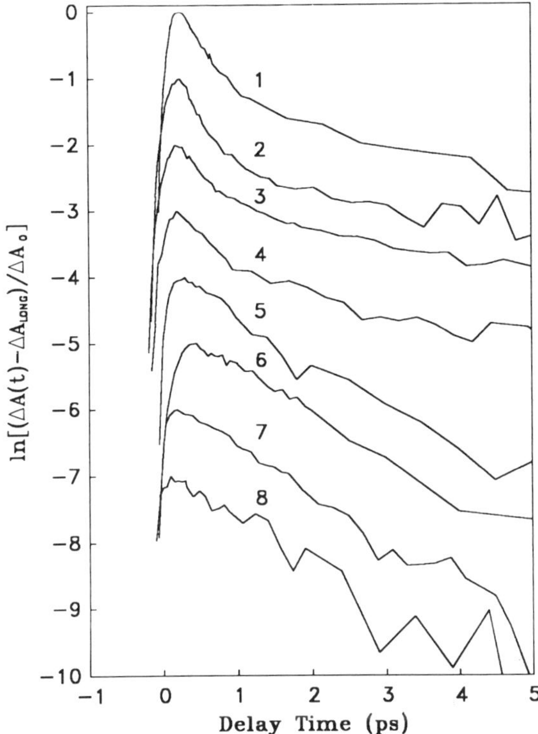

Fig. 29. Time dependence of the absorbance change due to the self-trapped excitons in several PDAs and PTs, after subtraction of the very long-lived component. The decay curves from the top to the bottom correspond to: 1. P3MT (observed at 10 K at the probe photon energy of 1.63 eV); 2. P3DT (10 K, 2.34 eV); 3. PDA-4BCMU (red-phase cast film, 10 K, 1.77 eV); 4. PDA-4BCMU (blue-phase oriented film, 10 K, 1.77 eV); 5. PDA-4BCMU (blue-phase oriented film, 290 K, 1.77 eV); 6. PDA-3BCMU (blue-phase cast film, 290 K, 1.77 eV); 7. PDA-3BCMU (oriented single crystals and amorphous domains prepared by evaporation on a KCl single crystal surface (100), 290 K, 1.77 eV); and 8. PDA-DFMP (297 K, 1.98 eV). The excitation pulse width and energy are 100 fs and 1.97 eV, respectively. The samples of PDA-4BCMU in red-phase are excited by the two-photon process.

P3DT, respectively, at 10 K. The decay becomes slower and slower at longer delay times because of the less-efficient tunneling due to the thicker barrier width the population of excitons must experience.

The lifetime of the self-trapped ^1B_u excitons in PDA-3BCMU is much shorter than the radiative life (~ 1 ns) estimated from the oscillator strength of the transition. This means predominant radiationless relaxation in the exciton decay process. Excitons in many semiconductors and organic molecular crystals have lifetimes between several hundred picoseconds and several hundred nanoseconds at low temperatures. More than an order of magnitude differences in the lifetime of these excitons are usually found between low temperatures and room temperature. The activation potential barrier or phonon scattering is very often invoked to explain the temperature dependence in most excitons in semiconductors, organic molecular crystals, and also in ionic crystals.

The relaxation of photoexcitations in polymers in general can be described as follows using PDAs and PTs as examples. The differences in the relaxation kinetics of excitons in the nonfluorescent PDA-3BCMU and PDA-4BCMU in the blue phase and fluorescent PDA-4BCMU in the red phase and P3DT and P3MT are discussed. The discussion is based on the Toyozawa's theory [115,116]. Figures 30 and 31 show the adiabatic potential curves of the ground state (G) and self-trapped exciton (STE), and free exciton (FE) band of the nonfluorescent (Fig. 30) and fluorescent (Fig. 31) polymers. The curvature radius of the potential of STE is larger than those of G and FE states in order to take into account the discussion in the reaction coordinate by Sumi *et al.* [114] and of the small Stokes shift observed for the fluorescent polymers. The electronic structure of the polymer systems are too complicated to be represented by the one-dimensional potential curve. The small Stokes shift may also be explained by the ground state potential having two minima with a small barrier between two configurations corresponding to the acetylenic and butatrienic configurations or to the aromatic and quinoid configurations in the cases of PDAs and PTs, respectively.

In Figs. 30 and 31, self-trapping process from FE (indicated by the number 1 in the figures) to "hot" STE (2) is the process from 1 to 2, followed by the emission of phonons of strongly coupled modes to the excitonic transition.

After the phonon emission process the STEs still remain in the nonthermal states (3). The nonthermal STEs (3) are coupled with phonon modes (intramolecular vibrations) of low frequencies and then thermalize to states in which temperature of the intramolecular vibration can be defined. The thermalization process (3 to 4) is observed as the spectral change at delay times from 0.5 ps to 5.0 ps. The time constant of the thermalization can be determined by the time dependence of absorbance change ratio among several different wavelengths. The obtained time constant of the thermalization process is about 1 ps in both red-phase PDAs and P3DT. However, the temperature of the STEs defined from equilibrated vibrational modes after this 1-ps thermalization process may be still higher than the bulk temperature of the polymer sample. Therefore, the temperature decreasing process

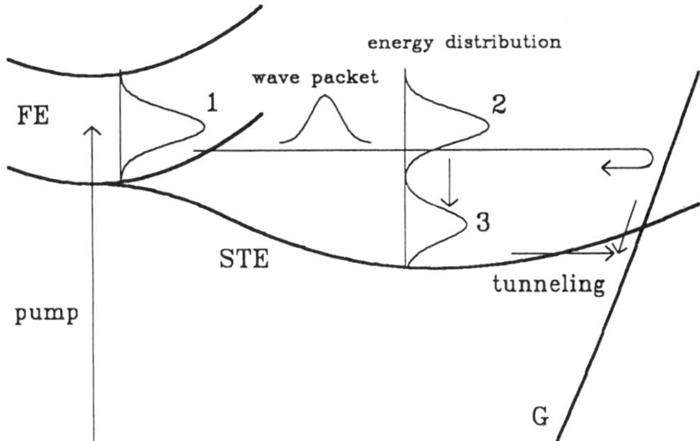

Fig. 30. Potential curves of the ground state (G), free exciton (FE), and self-trapped exciton (STE) of the nonfluorescent PDAs. The meaning of numbers 1, 2, and 3 are the same as in Fig. 28.

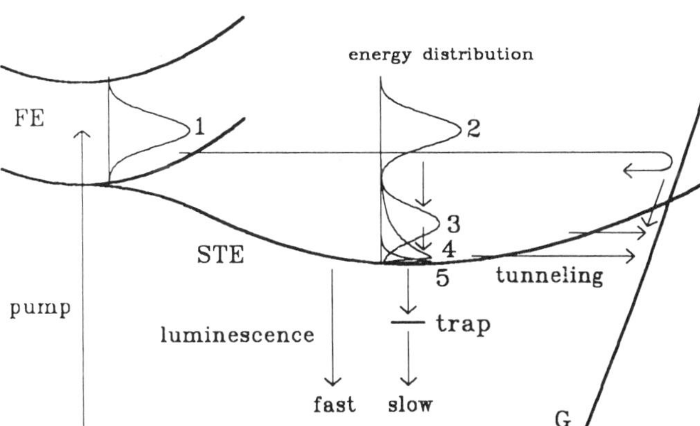

Fig. 31. Potential curves of the ground state (G), free exciton (FE) band, and self-trapped exciton (STE) of the fluorescent polymers. Numbers 1, 2, and 3 are the same as in Fig. 28. Number 4 means STE after thermal equilibration among intramolecular vibrational modes to a local temperature, which is higher than the bulk experimental temperature of the samples. Number 5 indicates the STE after the local temperature is lowered to the bulk experimental temperature.

(4 to 5) between several picoseconds and a few tens of picoseconds is expected to take place after the observed thermalization process. However, the spectral change due to the cooling-down process of the STEs could not be clearly detected in this study because of limited signal-to-noise ratio of the spectral data.

The change of the decay rate observed in the red-phase PDA-4BCMU and P3DT can be explained by the competition between the thermalization and tunneling processes. The STEs relax to the ground state mainly by tunneling through the barrier between the STEs and ground-state potentials. The initial fast decay is the relaxation from the unrelaxed and nonthermal STEs (indicated by 3 and 4, respectively, in Fig. 31) with the time constant of about 1 ps or slightly less. After the thermalization, the STEs come down close to the bottom of the potential and the loss rate by the tunneling becomes slower because of the increase in both the barrier height and thickness.

The decay time constant of about 5 ps is due to the tunneling from the thermal STE (indicated by 4 in Fig. 31). The thermalization process of the STE is estimated as about 1 ps from the spectral change. The wavelength dependence of the decay curve can be explained by the thermalization process. The absorbance change at 2.00 eV in the red-phase PDAs is mainly due to the nonthermal STEs, and the thermal STEs have the absorption peak at 1.6 eV. Therefore, the decay curve at 2.00 eV can be fitted to a single exponential function with the time constant of about 1 ps. The time constant of the absorbance change below 1.8 eV is longer than that of the bleaching, because the nonthermal STEs thermalize with the time constant of 1 ps, and the absorption peak due to the thermal STEs appears at 1.6 eV. The observed data for P3MT and P3DT can also be explained in the same way.

The decay of the nonthermal (but relaxed, namely after large-frequency phonon emission) STE in blue-phase PDAs is faster than in red-phase, because the crossing point is lower than the point in red-phase PDAs. The major part of the STEs relax to the ground state before thermalization. The decay rate from the nonthermal STEs to the ground state is not much slower than the decay rate from the unrelaxed STEs. Therefore, the decay curves in the blue-phase PDAs can be fitted to single-exponential functions. The wavelength dependence of the time constant is due to the mixed signals of the STE states. The bleaching signal exhibits the unrelaxed and relaxed (i.e., after large frequency phonon emission) but nonthermal STEs, while the absorption around 1.8 eV is due to the nonthermal STEs. Therefore, the observed decay time constant of the bleaching is shorter.

The difference between fluorescent and nonfluorescent polymers is the time constant of the tunneling from the nonthermal and thermal STEs to the ground state. The time constants in the blue-phase PDAs are between 1 ps and 3 ps, and they are shorter than that in the red-phase PDAs. Therefore the population of the remaining STE at the bottom of the STE potential curves in the blue-phase PDA is smaller, and the luminescence could not be detected. The fast component of the luminescence in the red-phase PDAs is considered to be due to the free excitons and/or STEs, and the slow component is due to the traps in the polymer chains. The detailed study using highly oriented samples is also consistent with the above model [117].

The proposed relaxation model predicts that the decay rate of the STE in conjugated polymers depends on the exciton energy. Conjugated polymers with large exciton energy are considered to have excitons with long lifetime and to be more fluorescent. For example poly[*p*-phenylenevinylene] (PPV), which has an absorption peak at 2.5 eV has strong luminescence with a quantum yield of several percent [118]. Recently, a Langumuir-Blodgett film of PDA-(12,8), which has an exciton peak at 1.88 eV, has been studied and the decay time constant of the STE is estimated as 1.3 ± 0.1 ps at 290 K [119]. It is shorter than that in other blue-phase PDAs, which have an exciton peak at 1.97 eV. These results are consistent with the model proposed above. However, PDA under the hydrostatic pressure has different decay kinetics. When the pressure increases, the absorption edge of red-phase PDA-4BCMU shifts to lower energy, but the decay kinetics becomes slower [120]. This pressure effect is probably due to a subsequent three-dimensional distortion of the polymer chain. The decay kinetics under the pressure may be different from that in the one-dimensional system.

Using the proposed model shown by Figs. 30 and 31, low fluorescence quantum efficiency of the one-dimensional polymers is systematically explained. The crossing points are higher than in PDA-3BCMU and PDA-4BCMU in the blue phase in fluorescent PDA-4BCMU in the red phase and P3MT and P3DT, because of the higher energy of free and ST exciton state. While in the red-phase PDA-4BCMU, a larger fraction (probably 10 to several tens of percent) of free excitons are relaxed to ST excitons with 3 ps lifetime at 10 K. This is represented in Figs. 30 and 31 as fast tunneling through and passing over the crossing point of the potential curve between the ST exciton and the ground state.

The ratio of the contribution of the tunneling and activation processes must be determined for each polymer by detailed experiments of the temperature dependence of the primary decay kinetics. The time constants of the various relaxation processes in three polymers are listed in Table 6 as examples.

The typical times of the relaxation processes in polymers are listed in Table 7 in comparison with those in the compound semiconductors such as GaAs in Table 8 made by Jagdeep Shah at AT&T Bell Labs.

Table 6.

Time Constants of the Relaxation Processes of Excitons at 290 K

Relaxation processes[a]	PDA-4BCMU (blue) (oriented film)	PDA-4BCMU (red) (oriented film)	P3DT
1. 2 (self-trapping)[b]	(10–20 fs)	(10–20 fs)	(10–20 fs)
2. 3 (phonon emission)	140 ± 40 fs	120 ± 60 fs	100 ± 50 fs
3. 4 (thermalization)	—[c]	1.1 ± 0.1 ps	1.0 ± 0.2 ps
2. G (tunneling from unrelaxed STE)	< 1.0 ps	0.7 ± 0.1 ps[d]	0.3 ± 0.1 ps[d]
3. G (tunneling from nonthermal STE)	1.6 ± 0.1 ps		
4. G (tunneling from thermal STE)	—[e]	4.4 ± 0.5 ps	4.7 ± 1.2 ps

[a] 1–4 and G are the states shown in Fig. 3.

[b] The time constants of the self-trapping processes could not be time-resolved in this study.

[c] The thermalization process could not be detected in the blue-phase PDA.

[d] The tunneling processes from the unrelaxed STE and from the nonthermal STE could not be separated.

[e] The tunneling from thermal STE in the blue-phase PDA-4BCMU could not be observed because the thermal STEs disappear before thermalization.

Table 7.

Relaxation Time Constants of Conjugated Polymers after Intrachain Excitation

1–2 fs	Creation of free exciton (FE)
10–20 fs	Formation of hot self-trapped exciton dephasing exciton (STE)
	Cooling down process by high-frequency phonon (intrachain) emission
100–200 fs	Decay of hot STE before thermalization
	Phonon emission from hot STE before thermalization
1–2 ps	Thermalization of STE, but with higher vibrational temperature than experimental temperature
5–10 ps	Cooling down of thermal STE with higher temperature than experimental temperature
< 5 ps	Tunneling before and during thermalization
5–30 ps	Tunneling after thermalization
0– < 30 ps (< resolution = 30 ps)	Hot luminescence from FE, hot and STEs before and after thermalization but before cooling down to the experimental temperature
> 30 ps	Luminescence from relaxed STE after cooling down
5 ps–10 ns	Disappearance of STE by being trapped after random walk
> 10 ns	Decay by radiation process

Table 8.

Time Scales of Relaxations in Semiconductor Physics

10–100 fs	Carrier dephasing Hole-optical phonon Carrier-carrier
100–500 fs	Electron-optical phonon Intervalley Exciton ionization Intersubband (E > L0)
0.5–2 ps	Thermalization of carriers Electron-hole thermalization
1–10 ps	Exciton dephasing Optical phonon decay
10–500 ps	Carrier cooling Intersubbance (E < L0)
>200 ps	Radiative recombination

Courtesy of Dr. Jagdeep Shah at AT & Bell Laboratories.

4. CONCLUSION

In the present paper we have clarified the mechanism of the low fluorescence quantum yield and no triplet exciton formation from the singlet exciton. This is because of the very short lifetime of the singlet exciton in the conjugated polymers. The short lifetime is caused by the exciton–phonon coupling in a one-dimensional system, which is explained as follows.

Since the conjugated polymers are one-dimensional, the spontaneous geometrical relaxation takes place without an activation barrier in the adiabatic potential between the free exciton and the self-trapped exciton. Because of this, the adiabatic potentials of the self-trapped exciton and the ground state cross at a point lower than ordinary three-dimensional systems.

The present paper has already been published partly in the publications of the authors group. The detailed study of the oriented film of PDA-3BCMU and PDA-4BCMU in the blue phase, difference in decay kinetics between amorphous cast film and the evaporated film on KCl of PDA-3BCMU, the triplet exciton formation in PDA-MADF, and relaxation in PMSPA will be described in [59], [108], and [63], respectively.

ACKNOWLEDGMENTS

The author would like to express thanks to Drs. M. Yoshizawa, K. Ichimura, S. Halle, and U. Stamm, M. Taiji, K. Minoshima, E. Tokunaga, and A. Yasuda for their collaboration in the femtosecond time-resolved experiment. He also is indebted to Prof. K. Yoshino for providing polythiophene samples and to Messrs. T. Uemiya and Y. Hattori for providing PDA-3BCMU/KCl samples. The author thanks Profs. Y. Toyozawa, E. Hanamura, and H. Sumi and Dr. S. Abe for their valuable discussion.

REFERENCES

1. N. Morita and T. Yajima, *Phys. Rev.* **A30**, 2525 (1984).
2. S. Asaka, H. Nakatsuka, M. Fujiwara, and M. Matsuoka, *Phys. Rev.* **A29**, 2286 (1984).
3. R. Beach and S. R. Hartmann, *Phys. Rev. Lett.* **53**, 663 (1984).
4. H. Nakatsuka, M. Tomita, M. Fujiwara, and S. Asaka, *Opt. Commun.* **52**, 150 (1984).
5. M. Fujiwara, R. Kuroda, and H. Nakatsuka, *J. Opt. Soc. Am.* **B2**, 1634 (1985).
6. S. R. Meech, A. J. Hoff, and D. A. Wiersma, *Chem. Phys. Lett.* **121**, 287 (1985).
7. T. Hattori and T. Kobayashi, *Chem. Phys. Lett.* **133**, 230 (1987).
8. A. M. Weiner, S. De Silvestri, and E. P. Ippen, *J. Opt. Soc. Am.* **B2**, 654 (1985).
9. A. Laubereau and W. Kaiser, *Rev. Mod. Phys.* **50**, 607 (1978).
10. S. M. George, H. Auwester, and C. B. Harris, *J. Chem. Phys.* **73** 5573 (1980).
11. S. M. George, A. L. Harris, M. Berg, and C. B. Harris, *J. Chem. Phys.* **80**, 83 (1984).
12. T. Hattori, A. Terasaki, and T. Kobayashi, *Phys. Rev.* **A35**, 715 (1987).
13. M. Tomita and M. Matsuoka, *J. Opt. Soc. Am.* **B3**, 560 (1986).
14. N. Morita, T. Tokizaki, and T. Yajima, *J. Opt. Soc. Am.* **B4**, 1269 (1987).
15. K. Kurokawa, T. Hattori, and T. Kobayashi, *Phys. Rev.* **A36**, 1298 (1987).
16. A. Mohktari and J. Chesnoy, *Europhys. Lett.* **5**, 523 (1988).
17. J. Chesnoy and A. Mokhtari, *Phys. Rev.* **A38**, 3566 (1988).
18. T. Hattori, A. Terasaki, T. Kobayashi, T. Wada, A. Yamada, and H. Sasabe, *J. Chem. Phys* **95**, 937 (1991).
19. B. J. Bern and R. Pecora, *Dynamical Light Scattering.* (Wiley, New York, 1976).
20. H. D. Dardy, V. Volterra, and T. A. Litovits, *J. Chem. Phys.* **59**, 4491 (1973).
21. J. F. Dill, T. A. Litovitz, and J. A. Bucaro, *J. Chem. Phys.* **62**, 3839 ((1975).
22. W. Danninger and G. Zundel, *Chem. Phys. Lett.* **90**, 69 (1982).
23. P.-A. Lund, O. Faurskov Nielsen, and E. Praestgaard, *Chem. Phys.* **28**, 167 (1978).
24. O. Faurskov Nielsen, *Chem. Phys. Lett.* **60**, 515 (1979).
25. M. Perrot, M. H. Brooker, and J. Lascombe, *J. Chem. Phys.* **74**, 2787 (1981).

26. J. Etchepare, G. Grillon, J. P. Chambaret, G. Hamoniaux, and A. Orszag, *Opt. Commun.* **63**, 329 (1987).
27. D. McMorrow, W. T. Lotshaw, and G. A. Kenney-Wallace, *IEEE J. Quantum Electron* **QE-24**, 443 (1988).
28. S. Ruhman, B. Kohler, A. G. Joly, and K. A. Nelson, *IEEE J. Quantum Electron.* **QE-24**, 470 (1988).
29. J. Etchepare, G. Grillon, G. Hamoniaux, A. Antonetti, and A. Orszag, *Revue Phys. Appl* **22**, 1749 (1987).
30. S. Ruhman, B. Kohler, A. G. Joly, and K. A. Nelson, *Chem. Phys. Lett.* **141**, 16 (1987).
31. C. Kalpouzos, D. McMorrow, W. T. Lotshaw, and G. A. Kenney-Wallace, *Chem. Phys. Lett.* **150**, 138 (1988); "Comment" in *Chem. Phys. Lett.* **155**, 240 (1989).
32. T. Kobayashi, A. Terasaki, T. Hattori, and K. Kurokawa, *Appl. Phys.* **B47**, 107 (1988).
33. T. Kobayashi, T. Hattori, A. Terasaki, and K. Kurokawa, *Revue Phys. Appl.* **22**, 1773 (1987).
34. S. Ruhman, A. G. Joly, B. Kohler, L. R. Williams, and K. A. Nelson, *Revue Phys. Appl* **22**, 1717 (1987).
35. Y.-X. Yan and K. A. Nelson, *J. Chem. Phys.* **87**, 6257 (1987).
36. S. Ruhman, A. G. Joly, and K. A. Nelson, *J. Chem. Phys.* **86**, 6563 (1987).
37. M. A. F. Scarparo, J. H. Lee, and J. J. Song, *Opt. Lett.* **6**, 193 (1981).
38. R. Trebino, C. E. Barker, and A. E. Siegman, *IEEE J. Quantum Electron.* **QE-22**, 1413 (1986).
39. C. E. Baker, R. Trebino, A. G. Kostenbauder, and A. E. Siegman, *J. Chem. Phys.* **92**, 4740 (1990).
40. K. Minoshima, M. Taiji, and T. Kobayashi, *Opt. Lett.* **16**, 1683 (1991).
41. M. J. LaGasse, D. Liu-Wong, J. G. Fujimoto, and H. A. Haus, *Opt. Lett.* **14**, 311 (1989).
42. D. Cotter, C. N. Ironside, B. J. Ainslie, and H. P. Girdlestone, *Opt. Lett.* **14**, 317 (1989).
43. N. Finlayson, W. C. Banyai, C. T. Seaton, and G. I. Stegeman, *J. Opt. Soc. Am.* **B6**, 675 (1989).
44. M. J. LaGasse, K. K. Anderson, H. A. Haus, and J. G. Fumimoto, *Appl Phys. Lett.* **54**, 2068 (1989).
45. G. R. Olbright and N. Peyghambarian, *Appl. Phys. Lett.* **48**, 1184 (1986).
46. S. Nomura and T. Kobayashi, *Solid State Commun.* **73**, 425 (1990).
47. R. J. Jain and R. C. Lind, *J. Opt. Soc. Am.* **73**, 647 (1983).
48. D. H. Auston, in *Ultrashort Light Pulses*, S. L. Shapiro, ed. (Springer, Berlin, 1977) p. 160.
49. E. Tokunaga, A. Terasaki, and T. Kobayashi, *Opt. Lett.* **17**, 1132 (1992).
50. E. Tokunaga, A. Terasaki, T. Wada, and T. Kobayashi, *Opt. Lett.* **18**, 370 (1993).
51. E. Tokunaga, A. Terasaki, and T. Kobayashi, *Phys. Rev. A*, to be published.

52. See, e.g. *Nonlinear Optics of Organics and Semiconductors*, edited by T. Kobayashi. (Springer-Verlag, 1989).
53. W.-S. Fann, S. Benson, J. M. J. Madey, S. Etemad, G. L. Baker, and F. Kajzar, *Phys. Rev. Lett.* **62**, 1482 (1989).
54. A. J. Heeger, S. Kivelson, J. R. Schrieffer, and W.-P. Su, *Rev. Modern Phys.* **60**, 781 (1988).
55. M. Sinclair, D. Moses, K. Akagi, and A. J. Heeber, *Phys. Rev.* **B38**, 10724 (1988).
56. T. Kobayashi, M. Yoshizawa, U. Stamm, M. Taiji, and M. Hasegawa, *J. Opt. Soc. Am.* **B7**, 1558 (1990).
57. M. Yoshizawa, M. Taiji, and T. Kobayashi, *IEEE J. Quantum Electron.* **QE-25**, 2532 (1989).
58. U. Stamm, M. Taiji, M. Yoshizawa, T. Kobayashi, and K. Yoshino, *Mol. Cryst. Liq. Cryst.* **182A**, 147 (1990).
59. M. Yoshizawa, A. Yasuda, and T. Kobayashi, *Appl. Phys.* **B53**, 296 (1991).
60. M. Yoshizawa, T. Kobayashi, K. Akagi, and H. Shirakawa, *Phys. Rev. B* **37**, 10301 (1988).
61. L. Rothberg, T. M. Jedju, S. Etemad, and G. L. Baker, *IEEE J. Quantum Electron.* **QE-24**, 311 (1988).
62. C. V. Shank, R. Yen, R. L. Fork, J. Orenstein, and G. L. Baker, *Phys. Rev. Lett.* **49**, 1660 (1982).
63. S. Takeuchi, M. Yoshizawa, T. Masuda, T. Higashimura, and T. Kobayashi, *IEEE J. Quantum Electron.* **28**, 2508 (1992).
64. Z. Vardeny, J. Orenstein, and G. L. Baker, *Phys. Rev. Lett.* **57**, 3229 (1986).
65. See, e.g. *Polydiacetylenes*, ed. by D. Bloor and R. R. Chance (Martinus Nijhoff Publishers, Dordrecht, Netherlands, 1985).
66. R. R. Chance, G. Patel, and J. D. Witt, *J. Chem. Phys.* **71**, 206 (1979).
67. K. S. Wong, W. Hayes, T. Hattori, R. A. Taylor, J. F. Ryan, K. Kaneto, K. Yoshino, and D. Bloor, *J. Phys.* **C18**, L843 (1985).
68. A. Kobayashi, H. Kobayashi, T. Tokura, T. Kanetake, and T. Koda, *J. Chem. Phys.* **87**, 4962 (1987).
69. T. Hattori and T. Kobayashi, *Chem Phys. Lett.* **133**, 230 (1987).
70. G. M. Carter, J. V. Hryniewicz, M. K. Thakur, Y. J. J. Chen, and S. E. Meyler, *Appl. Phys. Lett.* **49**, 998 (1986).
71. C. C. Hsu, Y. Kawabe, Z. Z. Ho, N. Peyghambarian, J. N. Polky, W. Krug, and E. Miao, *J. Appl. Phys.* **67**, 7199 (1990).
72. T. Hasegawa, K. Ishikawa, T. Kanetake, T. Koda, K. Takeda, H. Kobayashi, and K. Kubodera, *Chem. Phys. Lett.* **171**, 239 (1990).
73. J. Swiatkiewicz, X. Mi, . Chpra, and P. N. Prasad, *J. Chem. Phys.* **87**, 1882 (1987).
74. G. J. Blanchard and J. P. Heritage, *J. Chem. Phys.* **93**, 4377 (1990).
75. P. P. Ho, N. L. Yang, T. Jimbo, Q. Z. Wang, and R. R. Alfano, *J. Opt. Soc. Am.* **B4**, 1025 (1987).
76. F. Charra and J. M. Nunzi, *Organic Molecules for Nonlinear Optics and Photonics*, ed. by J. Messier *et al.* (Kluwer Academic Publishers, Netherlands, 1991), p. 359.

77. F. Kajzar, L. Rothberg, S. Etemad, P. A. Chollet, D. Grec, A. Boudet, and T. Jedju, *Opt. Commun.* **66**, 55 (1988).
78. B. I. Greene, J. Orenstein, R. R. Millard, and L. R. Williams, *Chem. Phys. Lett.* **139**, 381 (1987).
79. J. M. Huxley, P. Mataloni, R. W. Schoenlein, J. G. Fujimoto, E. P. Ippen, and G. M. Carter, *Appl. Phys. Lett.* **56**, 1600 (1990).
80. D. McBranch, A. Heys, M. Sinclair, D. Moses, and A. J. Heeger, *Phys. Rev.* **B42**, 3011 (1990).
81. Z. Vardeny, H. T. Grahn, A. J. Heeger, and F. Wudl, *Synth. Metals* **28**, C299 (1989).
82. I. D. W. Samuel, K. E. Meyer, R. H. Friend, J. Ruhe, and G. Wegner, *Proceedings of ICSM* preprint (1990).
83. Z. Vardeny, E. Ehrenfreund, O. Brafman, M. Nowak, H. Schaffer, A. J. Heeter, and F. Wudl, *Phys. Rev. Lett.* **56**, 671 (1986).
84. N. Colaneri, M. Nowak, D. Spiegel, S. Hotta, and A. J. Heeger, *Phys. Rev.* **B36**, 7964 (1987).
85. K. Kaneto, S. Hayashi, and K. Yoshino, *J. Phys. Soc. Jpn* **57**, 1119 (1988).
86. K. Kaneto, S. Hayashi, S. Ura, and K. Yoshino, *J. Phys. Soc. Jpn.* **54**, 1146 (1985).
87. K. Kaneto, S. Hayashi, and K. Yoshino, *Solid State Commun.* **65**, 783 (1987).
88. T. Kanetake, K. Ishikawa, T. Koda, Y. Tokura, and K. Takeda, *Appl. Phys. Lett.* **51**, 1857 (1987).
89. H. Murata, S. Tokito, T. Tsutsui, and S. Saito, *New Polym. Mater.* **2**, 11 (1990).
90. M. Joffre, D. Hulin, A. Migus, A. Antonetti, C. Benoit à la Guillaume, N. Peyghambarian, M. Lindberg, and S. W. Koch, *Opt. Lett.* **13**, 276 (1988).
91. J. P. Sokoloff, M. Joffre, B. Fluegel, D. Hulin, M. Lindberg, S. W. Koch, A. Migus, A. Antonetti, and N. Peyghambarian, *Phys. Rev.* **B38**, 7615 (1988).
92. C. H. Brito Cruz, J. P. Gordon, P. C. Becker, R. L. Fork, and C. V. Shank, *IEEE J. Quantum Electron.* **QE-24**, 261 (1988).
93. M. Lindberg and S. W. Koch, *J. Opt. Soc. Am.* **B5**, 139 (1988).
94. S. Hulle, M. Yoshizawa, H. Matsuda, S. Okada, H. Nakanishi, and T. Kobayashi, to be published.
95. B. I. Greene, J. F. Mueller, J. Orenstein, D. H. Rapkine, S. Schmitt-Rink, and M. Thakur, *Phys. Rev. Lett.* **61**, 325 (1988).
96. G. M. Carter, *J. Opt. Soc. Am. B* **4**, 1018 (1987).
97. A. M. Bonch-Bruevich and V. A. Khodovoi, *Usp. Fiz. Nauk.* **93**, 71 (1967) [*Sov. Phys. Usp.* **10**, 637 (1968)].
98. L. Y. Yang, R. Dornsville, Q. Z. Wang, W. K. Zou, P. P. Ho, N. L. Yang, R. R. Alfano, R. Zamboni, R. Danieli, G. Ruani, and C. Taliani, *J. Opt. Soc. Am. B* **6**, 753 (1989).
99. D. Frohlich, A. Nothem, and K. Reimann, *Phys. Rev. Lett.* **55**, 1335 (1985).
100. J. M. Nunzi and D. Frec, *J. Appl. Phys.* **62**, 2198 (1987).

101. A. Mysyrowicz, D. Hulin, A. Antonetti, A. Migus, W. R. Masselink, and H. M. Morkoc, *Phys. Rev. Lett.* **56**, 2748 (1986).
102. F. Kajzar, L. Rothberg, S. Etemad, P. A. Chollet, D. Grec, A. Boudet, and T. Jedju, *Opt. Commun.* **66**, 55 (1988).
103. D. Hulin, A. Migus, M. Joffre, and A. Antonetti, *Proceedings of Nonlinear Optics '90* **WI6**, p. 169 (1990).
104. W. M. Dennis, W. Blau, and D. J. Bradley, *Appl. Phys. Lett.* **47**, 200 (1985).
105. E. Tokunaga, A. Terasaki, T. Wada, and T. Kobayashi, to be published.
106. J. Orenstein, S. Etemad, and G. L. Baker, *J. Phys. C* **17**, L297 (1984).
107. L. Robins, J. Orenstein, and R. Superfine, *Phys. Rev. Lett.* **56**, 1850 (1986).
108. K. Ichimura, M. Yoshizawa, H. Matsuda, S. Okada, M. Osugi, S. Hourai, H. Nakanishi, and T. Kobayashi: to be published.
109. E. F. Steigmein, H. Auderset, and W. Kobel, *Synth. Met.* **18**, 219 (1987).
110. J. Jortner, Selected chapters in *Excitons*, E. I. Rashba and M. D. Sturge, ed. (North-Holland, Amsterdam, 1987), pp. 273–332.
111. C. X. Cui and M. Kertesz, *Phys. Rev.* **B40**, 9661 (1989).
112. G. Gustafson, O. Inganas, H. Osterholm, and J. Laakso, preprint.
113. K. Nasu, *J. Luminescence* **38**, 90 (1987).
114. H. Sumi, M. Georgier, and A. Sumi, *Rev. Solid State Science* **4**, 209 (1990).
115. Y. Toyozawa, *J. Phys. Soc. Japan* **58**, 2626 (1989).
116. Y. Toyozawa, *Tech. Rep. ISSP Institute for Solid State Physics*) 1963 (unpublished).
117. R. H. Friend, D. D. C. Bradley, and P. D. Townsend, *J. Phys.* **D20**, 1367 (1987).
118. M. Yoshizawa, K. Nishiyama, M. Fujihira, and T. Kobayashi, *Chem. Phys. Lett.*, to be published.
119. B. C. Hess, G. S. Kanned, Z. V. Vardeny, and G. L. Baker, *Phys. Rev. Lett.* **66**, 2364 (1991).

PART II

NONLINEAR OPTICS IN MOLECULAR MEDIA

Chapter 3

QUADRATIC MOLECULAR NONLINEARITIES AT THE AIR/WATER INTERFACE

Y. R. Shen

Department of Physics, University of California and Material Sciences Division, Lawrence Berkeley Laboratory, Berkeley, California

1. INTRODUCTION

The highly promising development in optoelectronics has greatly increased the demand for new nonlinear optical materials in recent years. In this regard, organic materials are most attractive owing to the known rich variety of organic compounds and the inherent flexibility in synthesizing such compounds with desired properties [1]. The first step in searching for highly

ISBN 0-12-784450-3

nonlinear organic materials is to find molecules that are highly nonlinear. For strong second-order nonlinearities, the molecules must be highly asymmetric and possess a high degree of electron delocalization [1,2]. Quantitative predictions of optical nonlinearities are usually difficult and one must often resort to experiments. Thus, a convenient and reliable method for measuring optical nonlinearities of molecules is needed.

In most cases, second-order nonlinear optical polarizabilities of molecules are determined by the EFISH (*dc* electric-field induced second harmonic generation) method [3]. Molecules to be measured are dissolved as solutes in a solvent. The applied *dc* field induces a molecular reorientation in the solution, breaks the inversion symmetry, and renders optical second harmonic generation (SHG) in the medium allowed. From the measured SHG output, the second-order polarizability $\alpha^{(2)}$ of the molecules can be deduced. The deduction, however, requires some knowledge of the third-order polarizability $\alpha^{(3)}$ and the static dipole moment μ of the molecules, as well as the assumption of negligible solute solvent interaction and local field correction. The values of $\alpha^{(2)}$ thus obtained may be subject to large uncertainty.

An alternative method adopted by some researchers is to spread a monolayer of molecules to be studied on a suitable substrate (namely, glass or water) and measure their $\alpha^{(2)}$ by surface SHG [4–9]. The symmetry requirement that SHG is only allowed in the electric-dipole approximation in a medium without an inversion center makes the process inherently surface-specific [7]. It has been demonstrated that SHG from a monolayer of molecules spread on water is easily detectable even when the molecular nonlinearity is relatively weak [6,10]. Frequently, the technique has the advantages of simplicity in the experimental arrangement, easy control of the surface density of molecules, and straightforward deduction of $\alpha^{(2)}$ from measurements. With a sufficiently low surface density, interaction between molecules in the adsorbed monolayer and the local field correction from the presence of neighboring molecules can be neglected.

Clearly, the adsorbed molecules must interact with the substrate, but in many cases the effect of such interaction on the molecular nonlinearity, including the local field correction, is also negligible [11]. The surface SHG measurement can then provide a rather realistic and accurate determination of $\alpha^{(2)}$ for individual molecules. It is of course important that the molecules under study can be spread uniformly with a well-defined polar orientation on the appropriate substrate. In order to have molecules adsorbed on water, they must be sufficiently hydrophobic. This can often be achieved by adding a sufficiently long hydrocarbon chain to each molecule via synthesis. The hydrocarbon chain has a very low second-order nonlinearity in the visible and near infrared range [6], and therefore has little effect on the value of $\alpha^{(2)}$ measured.

In many cases, the absorbed molecules do interact strongly with the substrate, resulting in a significant change in the molecular nonlinearities. This occurs, for example, with molecules that can be protonated [12]. The optical nonlinearities of such molecules on water can be very different from those of isolated ones, depending on the proton concentration at the air/water interface [13]. While in such cases surface SHG is no longer a valid technique to determine $\alpha^{(2)}$ for isolated molecules, it can actually be used to probe the environment the molecules are in, knowing the environmental effect on $\alpha^{(2)}$. The possibility of using optically nonlinear molecules as molecular probes to measure surface pH values (surface proton concentrations) can be of great interest to researchers in chemistry and biology. In other instances, surface SHG can be used to probe monolayer aggregation [14], crystallization [15], phase transition [10], and polymerization [16].

Even molecules soluble in water can form a polar-oriented equilibrium layer at the air/water interface if they have an amphiphilic character, depending on the balance between solubility and hydrophobicity of the molecules. Surface SHG is an ideal tool for studies of such interfacial layers [17,18]. That SHG can be used to measure adsorption isotherms, pressure-area isotherms, molecular nonlinearities, molecular orientations, and phase transitions of a soluble interfacial layer could lead to many exciting new research opportunities.

With tunable, ultrashort laser pulses, surface SHG can be extended to spectroscopic and dynamic studies of molecules at the air/water interface. Surface reactions, for example, are expected to be significantly different from those in a bulk because of the different molecular environment, and SHG allows time resolved in-situ monitoring of such surface processes [16,19]. This opens the door to a most interesting, but hardly explored, area of research on Langmuir films.

In the following, we first briefly describe the theory underlying the surface SHG and the experimental arrangement commonly used to study molecules at the air/water interface. Then, in the remaining sections, we discuss how SHG from molecular layers at the air/water interface is used to measure quadratic nonlinearities and average orientations of molecules, to probe the environmental effect on molecules, to monitor aggregations, phase transitions, polymerization, and surface reactions of monolayers, and to study soluble molecular monolayers adsorbed at the air/water interface.

2. THEORY OF SURFACE SHG

The theory of surface SHG was originally worked out by Bloembergen and Pershan [20] and later extended by others [21]. Consider the case of a laser beam of frequency ω incident from the air side into a nonlinear medium with a plane boundary surface at $z = 0$. SHG appears in both transmitting and reflecting directions. Here, we will focus on the reflected SH output. By solving the Maxwell equations with the proper boundary conditions, one can show that the reflected SH intensity is given by [22]

$$I(2\omega) = \frac{32\pi^3\omega^2 \sec^2\theta}{c^3}|\vec{e}'(2\omega)\cdot\overset{\leftrightarrow}{\chi}^{(2)}_{s,\,\mathrm{eff}}\cdot\vec{e}'(\omega)e'(\omega)|^2 I^2(\omega), \qquad (3.2.1)$$

where θ is the angle of incidence or reflection on the air side; $e'(\Omega) \equiv \overset{\leftrightarrow}{L}(\Omega)\cdot\hat{e}(\Omega)$; $\hat{e}(\Omega)$ is the unit vector describing the polarization of the field at Ω; $\overset{\leftrightarrow}{L}$ is a rank-2 tensor with diagonal elements L_{xx}, L_{yy}, and L_{zz} describing transmission Fresnel coefficients that relate the field components along $\hat{x}$, $\hat{y}$, and $\hat{z}$ in air to the corresponding ones in the surface layer ($\hat{z}$ is along the surface normal and $\hat{x} - \hat{y}$ the plane of incidence) and is the effective surface nonlinear susceptibility that includes the bulk contribution to the reflected SHG. Explicitly we have [22]

$$L_{xx} = 2k_{2z}/(\varepsilon_2 k_{1z} + k_{2z})$$

$$L_{yy} = 2k_{1z}/(k_{1z} + k_{2z})$$

$$L_{zz} = 2k_{1z}(\varepsilon_2/\varepsilon')/(\varepsilon_2 k_{1z} + k_{2z})$$

$$(\chi^{(2)}_{s,\,\mathrm{eff}})_{ijk} = (\chi^{(2)}_s)_{ijk} + (\chi^{(2)}_B)_{ijk}/[k_{2z}(2\omega) + 2k_{2z}(\omega)]f_i(2\omega)f_j(\omega)f_k(\omega). \quad (3.2.2)$$

In these expressions, k_{1z} and k_{2z} denote the z components of the wave vectors in air and in the medium, respectively, ε' and ε_2 refer to the dielectric constants of the surface layer and the medium, respectively, $\overset{\leftrightarrow}{\chi}^{(2)}_S$ is the true surface nonlinear susceptibility, $\overset{\leftrightarrow}{\chi}^{(2)}_B$ is the bulk nonlinear susceptibility, $f_j(\Omega) = \varepsilon_2(\Omega)/\varepsilon'(\Omega)$ for $j = z$, and $f = 1$ for $j = x, y$.

It is clear from Eqs. (3.3.1) and (3.3.2) that in order for SHG to be surface-specific, we must have $|P^s| \gtrsim |P^B/2k_2|$ or $|\chi^{(2)}_s| \gtrsim |\chi^{(2)}_B/2k_2|$. This can only be true, in general, if the medium has an inversion symmetry. In this case $\chi^{(2)}_{\mathrm{B}}$ vanishes under the electric-dipole approximation. If we assume that the surface is a simple termination of the bulk structure, and that both $\chi^{(2)}_s$ and $\chi^{(2)}_{\mathrm{B}}$ arise from the electric-quadrupole contribution that responds to the spatial variations of the fields, we should find $|\chi^{(2)}_s| \sim |\chi^{(2)}_B/k|$. This is because the field varies in the bulk on the scale of $1/k$, whereas it varies at the surface on the scale of surface layer thickness. More generally, $\chi_s^{(2)}$ may also have an electric-dipole contribution if the surface has the structure of a polar layer or a polar-oriented layer of molecular adsorbates as we shall discuss in this article. In the latter case, one often finds for a full monolayer of highly nonlinear molecules $|\chi_s^{(2)}| \gg |\chi_B^{(2)}/k|$.

If the input laser beam is in the form of a pulse with a pulsewidth T and a beam cross-section A, then the SH output, in terms of photons/pulse, is given by [15]

$$S(2\omega) = \frac{32\pi^3\omega \sec^2\theta}{\hbar c^3} |\mathbf{e}'(2\omega)\cdot\vec{\chi}_{s,\mathrm{eff}}^{(2)}\colon \mathbf{e}'(\omega)\mathbf{e}'(\omega)|^2 \mathbf{I}_1^2(\omega)AT, \qquad (3.2.3)$$

in which all quantities are in cgs units. We use Eq. (3.2.3) to give a crude estimate of the strength of surface SHG. The typical nonresonant value of $\chi_{s,\mathrm{eff}}^{(2)}$ for a polar monolayer of molecules adsorbed at an interface is 10^{-14} to 10^{-15} esu, as one would find from a perturbation calculation. If we take $|\chi_S^{(2)}| = 10^{-15}$ esu, $I(\omega) \sim 10$ MW/cm^2, $A \sim 0.2$ cm^2, $T \sim 10$ nsec (from a Q-switched laser), and $\theta_{2\omega} \sim 45^\circ$ in Eq. (3.2.3), the signal strength S is of the order of 10^4 photons/pulse. Since a good photon counting detection system is capable of detecting a minimum of 10^{-4} photons/pulse, one could measure a $|\chi_{s,\mathrm{eff}}^{(2)}|$ as small as 10^{-19} esu. Normally, $|\chi_{s,\mathrm{eff}}^{(2)}| \sim 10^{-18}$ esu is measurable without much excessive effort.

For a substrate with a well-defined bulk and interface structural symmetry, $\vec{\chi}_{s,\mathrm{eff}}^{(2)}$ possesses a definite symmetry form [21]. If the substrate is isotropic (such as water), the only nonvanishing (ijk) elements of $\vec{\chi}_{s,\mathrm{eff}}^{(2)}$ are $xxz = xzx = yzy = yyz$, $zxx = zyy$, and zzz. While $\vec{\chi}_{s,\mathrm{eff}}^{(2)}$ generally includes contributions from the substrate, we shall consider in this paper only cases where surface SHG is dominated by the adsorbed molecular monolayer (or the substrate contribution can be subtracted). We then have

$$\vec{\chi}_{s,\mathrm{eff}}^{(2)} \simeq \vec{\chi}_S^{(2)}$$

which describes the nonlinear response of the adsorbed monolayer with or without the characteristics changed by interaction with the substrate.

In the absence of any interaction between molecules in the adsorbed monolayer, $\vec{\chi}_S^{(2)}$ is related to the molecular nonlinear polarizability $\vec{\alpha}^{(2)}$ by the simple relation

$$\vec{\chi}_S^{(2)} = N_S\langle\vec{\alpha}^{(2)}\rangle, \qquad (3.2.4)$$

where N_S is the surface density of the molecules, and the angular brackets denote an average over the orientational distribution of the molecules. If the substrate is isotropic, the molecular distribution is usually azimuthally isotropic. In this case, as mentioned earlier, there are only three independent nonvanishing elements of $\vec{\chi}_S^{(2)}$ that can be measured by surface SHG. They allow the determination of three quantities in Eq. (3.2.4); for example, two elements of $\vec{\alpha}^{(2)}$ and one parameter characterizing the polar orientation of the molecules. For rodlike molecules, $\vec{\alpha}^{(2)}$ is dominated by a single element of $\alpha^{(2)}_{\xi\xi\xi}$ with $\hat{\xi}$ along the long molecular axis. We then find [4]

$$\left.\begin{aligned}(\chi_S^{(2)})_{xxz} &= (\chi_S^{(2)})_{yyz} = (\chi_S^{(2)})_{zxx} = (\chi_S^{(2)})_{zyy} \\ &= \tfrac{1}{2} N_S \alpha^{(2)}_{\xi\xi\xi} \langle \cos\theta \sin^2\theta \rangle \\ (\chi_S^{(2)})_{zzz} &= N_S \alpha^{(2)}_{\xi\xi\xi} \langle \cos^3\theta \rangle,\end{aligned}\right\} \tag{3.2.5}$$

where θ is the angle between $\hat{\xi}$ and $\hat{z}$. If we assume the molecular orientational distribution is a δ-function, then the measurement of $(\chi_S^{(s)})_{yyz}$ (or $(\chi_S^{(2)})_{zyy}$) and $(\chi_S^{(2)})_{zzz}$ permits the determination of $\alpha^{(2)}_{\xi\xi\xi}$ and θ. In other cases, if $\alpha^{(2)}$ has two dominant elements, one finds $(\chi_S^{(2)})_{zyy} \neq (\chi_S^{(2)})_{yzy}$, and both elements of $\vec{\alpha}^{(2)}$ and θ can be determined from the three independent measurements of $\vec{\chi}_S^{(2)}$.

In the presence of interaction between molecules in the monolayer, including the local field effect, $\vec{\chi}_S^{(2)}$ is no longer linearly proportional to N_S. The relation between $\vec{\chi}_S^{(2)}$ and $\vec{\alpha}^{(2)}$ becomes much more complicated even if the interaction between molecules is well specified and makes the determination of $\vec{\alpha}$ and θ very difficult. For that reason, it is preferable to have N_S small in the measurement in order to assure that Eq. (3.2.4) is valid. We should note from Eqs. (3.2.1) and (3.2.2) that the quantities measured by surface SHG are actually $(\chi_S^{(2)})_{ijk}/g_i(2\omega)g_j(\omega)g_k(\omega)$, where $g_j(\Omega) = 1$ for $j = x, y$, and $g_j(\Omega) = \varepsilon'(\Omega)$ for $j = z$. If ε' is significantly different from 1, then it also depends on N_S, and $(\chi_S^{(2)})_{ijk}/g_i(2\omega)g_j(\omega)g_k(\omega)$ with one or more gs different from 1 becomes nonlinear in N_S; the corresponding SH output should then have a dependence on the surface density other than $(N_S)^2$.

With either ω or 2ω approaching a resonant transition, $\vec{\alpha}^{(2)}$ is resonantly enhanced. This is reflected in the resonance enhancement of $\vec{\chi}_S^{(2)}$ [4]. In the limit where Eq. (3.2.4) is valid, the SHG measurement with a tunable laser should yield the dispersion of $\vec{\alpha}^{(2)}$.

3. EXPERIMENTAL ARRANGEMENT

We consider here only SHG measurements of molecular monolayers spread on water. The experimental setup is described in Fig. 1 [10]. A Langmuir trough is commonly used to prepare a system of adsorbed molecules at an

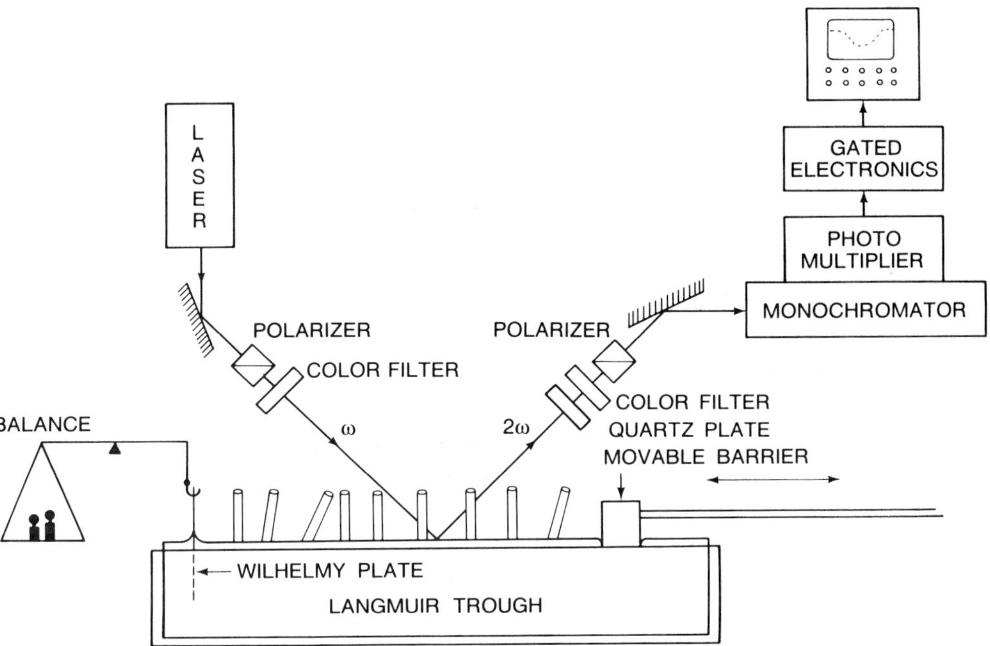

Fig. 1. Experimental arrangement for second-harmonic generation (SHG) and surface pressure studies of amphiphilic molecules at the air/water interface.

air/water interface. The adsorbed molecules change the surface tension, which can be easily measured by a Wilhelmy plate attached to a balance. A moveable barrier at the air/water interface allows a continuous variation of the surface molecular density N_S if the molecules are insoluble. With sufficiently low N_S, Eq. (3.2.4) should be valid. In preparing the system, care must be taken to prevent the interface from being contaminated.

One of the advantages of SHG as a surface analytical tool is the simplicity of the optical arrangement. As shown in Fig. 1, the basic elements in the setup are nothing but a laser and a photodetection system with proper spectral filtering. Mode-locked, Q-switched, and CW lasers can all be used. Which one is more advantageous depends on the system to be investigated. The SH signal from a smooth surface is coherent and highly directional; therefore, collection of the signal is a trivial matter.

Equation (3.2.3) shows that with a pulsed laser, the SH signal should increase with the increase of $I(\omega)$, A, and T; however, if the pump pulse energy $W(\omega) = I(\omega)AT$ is fixed, the signal should increase with the decrease of A and T. In practice, the laser fluence (energy/unit area) impinging on a surface is often limited by laser-induced breakdown. If we assume that the breakdown is due to laser-induced melting, and its threshold for a short-pulse laser excitation at 1.06 μm is $(IT)_{th}$, the maximum SH output is given by [23]

$$\zeta_{\max} \sim 10^{13} f |\chi_S^{(2)}|^2 (IT)_{th}^2 (A/T), \tag{3.3.1}$$

where f is the pulse repetition rate. Thus a Q-switched Nd:YAG laser with $T = 10$ nsec, $f = 10$ pulses/sec, and a pulse energy $W = 100$ mJ can yield an SH signal of 1 photon/sec assuming $(IT)_{th} = 1$ J/cm^2 and $|\chi_S^{(2)}| = 10^{-18}$ esu. The signal strength can be further improved with a Q-switched mode-locked laser. If we take $W = 10$ μJ, $T = 1$ psec, and $f = 10^4$ pulses/sec, we find $S \sim 100$ photons/sec. For typical values of $\alpha^{(2)}$ ranging from 10^{-29} to 10^{-31} esu, $|\chi_S^{(2)}| \sim 10^{-8}$ esu corresponds to a surface density of 10^{11} to 10^{13} molecules/cm^2. A fully packed molecular monolayer on water usually has a surface density larger than 10^{14}/cm^2.

4. MEASUREMENTS OF $\alpha^{(2)}$

As described in the previous sections, surface SHG can be used to measure $\alpha^{(2)}$ of molecules spread on a substrate in a monolayer form, in particular, long molecules with $\overleftrightarrow{\alpha}^{(2)}$ dominated by $\alpha_{\xi\xi\xi}^{(2)}$ along the long molecular axis. This technique has been applied in earlier studies of molecular monolayers spread or adsorbed on glass [4,5]. The disadvantages of preparing monolayers on solid substrates are the possibility of having nonuniform molecular distribution and the difficulty of continuously varying *in situ* the surface density of molecules. These disadvantages can be avoided if the molecules can be spread on water, which is possible for amphiphilic molecules. We discuss here such measurements on a group of related molecules [6].

Table 1 presents the results of surface SHG measurements at 0.53 μm on a number of molecules. With different input/output polarization combinations, $(\chi_S^{(2)})_{zzz}$, $(\chi_S^{(2)})_{yzy}$ and $(\chi_S^{(2)})_{zyy}$ could be deduced. In all cases, $(\chi_S^{(2)})_{yzy} = (\chi_S^{(2)})_{zyy}$, indicating that $\alpha_{\xi\xi\xi}^{(2)}$ is the only dominant element in $\overleftrightarrow{\alpha}^{(2)}$. By varying N_S, it was found that all $(\chi_S^{(2)})_{ijk}$ were linearly proportional to N_S. An example is shown in Fig. 2. Thus Eq. (3.2.5) is valid, and the local field effect due to the adsorbed molecules can be neglected. Moreover, the molecular orientation remains unchanged with different N_S. Assuming that the interaction between the molecules and the water substrate does not affect the molecular nonlinearity, Eq. (3.2.5) can then be used to find for the individual molecules and their average tilt angle θ. The results in Table 1 were obtained with N_S kept at 3×10^{14} molecules/cm^2 in all cases.

Table 1.

Second-order nonlinear polarizability of a number of organic molecules as determined by SHG from a monolayer using 532 nm excitation. The $(\chi_S^{(2)})_{yzy}$ for the monolayer is given for a molecular density of $N = 3 \times 10$ cm. θ is the angle between the molecular ξ axis and the surface normal.

	Molecule	$(\chi_S^{(2)})_{yzy}$ (10^{-16} esu) at $N_S = 3 \times 10^{14}\ cm^{-2}$	θ (°)	$\alpha^{(2)}_{\xi\xi\xi}$ (10^{-30} esu)
8CB	$C_8H_{17}(C_6H_4)_2CN$	11	71	25
9CB	$C_9H_{19}(C_6H_4)_2CN$	11	71	25
10CB	$C_{10}H_{21}(C_6H_4)_2CN$	11	71	25
12CB	$C_{12}H_{25}(C_6H_4)_2CN$	11	71	25
14FA	$C_{14}H_{29}COOH$	0.05	50	0.08
17FA	$C_{17}H_{35}COOH$	0.04	50	0.07
22FA	$C_{22}H_{45}COOH$	0.04	50	0.07
8BCA	$C_8H_{17}(C_6H_4)_2COOH$	2.8	63	6
7CPP	$C_7H_{15}(C_4N_2H_4)_2C_6H_4$	1.9	79	8
5CT	$C_5H_{11}(C_6H_4)_3CN$	3.5 (532 nm) 6 (586 nm)	60	7.5 (532 nm) 13 (586 nm)

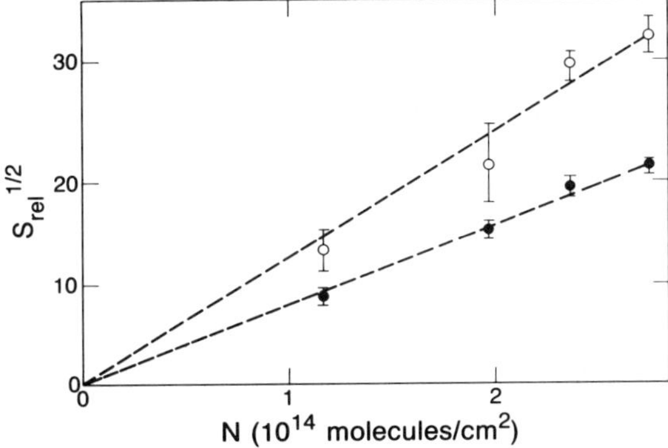

Fig. 2. The square root of the relative SHG intensity ($S_{rel}^{1/2}$) is plotted as a function of surface density for 5CT. The input laser field was polarized at 45° to the plane of incidence, for which both the *s*-polarized (open circles), and *p*-polarized (filled circles) SHG outputs were measured.

In Table 1, the values of $(\chi_S^{(2)})_{yzy}$, $\alpha^{(2)}_{\xi\xi\xi}$, and θ are listed for the various molecules. It shows that $\alpha^{(2)}_{\xi\xi\xi}$ of the cyano-biphenyl molecules are quite high (as compared to $\alpha^{(2)}_{\xi\xi\xi} \sim 2.5 \times 10^{-29}$ esu for the well known nonlinear molecule, 2-methyl-4-nitroaniline [MNA] measured at 1.06 μm). This high value results partially from the asymmetry induced in the delocalized π-electron system of the biphenyl rings by the presence of the CN group on one end and the hydrocarbon chain on the other end, and partially from resonance enhancement of the SH output at 0.265 μm. The hydrocarbon chain itself has very weak nonlinearity. Changing the chain length on a molecule has hardly any effect on $\alpha^{(2)}_{\xi\xi\xi}$. This is seen in the cyano-biphenyl $C_nH_{2n+1}(C_6H_4)_2CN)$ (NCB) case with n varying from 8 to 12 and in the fatty acid ($C_nH_{2n+1}COOH$) case with $n = 14$, 17, and 22. The weak but detectable nonlinearity of the fatty acid comes from the carboxylic acid (COOH) group.

Comparison of 8CB and $C_8H_{17}(C_6H_4)_2COOH$ (8BCA) in Table 1 shows that the CN group is a better electron acceptor than COOH, so that its connection to the biphenyl core leads to a stronger charge-transfer asymmetry in the delocalized electron distribution along the molecule and hence a larger second-order nonlinearity. Replacing a phenyl ring of the biphenyl core by a pyrimidine ring causes a significant reduction in $\alpha^{(2)}_{\xi\xi\xi}$ as seen in Table 1 from comparing $C_7H_{15}(C_4N_2H_2)(C_6H_4)CN$ (7CPP) with 8CB. This is due to an interruption of electron delocalization in the pyrimidine ring. Adding another phenyl ring to 8CB, as in $C_5H_{11}(C_6H_4)CN$ (5CT) surprisingly leads also to a decrease in $\alpha^{(2)}_{\xi\xi\xi}$, instead of an increase as one would expect from a longer delocalized π-electron system. This decrease is believed to be due to an actual nonplanar arrangement of the three phenyl rings in 5CT, reducing the effective length of electron delocalization.

The above discussion suggests that by measuring $\alpha^{(2)}_{\xi\xi\xi}$ systematically for a group of molecules with related structures one could learn how the molecular nonlinearity depends on the molecular structure. The SHG measurement of molecules spread on water is convenient and straightforward, but the method does have the drawback that only $\overleftrightarrow{\alpha}^{(2)}$ with a single dominant element, e.g., $\alpha^{(2)}_{\xi\xi\xi}$, can be conveniently measured. In other cases, additional information or assumptions about the other dominant elements of $\overleftrightarrow{\alpha}^{(2)}$ are needed. However, we notice that the same is true with the EFISH method [3].

With a tunable laser, surface SHG can be adopted for surface monolayer spectroscopy [4]. When either ω or 2ω approached a resonance, $\alpha^{(2)}$, and hence SHG, are resonantly enhanced. An example is presented in Fig. 3 for a monolayer of all-*trans* retinal molecules spread on water [24]. Two resonant peaks are discernible, one at 335 nm and the other at 360 nm. The former can be assigned to the main B_u-like absorption band of all-*trans* retinal, and the latter the one-photon forbidden A_g-like absorption band. Both bands appear to be blue-shifted from those of molecules in solution by approximately the same amount. Retinal molecules adsorbed in a membrane

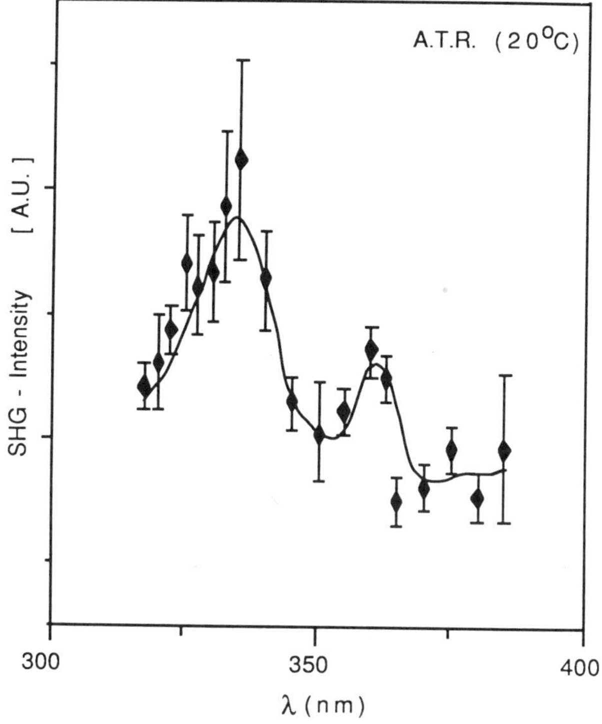

Fig. 3. SHG spectrum of all-*trans* retinal (ATR) at the air/water interface.

can also be studied with SHG, and how their properties are influenced by the surrounding environment can be investigated [24].

5. MOLECULAR ORIENTATIONS AT AN INTERFACE

The surface SHG measurement of long molecules yields both $\alpha^{(2)}_{\xi\xi\xi}$ and θ according to Eq. (3.2.5) [4,6]. Information on the molecular tilt is often useful in its own right. For example, surface adhesion and lubrication properties may well depend on the molecular orientation of surfactants adsorbed on the surface. On the water surface, the molecular tilt is determined by the molecular structure and its interaction with water. The SHG measurement of different molecules permits an investigation of this problem. The molecules in Table 1 all appear to orient with a large tilt angle from the surface normal. Note that the molecular orientation here refers to the orientation of a certain part of the molecule that gives dominant contribution to $\alpha^{(2)}_{\xi\xi\xi}$, for example, the phenyl core of a molecule. The large tilt angle then suggests that the phenyl rings would lie flat on the water surface if the hydrocarbon chain were not there to lift it up. The more water-soluble 7CPP molecules containing the pyrimidene ring have the largest θ. For 5CT, the tilt angle is significantly smaller, presumably because the nonplanar phenyl ring structure reduces the attractive interaction between the phenyl core and water.

SHG is a coherent optical process. The output is a coherent wave that carries phase information; the latter can be measured with an interference technique [25]. If a medium with a nonvanishing, electric-dipole-allowed $\vec{\chi}^{(2)}$ is inverted, we expect $\vec{\chi}^{(2)}$ to change sign and the SHG output to have a 180° phase shift. In this respect, surface SHG can be used to determine the polar orientation (i.e., head up or down) of adsorbed molecules if either by theory or by comparing with a reference molecular system, the sign of $\vec{\chi}_S^{(2)}$ is defined with respect to the polar orientation [26]. Such information cannot be obtained by any linear optical technique. As an example, consider cyanobiphenyl (8CB) monolayers spread on water and on a clean glass plate. The two cases yield $\vec{\chi}_S^{(2)}$ with the same phase indicating that the polar orientations of the two monolayers are the same [27]. Because hydrocarbons are hydrophobic, the 8CB molecules on water must have their CN end group facing water. The observed phase information asserts that 8CB must adsorb on glass with the CN group attached to glass.

By attaching a radical group with high nonlinearity to a specific part of a molecule, one can use SHG to learn about orientation of a local section of the molecule. In a recent experiment, Hsiung *et al.* [28] studied the orientations of two nearly identical amphiphilic molecular species adsorbed on water except that the hydrocarbon chains of the two have their numbers of methylene (CH_2) groups differ by 1. One would expect that the chain orientations of the two are the same, but the orientations of the terminal methyl groups differ because of the even-odd effect of the C—C bond orientation. By replacing the terminal methyl group by a *p*-nitroaniline (*p*NA) group, SHG from a monolayer of such molecules spread on water is dominated by *p*NA. Its polarization dependence then yields the orientation of the *p*NA group which should be very different for the two amphiphilic molecules (see Fig. 4). This was indeed what was observed by Hsiung *et al.*

A monolayer of amphiphilic molecules spread on water could undergo an orientational phase transition when the surface density is varied. This can also be monitored by surface SHG through its polarization dependence, as was first demonstrated with a fatty acid monolayer [10]. In the phase coexistence region, domains of one phase grow out of the other. They drift slowly on water and can be observed as slow fluctuations in the SHG signal

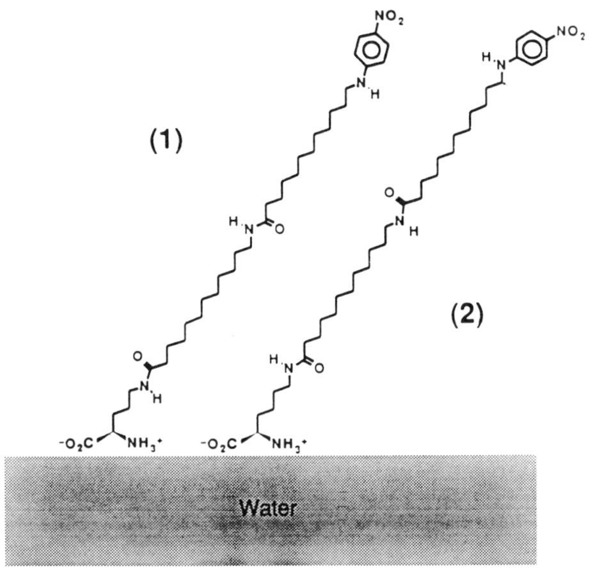

Fig. 4. Schematic showing the orientations of two amphiphilic molecules, $O_2NC_6H_6$—$NH(CH_2)_{11}(CO)NH(CH_2)_{11}(CO)NH(CH_2)_nCH(NH_3 + COO^-)$ with (1) $n = 3$ and (2) $n = 4$, at the air/water interface. The orientation of the *p*NA group ($O_2NC_6H_6$) at the end of the alkyl chain shows the even-odd effect of n (from Ref. 28 with permission from Elsevier Science Publishers BV).

if the laser beam spot is smaller than or comparable to the domain size [29]. The same can be observed by reflection ellipsometry [30].

6. ENVIRONMENTAL EFFECT ON MOLECULAR NONLINEARITY

Nonlinear optical properties of molecules are easily affected by the local environment. First, there is the usual local field effect. Then, interaction of the molecules with their surroundings can result in a change of their electronic properties. The latter can be labeled as the chemical effect and is often most significant for molecules with strong electron delocalization. For example, protonation or deprotonation of these molecules can alter the charge-transfer asymmetry and drastically modify the quadratic nonlinearity [12]. Interaction with neighboring molecules may shift the dissociation equilibrium of the molecules, lead to aggregation or bond cleavage, and thereby change their optical properties [13,31]. It is convenient to study the environmental effect with molecules that can be spread on water and form a monolayer. The environment can then be changed by film compression, matrix dilution, variation of proton or other ion concentration in the subphase layer, and coadsorption of different molecular species on water. The resulting change in the molecular hyperpolarizability can be probed by SHG from the monolayer.

The local field effect is known to be a complex physics problem in general. This is particularly true for systems involving large molecules because dominant contributions to linear and nonlinear polarizabilities often come from different parts of the molecules. Then the usual point-dipole approximation [11,32] in the local field calculation has to be extensively modified. In the case of a molecular monolayer on water, the problem can be greatly simplified. As discussed earlier, at a sufficiently low surface density, the local field effect due to molecular interaction within the monolayer is negligible. The one due to water substrate is also unimportant if the molecules do not lie flat on the water surface as judged from the point-dipole calculation [11].

The chemical effect can be much more significant as it perturbs the electron distribution in the molecules. An example is protonation of molecules which often occurs with molecules having end groups that can share electrons with neighboring protons. Consider the two dye molecules, hemicyanine (HC) and nitrostilbene (NS), listed in Fig. 5. Both molecules are known to have

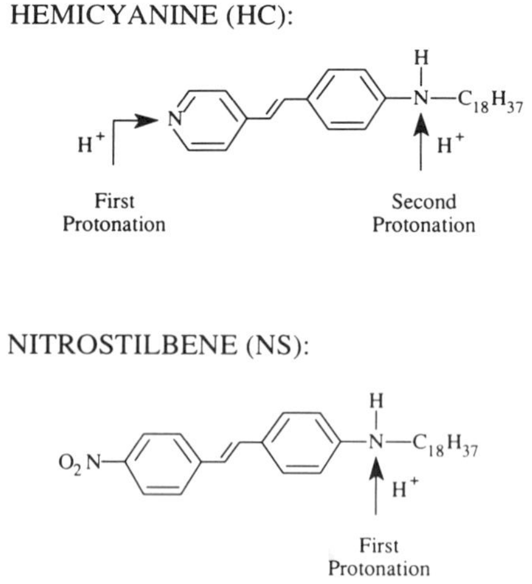

Fig. 5. Molecular structures of hemicyanine (HC) and nitrostilbene (NS) dyes. The arrows indicate the positions and the orders of protonation of the electron donor and acceptor groups.

large quadratic optical nonlinearities because of extensive electron delocalization and large charge transfer asymmetry along the molecules [1]. Assuming that the two-level model is valid for such charge-transfer molecules, we can write the dominant second-order nonlinear polarizability of the molecules as [1]

$$\alpha^{(2)}_{\xi\xi\xi}(2\omega = \omega + \omega) = \frac{3e^2}{m\hbar} \frac{\omega_{ng}}{(4\omega^2 - \omega_{ng}^2)(\omega^2 - \omega_{ng}^2)} f\delta,$$

where ω_{ng} and f are the resonant frequency and oscillator strength of the charge-transfer absorption band, respectively, and δ is the difference of the permanent dipole moments of the two levels. Protonation of the molecules affects the electron delocalization and charge-transfer asymmetry, shifts the resonant frequency ω_{ng}, and changes δ. It is interesting to see how the nonlinearities of two seemingly similar molecules, HC and NS, respond to protonation [13].

The two molecules have the same electron donor group, i.e., the amino (NH) group, but different acceptor groups. The NS has a stronger electron acceptor group (NO_2) than the HC (pyridine N). The electron delocalization length is also somewhat longer in NS. Correspondingly, the $\pi - \pi^*$ charge-transfer transition of NS is at a longer wavelength than that of HC (430 nm as compared to 385 nm). As a result, NS has a significantly larger $\alpha^{(2)}_{\xi\xi\xi}$ at 1.06 μm than HC. Consider now the effect of protonation on these dye molecules. For HC, both the pyridine nitrogen and the amine can be protonated. The protonated pyridine nitrogen becomes a stronger electron acceptor, and the charge-transfer band is red shifted. The result is an increase of $\alpha^{(2)}$. Following the protonation of the pyridine end, the amino nitrogen of HC can also be protonated. This protonation reduces the ability of the p_z electrons of the nitrogen to participate in the delocalized π-electron system. It causes a blue shift of the charge-transfer band, a reduction in the electron donor strength of the amino group, and a resultant decrease of $\alpha^{(2)}$. For NS, only the amino nitrogen is expected to be protonated. Again, the protonation reduces the donor strength of the amino group, hence induces a blue shift of the charge-transfer band and decreases the value of $\alpha^{(2)}$. The nitro group of NS may also be "protonated" (or associated with a proton without forming a covalent bond) when the surrounding proton concentration is sufficiently high. This "protonation" would increase the acceptor strength on the nitro group and increase the value of $\alpha^{(2)}$ for NS.

The above prediction has been substantiated by experiment [13]. The quadratic nonlinearities $\alpha^{(2)}_{\xi\xi\xi}$ of HC and NS were measured via SHG by spreading the molecules on water. To see the effect of protonation, the pH

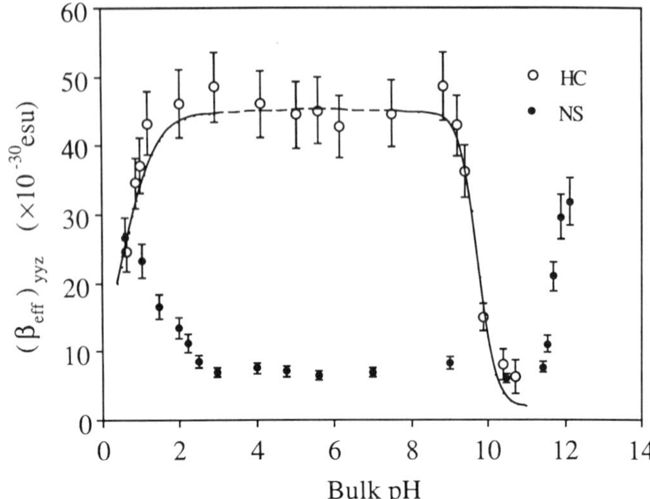

Fig. 6. Effective hyperpolarizability $(\beta_{\text{eff}})_{yyz}$ $[\equiv(\alpha^{(2)}_{\text{eff}})_{yyz}]$, versus bulk pH for HC (○) and NS (●) dispersed in a monolayer of stearic acid. The molar ratios of dye/stearic acid for the two cases are 1:15 and 1:6, respectively. The curve for HC is obtained from theoretical calculation.

value of the bulk water was varied by adding NaOH or HCl. Since the dipole–dipole interaction among the dye chromophores might lead to aggregation, the HC or NS dye molecules were diluted in a monolayer matrix of stearic acid (C18). The nonlinearity of C18 is negligibly small compared to either HC or NS so that the C18 matrix would not affect the measured values of $\alpha^{(2)}$ for HC or NS. The experimental results of average nonlinear polarizability $\alpha^{(2)}_{\text{eff}}$ for these two dye molecules versus the bulk pH value of water are presented in Fig. 6. They were obtained from a dye/C18 molar ratio of 1:15 for HC and 1:6 for NS. As shown in Fig. 6, the two dye molecules show opposite behaviors. HC has a large $\alpha^{(2)}_{\text{eff}}$ in the intermediate range of bulk pH values and drops off at both high and low pH values, while the opposite is true for NS. Obviously, the variations of $\alpha^{(2)}_{\text{eff}}$ with bulk pH come from protonation of the dye molecules. The large, nearly constant $\alpha^{(2)}_{\text{eff}}$ for HC with pH between 1.5 and 9 indicates that all the HC molecules in the monolayer have been converted to the singly protonated form. This gives $(\alpha^{(2)}_{\text{eff}})_{yyz} = \langle\alpha^{(2)}_{\text{HC}-\text{H}^+}\rangle_{yyz} = 44 \times 10^{-30}$ esu. The large value of $\langle\alpha^{(2)}_{\text{HC}-\text{H}^+}\rangle_{yyz}$ is due to enhanced charge transfer resulting from protonation of the pyridine end of the molecule as we discussed earlier. For pH > 10, the low value of $\alpha^{(2)}_{\text{eff}}$ indicates that the HC molecules are mainly in the neutral form with

$(\alpha_{\text{eff}}^{(2)})_{yyz} = \langle \alpha_{\text{HC}}^{(2)} \rangle_{yyz} < 6 \times 10^{-30}$ esu. For pH < 1.5, part of the HC molecules become doubly protonated with $\langle \alpha_{\text{H}^+-\text{HC}-\text{H}^+}^{(2)} \rangle_{yyz} < 24 \times 10^{-30}$ esu. From the fact that measured values of $(\alpha_{\text{eff}}^{(2)})_{yyz}/(\alpha_{\text{eff}}^{(2)})_{zzz}$ remained unchanged for all pH values, it was concluded that the average molecular tilt angle is $\theta = 65 \pm 5^\circ$ for HC in all protonation stages. For NS, the singly protonated molecules have a weaker charge transfer and hence a smaller $\alpha^{(2)}$. The data in Fig. 6 indicate that the molecules are partially kept in the neutral form for pH > 11, essentially all converted into the singly protonated form for 11 > pH > 2.5, and then partially converted into doubly protonated form for pH < 2.5. They have an average tilt angle of $\theta = 68^\circ \pm 5^\circ$ in all stages.

The proton concentration at an air/water interface is not necessarily the same as that in the bulk. They are clearly different if the interface is charged [33]. This is the case with a stearic acid monolayer on water. The molecules can have their head groups deprotonated and result in a negative surface charge layer. The negative charges attract protons to the interfacial region and render the surface pH value (pH_s) significantly larger than the bulk pH value (pH_b). The knowledge of surface pH_s (or local pH values) is of great importance in many chemical and biological problems. The measurements of pH_s require a molecular probe. Here, through the protonation effects on $\alpha^{(2)}$ of dye molecules, we may actually have such a probe [13].

Consider single protonation of HC molecules. Let the process be described by

$$\text{H}^+ + \text{HC} \rightleftarrows \text{HC} - \text{H}^+, \tag{3.6.1}$$

which is governed by an equilibrium constant

$$\text{K}_a = [\text{H}^+][\text{HC}]/[\text{HC} - \text{H}^+], \tag{3.6.2}$$

where $[X]$ denotes the molar concentration of the X species. With the definitions of $\text{pK}_a = -\log \text{K}_a$, $\text{pH} = -\log[\text{H}^+]$, and $n = [\text{HC} - \text{H}^+]/([\text{HC}] + [\text{HC} - \text{H}^+])$, Eq. (3.6.2) can be written as

$$\log\left(\frac{n}{1-n}\right) = \text{pK}_a - \text{pH}. \tag{3.6.3}$$

Now that we have

$$(\alpha_{\text{eff}}^{(2)})_{ijk} = n\langle \alpha_{\text{HC}-\text{H}^+}^{(2)} \rangle_{ijk} + (1-n)\langle \alpha_{\text{HC}}^{(2)} \rangle_{ijk}, \tag{3.6.4}$$

as long as $[\text{H}^+ - \text{HC} - \text{H}^+]$ is negligible (pH > 2), the mole fraction n can be directly deduced from the measured $(\alpha_{\text{eff}}^{(2)})_{ijk}$, knowing $\langle \alpha_{\text{HC}-\text{H}+}^{(2)} \rangle_{ijk}$ and $\langle \alpha_{\text{HC}}^{(2)} \rangle_{ijk}$. Equation (3.6.3) then allows us to find (pK_a − pH). Thus the pH value seen by HC can be obtained if pK_a for the molecules is known.

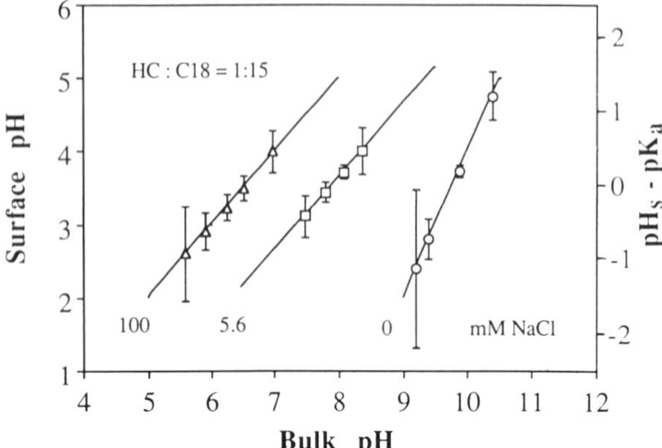

Fig. 7. Surface pH_s versus bulk pH_b and $pH_s - pK_a$ versus pH_b for a system of an HC/C18 monolayer in 1:15 molar ratio on water with three different salt concentrations. The solid lines were calculated from the Gouy-Chapman theory and the data points were deduced from SHG measurements with $pK_a = 3.5$ for the protonation HC. The pH value of the solution was adjusted by the addition of HCl and NaOH, respectively.

The above idea was tested with the 1:15 HC/stearic acid monolayer mixture on water [13]. Solutions with three different salt (NaCl) concentrations were used. The salt solutes provided additional counter ions in water that could compete with protons in their attempt to neutralize the negative surface charges of stearic acid molecules at the surface. The relations between pH_s and pH_b in the three cases are therefore expected to be different. The experimental results of $pH_s - pK_a$ versus pH_b deduced from the measurement of $(\alpha^{(2)}_{\text{eff}})_{yyz}$ are shown in Fig. 7. Unfortunately, pK_a for protonation of HC on water was not known experimentally. One must then resort to theory to find a value for pK_a.

The negative charge on a deprotonated stearic acid molecule is fairly localized and can be approximated by a point charge. One can then use the Gouy-Chapman model [33] (similar to the Debye model for charge screening) to relate pH_s and pH_b. If the interface is charged, then there exists an interfacial potential ψ that causes the proton concentration to vary from the bulk to the surface. We have

$$[H^+](\vec{r}) = [H^+]_{\text{Bulk}} \exp[-|e|\psi(\vec{r})/kT], \tag{3.6.5}$$

with $\psi = 0$ in the bulk. This leads to

$$pH_s = pH_b + \frac{|e|\psi_s}{2.3kT}, \tag{3.6.6}$$

where ψ_s is the surface potential. From the Gouy-Chapman model, ψ_s can be found in terms of the surface charge density and the bulk concentration of total counter ions. Equation (3.6.6) was used to fit the data in Fig. 7. It is seen that by choosing $pK_a = 3.5$, the agreement between theory and experiment for all three salt concentrations is good. This means that we can now use SHG with HC as probe molecules and with $pK_a = 3.5$ for protonation of HC to measure surface pH_s of an air/water interface.

The theoretical fit of the data for the case of 0 – mM NaCl in Fig. 7 can be employed to calculate from Eqs. (3.6.3) and (3.6.4) $\alpha_{\text{eff}}^{(2)}$ for $pH_b > 2.0$. The good agreement with the experimental result is depicted in Fig. 6. In the low pH_b limit, few stearic acid molecules could be deprotonated and hence $\psi_s = 0$. A similar calculation on the double protonation process $H^+ + HC - H^+ \rightleftarrows H^+ - HC - H^+$ with the corresponding pK_a value chosen to be 0.5 fits the data of $pH_b < 2$ in Fig. 6 satisfactorily [13].

Stearic acid and HC form homogeneously mixed monolayers at an air/water interface. Therefore, stearic acid can be used as matrix molecules to vary the mean distance between dye chromophores [9,34]. Because stearic acid (C18) molecules on water tend to be deprotonated, increasing the C18/HC molar ratio decreases the surface pH_s. This could affect protonation of HC and change the measured $\vec{\alpha}_{\text{eff}}^{(2)}$. In other cases, the dye molecules such as NS may aggregate via dipole-dipole attraction between chromophores. Protonation of these molecules could break the aggregation by introducing Coulomb repulsive interaction between protonated molecules to partially cancel the attractive dipole-dipole interaction [13]. For this reason, dilution of dye monolayers by stearic acid can be used to prevent dye aggregation in a monolayer. In a recent experiment, Hayden *et al.* [9,35] observed a significant change in the nonlinear susceptibility of a hemicyanine monolayer upon dilution by stearic acid and attributed the effect to local field variation. The hemicyanine dye used for that study cannot be protonated, and therefore the observed change cannot be due to protonation. In a study by Schildkraut *et al.*, [34] however, the same dye was found to form aggregates in the pure dye monolayer. Breaking the aggregates to monomers by stearic acid dilution led to a red-shifted absorption band. This should yield a change in the second-harmonic response, and could explain the observation of Hayden *et al.*

Surface SHG can also be used to monitor aggregates in a monolayer at an air/water interface [13,14]. Molecular aggregates are usually anisotropic in the surface plane. They can be recognized from the symmetry of the $\vec{\chi}_S^{(2)}$ tensor responsible for SHG. For example, if an aggregate has a C_{1v} symmetry, then $\chi_{xxx}^{(2)}$, $\chi_{xyy}^{(2)}$, $\chi_{xzz}^{(2)}$, and a few others all become nonzero.

Phase transitions are also manifestations of molecular interactions, and SHG can be used to probe phase transitions of a molecular monolayer. As mentioned in Sec. 5, it has been employed to identify the liquid expanded $\rightleftarrows$ liquid condensed transition of a fatty acid monolayer on water as an orientational phase transition [10].

Polymerization results from chemical reactions between monomers. Two-dimensional polymerization is a problem of special interest to many researchers. Again it can be studied by surface SHG with monolayers of monomers spread on water. This has the advantage that the density of monomers can be varied by film compression. The UV irradiation initiates the polymerization process, which can be monitored *in situ* by the change of SHG from the monolayer. That such an investigation is possible has been demonstrated with vinyl stearate and octadecyl methacrylate [16].

Molecules spread on water can interact with soluble molecules in the water subsurface. The interaction can cause the orientation of the insoluble molecules to change and the soluble molecules in the subsurface to aggregate or crystallize. This has been demonstrated by Berkovic and coworkers using SHG as a probe [36].

7. SOLUBLE MOLECULES ADSORBED AT THE AIR/WATER INTERFACE

Amphiphilic molecules soluble in water may prefer to stay near the surface. Adsorption of such molecules from a bulk solution to an air/water interface is a problem of great interest in colloidal and surface science [37]. To understand this type of adsorption, we need to know the adsorption isotherm and the $\pi - A$ isotherm of surface pressure (π) versus area per adsorbed molecule (A). One can easily measure the bulk concentration of the adsorbates and the surface pressure but not the surface density of excess adsorbate molecules. The existing techniques for surface density measurements [38] suffer from poor spatial resolution separating the surface from the bulk. With SHG, it is now possible to measure directly the surface density as well as deduce information about the polar orientation of the adsorbed molecules.

Consider the water-soluble molecules alkyl sodium naphthalene sulforates, C_nH_{2n+1}—$C_{10}H_8$—$SO_3Na(C_nNS)$ with $n = 6$ as an example [18,39]. The surface nonlinear susceptibility component $\chi^{(2)}_{pm}$ (SH output p-polarized; fundamental input at 45° from the incident plane) deduced from the surface SHG measurements at the air/water interface as a function of bulk concentra-

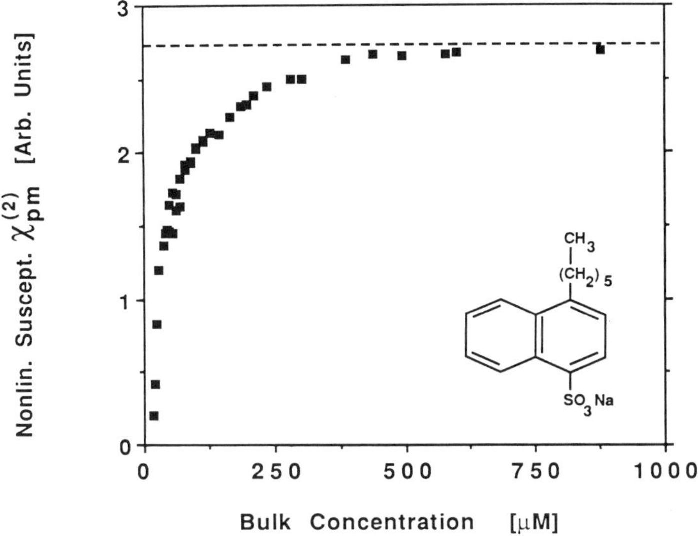

Fig. 8. Second-order susceptibility $\chi^{(2)}_{pm}$ versus the bulk concentration of the soluble C6NS molecules (0.35 M NaCl solution with pH = 5.6 at 20°C). $\chi^{(2)}_{pm}$ of a saturated adsorption layer is indicated by the broken line.

tion of C6NS is shown in Fig. 8. It was also found that the ratio $\chi^{(2)}_{sm}/\chi^{(2)}_{pm}$ (subindex s denotes SH output s-polarized) remains the same in the range of bulk concentration measured. These results indicate that:

1. There exists an adsorbed layer of polar-oriented C6NS molecules at the air/water interface.
2. The orientation of these adsorbed molecules remains unchanged with the surface density.
3. Eq. (3.2.4) is valid.
4. The surface density of C6NS approaches saturation for large bulk concentration.

In this case, the surface density of C6NS can actually be calibrated against that of C18NS [18,39]. The latter, with the longer alkyl chain, is insoluble in water. It can be spread on water with a prescribed surface density that can be directly measured by SHG. Because the alkyl chain contributes little to the nonlinearity $\vec{\alpha}^{(2)}$ of C18NS is the same as that of C6NS. Measurements also showed that $\chi^{(2)}_{sm}/\chi^{(2)}_{pm}$ is the same for the two cases, indicating that the two molecules have the same orientation. Thus, by using C18NS as a reference, $\vec{\alpha}^{(2)}$ becomes known and the surface density N_S of C6NS can be obtained directly from the measured $\chi^{(2)}_{pm}$ using Eq. (3.2.4). With $\chi^{(2)}_{pm} \propto N_S$,

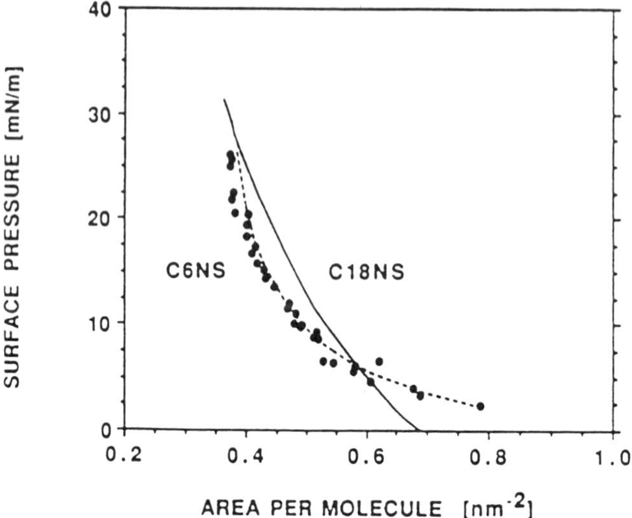

Fig. 9. Surface pressure versus the mean area per molecule for the insoluble monolayer of C18NS (——) and the soluble C6NS film (•) (0.35 M NaCl, 20°C). The former isotherm is governed by compression of the insoluble C18NS monolayer; the latter by adsorption of C6NS molecules to the air/water interface for various bulk concentrations. In the adsorption study, the area per molecule is obtained from the SHG measurement and the dotted curve is calculated from theory.

Fig. 8 is then nothing but the adsorption isotherm. As seen in Fig. 1, the experimental setup allows the simultaneous measurement of surface pressure π. One can then also find π versus N_S, or π versus $A = 1/N_S$ as depicted in Fig. 9.

In principle, the $\pi - A$ isotherm can be derived from the Gibbs equation [33]

$$(\delta\pi/\delta\mu)_T = N_S,$$

derived from the second law of thermodynamics, where μ is the equilibrium chemical potential which can be expressed as

$$\mu = \mu_b^\circ + kT \log \alpha_b,$$

with μ_b° being a constant and α_b the activity coefficient of the solute in the solution. If $(\delta\pi/\delta\mu)_T$ or $(\delta\pi/\delta \log \alpha_b)_T$ is known, then N_S can be obtained [17]. Unfortunately, this is usually not the case. The possibility that SHG can provide a direct measurement of N_S is therefore most helpful.

The value of $\chi_{pm}^{(2)}$ for the absorbed C6NS layer at saturation was found to be equal to that of a C18NS closed-packed monolayer. This indicates that the interfacial layers in the two cases contain the same number of polar-oriented adsorbed CNNS molecules, although one is soluble and the other insoluble in water. It then leads to the conclusion that the soluble C6NS molecules must also form a single polar-ordered close-packed monolayer at the air/water interface when the bulk concentration of C6NS approaches saturation [18]. One may argue that by coincidence, the interfacial layer can be composed of a partially polar-oriented C6NS surface monolayer and some polar-oriented C6NS molecules in the subsurface region. Further experiments with SHG and surface pressure measurements on mixed soluble and insoluble surface molecule layers have ruled out this possibility [18].

By monitoring excess soluble molecules at the air/water interface, SHG can be used to study how molecular structure can affect molecular solubility in water and the surface-bulk equilibrium [17]. For example, increasing the hydrocarbon chain length of amphiphilic molecules will drive more molecules to the surface. Charging the head group, on the other hand, will increase the solubility and drive the molecules into the bulk water. Using SHG, Eisenthal and coworkers [17] measured the surface density of alkylphenolate ions ($C_nH_{2n+1}C6H_5O^-$) and showed that it takes 4 CH_2 groups attached to a phenolate ion to balance the repulsion of the ion from the interface. In another experiment [17], it was seen that converting *p*-nitrophenol (NO_2C6H_6OH) to anions ($NO_2C6H_6O^-$) via increasing the bulk pH value drives the molecules into the bulk water. SHG measurements of the adsorption isotherms of charged and neutral solute molecules at the air/water interface can yield the adsorption-free energies for the respective species if the adsorption can be approximated by simple Langmuir kinetics [17].

Adsorption dynamics of soluble molecules can be studied by time-dependent SHG. One measures by SHG how the molecules appear and approach the equilibrium surface density at an initially clean air/water interface. This was carried out from a freshly swept water surface of a solution in a Langmuir trough [40] or from the surface of a water (solution) jet [41]. In principle, from the polarization dependence of SHG, the time variation of molecular orientation at the interface can also be measured. The result of the water jet experiment suggests that the orientation of phenol adsorbing to the air/water interface reaches equilibrium on a time scale much faster than milliseconds [41].

8. ULTRASHORT TIME-RESOLVED STUDIES

SHG is limited in time only by the electronic response of the medium. Therefore, with ultrashort laser pulses, the technique permits studies of surface dynamics with an extremely high time resolution. This is most exciting as it opens the door to a virtually unexplored research area. Reaction dynamics, for example, can be very different at the air/water interface than in a solution. Sitzmann and Eisenthal [19] first carried out such an experiment. They studies the *cis-trans* photoisomerization rate of 3.3′-diethyloxadicarbocyanine iodide and found that it is significantly faster at the air/water interface than in the bulk solution ($\tau = 220$ ps as compared to $\tau = 520$ ps).

With tunable lasers, it is possible to use resonant SHG (or, more generally, resonant sum-frequency generation) to probe excitation dynamics of surface molecules. For example, both longitudinal and transverse relaxations of a surface molecular vibration can be measured [42]. Such experiments at an air/water interface, however, have not yet been reported.

9. SUMMARY

Being highly surface-specific, surface SHG is an effective surface analytical tool. The submonolayer sensitivity and *in situ* measurement capability make it ideal for investigation of molecules at gas/liquid interfaces. Spreading molecules at an air/water interface allows us to measure the quadratic nonlinearity of the molecules without complication from the local-field correction. Molecular orientations at the air/water interface can also be deduced from the measurement. The nonlinearities can however be significantly affected by environment. Protonation of a molecule, for example, can drastically change the delocalized electron distribution in the molecule and alter its nonlinearity. Measurements of molecule nonlinearities can, in turn, be used to probe proton concentrations or pH values. Aggregation, polymerization, and phase transitions of molecular monolayers are also reflected in changes of molecular nonlinearities. Surface reactions in general can be monitored by time-dependent surface SHG. With tunable picosecond or subpicosecond laser pulses, the surface SHG technique allows many exciting opportunities of research in surface spectroscopy and ultrafast surface dynamics.

REFERENCES

1. See, for example, *Nonlinear Optical Properties of Organic and Polymeric Materials*, ed. D. J. Williams (Am. Chem. Soc., Washington, 1983); J. Zyss, *J. Mol. Electron.*

1, 25 (1985); *Nonlinear Optical Properties of Organic Molecules and Crystals*, ed. D. Chemla and J. Zyss (Academic, New York, 1987); *Organic Materials for Nonlinear Optics*, ed. R. A. Hann and D. Bloor (Royal Soc. Chemistry, London, 1989); Langmuir-Blodgett Films 3, ed. D. Möbius, in *Thin Solid Films* 159–160 (1988); Langmuir-Blodgett Films 4, ed. K. Fukuda and M. Sugi, in *Thin Solid Films* 179–180 (1989).
2. A. Dulcie, C. Flytzanis, C. L. Tang, D. Pepin, M. Fetizon, and Y. Hoppilliard, *J. Chem. Phys.* **74**, 1559 (1981); J. L. Oudar and D. S. Chemla, *J. Chem. Phys.* **66**, 2664 (1977); J. L. Oudar, *J. Chem. Phys.* **67**, 446 (1977).
3. B. F. Levine and C. G. Bethea, *Appl. Phys. Lett.* **24**, 445 (1974); *J. Chem. Phys.* **63**, 2666 (1975).
4. T. F. Heinz, C. K. Chen, D. Ricard, and Y. R. Shen, *Phys. Rev. Lett.* **48**, 478 (1982); T. F. Heinz, H. W. K. Tom, and T. R. Shen, *Phys. Rev.* **A28**, 1883 (1983).
5. I. R. Gisling, N. A. Cade, P. V. Kolinsky, and C. M. Montgomery, *Electron. Lett.* **21**, 169 (1985); I. R. Girling, N. A. Cade, P. V. Kolinsky, J. D. Earls, G. H. Cross, and I. R. Peterson, *Thin Solid Films* **132**, 101 (1985).
6. Th. Rasing, G. Berkovic, Y. R. Shen, S. G. Grubb, and M. W. Kim, *Chem. Phys. Lett.* **130**, 1 (1986); G. Berkovic, Th. Rasing, and Y. R. Shen, *J. Opt. Soc. Am.* **B4**, 945 (1987).
7. G. Marowsky, A. Gierulski, R. Steinhoff, D. Dorsch, R. Eidenschnik, and B. Rieger, *J. Opt. Soc. Am.* **B4**, 956 (1987).
8. D. Lupo, W. Prauss, U. Scheunemann, A. Laschewsky, H. Ringsdorf, and I. Ledoux, *J. Opt. Soc. Am.* **B5**, 300J (1988).
9. L. M. Hayden, *Phys. Rev.* **B38**, 3718 (1988).
10. Th. Rasing, Y. R. Shen, M. W. Kim, and S. Grubb, *Phys. Rev. Lett.* **55**, 2903 (1985).
11. P. Ye and Y. R. Shen, *Phys. Rev.* **B28**, 4288 (1983).
12. G. Marowsky, L. F. Shi, D. Möbius, R. Steinhoff, Y. R. Shen, D. Dorsch, and B. Rieger, *Chem. Phys. Lett.* **147**, 420 (1988).
13. X. D. Xiao, V. Vogel, and Y. R. Shen, *Chem. Phys. Lett.* **163**, 555 (1989); X. D. Xiao, V. Vogel, Y. R. Shen, and G. Marowsky, *J. Chem. Phys.* **94**, 2315 (1991).
14. V. Mizrahi, G. I. Stegeman, and W. Knoll, *Phys. Rev.* **A39**, 3555 (1989).
15. I. Weissbuch, G. Berkovic, L. Leiserowitz, and M. Lahav, *J. Am. Chem. Soc.* **112**, 5874 (1990).
16. G. Berkovic, Th. Rasing, and Y. R. Shen, *J. Chem. Phys.* **85**, 7374 (1986).
17. J. M. Hicks, K. Kemnitz, K. B. Eisenthal, and T. F. Heinz, *J. Phys. Chem.* **90**, 560 (1986); K. Kemnitz, K. Battacharyya, J. M. Hicks, G. R. Pinto, and K. B. Eisenthal, *Chem. Phys. Lett.* **131**, 285 (1986); K. Bhattacharyya, E. V. Sitzmann, and K. B. Eisenthal, *J. Chem. Phys.* **87**, 1442 (1987); K. Bhattacharyya, A. Castro, E. V. Sitzmann, and K. B. Eisenthal, *J. Chem. Phys.* **89**, 3376 (1988).
18. V. Vogel, C. S. Mullin, and Y. R. Shen, *Langmuir* **7**, 1222 (1991).
19. E. V. Sitzmann and K. B. Eisenthal, *J. Phys. Chem*, **92**, 4579 (1988).

20. N. Bloembergen and P. S. Pershan, *Phys. Rev.* **128**, 606 (1962).
21. C. C. Wang, *Phys. Rev.* **178**, 1457 (1969) T. F. Heinz, Ph.D. Thesis, Univ. Calif. Berkeley, 1982 (unpublished); P. Guyot-Sionnest, W. Chen, and Y. R. Shen, *Phys. Rev.* **B33**, 8254J (1986); J. E. Sipe, D. J. Moss, and H. M. Van Driel, *Phys. Rev.* **B35**, 4420 (1987); P. Guyot-Sionnest and Y. R. Shen, *Phys. Rev.* **A38**, 7985 (1989).
22. See, for example, Y. R. Shen, *The Principles of Nonlinear Optics.* (J. Wiley, New York, 1984), Ch. 25.
23. G. T. Boyd, Y. R. Shen, and T. W. Hansch, *Optics Lett.* **11**, 97 (1986).
24. Th. Rasing, J. Huang, A. Lewis, T. Stehlin, and Y. R. Shen, *Phys. Rev.* **A40**, 1684 (1989); J. Huang, A. Lewis, and Th. Rasing, *J. Chem. Phys.* **92**, 1756 (1988).
25. H. W. K. Tom, T. F. Heinz, and Y. R. Shen, *Phys. Rev. Lett.* **51**, 1983 (1983).
26. K. Kemnitz, K. Bhattacharyya, J. M. Hicks, G. R. Pinto, K. B. Eisenthal, and T. F. Heinz, *Chem. Phys. Lett.* **131**, 285 (1986).
27. C. Mullin, P. Guyot-Sionnest, and Y. R. Shen, *Phys. Rev.* **A39**, 3745 (1989).
28. H. Hsiung, G. R. Meredith, H. Vanherzeele, R. Popovitz-Biro, E. Shavit, and M. Lahav, *Chem. Phys. Lett.* **164**, 539 (1989).
29. X. Zhao, M. C. Goh, and K. B. Eisenthal, *J. Phys. Chem.* **94**, 2222 (1990).
30. Th. Rasing, H. Hsiung, Y. R. Shen, and M. W. Kim, *Phys. Rev.* **A37**, 2732 (1988).
31. G. Gillberg, R. Keosian, J. L. Pruksarnukul, and D. W. Lupo, *Mol. Eng.* (to be published).
32. A. Bagehi, R. G. Barrera, and B. B. Dasgupta, *Phys. Rev. Lett.* **44**, 1475 (1980); A. Bagehi, R. G. Barrera, and R. Fuchs, *Phys. Rev.* **B25**, 7086 (1982).
33. See, for example, J. T. Davis and E. K. Rideal, *Interfacial Phenomena* (Academic Press, New York, 1963); A. A. W. Adamson, *Physical Chemistry of Surfaces*, 4th ed. (J. Wiley, New York, 1982).
34. J. S. Schildkraut, T. L. Penner, C. S. Willand, and A. Ullman, *Optics Lett.* **13**, 134 (1988); I. R. Girling, N. A. Cade, P. V. Kolinsky, R. J. Jones, I. R. Peterson, M. H. Ahmad, D. B. Neal, M. C. Petty, G. G. Roberts, and W. J. Feast, *J. Opt. Soc. Am.* **B4**, 950 (1987).
35. L. M. Hayden, B. L. Anderson, J. Y. S. Lam, B. G. Higgins, P. Stroeve, and S. Kowel, *Thin Solid Films* **160**, 379 (1988).
36. I. Weissbuch, G. Berkovic, L. Leiserowitz, and M. Lahav, *J. Am. Chem. Soc.* **112**, 5874 (1990); G. Berkovic and I. Weissbuch, *Mol. Cryst. Liq. Cryst.* (to be published).
37. See, for example, P. Schaef and J. Talbot, *Phys. Rev. Lett.* **62**, 175 (1989); E. M. Lee, R. K. Thomas, J. Penfold, and R. C. Ward, *J. Phys. Chem.* **93**, 381 (1989); G. W. Woodbury and L. A. Noll, *Colloids and Surfaces* **33**, 301 (1989).
38. W. McBain and R. C. Swain, *Proc. Roy. Soc.* **A154**, 608 (1936); A. W. Adamson, *Physical Chemistry of Surfaces* (J. Wiley, New York, 1990) and references therein.
39. V. Vogel, C. S. Mullin, and Y. R. Shen, *J. Chem. Phys.* (to be published).
40. Th. Rasing, T. Stehlin, Y. R. Shen, M. W. Kim, and P. Valint, *J. Chem. Phys.* **89**, 3386 (1988).

41. A. Castro, S. Ong, and K. B. Eisenthal, *Chem. Phys. Lett.* **163**, 412 (1989).
42. A. L. Harris and N. J. Levinos, *J. Chem. Phys.* **90**, 3878 (1989); A. L. Harris, L. Rothberg, L. Dubois, N. J. Levinos, and L. Dhar, *Phys. Rev. Lett.* **64**, 2086 (1990); P. Guyot-Sionnest, P. Dumas, Y. J. Chabal, and G. S. Higashi, *Phys. Rev. Lett.* **64**, 2156 (1990); X. D. Zhu and Y. R. Shen, *J. Appl. Phys.* **B50**, 535 (1990); P. Guyot-Sionnest, *Phys. Rev. Lett.* **66**, 1489 (1991).

Chapter 4

ADVANCES IN MOLECULAR ENGINEERING FOR QUADRATIC NONLINEAR OPTICS

Joseph Zyss and **Isabelle Ledoux**

Centre National d'Etudes des Télécommunications, Département d'Electronique Quantique et Moléculaire, Bagneux, France

and

Jean-François Nicoud

Groupe des Matériaux Organiques, Institut de Physique et Chimie des Matériaux de Strasbourg, Strasbourg, France

ISBN 0-12-784450-3

1. INTRODUCTION

Over the last decade, intensive activity in the field of organic synthesis has come to implement molecular engineering guidelines established in the late seventies to early eighties: these efforts have focused on relatively simpler organic systems with particular emphasis on the paranitroaniline family [1,2]. The urgent task was then to demonstrate at both molecular and crystalline levels the relevance of organic media for quadratic nonlinear optics: confirmation of the adequacy of earlier models, such as the two-level quantum description of charge-transfer molecules [3], required experimental confrontation with relatively simple molecular structures. Beside, difficulties in crystallization of large-size complex molecules restricted investigations to smaller systems of limited chemical originality. The situation gradually changed in the mid-eighties under the influence of a number of factors of different origins: while previous exploration of the domain has been confined to a relatively small group of chemists working closely with physicists, convincing reports of early breakthroughs reached the chemical community by and large, and many synthetic possibilities were henceforth proposed and actually developed. Furthermore, a rather systematic exploration of the possibilities of the paranitroaniline family undertaken in the early eighties had led to the development of a number of interesting single crystals, and the bulk of the research effort for that class of materials progressively shifted from molecular engineering to optical physics and device demonstrations. A symbolic achievement in this field is the recent operation of an optical parametric oscillator [4] based on the optimized paranitroaniline-like NPP crystal [5]. Chemists have then increasingly extended, over the last five years, their field of investigation, which now encompasses newer generations of organic molecules of greater complexity and sophistication than for earlier examples. Such diversification is demanded by applications where "yellow" materials, satisfying the spectral requirements associated with traditional YAG laser sources for second harmonic generation or parametric amplification, are to be complemented to satisfy other needs: more coloured materials are of interest for near infrared electro-optic modulation where diminishing the spectral window can be afforded with the additional benefit of a considerable preresonant enhancement of the efficiency. Conversely, extension of the transparency to the UV, so as to encompass the blue spectral range, has been spurred by applications to high-density optical data storage. The subtler concept of transparency-nonlinearity tradeoff has thus led to a more adequate classification of compounds [6]. Functionalized polymers achieved a major breakthrough in the early eighties [7], to become widely adopted in the mid-eighties [8], by permitting the bottleneck of single crystal growth to be bypassed. The potential of organics for applications was subsequently considerably broadened to encompass waveguided optoelectronics. As far as molecules are concerned, poled polymers whereby nonlinear chromophores are introduced as guests or attached to the polymeric backbone allow to consider, even for applications, advanced molecules that would have hardly been considered for crystallization, at least in usable format. No molecular system, in this renewed context, can now be considered as unduly exotic in view of the various alternatives now available to overcome the barrier of single crystal formation. Other ways of assembling molecules for nonlinear optics have also matured, including Langmuir-Blodgett (LB) films, mesogenic phases, or self-assembly. It is therefore our purpose, in this paper, to highlight extensions of classical ways in molecular engineering that refine over previously general synthetic directions, as well as radically new orientations that have emerged more recently.

The blueprint of molecular systems for quadratic nonlinear optics must incorporate features pertaining to purely molecular properties, such as the transparency-efficiency tradeoff, as well as to macroscopic structural organization, be it in single crystalline or statistically poled media. Although such a division may be too crude to account for the incidence of structural features on the optical spectrum, as would result from excitonic effects in single crystals, or aggregation processes in dispersed media, it is of great help as a first iteration in the molecular design process.

Chemists have to find rational ways of exploring the wealth of molecular structures and synthetic pathways: they will be generally guided by a combination of their own experience or specialty and of previously established engineering guidelines to be ultimately refined or extended by their research. A physical approach to the molecular design will tend to be guided as the first step by more abstract considerations where, for example, symmetry or quantum-mechanical views will play a predominant role. Progress in the field results from a permanent confrontation between the two approaches and it is our ambition, in this chapter, to propose a dual perspective whereby chemical and physical approaches reinforce each other.

This chapter is organized along the following lines: original ways to locate materials at relevant positions on the transparency-efficiency trade-off scale are detailed in Sec. 2; the key parameter for classifying materials within this scale is the hybridization geometry of polarizable atoms at strategic sites in the conjugated system. The quadratic and cubic nonlinear optical properties of polyene oligomers are reviewed in Sec. 3, in view of the record high nonlinearities achieved in these molecules and for their fundamental interest as polarizable one-dimensional objects of variable finite extensions. Various ways to influence, at the molecular level, the nature and the symmetry of the macroscopic assemblies are reviewed in Secs. 4 and 5: examples in the fields of single crystals, either purely organic or organomineralic, inclusion compounds, Langmuir-Blodgett films, and mesogenic materials illustrate the considerable diversification of the field over recent years. In Sec. 5, the originally fruitful concept of dipolar conjugated molecules is revisited, and a new direction, initially based on symmetry considerations, is shown to open up a wealth of opportunities in a 3-dimensional chemistry based on the concept of multipolar nonlinear systems.

2. TRANSPARENCY/NONLINEARITY TRADEOFF

2.1. Transparency tuning

It is necessary to characterize a quadratic NLO material with reference to its wavelength transparency range [6], a minimal information being the wavelength absorption cutoff in solution with an appropriate solvent, knowing that a red shift of the shortest cutoff wavelength by as much as 100 nm may occur in the condensed phase (see Sec. 5.4 for new perspectives in that field).

The conjugated paranitroaniline (*p*-NA) system has led to several nearly optimized NLO materials, such as NPP and so on. These materials however display red-shifted cutoff, which precludes their utilization for blue-light SHG from current laser diodes. The need for a better control of the material transparency has thus emerged. Several attempts to address the transparency-tuning issue have been proposed. The use of Dewar's rules has been exposed previously [2], and need not be recalled here. These rules pointed out the more favorable transparency/nonlinearity (T/NL) tradeoff of the 2-amino-5-nitropyridine (2A5NP) system in comparison with the classical *p*-NA system. Several 2A5NP derivatives are now currently studied (see Secs. 4 and 5.2).

A mean for improving the T/NL trade off was to compensate the unavoidable bathochromic effect due to the lengthening of the conjugated system, by using donor-acceptor combinations of lesser efficiency than the classical amino-nitro one. By this way some efficient stilbenes, like 3-methyl-4-methoxy-4′-nitrostilbene (MMONS, **1**) have been discovered [9,10].

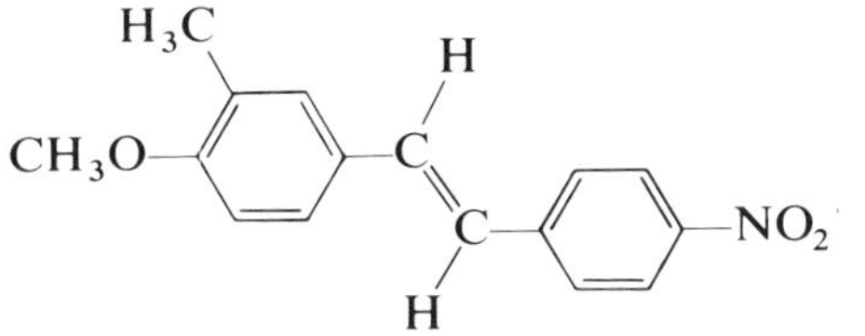

1 MMONS

Nevertheless the results are almost invariably the same: the compounds are always more or less colored. It seems possible to get a relatively easy tuning between red and yellow colored materials, but getting a still highly SHG efficient *and* uncolored material remains a somewhat problematic challenge. Two recent results, which could open the way to blue-light SHG, seem to show that improvements in that direction are now conceivable.

First is the use of five-member nitrogenous heterocycles as donor moieties in 1-substituted-4-nitrophenyl derivatives, by researchers of Fuji Photo Film Co. in Japan. They discovered two efficient materials: 3,5-dimethyl-1-(4-nitrophenyl)-pyrazole (DMNP, **2**) which has a wavelength cutoff at 402 nm (in solution) and an SHG signal of 16 × urea at 1.06 μm, and 3,5-dimethyl-1-(4-nitrophenyl)-1,2,4-triazole (DMNT, **3**), which has a wavelength cutoff at 390 nm and an SHG signal at 12.9 × urea at 1.06 μm [11]. The nonlinear coefficients d_{32} and d_{33} are 90 ± 10 pm/V and 29 ± 4 pm/V, respectively.

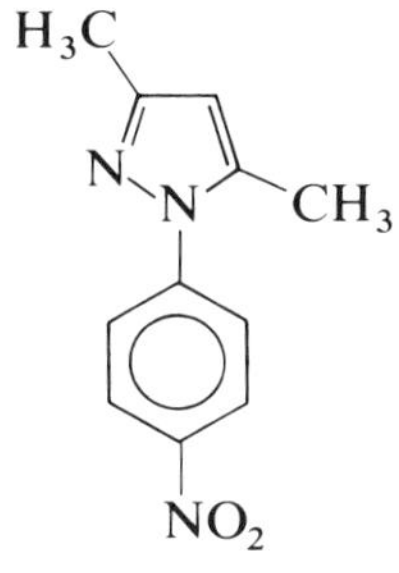

2 DMNP

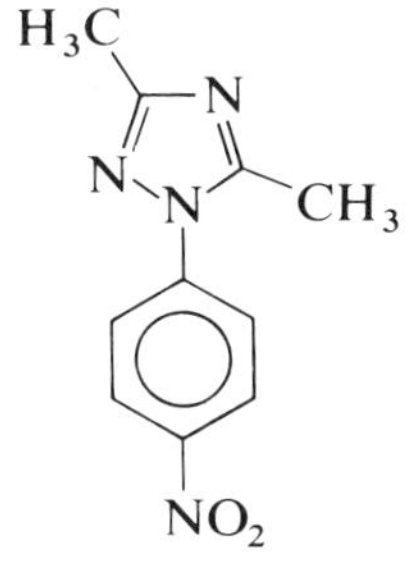

3 DMNT

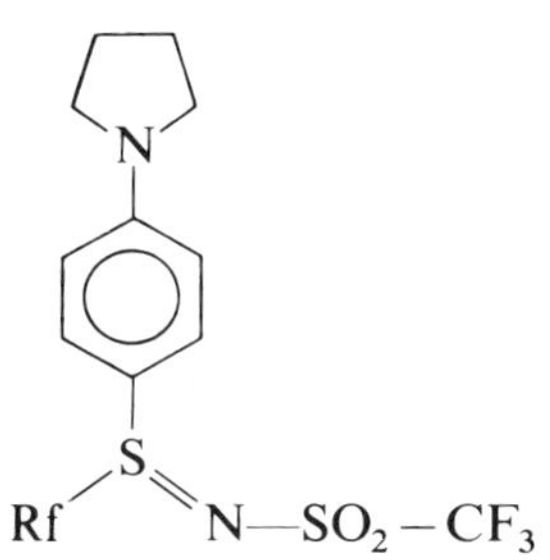

4

Efficient frequency doubling of a continuous-wave semiconductor laser was attained by using an optical fiber with a single crystal core of DMNP [12].

Second is the use of new substituted sulfone derivatives as acceptor. From the examination of its Hammett's (σ_p) constant, it has been expected that the perfluorinated alkylsulfone ($-SO_2-Rf$) would be a more effective acceptor than the nitro group [$\sigma_p(NO_2] = 0.81$, $\sigma_p(—SO_2—CF_3) = 0.93$ [13]. In analogous systems the β were comparable, but the λ_{max} of the fluorinated sulfone derivative is blue-shifted in comparison with its nitro analog. The T/NL tradeoff of these new materials seems very favorable and should be further investigated [14]. Moreover the sulfonylsulfimide (SSI) group [$S(R_f)$=N—SO_2—CF_3] presents even more interesting accepting properties as it leads to the overall effect of two nitro groups, by inductive contribution essentially. The SSI σ_I coefficient (field effect contribution to σ_p) is 1.5 times larger than the σ_I coefficient of the SO_2CF_3 group. The resulting effect on the favorable T/NL tradeoff is shown on compound **4**, which exhibits a $\beta = 13 \times 10^{-30}$ esu and $\lambda_{max} = 336$ nm.

2.2. *sp*^{3} hybridized hetero-atoms: Silicon

Another method leading to a strong increase of the UV transparency consists of using sp^3 hybridized hetero-atoms of the third row of the periodic table. These atoms may be placed either at one end of the conjugated path (and then be used as electroactive substituents) or between two fully conjugated moieties such as phenyl rings. The expected blue shift should be related to the electron isolating property of σ bonds as compared to π ones and to the relatively poor overlap of the external 3*s* and 3*p* orbitals of the hetero-atoms with the 2*s* and 2*p* ones of the carbon. On another hand, owing to the increased polarizability of the external 3*s* and 3*p* electrons of the hetero-atoms as compared to that of the 2*s* and 2*p* ones in carbon, a significant nonlinear response could be expected. Therefore, there is a possibility of improving the nonlinearity/transparency tradeoff using this delicate balance between the polarizability of 3*s* and 3*p* orbitals and the isolating character of the corresponding σ bonds. Sulfur-containing systems, where S is used either as an electron donor group or as part of the conjugated path, evidenced a significant improvement of the transparency/nonlinearity trade-off as compared to *p*-NA derivatives [15–18].

However, a more systematic study could be carried out on silicon derivatives. Like sulfur, silicon may be used in quadratic nonlinear molecules, either as a part of an electron donor or acceptor group, or inserted within the conjugation path [19–21]. The specific interest of silicon as compared with sulfur lies in higher connectivity (4 instead of 2) to possible adjacent moieties, opening the way to an increased "multifunctional" character of the nonlinear unit, which can be more easily associated to other structures such as polymer backbones, sol-gels compounds, or inorganic silicon-based substrates for microelectronics.

2.2.1. $Si(CH_3)_3$ *as an electron donor or acceptor group*

As evidenced in Table 1, the trimethylsilane substituent weakly interacts with donor groups such as dimethylamino or methoxy ones. Both dipole moments and static first-order hyperpolarizabilities are quite low. Trimethylsilane is markedly more effective as an electron releasing group. The mechanism involved is the interaction between the σ silicon-carbon bond and adjacent π^* orbitals of the benzene ring [22–23]. The nonlinear polarizability results primarily from an electronic intramolecular charge-transfer (ICT) between the donor $Si(CH_3)_3$ and the acceptor dicyanovinyl groups through the conjugated aromatic ring. The trimethylsilyl group seems to be approximately as active as the methoxy group.

There is a huge blue shift of the maximum absorption wavelength if one compares **8** (λ_{max} = 440 nm) to **7** (λ_{max} = 322 nm). However, the magnitude of the σ—π interaction involved in the ICT process is strongly reduced as

Table 1.

Maximum Absorption Wavelength λ_{max}, Dipole Moment μ and Static First-Order Hyperpolarisability $\beta(0)$ of Methylsilane Derivatives.

Compound	Number	λ_{max} (nm)	μ (Debye)	$\beta(0)$ (10^{-30} esu)
$Me_3Si-C_6H_4-OMe$	**5**	281	1.1	0.3
$Me_3Si-C_6H_4-NMe_2$	**6**	266	1.8	4.0
$Me_3Si-C_6H_4-C(H)=C(CN)-CN$	**7**	322	5.0	9.0
$Me_2N-C_6H_4-C(H)=C(CN)-CN$	**8**	440	7.0	16.0

compared to π—π conjugation. This leads to a severe reduction of the static hyperpolarizability values (by a factor of 3), which limits the interest of the trimethylsilyl substituent for obtention of highly efficient molecules for frequency doubling towards short wavelengths.

2.2.2. *Si as part of the conjugation path*

The Si atom has also been tested as an electron bridge connecting separate phenyl rings [24]. In Table 2 are reported dipole moment, maximum absorption wavelength, and hyperpolarizability values of the series:

$$(CN)_2C{=}CH{-}\Phi{-}(-\overset{|}{\underset{|}{Si}}-)_n{-}\Phi{-}N(CH_3)_2, \qquad \text{where } n = 1, 2, 6.$$

The experimental $\beta(0)$ values are compared to these obtained by use of an additive model where β_{add} is estimated as the vector sum of the quadratic hyperpolarizabilities of each subunit:

$$(CH_3)_2N{-}\Phi{-}\overset{|}{\underset{|}{Si}}{-}(CH_3)_3$$

and $$(NC)_2C{=}CH{-}\Phi{-}Si(CH_3)_3$$

The relatively weak $\beta(0)$ value for $n = 1$ and the absence of any ICT between the electron donating and electron attracting units, clearly evidences that one Si atom cannot be considered as a polarizable electron connection between the phenyl rings. The ICT processes involved here remain located within each subunit. It should be pointed out that, for $n = 1$, the two individual ICT axes of each subunit are not aligned but make an angle of approximately 109°, due to sp^3 hydridization of Si. Such a bent structure is not favorable to ICT through the whole molecule, the steric intramolecular hindrances making the two benzene rings not coplanar.

For $n = 2$, the possible ICT contribution to the hyperpolarizability remains within the experimental error range. However, the $\beta(0)$ value is significantly higher; this is mainly due to the all-*trans* configuration of the molecule along the Si–Si bond, as confirmed by crystallographic data. This geometry allows for a better coplanarity between the two benzene rings (the corresponding dihedral angle being only 11°) in the solid state [19] and a good parallelism between the vector parts of the $\beta(0)$s of each subunit, then maximizing the cosine projection factors.

For $n = 6$, the optical first-order hyperpolarizability is significantly improved with respect to smaller n values. This is due to the delocalization of

Table 2.

Dipole Moment μ, Maximum Absorption Wavelength λ_{max}, Static First-Order Hyperpolarizability Deduced from Experimental Data ($\beta\,exp(0)$), from an Additive Model ($\beta_{add}(0)$) and Intramolecular Charge-Transfer Contribution $\beta_{CT}(0)$ to $\beta\,exp(0)$, for Silane Derivatives with Various Numbers (1, 2 and $n = 6$) of Dimethylsilane Units between the Electron-Donating and the Electron-Attracting Parts of the Molecule.

Compound	Number	μ (D)	λ_{max} nm	$\beta_{exp}(0)$ (10^{-30} esu)	$\beta_{add}(0)$ (10^{-30} esu)	$\beta_{CT}(0)$ (10^{-30} esu)
$Me_2N-C_6H_4-SiMe_2-C_6H_4-CH=C(CN)_2$	**9**	6	320	11	11	0
$Me_2N-C_6H_4-SiMe_2-SiMe_2-C_6H_4-CH=C(CN)_2$	**10**	7	334	16	13	3 ± 2
$Me_2N-C_6H_4-[SiMe_2]_n-C_6H_4-CH=C(CN)_2$	**11**	6.8	385	26	13	13

σ electrons along the silicon–silicon backbone. An important ICT (50 percent of the $\beta(0)$ value) is clearly pointed out by comparing the experimental value to that obtained from an additive model. The compound shows an interesting nonlinearity/transparency compromise: the $\beta(0)$ value is three times larger than the most recent $\beta(0)$ values reported in the literature for paranitroaniline ($\lambda_{max} = 370$ nm and $\beta(0) = 8 \times 10^{-30}$ esu in dioxane [25]) for a similar transparency. The σ delocalization over the Si–Si chain clearly evidences the ability of silicon-based molecules to bring a better optimization of the transparency-efficiency tradeoff.

2.2.3. *Role of substituents in ICT through disilane units*

We now concentrate on disilane derivatives where the two silicon atoms ensure the connection between two conjugated moieties: one containing an electron donating group ($N(CH_3)_2$, OCH_3, F ...) and the other one an electron accepting group (vinyl dicyano). It is clearly established, from NLO experiments and from various other chemical or physical properties [2], that the scale of the electron donating strength is

$$H < F < OCH_3 < N(CH_3)_2.$$

A quantitative estimate of these electron donating properties may be given by Hammett's constants. Therefore, we could expect a better nonlinear optical response for compounds bearing the strongest electron donor substituents.

Figure 1 plots the $\beta(0)$ of four disilane derivatives bearing donor groups with various strengths, as a function of the corresponding Hammett's constants [21]. One may notice the significant decrease in $\beta(0)$ when replacing the methoxy group by an N,N dimethylamino one, which is supposed to be a stronger donor. One should also note that there is a strong correlation between the experiment and the additive model for donor substituents such as H, F or N,N dimethylamino, but that there is no such link for the corresponding methoxy derivative.

The behavior is close to the predicted one depicted by Beratan [26]. The remarkably similar $\beta(0)$ values for molecules substituted with H or F atoms clearly evidence that, for an identical attractor group, using various substituents having a poor electron donor character is not worthwhile for an increase of the transparency/efficiency tradeoff.

This experimental behavior also evidences the possibility of a decrease of the quadratic nonlinear response of an electronic system substituted by quite strong electron donor or acceptor units, as already predicted for inorganic compounds [27]. The use of substituents with a strong electropositive

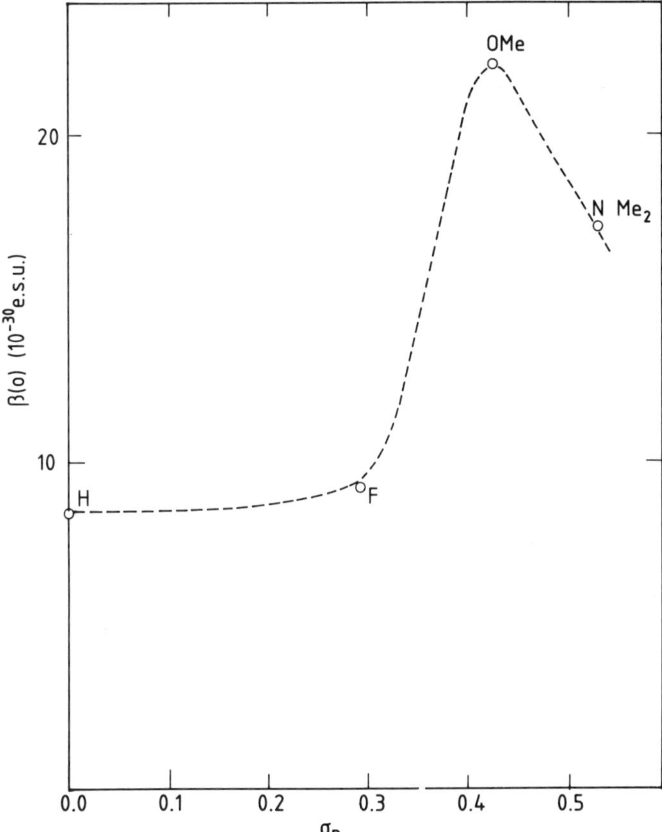

Fig. 1. Static first-order hyperpolarizabilities $\beta(0)$ of vinyldicyano-disilane derivatives substituted by various electron donor groups H, F, MeO and Me_2N, as a function of the Hammett's constant σ_R of these substituents.

or electronegative character (in inorganics) or with strongly positive or negative Hammett's constants (in organics) leads to a large decrease of the polarizability of the electrons involved in the NLO process, and is therefore reponsible for a significant lowering of $\Delta\mu$ and then of β values.

2.3. *sp*^{2} hybridized carbon atoms

2.3.1. *Carbon atoms: polyaryls*

An efficient way to increase the transparency, in keeping with a reasonably large molecular β value, has been reported in the case of polyphenyls

[28–33]. Theoretical studies, performed on hypothetic planar structures [34], have first evidenced a large enhancement of β when increasing the number n of phenyl rings in the series:

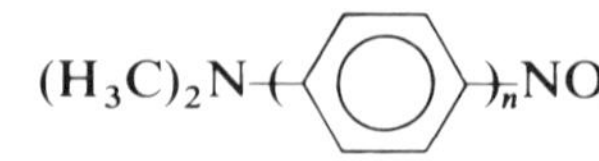

$$(H_3C)_2N\text{-}(C_6H_4)_n\text{-}NO_2$$

According to these estimates, the β value stabilizes beyond $n = 3$, in the range of 150×10^{-30} esu for SHG at 1.06 μm. However, this work, assuming coplanar phenyl rings, overestimates the λ_{max} values of the calculated molecules. In fact, the good transparency of polyphenyls in the blue or UV range can be partially accounted for by the limitation of the ICT between the donor and the acceptor group, due to the important torsion angle ($\sim 40°$ for dimethylamino-cyano-benzene (DMACB)) between the phenyl rings.

Nonlinear optical studies performed in the solid state of DMACB evidenced an additive factor enhancing the UV transparency of these materials: in crystalline DMACB, the almost perfect alignment of the molecules parallel to each other within the crystalline frame leads to an excitonic gas to crystal blue shift of the absorption edge resulting from molecular interactions in the lattice [32,35]. Adequate packing geometry may then combine its effect with that of intramolecular torsion angles to shift the transparency of the crystal towards the blue.

Comparison of solution and solid-state absorption spectra of DMACB and POM is illustrated by Fig. 2: the value of the λ_{max} of DMACB in chloroform is slightly red-shifted as compared to that of POM. On the contrary, the solid-state absorption cutoff of DMACB is considerably extended towards the UV (by 50 nm) as compared to POM. In addition, the quadratic nonlinear susceptibility coefficient d_{33} of DMACB (deduced from an oriented gas model) appears to be close to 10^{-6} esu, a value surpassing by a factor of two the best reported ones within this optical frequency range.

The case of DMACB clearly evidences the considerable interest of polyphenyls for quadratic nonlinear optics in terms of transparency-nonlinearity tradeoff. A more extensive study of cyanopolyphenyls of the general formula:

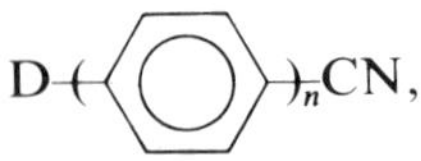

$$D\text{-}(C_6H_4)_n\text{-}CN,$$

where D is an electron donor group ($-N(CH_3)_2$, $-S-CH_3$, $-O-CH_3$, CH_3), allows for a more extensive screening of ICT polyphenyls for nonlinear optics, with possible combinations to other structural properties such as

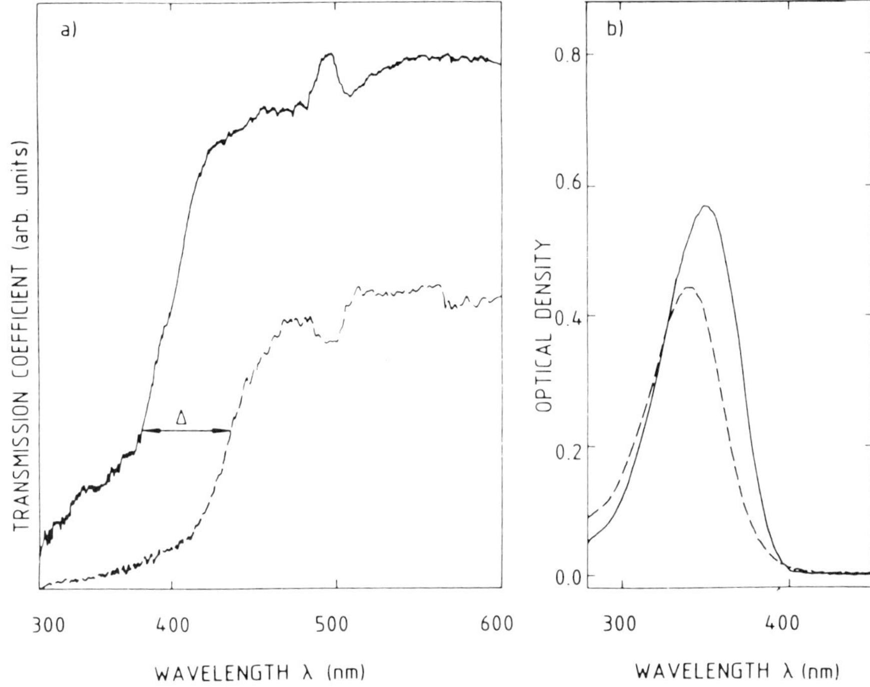

Fig. 2. (a) Optical density versus wavelength of 100 μm thick powder samples (grain calibration between 50 and 100 μm) of POM (dashed line) and DMACB (solid line). The corresponding spectra in chloroform are given in (b). Δ is the spectral shift observed in the solid state when going from POM to DMACB.

liquid crystals, the group

$$-\left(\mathrm{C_6H_4}\right)_n\mathrm{CN}$$

being a classical mesogenic moiety for $n > 1$.

2.3.1.1. Linear optical properties. In Fig. 3 are sketched the λ_{max} values as a function of n for four different electron-donating groups with increased electron-releasing character when going from methyl to dimethylamino one. In terms of relative UV transparency, λ_{max} increases with the polarizability of the donor groups.

For a given donor substituent, the large increase of λ_{max} when going from

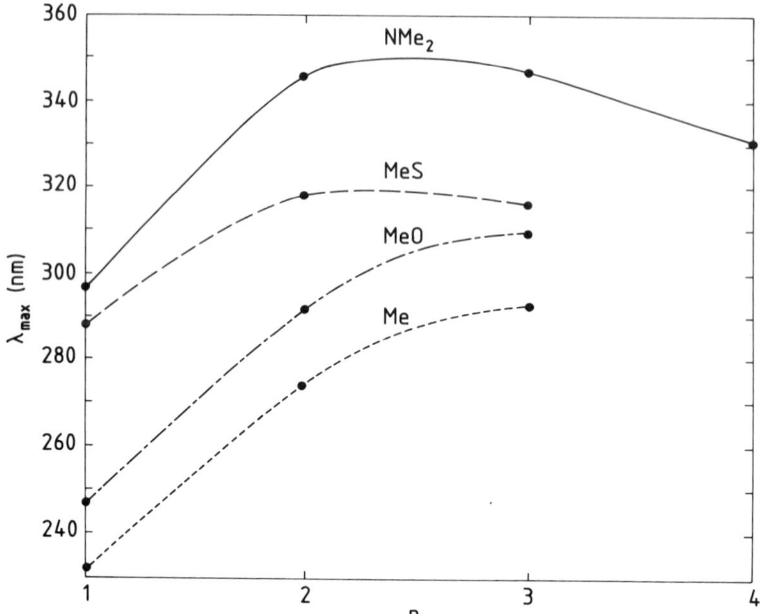

Fig. 3. Plot of the maxima absorption wavelengths of four families of polyphenyls exhibiting the same electron accepting group (cyano) but different electron donating groups, such as dimethylamino (continuous line), methoxy (dashed-dotted line), methylthio (dashed line), and methyl (dotted line), versus the number of phenyl rings.

$n = 1$ to $n = 2$ results from the additional number of π electrons involved in the ICT process, which largely overcomes the electron polarizability hindrance due to the twist angle between the two phenyl rings. For $n = 3$, the increase of the number of π electrons due to the third additional phenyl contributes less to the ICT as a result of the presence of two torsion angles between the benzene rings. The λ_{max} saturates or even slightly decreases in some cases. It should be pointed out that, for molecules of these series, even for the strongest donor group and the largest π-electron system, λ_{max} is significantly smaller than for *p*-NA. All compounds are white or very pale yellow in the solid state.

2.3.1.2. Nonlinear optical properties. $\beta(0)$ values of the four molecular series are plotted on Fig. 4 as a function of *n*. It points out that the electron-donating character of thiomethyl is slightly weaker than that of methoxy, although the polarizability of the surrounding nonbonding electrons is higher in the case of the sulfur atom. This is in keeping with dipole

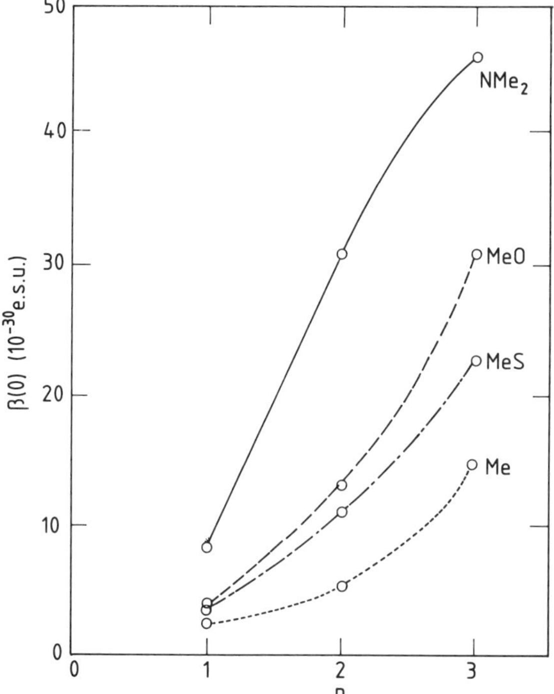

Fig. 4. Plot of $\beta(0)$ of three families of polyphenyl exhibiting the same electron accepting group (cyano) but different electron donating groups such as dimethylamino (continuous line) versus the number of phenyl rings.

measurements, which exhibit a similar relative behavior in these two molecular families. In fact, owing to the larger atomic volume of sulfur, the orbital overlap integral between the neighboring sp^2 hybridized carbon atom and the sulfur one is lower than in the case of the oxygen-carbon pair. In the later case, steric constraints are relaxed and allow for an improved mesomeric interaction between the nonbonding electrons and the aromatic phenyl ring as compared to the case of a sulfur atom. The ICT is then significantly stronger for the methoxy substituent.

Saturation of $\beta(0)$ values can only be observed for a *ter*-phenyl molecule containing the most strongly interacting pair of donor-acceptor substituents (—CN, —$N(CH_3)_2$). This phenomenon corresponds to a saturation of the transmitting capacity of the conjugated channel linking the donor and acceptor groups. The limitation of the ICT has a purely electronic origin, predicted even for idealized fully planar systems [34] and an additional steric origin, related to the torsion angles between the adjacent phenyl rings. The two effects are self-consistently related, in that the existence of torsion angles has a noticeable influence on the electronic properties of the whole π-electron system of the phenyl rings while, conversely, torsion angles between neighboring phenyl rings will depend on the strength of the donor-acceptor interaction. Such a "saturation" of the $\beta(0)$ values is not observed for the other weaker electron donating groups. This behavior is probably linked to the electronic properties of ICT molecules described in the models proposed by Choy [27] and Beratan [26], in the case of chemical bonds connecting strong electronic donor and acceptor units.

2.3.2. The Enone System, a "Cross-Conjugated" System: Chalcones and Related Compounds

In a "cross-conjugated" system, three groups are present, two of which are not conjugated with each other, although each of them is conjugated with the third one. The representative system is given by 3-methylene-1,4-pentadiene molecule (**12**). Five resonance structures must be written for this pentadiene molecule, making it difficult to treat such a system by the valence-bond method (Scheme 1). In most cases, it has been shown that it is easier to treat cross-conjugated molecules by the molecular orbital method [36]. In the case of **12**, it is shown that a resonance energy does exist.

Scheme 1

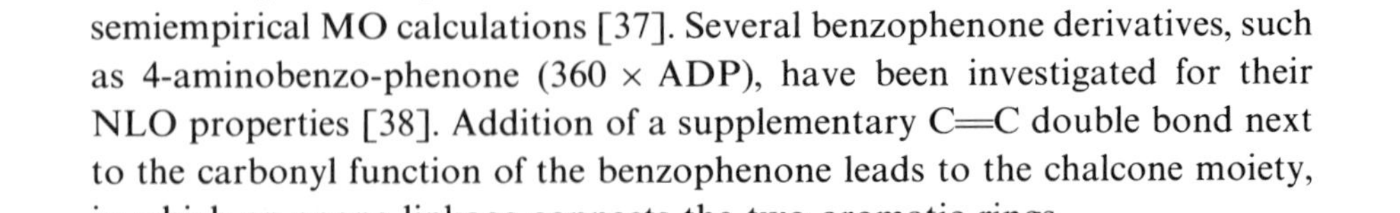

$$H_2C{=}CH{-}C({=}CH_2){-}CH{=}CH_2 \longleftrightarrow H_2C{=}CH{-}C(\overset{\oplus\ \text{or}\ \ominus}{C}H_2){=}CH{-}\overset{\ominus\text{or}\oplus}{C}H_2 \longleftrightarrow \overset{\ominus\text{or}\oplus}{H_2C}{-}CH{=}C(\overset{\oplus\ \text{or}\ \ominus}{C}H_2){-}CH{=}CH_2$$

12

The benzophenone system is also cross-conjugated and has been studied by semiempirical MO calculations [37]. Several benzophenone derivatives, such as 4-aminobenzo-phenone (360 × ADP), have been investigated for their NLO properties [38]. Addition of a supplementary C=C double bond next to the carbonyl function of the benzophenone leads to the chalcone moiety, in which an enone linkage connects the two aromatic rings.

Another example is illustrated by chalcones: several of them, bearing the classical nitro- and cyano-accepting groups, have been reported [39,40] as efficient $\chi^{(2)}$ materials. We have shown that these "push-pull" materials are generally more efficient when the π-acceptor group is placed on the benzoyl (Ph—CO—) group (**13a**) rather than on the other end (**13b**). 3-(4-dimethyl-aminophenyl)-1-(4-cyanophenyl)-phenyl)-2-propene-1-one, (DACC, **13c**, for Dimethyl-Amino-Cyano-Chalcone), is a red crystalline material with a SHG powder test close to that of NPP at 1.34 μm. Its β value is 191×10^{-30} esu measured by the EFISHG technique at the same wavelength in acetone solution. The maximum wavelength of the charge transfer transition is $\lambda_{max} = 436$ mm in the same solvent, the corresponding $\beta(0)$ value is then of 95×10^{-30} esu, a rather high value surpassing that of the corresponding stilbene [41]. The chalcone cross-conjugated path appears thus as a powerful electronic system for the molecular engineering of a highly hyperpolarizable molecule.

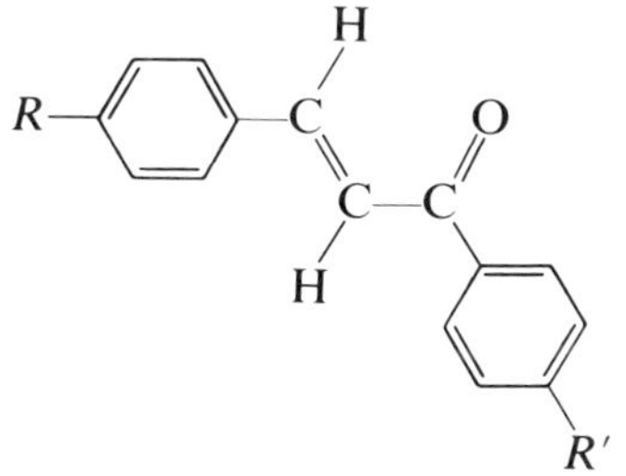

	R	*R′*	
13a	donor	acceptor	
13b	acceptor	donor	
13c	Me_2N	CN	(DACC)
13d	Br	OCH_3	(BMC)
13e	OCH_3	OC_2H_5	(MEC)

Chalcone chemistry has been extensively exploited in Japan, especially in the search of materials possessing a wide transparency range in the visible light. Along this direction, it was shown that the addition of a powerful π-acceptor group such as NO_2 or CN on the benzoyl group was unnecessary, the withdrawing effect of the sole carbonyl function being sufficient [42]. It is even possible to place one donating group on each phenyl ring. Two corresponding crystalline chalcone derivatives were recently investigated: 4-bromo-4′-methoxychalcone (BMC, **13d**) and 4-methoxy-4′-ethoxychalcone (MEC, **13e**). BMC presents an SHG signal of $11\times$ urea, and has a cutoff wavelength at 440 nm in the solid state [43]. The SHG signal of MEC was found to be $5\times$ urea, when its cutoff wavelength is 430 nm in the solid state. The effective NLO coefficients d_{eff} are in the range from 3.5 to 5.7 pm/V. The last improvements in the search of chalcone NLO materials come from the investigation of thienylchalcone derivatives [44]. Of special interest is 1-(2-thienyl)-3-(4-tolyl)propene-1-one (T-17, **14**) which shows an SHG signal intensity of $15\times$ urea and has a high blue-light region transparency with a cutoff wavelength of 390 nm. The effective nonlinear optical coefficient is 7.1 pm/V. This demonstrates the attractive T/NL tradeoff of such chalcones and their possible use for green and blue laser light generation [45].

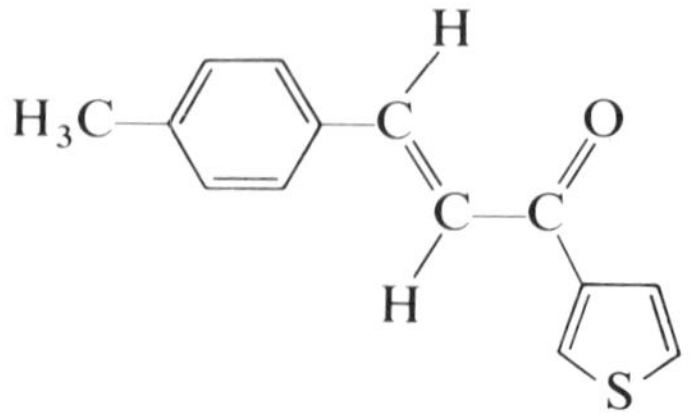

14 T-17

The cyclobutenedione system, built up from four sp^2 carbons and two sp^2 oxygens is related to the previous one in that it contains an enone moiety. Two positions are available on this group to attach specific substituents endowed with donor properties, or chirality, or both. For that reason the corresponding bisubstituted molecules can be considered as bearing a "crossed" bichromophoric system, contrarily to those described in Sec. 4.3.2 or 4.3.3. Several new SHG active materials containing cyclobutenedione units were reported by Pu [46]. In particular, the optically pure amino-alcohol derivative (−)-4-(4′-dimethylamino-phenyl)-3-(2′-hydroxypropylamino)cyclobutene-1,2-dione, (DAD, **15**), shows an SHG powder signal of 64 × urea at 1.06 μm. DAD crystallizes in the rather rare triclinic noncentrosymmetric space group P1. Its x-ray crystal structure shows that the direction of the dominant charge transfer axis is perfectly aligned in one dimension, which seems more adequate for electro-optical properties or frequency doubling in optical waveguides. Two hydrogen bonds between adjacent molecules, involving the carbonyl groups and NH and OH functionalities account for this perfect alignment [47].

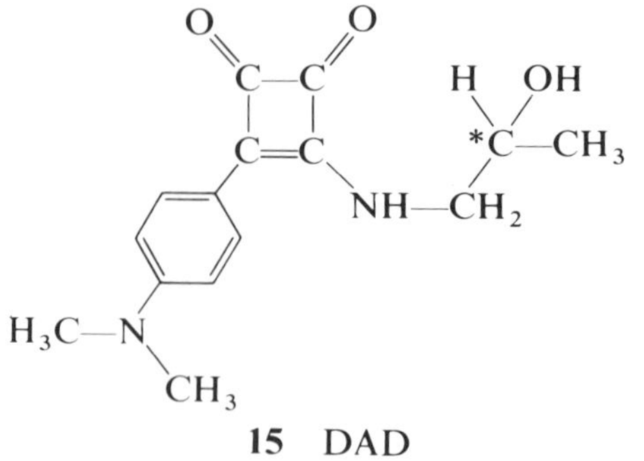

15 DAD

2.3.3. Benzylidene-aniline derivatives: Schiff bases

The benzylidene-aniline moiety contains a $Csp^2{=}Nsp^2$ double bond as part of its conjugated system. It constitutes a dissymmetric aza analog of the stilbene, that leads to two kinds of donor-acceptor derivatives **16a** and **16b**.

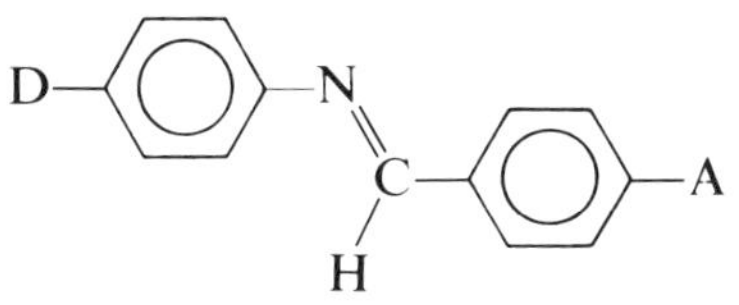

16a 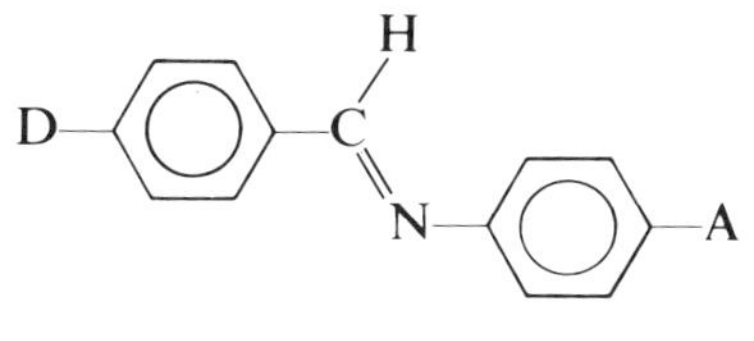 **16b**

These compounds are the well-known Schiff bases, easily prepared by condensation of the appropriate aniline and benzaldehyde starting materials.

Early studies of the hyperpolarizabilities of these compounds were reported by Nicoud and Twieg [2] as an illustration of the transparency tuning by use of Dewar's rules. The position of the imine nitrogen induces a dramatic position change of the maximum ICT absorption. The studies were resumed by Toray Industries Research Laboratories, where the molecular geometries and β and μ values of several donor-acceptor bisubstituted benzylidene-aniline derivatives were calculated [48]. Some selected results are shown in Table 3. The calculations show that derivatives of type **16a** have a small molecular dipole moment and a large hyperpolarizability β. A favorable T/NL tradeoff seems to be obtained with a CH_3O/NO_2 donor-acceptor pair in the structure of **16a** (**19**), as with stilbene derivatives (cf: MMONS, **1**). A number of chemical modifications of the basic structure **16**, especially by introducing a hydrogen-bond forming system, have led to a series of highly efficient $\chi^{(2)}$ crystals. One of them, 4′-nitrobenzylidene-3-acetamido-4-methoxy-aniline, (MNBA, **20**), has a SHG signal of 230 × urea. Strongly polar noncentrosymmetric molecular packing has been achieved in this yellow crystal, of monoclinic structure, space group Cc, by the presence of an intermolecular hydrogen bond between neighboring acetamido groups. The wavelength cutoff in the crystal is

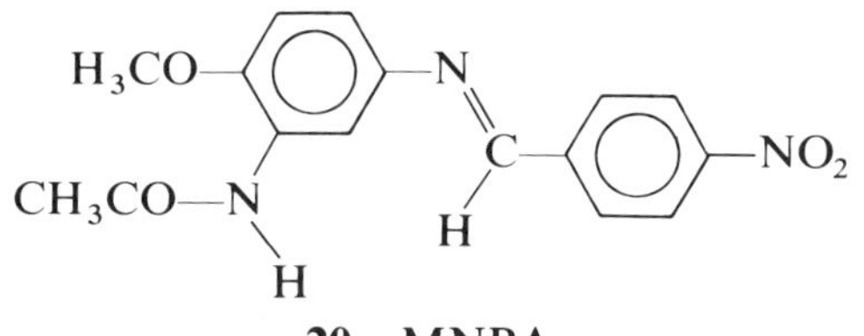

20 MNBA

Table 3.

Linear and Nonlinear Optical Data of Benzylidene-Aniline Derivatives. λ_{max} is Given in nm, β_{zzz} in 10^{-30}esu.

Compound	μ (D)		λ_{max}		β_{zzz} at 1.17 eV	
	a	b	a	c	a	b
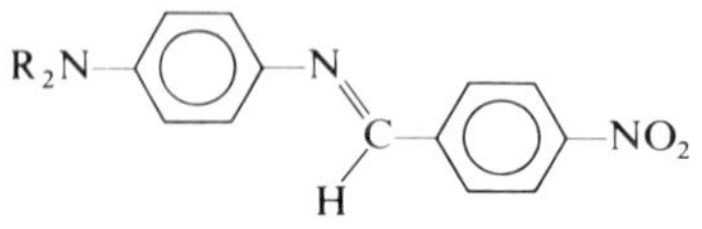 $R_2N-C_6H_4-N=CH-C_6H_4-NO_2$ **17**	6.0	7.1	410	371.0	135.5	135.3
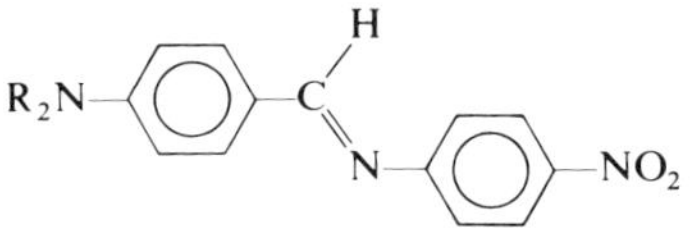 $R_2N-C_6H_4-CH=N-C_6H_4-NO_2$ **18**	13.3	9.7	335	387	77.0	31.8
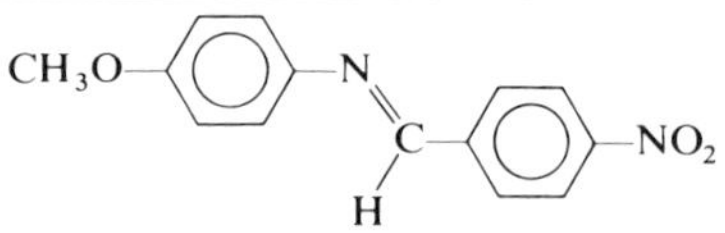 $CH_3O-C_6H_4-N=CH-C_6H_4-NO_2$ **19**	—	5.9	—	—	—	105.3

[a] Calculated values, R = H; [2].
[b] Calculated values, R = H; [49].
[c] Experimental values, R = CH_3; [50]

505 nm, which appears inadequate for blue-light generation, but the value of the d_{11} coefficient as measured by the Maker fringe method is as high as 494 pm/V [49].

2.3.4. *Hydrazones*

Hydrazones are usually obtained by coupling a substituted hydrazine to a carbonyl functionality. As a consequence, a nitrogen atom with three single σ bonds and bearing an electronic lone pair is conjugated with the π electron system of the neighboring $sp^2N{=}sp^2C$ azomethine double bond. The resulting configuration is a planar trigonal nitrogen. Thus the 1,3-diphenylhydrazone system offers a new example of a conjugation path by introduction of an adjacent filled p orbital to a benzylidene-aniline system. Here again, as in Secs. 2.32 and 2.33, two kinds of "push-pull" derivatives are possible, according to the disposition of the donor-acceptor pair, leading to derivatives of type **21a** and **21b**, β values of various 1,3-diphenylhydrazone derivatives

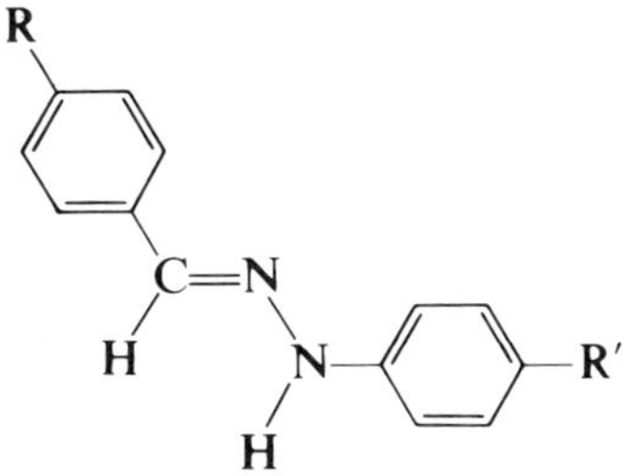

R

C=N

H N—R′

H

21

21a R = donor, R' = acceptor
21b R = acceptor, R' = donor

related to type **21a** were obtained from EFISHG experiments and from SHG measurements carried out on Langmuir-Blodgett monolayers [51–53]. The main linear and nonlinear optical data of these materials are given in Table 4. The acceptor moieties are 4-nitro- and 2,4-dinitrobenzene; the strength of the donor groups can be weak ($H_{33}C_{16}OCO$—), or quite strong (dialkylamino). As expected, on the electron-donating part of the molecule, grafting one or two additional donor groups in *meta* position with respect to the main molecular axis greatly weakens the nonlinear response of the molecule. On the contrary, on the electron accepting side, the presence of an additional nitro group in *ortho* position with respect to the molecular axis significantly improves the nonlinearity of the molecule.

Focusing on compound **24**, the transparency is similar to that of **32** [54], for a significantly larger static first-order hyperpolarizability $\beta(0)$. 1,3-diphenylhydrazones seem then to offer a slight improvement of the T/NL tradeoff as compared to the corresponding diene derivative. In addition, the significant reduction of the molecular size when going from the diene to the corresponding diphenylhydrazone could be another advantage in terms of the "hyperpolarizability density" and therefore could lead (in the case of an adequate packing) to higher $\chi^{(2)}$ values in the solid state.

It should be pointed-out that the large $\beta(0)$ values observed in the above diphenylhydrazones (almost six times higher than that of 4-NA) cannot be accounted for by a partial ICT, for example between the nitro group and one of the nitrogen atoms of the molecular axis. The conjugation should extend over all the π and the nonbonding electrons of the molecular backbone. The diphenylhydrazone moiety represents a new kind of "cross-conjugation" involving the nonbonding electron pair of the planar trigonal nitrogen atom in position 3. Some chiral and achiral SHG active crystalline derivatives of type **21a** have already been reported [2]. The other kind of 1,3-diphenylhydrazone derivatives (**21b**), to the best of our knowledge, has not been studied.

Table 4.

Static First-Order Hyperpolarizabilities Deduced from EFISH Measurements at 1.34 μm, Maximum Absorption Wavelengths of the Electronic Transition (in Solution), and $\beta(0)$ Values Deduced from SHG Experiments in LB Monolayers, for Various Phenylhydrazone Derivatives.

Molecule	No.	$\beta_{(0)}$ EFISHG 10^{-30} esu[a]	λ_{max} (nm)	$\beta_{(0)}$ *LB* 10^{-30} esu[b]
n-$C_{18}H_{37}$—O n-$C_{18}H_{37}$—O—⟨⟩—CH=N—NH—⟨⟩—NO_2 n-$C_{18}H_{37}$—O	**22**	17	390	17.6
n-$C_{16}H_{33}$—O n-$C_{16}H_{33}$—O—⟨⟩—CH=N—NH—⟨⟩—NO_2	**23**	46.3	390	
n-$C_{18}H_{38}$—O—⟨⟩—CH=N—NH—⟨⟩—NO_2	**24**	67	400	21
n-$C_{12}H_{25}$—O n-$C_{12}H_{25}$—O—⟨⟩—CH=N—NH—⟨⟩—NO_2 n-$C_{12}H_{25}$—O O_2N	**25**	22	402	23.5
n-$C_{16}H_{33}$—O n-$C_{16}H_{33}$—O—⟨⟩—CH=N—NH—⟨⟩—NO_2 O_2N	**26**	38	400	22.4
CH_2=CH—C_9H_{18}O—⟨⟩—CH=N—NH—⟨⟩—NO_2 O_2N	**27**	67.8	390	
$(H_{33}C_{16})_2$N—⟨⟩—CH=N—NH—⟨⟩—NO_2 O_2N	**28**	80	460	
n-$C_{16}H_{33}$O n-$C_{16}H_{33}$O—⟨⟩—CH=CH—CH=N—NH—⟨⟩—NO_2 O_2N	**29**	56	417	20.2
O C—⟨⟩—CH=N—NH—⟨⟩—NO_2 $H_{35}C_{17}$—O	**30**	13.5	388	
O C—⟨⟩—CH=N—NH—⟨⟩—NO_2 $H_{35}C_{17}$—O O_2N	**31**	24.6	373	
H_3CO—⟨⟩—C(H)=C(H)—C(H)=C(H)—⟨⟩—NO_2	**32**	37	397	

[a]Ledoux [52].
[b] Bubeck [53].

2.4. *sp*-Hybridized Carbon Atoms: Tolans and Diaryl-1,2-Acetylenes

The π electron system present in the carbon–carbon triple bond can be part of the conjugated system linking a donor–acceptor pair in a NLO chromophore. Diphenyl-1,2-acetylene is often called *tolan*. The tolan skeleton presents the advantage of avoiding the chemically and photochemically readily induced Z-E (*cis-trans*) isomerism that can occur within corresponding stilbenes. In addition, the acetylenic triple bond induces an hypsochromic shift in comparison with the same molecule bearing a double bond. This led several groups to investigate the NLO properties of 4,4′-disubstituted donor-acceptor diphenyl-1,2-acetylenes (**33**), also called "push-pull" tolans.

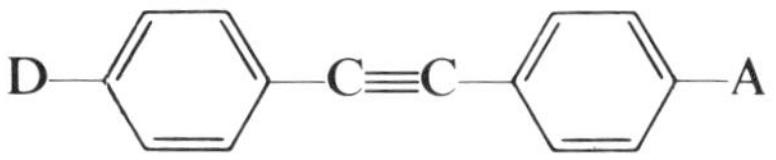

D—C₆H₄—C≡C—C₆H₄—A

33

The first report of an SHG active tolan derivative originates from NTT-Laboratories where the NLO properties of crystalline 4-methoxy-4′-nitrotolan (MONT, **34**) were studied [55,56]. Researchers from the Jet Propulsion Laboratory (JPL) screened several push-pull tolans and found them to be highly SHG efficient, like 4-amino-4′-carbomethoxy-tolan (**44**) and 4-methylthio-4′-nitro-tolan (**36**), with respectively 120× urea and 65× urea at 1.06 μm (Table 5) [57,58].

At the same time a series of push-pull tolans have also been synthesized and studied for their NLO properties by IPCMS groups in Strasbourg. Classical considerations about conjugated systems reveal that the most stable conformation of each push-pull tolan is obtained when the phenyl rings both lie in the same plane. Contrarily to biaryl or stilbene analogs, no steric interactions between the ortho-hydrogens occur here. Calculations of the molecular structure and of the heat of formation by the AM1 semi-empirical method, indicate that the resonance energy difference between an all-planar geometry, and an orthogonal orientation of the aryl rings is less than 0.3 kcal.mol^{-1} [59]. The consequence is a quasi-free rotation around each aryl-ethynyl carbon–carbon single bond, and the coexistence of all conformers at room temperature. Any value of the interplanar angles in the crystalline state can thus be reached as environmental effects can induce such torsional changes easily. This had been actually observed by Desiraju and Krishna [60] from a study of unsymmetrically substituted tolans with moderate dipole moments. Their results show that these compounds are likely to adopt

Table 5.

Linear and Nonlinear Optical Data of Push-Pull Tolans.

			λ_{max} (in $CHCl_3$)		β^e (10^{-30} esu)		SHG effic. (× urea at 1.06 μm)	
A	D	No.	a	b	a	b	c[f]	b
NO_2	OMe	**34**	357	347	25	14	++	MONT[d]
NO_2	NMe_2	**35**	415	416	42	46	0	—
NO_2	SMe	**36**	362	358	29	20	+++	65
NO_2	Me	**37**	342.5	—	21	—	++	—
NO_2	Br	**38**	335	—	21	—	++	—
CN	NMe_2	**39**	373	373	32	29	0	—
CN	OMe	**40**	328	—	13	—	0	—
CN	SMe	**41**	334	326	15.5	15	0	0
CN	Me	**42**	314	—	10	—	+	—
CN	Br	**43**	310	—	9	—	+	—
COOMe	NH_2	**44**	—	337	—	15	—	120
COMe	NH_2	**45**	—	336	—	12	—	39
SO_2Me	NH_2	**46**	—	338	—	13	—	0

[a] Barzoukas [58].
[b] Stiegman [57].
[c] Nicoud [59].
[d] Kurihara [55].
[e] β stands for $\beta(0)$ in (a) and at 1.91 μm in (b).
[f] Given the broad range of sizes of crystalline particles in the samples, and possible preferential orientation, the powder tests were not quantified: + denotes a signal comparable to or a few times greater than that of urea, whereas ++ and +++ refer to one order and two orders of magnitude greater signals respectively. 0 denotes no eye-detectable SHG signal.

chiral nonplanar conformations in the solid state, which seems to lead to a high proportion of noncentrosymmetric crystal structures. The effect of the free rotational distortion of the diaryl-acetylene backbone on the molecular hyperpolarizability has been calculated by a Finite Field MNDP method

[5]. The results showed that even when the two aryl rings are perpendicular to each other, the charge-transfer interaction is still present, the residual hyperpolarizability being only half that of the maximum obtained for a full planar conformation. We can then conclude that whatever the conformation of the push-pull tolan in the solid state is, there always remains a significant hyperpolarizability. The β values of ten different push-pull tolans were measured by the EFISH technique, pointing out that the methylthio (MeS) group leads to a noticeable increase of β without significant loss of transparency in comparison with the methoxy (MeO) group and that the bromo substituent, though electronegative, can allow for a relatively large β and good transparency when opposed to the cyano and nitro groups. Some selected results are reported in Table 5. Several SHG powder active compounds, which correspond to a high proportion of noncentrosymmetric crystal structures and seem to confirm the analysis of Desiraju and Krishna [60], have been observed.

In order to investigate a possible tuning of the linear and nonlinear optical properties of donor-acceptor tolan analogs, several new push-pull diaryl-1,2-acetylenes of type **47**, where the aryl ring is a pyridyl or a phenyl ring, were synthesized [61]. All the compounds prepared present a blue-shifted charge transfer band as expected according to Dewar's rules when the pyridyl ring is bearing the donor group, but a surprising red shift, against Dewar's rules, when the pyridyl group is placed on the acceptor side. The β values, measured by the EFISHG technique at IBM-Almaden Laboratories, are summarized in Table 6 [62]. The results confirm that the ethynyl bond ensures the conjugated linkage between the donor and the acceptor groups, and that the two aryl rings behave relatively independently of each other. The nitro-phenyl ring appears to be a lesser acceptor group than the nitro-pyridine one. Several chiral derivatives of **47** have been prepared, while not leading to highly SHG efficient crystalline materials.

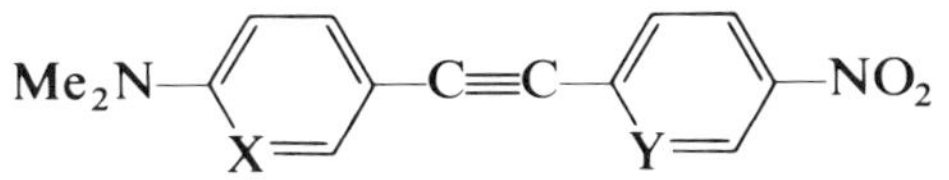

Me_2N–(X-ring)–C≡C–(Y-ring)–NO_2

47

In conclusion, the push-pull diaryl-1,2-acetylene derivatives, in which two *sp*-hybridized carbon atoms ensure the communication between a donor and an acceptor moiety, have been extensively studied. The tolan skeleton, and more generally the diaryl-1,2-acetylene moiety remain interesting building blocks for introducing hyperpolarizable chromophores in

Table 6.

Linear and Nonlinear Optical Data of Pyridine-Containing 1,2 Diaryl-acetylenes.[a]

X	Y	Number	μ_g (D)[b]	λ_{max} (nm) (in EtOH)	ε_{max} ($M^{-1}\,cm^{-1}$)	$\beta(0)$ (10^{-30} esu)
CH	CH	**35**	7.09	404	25.000	37
N	CH	**48**	7.15	384	23.200	27
CH	N	**49**	7.16	424	24.500	34
N	N	**50**	6.11	400	27.500	27

[a] Data from Nicoud, Twieg, Moylan, and Burland, [62].
[b] All μ_g in dioxane except $X = Y = N$ in $CHCl_3$.

side-chain $\chi^{(2)}$ NLO polymers [63] or in polymeric chains for $\chi^{(3)}$ effects [64].

3. OPTIMIZATION OF QUADRATIC AND CUBIC NONLINEARITY: POLYENES

3.1. Introduction

Early theoretical and experimental research had pointed out the interest of increasing the extension of the conjugated systems in order to enhance the molecular β and γ tensors [65]. Within this domain of investigation, polyene oligomers are of particular interest as model systems of one-dimensional conjugated chromophores [66–72]. This family of molecules displays the largest first-order hyperpolarizabilities reported to-date [67–68] and could then open interesting perspectives for improving the present performances of devices based on the linear electro-optic (Pockels) effect, such as modulators and switches. In view of the wavelength values currently used in optical communication devices (typically 1.5 μm), there is no limitation of the nonlinear optical performances due to the proximity of electronic resonances, which are mostly located within the visible spectral range. Polyenes also exhibit significant third-order polarizabilities, leading to various applications in the field of all-optical signal processing.

The main questions arising from experimental studies and related models are the following:

- Is there an "asymptotic" limit for β and γ when the number of double bonds n extends infinitely? If such is the case, there should be an optimal conjugated system dimension beyond which extending the chain length would not only be pointless, but even detrimental in terms of hyperpolarizability density. Such a saturation is expected for asymmetric conjugated systems exhibiting a strong intramolecular charge transfer between an electron donor and an electron acceptor group through the π electron linkage; the ability of substituents to communicate must break down when the length of the connecting conjugated pathway rejects them too far away from each other to establish the charge transfer connection. This trend has been rationalized in the case of cubic properties of conjugated polymers by the introduction of the electron delocalization length defined as the quadratic mean deviation of the electron position in a one-electron model as derived from the chain bond alternation [73]: extension of the chain length beyond the delocalization length, which is found to be of the order of a few repeat units in usual cases, will not increase the oligomer susceptibility owing to such saturation effects.
- The precise conjugated chainlength dependence for β and γ before the onset of saturation effects and for various conjugated systems as well as substituent groups is of considerable interest in view of both applied and fundamental types of issues. Earlier models based on a dispersionless one-electron [65] description of light-molecule interactions may be too crude. In addition, the issue of asymmetric molecules and related β tensors has not been addressed within this framework.

A more extended study of the second- and third-order polarizabilities will include various types of symmetrical or unsymmetrical carotenoids, with increasing chain length up to 11 double bonds in the conjugated backbone. This allows for a more systematic investigation of the influence of the conjugated length and of the nature of the end substituents on the quadratic and cubic nonlinear response of these exceptionally active molecules.

3.2. Polyenes for Quadratic Nonlinear Optics

A wide choice of ICT polyenes for quadratic nonlinear optics is represented on Fig. 5. The molecules bear various donor and acceptor groups, linked

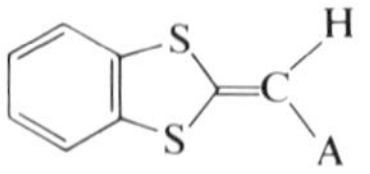

Fig. 5. Structural formulae of the asymmetric polyene series.

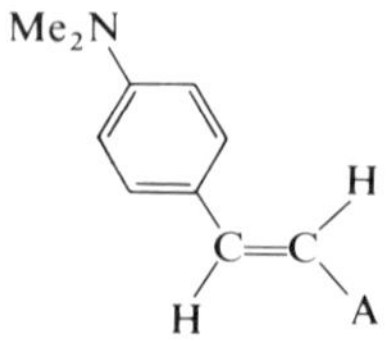

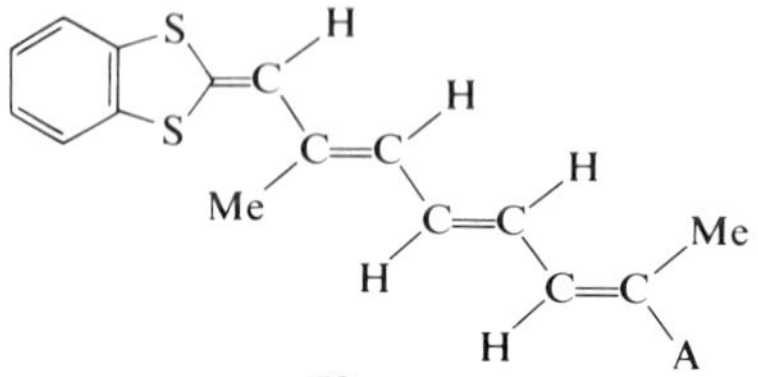

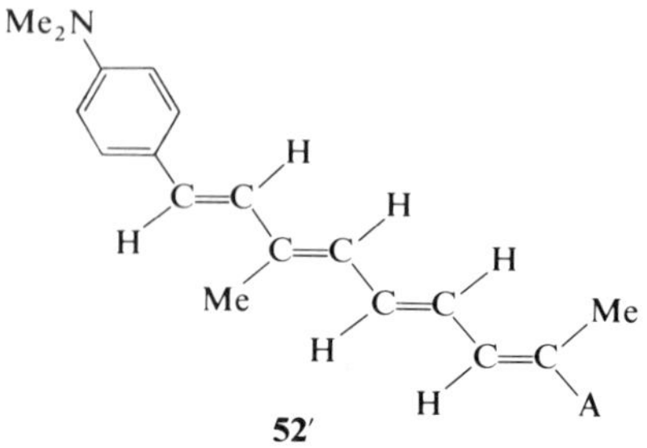

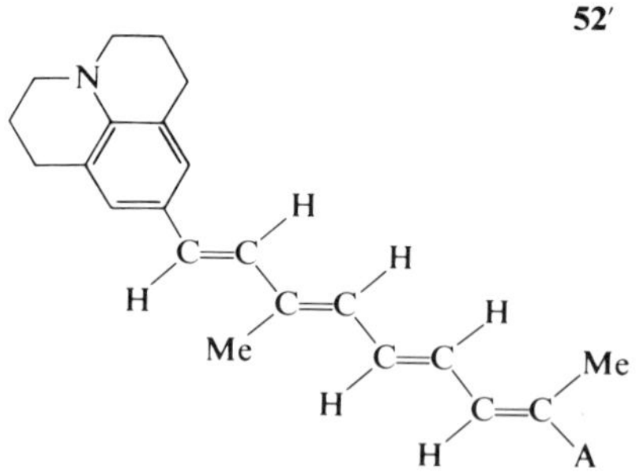

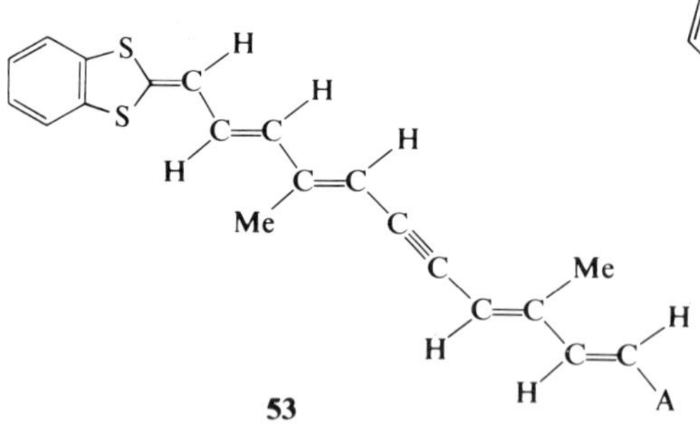

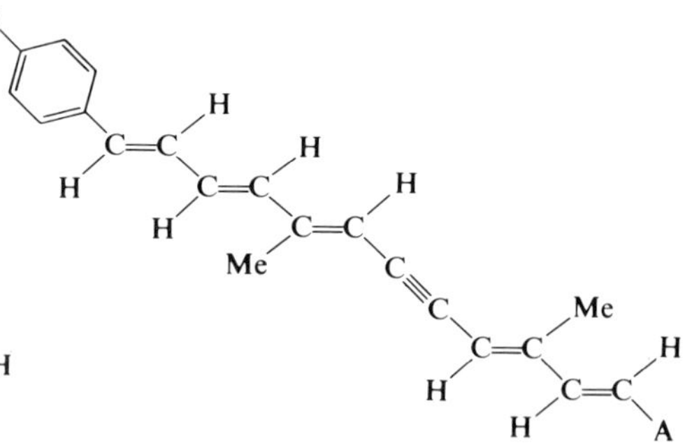

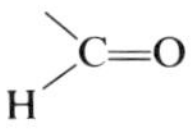

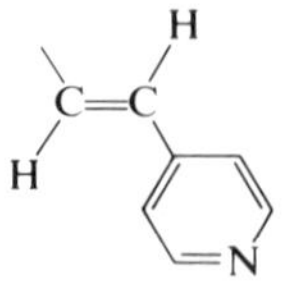

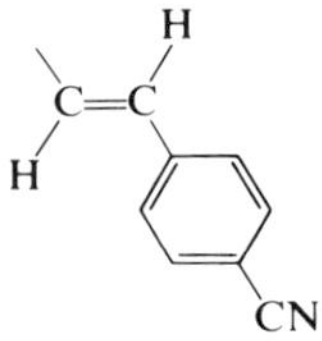

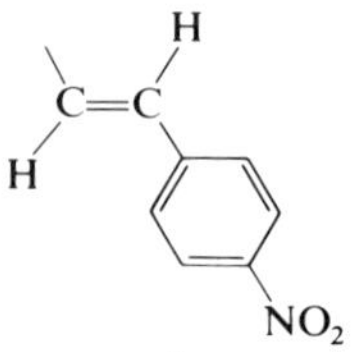

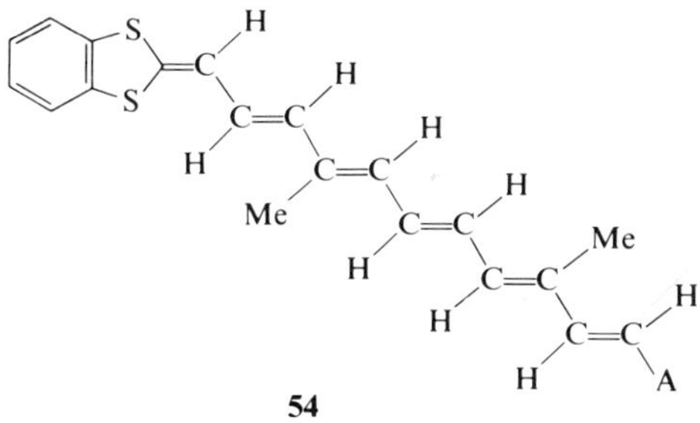

Fig. 5. *continued*

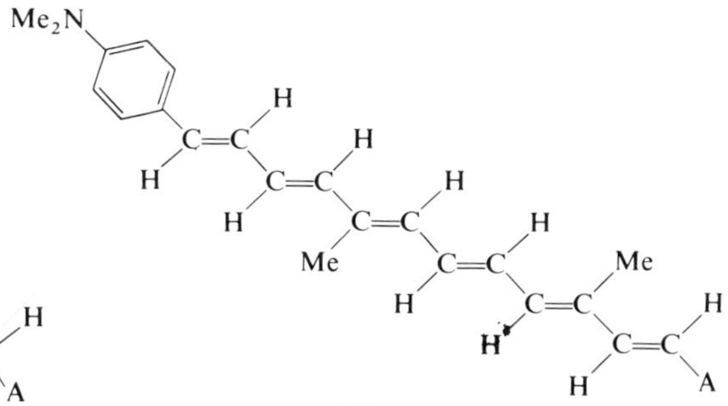

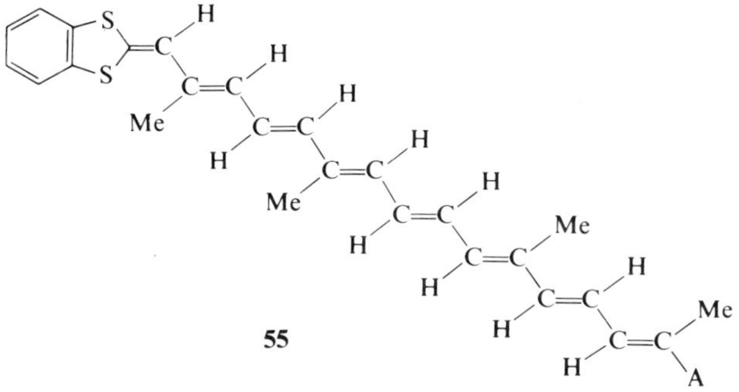

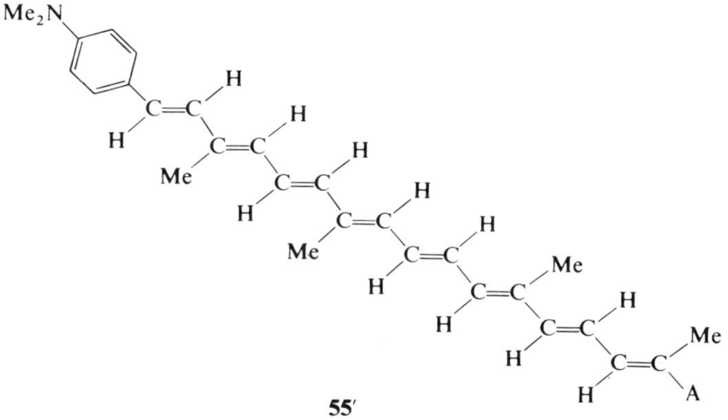

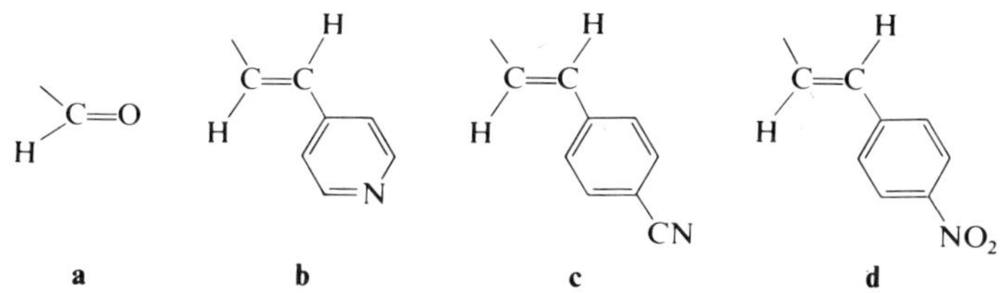

together by polyenic chains of different length. The donor groups are benzodithia (compounds **51**–**55**), *N*,*N*-dimethylanilino (compounds (**51**′–**55**′) and julolidino (compound **52**″). The chemical formulae of the accept groups are given as **a**, **b**, **c**, **d**. The effect of the triple bond is compared to that of the double bond.

Semi-empirical computations carried out on dimethylaminopolyenals and dimethylaminopolyenes [34] evidenced a rapid increase of β with the number of double bonds n, with a saturation around $n = 20$. The polymers display an efficiency more than one order of magnitude higher than that of the polyphenyl system. Synthesis and nonlinear optical characterization of asymmetric carotenoids should bring a confirmation of these predicted trends, along with a fine analysis of the influence of the strength of donor and acceptor groups.

All the experimental results, obtained by EFISH measurements at 1.34 μm, are collected in Table 7 [67,68]. The β tensor is supposed to be one-dimensional along the CT axis. The EFISH experiment allows for the determination of γ_0:

$$\gamma_0 = \gamma_e(-2\omega; \omega, \omega, 0) + \frac{\mu\beta(-2\omega; \omega, \omega)}{5kT}.$$

The first term is the scalar part of the cubic hyperpolarizability tensor γ_{ijkl}, whereas the second term originates from the partial orientation of the permanent dipole moment μ in the static field. The orientational contribution is usually assumed to be predominant over the electronic term γ_e for small molecules; this could be different for large molecular systems, such as polyenes with a large conjugation length; in that case, the contribution of the cubic hyperpolarizability could be significant (larger than 30 percent of the total value). Then we will assume that the EFISH results are fully valid, for determination of the $\mu\beta$ values, only for a limited number of double bonds (not more than four).

Table 7 reports the values of the product $\mu\beta$, the measurement of the dipole moments being not possible, owing to the poor solubility of the longest compounds. However, it appears from theoretical considerations on benzodithia derivatives [74], that the dipole moment (estimated to lie in the 6 to 8 D range) [75] is almost constant with respect to n ($\mu \propto n^{0.1}$).

For each series of molecules, a red shift of the CT band is observed together with a substantial increase of the quadratic hyperpolarizability with n. The estimated $\beta(0)$ reaches a value as high as 50 times that of 4-nitroaniline. The dependence of $\beta(0)$ on n for both series can be approximated by $\beta(0) \propto n^{2.4}$.

For compounds **51′a**–**55′a** a similar trend is taking place, although the

Table 7.

Maximum Absorption Wavelength λ_{max} (in nm), and $\mu\beta(2\omega)$ (resp. $\mu\beta(0)$) Values at 1.34 μm (resp. at Zero Frequency) (in 10^{-48} esu) for Various Asymmetrical Polyenic Molecules.

Molecule	λ_{max}	$\mu\beta$ (2ω)	$\mu\beta(0)$
51a	372	30	20
52a	456	1200	570
53a	466	2200	1000
54a	485	2700	1100
55a	500	7250	2800
51′a	384	320	200
52′a	450	2000	1000
53′a	461	4200	2000
54′a	498	8900	3400
52″a	480	2900	1200
52b	457	1500	715
51c	410	250	140
52c	465	1950	900
51d	452	1000	480
52d	488	2200	900

value for molecule **51′a** falls out of the line. It should be noted that a phenyl ring has been arbitrarily included in the donor group, an approximation not valid for short chain lengths. These results are in good agreement with earlier experimental data reported by Dulcic *et al.* [76], where a nearly quadratic dependence of β is shown, and also with computed results given by Morley *et al.* [34].

The rise in $\beta(0)$ is steeper for the series bearing the less-efficient donor group; the difference in efficiency between donor substituents reduces as the chain length increases. This result may also be applied to series substituted with different acceptor groups, as illustrated by the $\mu\beta(0)$ values obtained for molecules **51c**–**52c** and **51d**–**52d**. The difference in donating or accepting power between donor or acceptor groups is smeared out at longer chain lengths. The efficiency of the donor-acceptor interaction appears to level off as n increases.

On the other hand, experimental results on molecules **53a** and **54a** show a decrease in the static hyperpolarizability, when a triple bond is included in the conjugated path instead of the double bond. The hypsochromic effect of the triple bond and the drop in the quadratic nonlinearity indicate a reduction in the electronic ICT.

The exceptionally large $\mu\beta(0)$ values of long polyenic systems evidence the potential of such compounds as candidates for inclusion in thin films such as LB or poled polymer films. However, the crucial problem of chemical stability must be solved by using specific synthetic strategies, in order to take full advantage of the huge nonlinearities of these compounds for applications in active optical devices.

3.3. Polyenes for cubic nonlinear optics

Polyenic molecules appear to exhibit the largest third-order optical nonlinearities among the various organic structures presently available for this type of application. $\chi^{(3)}$ up to 3×10^{-6} esu has been reported for thin films of polyacetylene [77]. These values lie far beneath these displayed by semiconductor-based multiple quantum wells; however, within the transparency domain, organic materials exhibit a much faster response, linked to the purely electronic character of the optical susceptibilities. It seems to be difficult to increase further the third-order nonlinear optical response of organic materials without performing a systematic study of the various parameters underlying the optimization of the corresponding susceptibility, including the nature of the conjugated unit, influence of the chain length, and role of substituents. For example, it would be of crucial interest to verify experimentally the possible saturation of the hyperpolarizabilities above a given value of the number of double bonds n for polyenic moieties. Another point would be to compare the role of donor and acceptor substituents, symmetrically or asymmetrically substituted.

In the following the influence of these parameters is investigated, using a wide choice of polyenic molecules with various chain lengths or substituents.

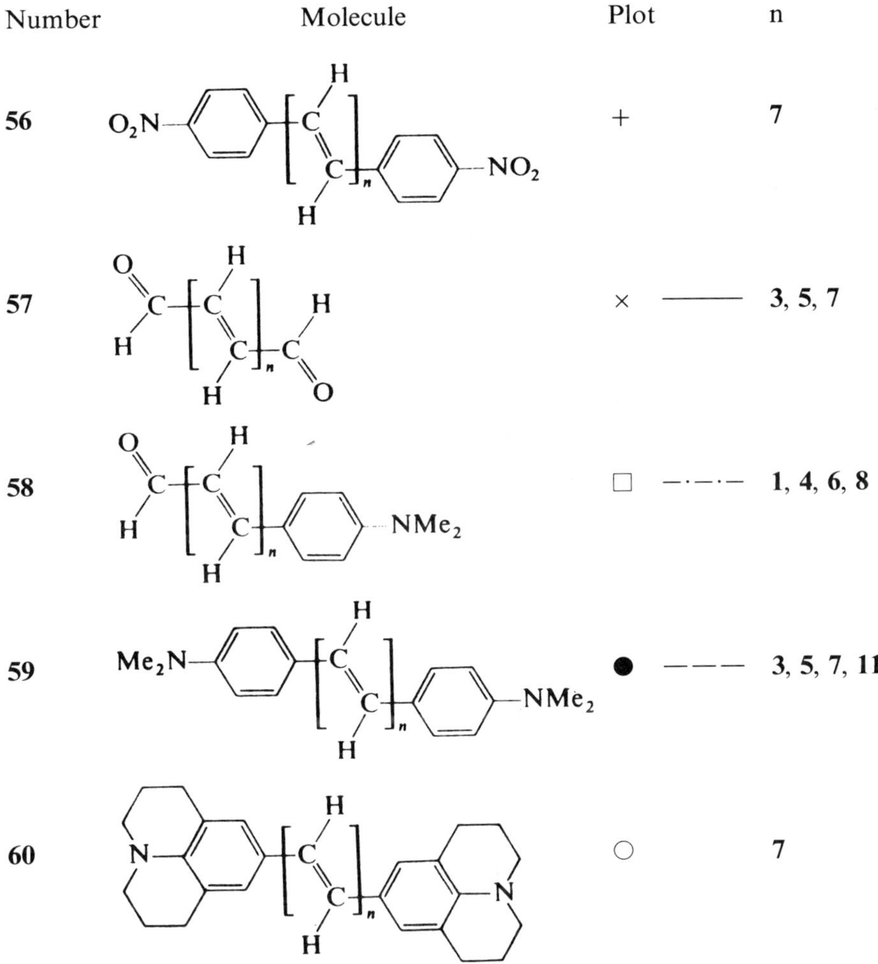

Fig. 6. Structural formulae of symmetrical and asymmetrical polyenes for γ measurements.

A list of the molecules to be studied is given in Fig. 6. The typical donor (resp. acceptor) substituent is *N*,*N*-dimethylamino (resp. formyl) group. For comparison, a strong electron donor (resp. electron acceptor) group julolidino (resp. nitro) is used for the same conjugation path $n = 7$. A comparison between the behavior of the γ values as a function of n is made for different pairs of donor-donor, acceptor-acceptor and donor-acceptor substituents [78].

γ measurements are performed using the EFISH experiment for symmetrical molecules, whereas only third harmonic generation (THG) is able to infer the second-order susceptibility values for asymmetrical structures, in order to avoid any $\mu\beta$ contribution. EFISH measurements use $\lambda = 1.34\ \mu m$ as a fundamental wavelength; THG experiments are performed at 1.91 μm; in both cases, the harmonic frequencies are sufficiently far away from the resonance to allow for neglecting the imaginary part of the third-order nonlinearity.

The dependence of the third-order polarizability γ as a function of n, for three couples of substituents: A–A (A = CHO); D–D (D = dimethylamino) and the corresponding D–A compound, has been studied. For comparison, the γ values of the polyenes substituted by the more electron accepting group nitro (resp. more electron withdrawing group julolidino) were also measured (for $n = 7$) and compared to the corresponding values of symmetrical formyl and dimethylamino derivatives.

It may be relevant to plot the different γ values as a function of N, where N takes into account the contribution of the phenyl rings to the conjugated path. A crude approximation could be to consider that one phenyl ring is equivalent to 1.5 double bonds [79]. Therefore, we will replace, for D-A (resp. D–D) substituents, n by $N = n + 1.5$ (resp. $N = n + 3$). The same procedure will be applied to the dinitro and julolidino compounds, with $N = n + 3$. Fig. 7 illustrates the corresponding behavior of the various polyenic molecules after these corrections. The slopes are 2.3, 3.4, and 4.6 for A–A, D–A and D–D compounds, respectively. For the D–A compound, it must be pointed out that the slope of $\mathrm{Ln}\,\gamma = f(\mathrm{Ln}\,N)$ comes quite close to that (3.5) observed by Messier *et al.* [72] with the same acceptor group, the donor substituent being benzodithia. In addition, the dinitro compounds come close to, although still slightly above, the line corresponding to the formyl derivatives. The julolidino derivative is found to be much more active than the dimethylanilino one, owing to the much stronger donor character of the fully planar amino moiety in the julodinine group.

The slope of the curve $\mathrm{Ln}\,\gamma = f(\mathrm{Ln}\,N)$ increases when going from electron-accepting to electron-withdrawing substituents. This is consistent with the fact that donor groups tend to push the external electrons away from the molecular backbone, then increasing the (hyper)polarizability of the conjugated system. The γ values increase much faster with N with electron donor substituents than with acceptor ones. However, it should be pointed out that the lines describing the behavior of D–D and A–A derivatives intersect for $N = 5.6$. Below this value, γ of diacceptor derivatives surpass those of didonor ones. A similar phenomenon was observed by Spangler *et al.* [54], where γ values of dinitro polyenic derivatives were found to be larger

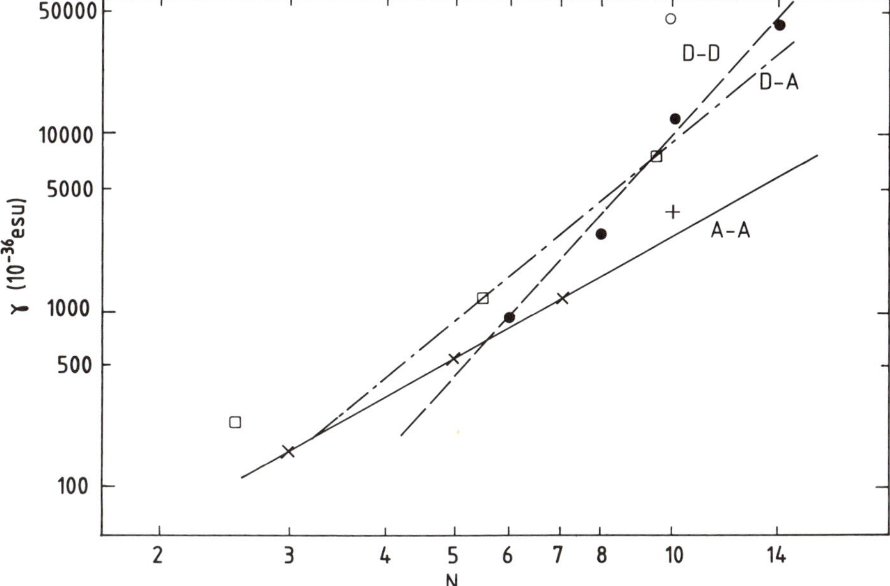

Fig. 7. Third-order polarizabilities of *A–A*, *A–D* and *D–D* polyenic derivatives (where *A* is a formyl group and *D* is a dimethylamino one), as a function of the "equivalent" number of double bonds N ($N = n + 1.5$ for *D–A* compounds, and $N = n + 3$ for *D–D* compounds) (logarithmic scale). + and o correspond respectively to the dinitrophenyl and the di-julolidino derivatives, for $n = 7$.

than these of dimethylamino ones. In addition, preliminary *ab initio* calculations carried out on these compounds seem to evidence a similar behavior, at least for small N values. Asymmetric compounds also exhibit much larger γ values as compared to the other ones, the inversion of this tendency occurring only for $N = 9.5$.

No saturation of γ could be observed up to $N = 14$. Huge synthesis and solubility problems prevents the availability of large conjugation lengths for this type of molecule.

More recently, an attempt to further increase the conjugation length, using triblock synthesis [80], towards elaboration of polyenes substituted by norbornene groups, led to availability of 13 conjugated double bonds. EFISH measurements of the γs of these compounds at 1.34μμm, for various n values, led to a dependence in $n^{3.6}$ [81]. This exponent is close to that of *D–A* polyenic derivatives.

4. INFLUENCE OF SHAPE FACTORS

4.1. Chirality

The oldest strategy in the so-called "molecular and crystal engineering" for $\chi^{(2)}$ NLO materials is chirality. As is well known, it provides the synthetic chemist with a means of guaranteeing that crystallization of a pure enantiomer will occur in a noncentrosymmetric point group. Such effect can be considered as the consequence of geometry, although chirality itself is not the only parameter but is often associated to other effects such as hydrogen bonding [5]. However, the fact that a molecule is optically pure does not guarantee that the molecular packing will be adequate for NLO effects, but only that we have a 100 percent probability of getting a noncentrosymmetric crystal structure. It is essentially the chirality of sp^3 tetrahedral carbon atom(s) that has been widely utilized so far. Methyl-(2,4-dinitrophenyl)-aminopropanoate (MAP, **61**) was the first deeply studied $\chi^{(2)}$ NLO material that exploited the chirality strategy [82]. Several other chiral amino-nitro-aromatics have been screened for their SHG activity, and it is not our purpose to make a comprehensive review [2,61].

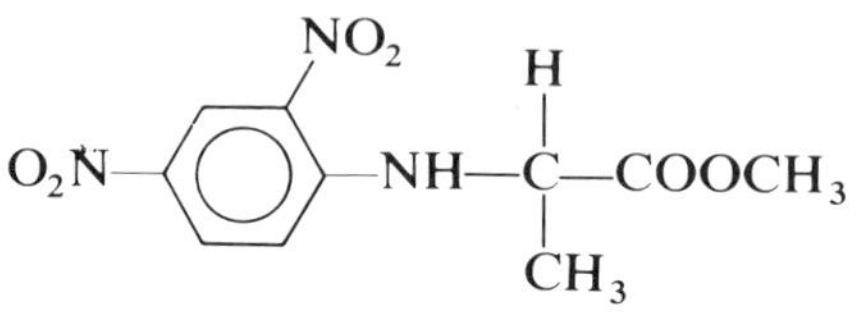

NO_2 H

O_2N—⟨benzene ring⟩—NH—C—COOCH$_3$

CH_3

61 MAP

4.2. Inactive bulky groups

One-dimensional NLO chromophores are very often highly polar in their ground states, since the charge transfer between the donor and the acceptor groups already exists in the ground state. An exception to this rule is given by using a push-pull substituent, such as in 4-nitropyridine-1-oxide derivatives in which the ground state dipole moment may vanish without cancelling the intramolecular charge transfer. 3-methyl-4-nitropyridine-1-oxide [83] (POM, **62**) was the first organic $\chi^{(2)}$ NLO crystalline material that exemplifies with success this strategy. It was indeed supposed that dipole–dipole interactions between highly polar molecules would favor head-to-tail packing, and hence lower $\chi^{(2)}$ properties even in the case of an acentric packing. This classical idea, however, should be used cautiously according to a recent study [84]. In order to reduce the probability of antiparallel stacking, amphiphilic molecules in which a neutral bulky group alternates with the hyperpolarizable polar moiety have been prepared. 2-cyclooctylamino-5-nitropyridine (COANP, **63**), space group $Pca2_1$, illustrates this crystal engineering. The molecular amphiphilic property gives rise to a packing where the polar and neutral parts alternate in layers perpendicular to the *a* axis [85,86]. The new *N*-(5-nitro-2-pyridyl)-(S)-phenylalaninol (NPPA, **64**), whose highly efficient SHG activity had been reported by Nicoud [61] can be classified in that category of materials. In the crystal the conformation of the molecules is such that the benzene ring bends itself over the pyridine ring to form a scorpionlike structure. Thus the benzyle moiety can be considered as an inactive bulky group that contributes in inducing a favorable crystal packing. In fact, two intermolecular hydrogen bonds N—H···O(OH) and O—H···$O(NO_2)$ tightly hold the molecules in each layer [87,88].

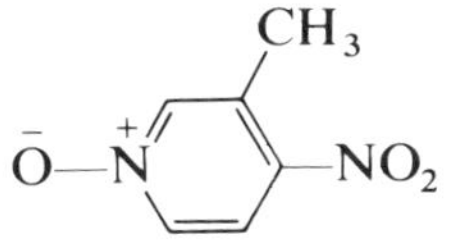

62 - POM

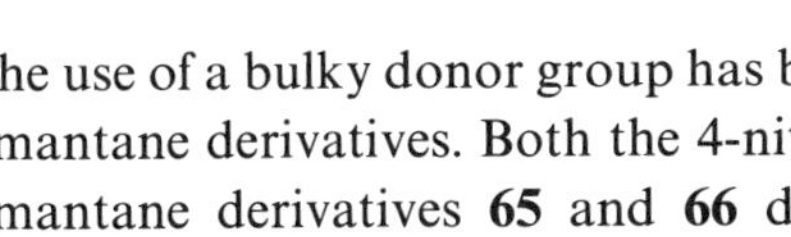

63 COANP

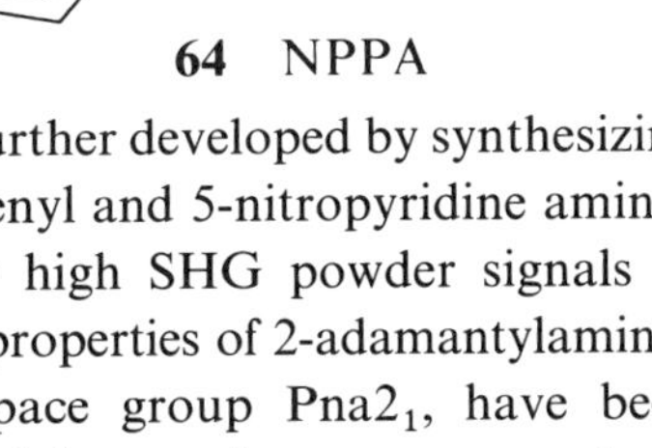

64 NPPA

The use of a bulky donor group has been further developed by synthesizing adamantane derivatives. Both the 4-nitrophenyl and 5-nitropyridine amino-adamantane derivatives **65** and **66** display high SHG powder signals at 1.06 μm [61]. The efficient nonlinear optical properties of 2-adamantylamino-5-nitropyridine (AANP, **66**), crystals of space group $Pna2_1$, have been confirmed by NTT-Laboratories. By second harmonics measurements at 1.06 μm, the d_{31} and d_{33} coefficients were determined to be 80 and 60 pm/V, which compares to NPP [89]. The molecular orientation of the charge transfer axis is 58°, which comes close to the optimum angle of 54.74° for the mm2 crystalline point group. Thus the introduction of a bulky inactive group in a nonlinear chromophore, though diluting the density of active

NLO moieties in the final crystal, has proved to be an efficient crystal engineering strategy.

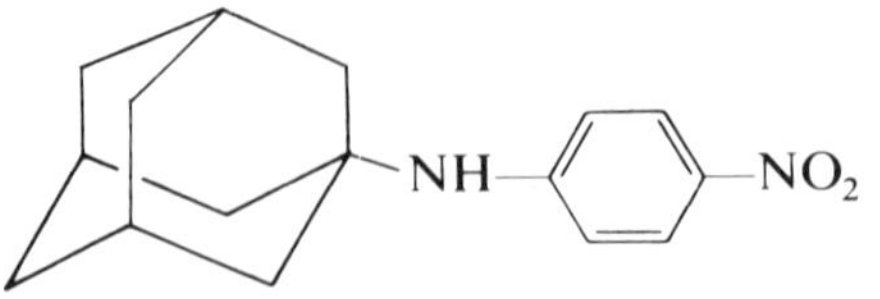

NH—⟨⟩—NO_2

65

NH—⟨N⟩—NO_2

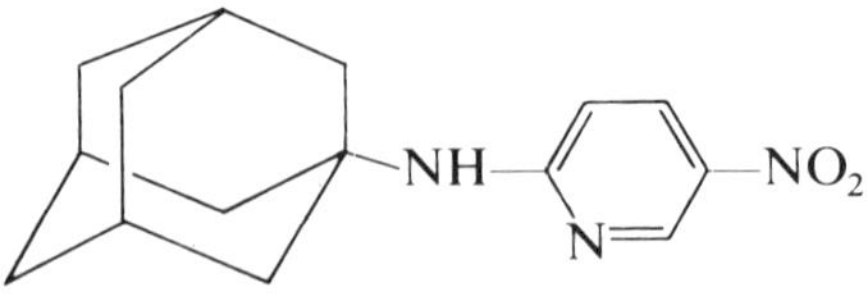

66 AANP

4.3. Nonplanar polychromophoric systems

4.3.1. Polysiliciated compounds

The interaction of more than two aromatic systems via a silicon atom and the resulting nonlinearity has been reported by Mignani *et al* [19] whereby compounds **67** and **68** in Table 8 have been compared to their monoaromatic counterparts **6** and **7**. The nonplanar geometry resulting from the silicon sp^3 hybridized orbitals almost completely breaks down the interaction between the donor dimethylamino and acceptor dicyanovinyl substituent groups as demonstrated in Table 8 by the validity of a vectorially additive model.

In this model, the vectorial component of the β tensor for each monoaromatic submoiety of **67** and **68** has been replaced by that of the equivalent trimethylsilyl monosubstituted molecule **6** and **7**, as independently measured by EFISH. β^Z_{add} is projected along the vectorially added dipole moments of **6** and **7** and defined as

$$\beta^Z_{add} = \frac{(\vec{\beta}_1 + \vec{\beta}_2)\cdot(\vec{\mu}_1 + \vec{\mu}_2)}{\|\vec{\mu}_1 + \vec{\mu}_2\|}.$$

The good agreement between calculated and experimental values confirms the low level of interaction between donor and acceptor moieties. Although the Si ← A contribution is predominant, the Si ← D one is nonnegligible, and both remain confined within each moiety.

4.3.2. The lambda-shaped concept

All the previous ONL molecules that have led to $\chi^{(2)}$ materials as crystals or in any other condensed phases are one-dimensional charge-transfer molecules. In that case it is well known that even when an optical noncentrosymmetric molecular orientation for phase-matching occurs in the bulk phase, only 38 percent of β_{xxx} can be utilized in the phase-matched process. This loss of NLO activity has led researchers to design a new molecular shape that could more easily induce noncentrosymmetric structures

Table 8.

Dipole Moment (in Debyes), Experimental Hyperbolarizability β_Z (at 1.17 eV) and Extrapolated at Zero Frequency by a Two-Level Model $\beta(0)$) and Theoretical β_{add} as from a Vectorially Additive Model, for Various Mono- and Trisubstituted Silane Derivatives. Vectors alongside Molecular Structures Represent Dipole Moments and Charge Transfer Orientations with usual Conventions (from Donor to Acceptor or from Electronic Charge Vacancies to Charge Accumulation).

Molecule	No.	μ (D)	β_Z (1.17 eV) (10^{-30} esu)	$\beta_{(0)}$ (10^{-30} esu)	β_{add} (10^{-30} esu)
$(CH_3)_2N-C_6H_4-SiMe_3$	**6**	1.76	3.8 ± 1	4	—
$(CN)_2C=CH-C_6H_4-SiMe_3$	**7**	5.0	12 ± 3	9	—
$Me_2N-C_6H_4-Si(Me)(C_6H_4-CH=C(CN)_2)_2$	**67**	7.1	22.5 ± 5	13	12
$(NC)_2C=CH-C_6H_4-Si(Me)(C_6H_4-NMe_2)_2$	**68**	6.6	18 ± 4	16	13

together with an optimization of the phase-matching properties. Miyata and coworkers recently proposed the lambda (Λ)-shaped molecular concept for that aim [90]. In a Λ-shaped molecule, two independent hyperpolarizable moieties are linked by σ bonds (for example a methylene bridge CH_2), so that the intramolecular charge transfer contributing to β has almost a two-dimensional character. Therefore the largest molecular hyperpolarizability tensor β_{ijk} is a nondiagonal component. It has been shown that Λ-shaped molecules stack easily along one direction and often crystallize into noncentrosymmetric space groups. In addition, the Japanese group found that in their most stable conformation such molecules display an internal angle of about 120° between the two charge transfer axes, coming close to an optimal molecular orientation for phase-matching. Therefore, when Λ-shaped molecules crystallize in a noncentrosymmetric crystal structure, it is claimed that the full β_{xyy} coefficient can be utilized. The simplest molecule based on this concept, in which two 4-nitroaniline moieties are bonded together via a methylene bridge has led effectively to an efficient NLO $\chi^{(2)}$ material: *N*,*N*′-bis-(4-nitrophenyl)-methanediamine (NMDA, **69**) takes a Λ-shaped conformation and seems to form a noncentrosymmetric crystal structure as it easily stacks along the same direction. Several other materials based on Λ-shaped molecules exhibit efficient $\chi^{(2)}$ NLO properties. Among these, *N*,*N*′*-bis*-(4-ethylcarboxyphenyl)-methanediamine (ECPMDA, **70**) possess large $\chi^{(2)}$ coefficients for the unusual tetragonal $I4_1cd$ space group (25 × urea by SHG powder test), a low cutoff wavelength of 330 nm and the possibility of noncritical phase-matching. This new crystal should appear as a good NLO material for blue-light emission by frequency doubling of semiconductor laser light.

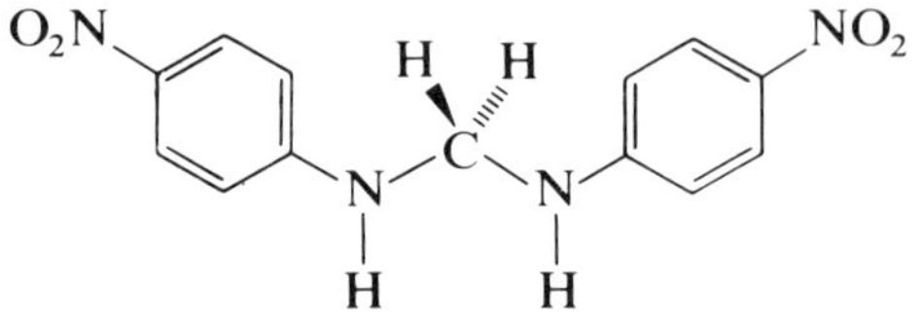

O_2N H H NO_2 C N N H H

69 NMDA

H H C N N H H O C O H_5C_2 O C O C_2H_5

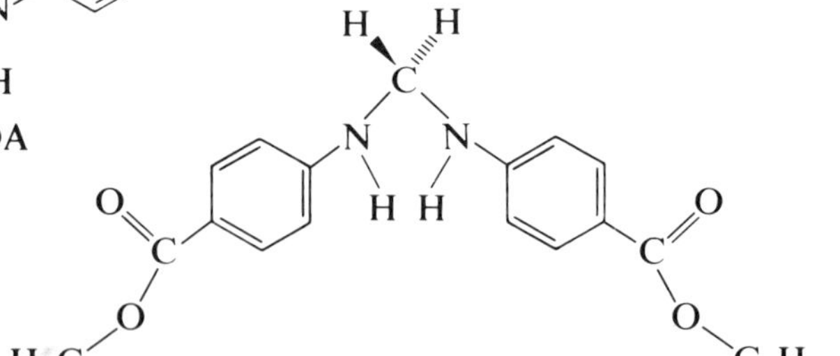

70 ECPMDA

4.3.3. The roof-shaped concept

Another Japanese group has recently developed a new strategy in NLO molecular engineering in order to control the molecular alignment of the NLO chromophores leading to optimized phase-matching properties. Dihedral rigid molecules bearing two 4-nitroaniline moieties rigidly connected through a bicyclic skeleton having a *cis* C*sp*^{3}-C*sp*^{3} junction have been designed. Thus the molecule consists of two rigid planar parts making a theoretical angle of 109.5°. Such dihedral rigid molecules understandably bear the nickname of "roof-shaped" molecules. Once again, one of the simplest molecules synthesized according to the roof-shaped concept, 3,9-dinitro - 5*a*, 6, 11*a*, 12 - tetrahydro[1, 4]benzoxazino[3, 2 - *b*][1, 4]benzoxazine (DNBB, **71**), led to a new highly efficient nonlinear crystal that exhibits, according to the authors, larger $\chi^{(2)}$ properties than that of NPP [91]. However, the cutoff wavelength of DNBB material is 500 nm, probably owing to the bathochromic effect induced by the ortho oxygens, as can be anticipated by Dewar's rules. It may be noticed that the DNBB molecule is chiral, with two asymmetric sp^3 carbon atoms providing the *cis* junction, which is symmetric with respect to a C2 axis.

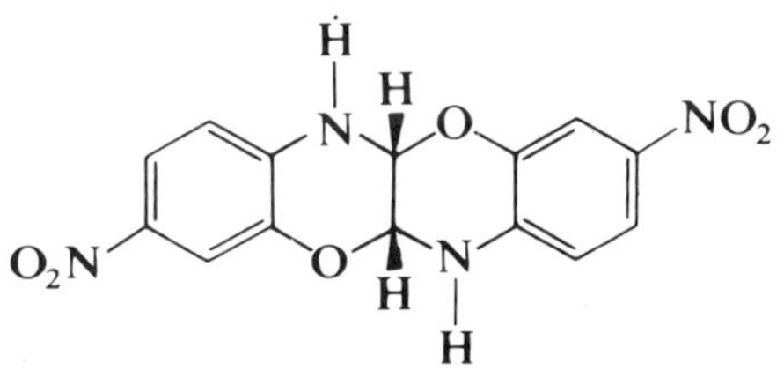

H H N O NO_2 O_2N O N H H

71 DNBB

New molecules, combining the shape factors reviewed here, i.e., C2 symmetry, Λ- and roof-shaped concepts, have been reported recently by AT&T-Bell-Laboratories [92]. The author's aim was to prepare NLO chromophores that could lead to an increased polar order achievable by applying an electric field. A 2,5-endo-disubstituted norbornane bearing two 4-nitroaniline moieties (**72**) was synthesized stereospecifically, in the anticipa-

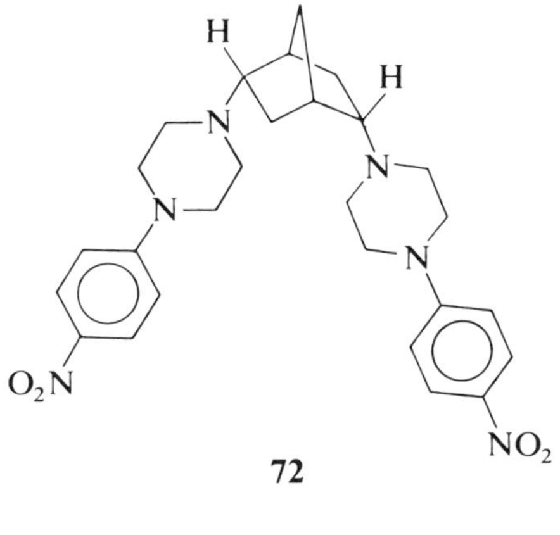

H H N N N N O_2N NO_2

72

tion that the rigid molecular backbone would force the two chromophores to adopt parallel directions. The x-ray structure of the analog without the nitro groups shows a deviation from the expected parallelism leading to an angle of almost 90° between the two amino-phenyl moieties. The molecule then adopts a Λ-shaped conformation built on a roof-shaped norbornane skeleton and accommodates a C2 symmetry axis. Here, too, this C2 symmetric molecule is chiral.

4.4. Mesogenic materials

Mesogenic properties of some molecules leading to liquid crystalline (LC) materials can be considered as the consequence, among other features, of their shape. The NLO properties of LC materials have been less investigated than those of other condensed phases. Early studies in that field were well reviewed by Arakelyan *et al.* in 1980 [93]. In the past ten years, only a few papers reported the study of LC phases specifically engineered for second-harmonic generation. As far as small mesogenic molecules are concerned, highly polarizable smectogens have been specifically synthesized for NLO studies, such as **73** and **74** [94] and **75** [95]. However, since smectic mesophases are difficult to organize in single domains, these materials were not further studied.

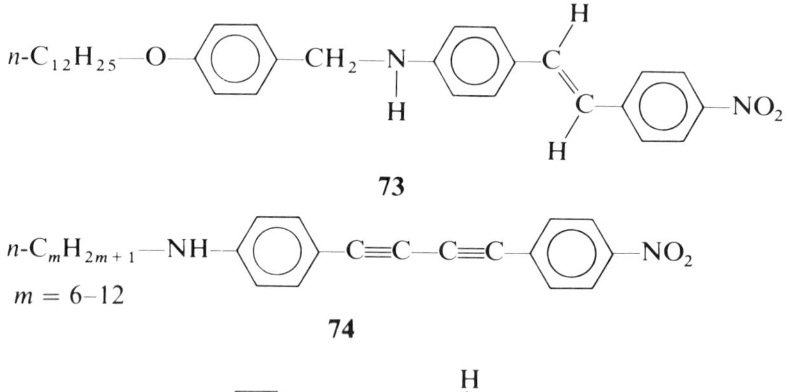

n-$C_{12}H_{25}$—O—(C₆H₄)—CH_2—NH—(C₆H₄)—CH=CH—(C₆H₄)—NO_2

73

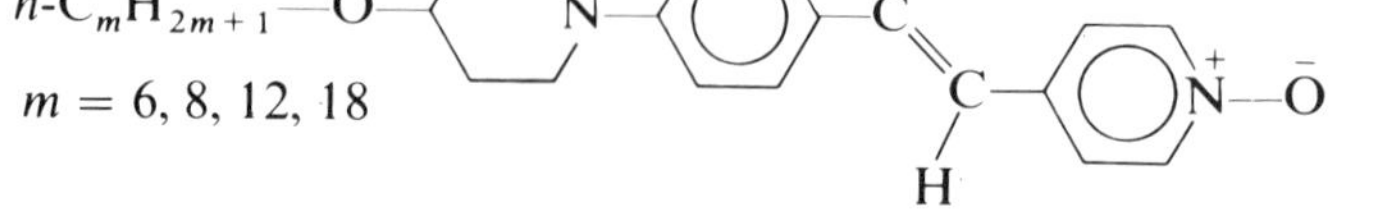

n-C_mH_{2m+1}—NH—(C₆H₄)—C≡C—C≡C—(C₆H₄)—NO_2

$m = 6$–12

74

n-C_mH_{2m+1}—O—(piperidine)N—(C₆H₄)—CH=CH—(pyridine)$\overset{+}{N}$—$\overset{-}{O}$

$m = 6, 8, 12, 18$

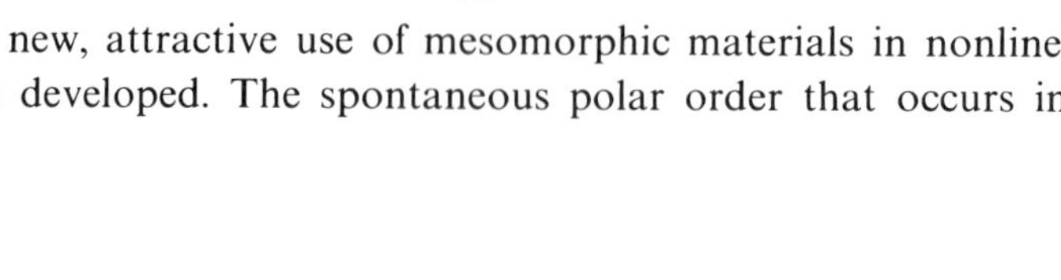

75

A new, attractive use of mesomorphic materials in nonlinear optics has been developed. The spontaneous polar order that occurs in ferroelectric liquid crystals (FLCs) is being used to control the polar orientation of an NLO active moiety, for the design of new $\chi^{(2)}$ materials. Taguchi *et al.* [96] have evidenced second harmonic generation for several FLCs in response to the applied electric field, which confirmed that the second harmonic intensity has a correlation with the spontaneous polarization. Another group, from Osaka University, reported recently a detailed study of SHG in FLCs as a function of electric field strength, rotating angle, temperature, cell thickness, molecular structure and so on [97]. In 3M2CPOOB (**76**) the authors observed an intense phase-matched SHG in the chiral smectic C* phase, in which a good alignment has been established by cooling from the isotropic phase.

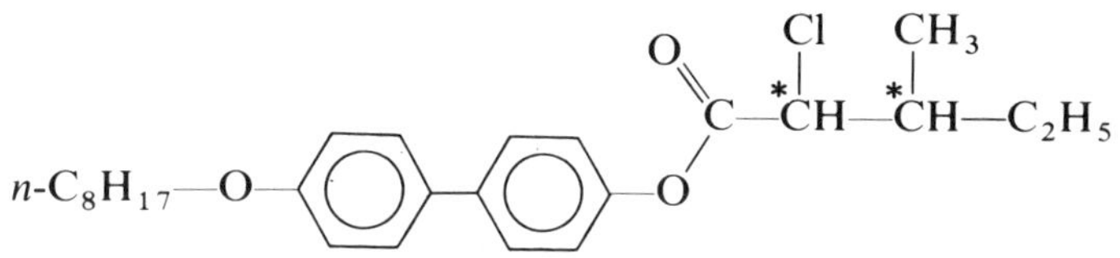

76 3M2CPOOB

The Boulder groups suggested a reason for the small $\chi^{(2)}$ values encountered so far in FLSs: though the molecules may indeed possess large β, the structure of the ferroelectric phase is not good for producing large $\chi^{(2)}$ when the hyperpolarizable chromophores are oriented along the director. Nonlinear molecules must subsequently be oriented in a polar fashion along the polar axis. Walba and coworkers succeeded in synthesizing the first NLO–FLC mesogen (W314, **77**) bearing a 2-nitroalkoxyphenyl unit as hyperpolarizable moiety, that shows $d_{22} = d_{23} = 0.6$ pm/V ($d_{eff} = 0.23$ pm/V) and λ_{max}(EtOH) = 308 nm, in a monotropic SmC* phase at 65°C [98]. The Boulder group's studies shows that $\chi^{(2)}$ is not proportional to the spontaneous polarization **P** and may increase much faster than **P** with appropriately designed FLCs, and suggest that very large increases in $\chi^{(2)}$ in FLC films should be possible.

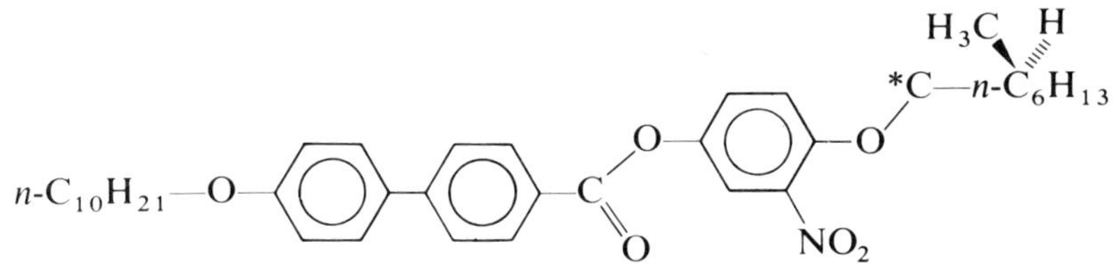

77 W314

5. INTERMOLECULAR STRUCTURES: PACKING AND ELECTRONIC FEATURES

5.1. Cocrystals and H-Bonding Networks

Besides chirality or other shape factors, several crystal engineering approaches have been proposed in order to obtain an acentric organization of NLO chromophores in a crystal lattice. One of them is the cocrystallization of two (or more) organic materials that can be individually SHG active or inactive: the mixture in adequate proportions may then induce a noncentrosymmetric packing of each NLO chromophore. Success here is not guaranteed, unless at least one of the compounds is chiral. It should be mentioned that in most cases, intermolecular hydrogen-bonding is involved in such mixed crystals. Besides results from earlier studies [2,99–101], mixed crystals have been recently investigated by a Japanese group: Okamoto *et al.* [102], for example, have studied crystalline mixtures of 4-nitroaniline (4-NA) and some of its *N*-alkyl derivatives. The SHG powder test shows that the mixtures of 4-NA and *N*-isopropyl-4-NA, *N*-*n*-propyl-4-NA, or *N*-*n*-butyl-4-NA exhibit large SHG activities, although each individual compound has relatively small or no SHG activity. More recently, using 4-NA as guest and various substituted benzenes as host in order to get mixed crystals, Matsushima *et al.* [103] found several new SHG efficient materials. By varying the crystallization conditions as well as host and mixing ratios, the second harmonic intensities at 1.06 μm were shown to be as high as 100 to 189 times that of urea with 1,4-dihydroxybenzene or 1,4-dicyanobenzene as host. Provided that mixed crystals are stable enough and of sufficient optical quality, these results indicate that cocrystallization of organic molecules seems promising to generate highly efficient NLO crystals.

Since intermolecular hydrogen bonds are an efficient tool for organizing molecules in condensed phase, it is of major interest to be able to predict the geometry of such arrangements. The packing patterns of a series of nitroaniline compounds, which associate via intermolecular hydrogen bonding between amino groups and nitro groups, were systematically analyzed in order to find possible hydrogen-bond rules with the potential of predicting the orientations of neighboring molecules [104]. A first step in the design of acentric organic materials of interest in nonlinear optics emerged from these studies [105,106]. The establishment of intermolecular hydrogen bonding between the components of a mixture has been the most widely exploited in material engineering. A recent example is provided by Etter *et al.* [107], who found that cocrystallization of 4-aminobenzoic acid and 3,5-dinitrobenzoic

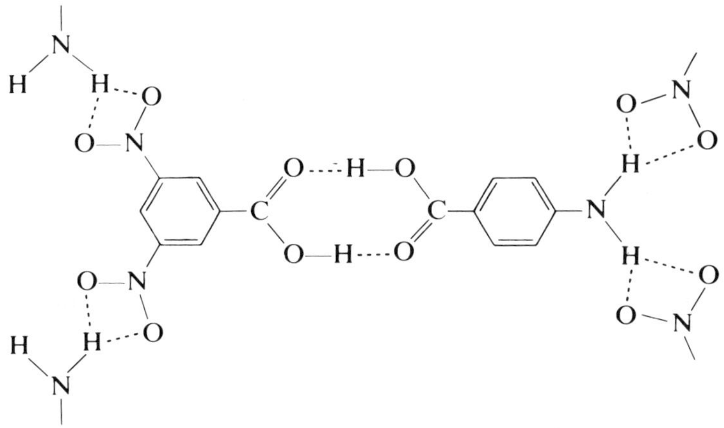

Fig. 8. Representation of an NLO active hydrogen-bonding network obtained by cocrystallization: 1-1 complex between 4-aminobenzoic acid and 3,5-dinitrobenzoic acid.

acid leads to an SHG active crystal of acentric structure Fdd2 (Fig. 8). This hydrogen-bonding strategy should be developed so as to introduce acentricity tentatively in crystalline phases by the combination of several NLO molecular partners. Furthermore, the nonlinearity of the resulting assembly could be enhanced by the establishment of the hydrogen-bonding network. Some promising results in that direction have been recently reported [108].

5.2. Organomineral structures

Some difficulties are inherent in organic crystals as a result of their often somewhat loose intermolecular cohesion ensured by Van Der Waals interactions: the additional crystalline binding brought about by hydrogen bonding, such as in NPP [5] or nitrouracil [109], has been, however, invoked to account for their significantly improved crystalline stability. An interesting concept consists of combining organic and mineral moieties in a mixed organomineral crystalline lattice that would jointly benefit from the respective qualities of both entities, the former contributing to its higher polarizabilities and its molecular engineering potential while improved structural stability would be expected from the latter. While paranitroanilinelike molecules are obvious candidates on the organic side, various assets, such as recalled hereafter, recommend the selection of the $H_2PO_4^-$ anion as the mineral counterpart [110]:

- $H_2PO_4^-$ connects to neighboring anions by short hydrogen bonds, thus forming quasi-polymeric mineral polyanionic strands.
- The oxygen atoms in the polyphosphate strands are liable to behave as eager acceptors of external hydrogen atoms released by surrounding organic cations.
- Dihydrogenophosphate tetrahedrons are distorted and bear a dipole moment capable of interacting with the organic polar cation.
- The first acidic function of H_3PO_4 is easy to handle and allows for the relatively easy growth of large size $(H_2PO_4^-)^n$-containing single crystals.

Various organic dihydrogen phosphates have been reviewed or proposed [111], some of which were obtained from a variety of alkyl ammonium ions as well as from chiral and achiral amino acids. A most prominent example is that of *L*-Arginium Dihydrogen Phosphate (LAP), an SHG active $P2_1$ organomineral structure [112,113]: its moderate nonlinear efficiency is believed to originate from a quasi-trigonal guanidiniumlike fragment leading to an "octupolar" β [114] (see also Sect. 6). A systematic strategy deliberately targeting at the design of efficient organomineral crystals is developed by Masse and Zyss [110] and exemplified in the case of 2-amino-5-nitro-pyridinium-dihydrogen-phosphate (2A5NPDP) extensively characterized by Kotler *et al* [115]. The specific choice of a pyridine rather than a benzenoid ring is meant to introduce a proton heteroatom and thus potentially increase the potential of the molecule for clipping via hydrogen bonding on to the phosphate anions while the amino hydrogens may be partially released so as to jointly contribute to a tightly bonded H-bonded structure as shown in Fig. 9.

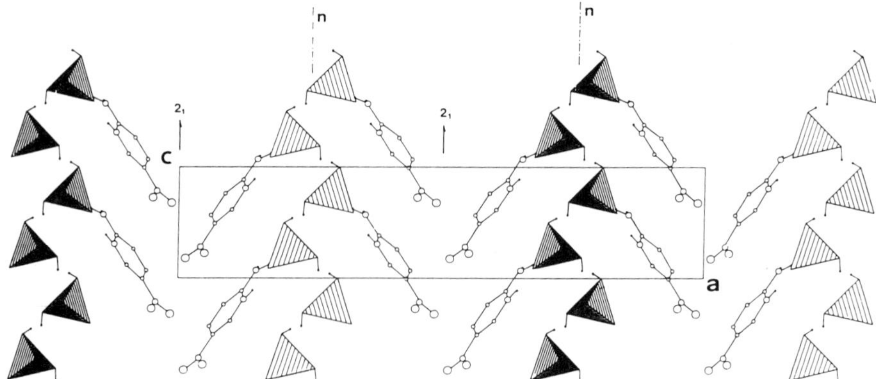

Fig. 9. Polar arrangement of the 2A5NPDP crystal structure evidenced in the (a, c) projection).

The crystalline structure of 2A5NPDP evidences a herringbonelike $Pna2_1$ structure whereby rows of organic cations are inclined at an angle of 36.7° with respect to the two-fold crystalline symmetry axis. The mineral sub-network consists of a layered structure tightly connected by multidirectional internal hydrogen bonds. Phosphate "walls" are seen periodically to segregate organic and mineral moieties: they may also play a role in screening the strongly dipolar double strands of organic cations, thus favoring the formation of noncentrosymmetric structure. The herringbone arrangement seems to be a fairly general feature of 2-amino-5-nitropyridine derivatives, either in crystalline phases such as COANP [85], an adamantane derivative [89], and MBANP [116], as well as in Langmuir-Blodgett structures such as the unusual X-type DCANP multilayers [117]. The absorption edge of 2A5NPDP is located at 420 nm. Protonation of the nitrogen heteroatom leads, as pointed out by solution absorption spectroscopy, to a pronounced blue shift ($\lambda_{max} = 347$ nm in acetone and 305 nm on phosphoric acid, the latter in protonation conditions closer to that in the actual crystalline structure). The following values have been reported for the nonlinear properties of 2A5NPDP [115]: $d_{33} = 12$ pm/V at 1.34 pm/V, $d_{15} = 7.2$ pm/V at 1.06 μm, and $d_{24} = 1.3$ pm/V at 1.06 μm.

A variant of 2A5NPDP has also been proposed [118]: tartaric acid is used as the polyanionic network embedding 2A5NP, thus leading to 2-amino-5-nitropyridinium-*L*-monohydrogenate tartrate (2A5NPLT). Although the resulting structure is an organic salt, it may be considered as falling into the same category as previously mentioned organomineral structures.

A preliminary attempt to host octupolar guanidinium entities in a *L*-monohydrogentartrate lattice so as purposely to optimize the LAP structure has been so far only partially successful [118]: a $P2_12_12_1$ structure has been obtained, evidencing strong interlocking of anionic and cationic lattices by a multidirectional H-bonding network. The limited powder SHG efficiency of the compound, of the order of that of urea, has been ascribed to the small dihedral angle of 8° between the guanidinium and the (0, 1, 0) crystalline plane, and a subsequently too limited departure from centrosymmetric packing.

The nonlinear properties of an interesting class of 1:3 trigonal triiodide adducts of general formula TI_3^*3D have been reported [119,120] with R = CH and D = S_8 (triiodomethane sulfur, crystal space group R3m), R = CH and D = C_9H_7N (triiodomethane quinoline, crystal space group R3), R = Sb and D = S_8 (Antimony triiodide-sulfur, crystal space group R3m). All these crystals exhibit a higher site symmetry at the triiodide molecule location with its three-fold symmetry axis coinciding with that of the crystal. These structures exemplify a class of "optimal" octupolar crystalline structures as defined in Sec. 6. Significant nonlinearities have been evidenced from SHG and electrooptic assessments and believed to originate mainly from the triiodide moiety: the major part of the nonlinearity is found to be perpendicular to its three-fold symmetry polar axis and is thus of octupolar origin.

5.3. Inclusion compounds

Besides cocrystallization, another approach involving intermolecular structures to promote noncentrosymmetry in the crystalline state is the use of inclusion complexes. It is known that complexation of a substrate by a host can markedly influence the chemical and physical properties of the guest. Several hosts of various sizes and shapes, most often chiral, are available for inducing an acentric molecular alignment. This is particularly the case of cyclodextrins (CD). CD molecules are cyclic oligomers of glucose and appear to be like a lampshade, with a large conical empty space allowing for insertion of various molecules, inasmuch as their size is compatible with the internal volume of the cyclodextrin host. A review dealing with CD complexation as a probe of molecular photophysics appeared recently [121]. Inclusion compounds of 4-nitroaniline and derivatives in β-cyclodextrin were the first to be reported as SH-generating materials [122–124]. Other organic hosts have been tested such as thiourea, *tris*-o-thymotide (TOT) and deoxycholic acid (ADC, **78**) [125,126]. This method has been extended to inorganic hosts such as molecular sieves [127,128]. However, the SHG efficiencies of these inclusion materials are not very high, and the method is limited by the size of the guest molecule. This approach requires that the selected guest nonlinear molecule possesses a large hyperpolarizability, so as to compensate for its dilution in the resulting inclusion material. Some success in that direction has been achieved by choosing 4-dimethylaminostilbazole-*N*-oxide

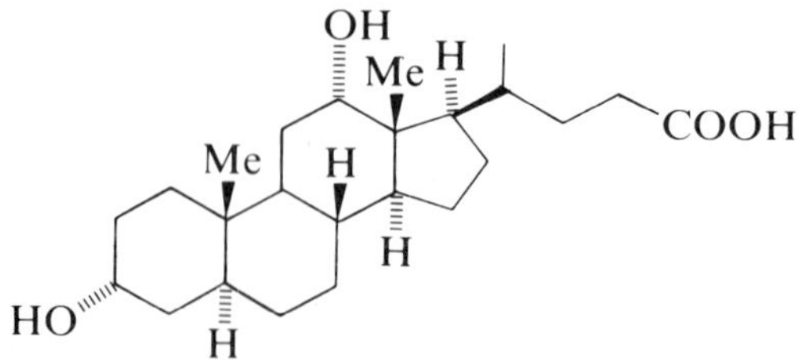

OH
Me H
COOH
Me H
H
HO
H

78 ADC

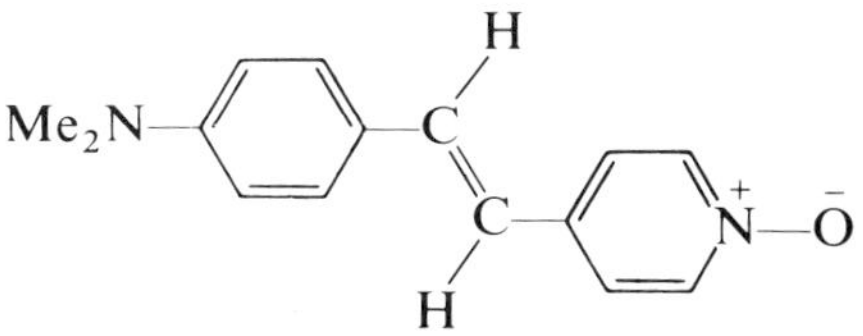

79 DASNO

(DASNO, **79**) as highly hyperpolarizable NLO guest chromophore and deoxycholic acid as the host. When a solution of DASNO and DCA in ethanol is slowly evaporated, a bright yellow crystalline powder is formed that shows an SHG efficiency a few times larger than that of urea at 1.06 μm.

Another interest of CD's hosts lies in their ability to incorporate, in a noncentrosymmetric manner, strongly elongated molecules. The rodlike shape of several ICT molecules may be responsible for various drawbacks in terms of molecular organization and collective electronic response. The rigid elongated structure of molecules with extended conjugation, such as polyenes, betaines, and hemicyanines, limits their ability towards further organization in such structures as single crystals or conventional LB films. The aggregates form islands on the water–air interface, leading to a multidomain microscopic structure of the films after transfer onto a solid substrate; these domains are separated by grain boundaries, which are responsible for severe scattering phenomena limiting the practical use of LB layers in waveguiding applications. In addition, aggregation occurring in LB films made of rodlike molecules leads usually to a strong blue shift of the linear absorption spectrum, therefore lowering the resonance enhancement of the quadratic nonlinear response of the films. Attempts to isolate a nonlinear molecule from the neighboring ones by use of proper techniques should lead to a noticeable enhancement of the quality of mono- and multilayers.

For all these reasons, the aggregation process was tentatively contradicted by way of inclusion in cyclodextrins. An amphiphilic β cyclodextrin presents interesting properties of building up ordered and high-optical-quality LB films. The geometry of the molecule favors a two-dimensional, in-plane hexagonal packing in each monolayer, preventing the formation of grain boundaries. Interesting results were obtained by mixing β-CyD's with donor-acceptor polyenic molecules [129]. Using cyclodextrin as host medium leads to a significant decrease of the UV aggregate peak and on the onset of another peak around 500 nm, which corresponds exactly to the solution spectrum.

SHG studies on mono- and multilayers of polyenic molecules evidence the crucial role of cyclodextrins in the improvement of the film quality. For example, when mixing the pyridine derivative **55b** with fatty acid, the measured apparent β value in 10 multilayers of this mixture alternated with pure fatty acid is half that of the corresponding monolayer. On the contrary, similar comparisons for mono- and multilayered films made of pyridinium polyenic molecules derived from **55b** mixed with amphiphilic cyclodextrin **80** evidence a large increase of the molecular order when building up non-centrosymmetric multilayers (mixed cyclodextrin/polyene alternated with pure cyclodextrin layers) as compared with the monolayer: the enhancement factor is higher than two, for the same surface density in active molecules [130]. A similar behavior is observed in betaine mono- and multilayers.

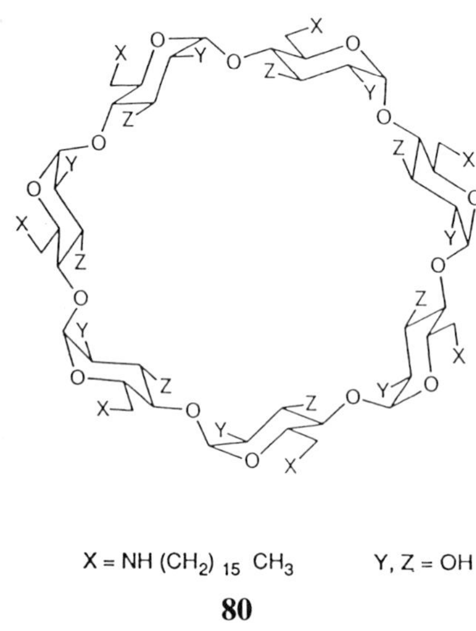

X = NH $(CH_2)_{15}$ CH_3 Y, Z = OH

80

5.4. Excitonic effects

The nonlinear behavior of organic crystals, in the parametric regime, can be satisfactorily accounted for by the oriented gas model whereby molecules are assumed to react independently and their nonlinear responses merely added, mutual polarization effects being taken care of by local field corrections. Little attention has been devoted so far to resonant configurations in organic nonlinear crystals, although precise identification and assignment of their electronic spectra would by highly needed. It has been shown [131,132] that the nature of the gas-to-crystal shift is governed by electrostatic dipole–dipole interactions strongly dependent on the lattice geometry. Similar molecules packed in different geometries as a result of sometimes structural differences may then lead to tunable shifts either to the red or blue part of the spectrum. The main problem in performing crystalline spectroscopic measurements in the visible and near-UV portions of the spectrum will then lie in the strong absorption attached to the chromophore units, which is essential to ensure a significant nonlinearity. Based on reflection spectroscopy at room (300°K) and liquid Helium (2°K) temperatures and subsequent Krämer-Krönig transformation, it has proved possible to circumvent this difficulty in the case of NPO (4-nitropyridine-1-oxyde), a molecule analogous to POM, except for the absence in NPO of a methyl group: its simpler geometry considerably eases the interpretation of spectra as compared to POM, while their electronic spectra and molecular (hyper)-polarizabilities are expected to be similar [133]. The crystal structure of NPO is Pnma with four molecules per unit cell all parallel to the (101) crystallographic plane. Reflection and transmission spectra are represented on Fig. 10.

The intramolecular charge-transfer (ICT) transition gives rise to the reflexion maxima at wavelengths shorter than 400 nm. The reflectivity is more pronounced for light parallel to the *c* axis and presents a marked vibronic structure. The *a* component is broader and less intense. In transmission, the crystal becomes opaque at the vicinity of the ICT transition, and the first sharp peak assigned to a $n\pi^*$ transition.

The maximum of the *c* (resp. *a*)-polarized crystal spectrum is red (resp.

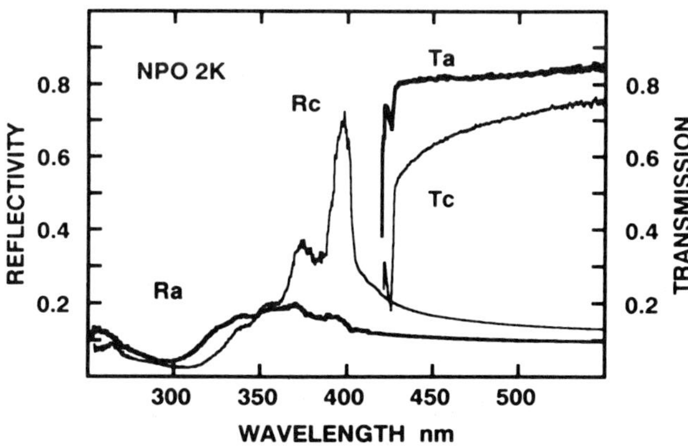

Fig. 10. Reflection and transmission spectra of NPO crystals at 2°K. Suffixes *a* and *c* refer to the polarizations of the incoming beam for both reflection and transmission. The transmission spectra are recorded on crystal plates of 80 μm thickness. Data are from Pierre *et al.* [133].

blue)-shifted by 3580 cm^{-1} (resp. 800 cm^{-1}) at 2°K. These two polarized spectral components belonging to the same molecular transmission reflect the exchange interaction. The so-called Davydov splitting of 4380 cm^{-1} is much larger than that observed in the $S_1 \leftarrow S_0$ transition of anthracene with an (a, b) splitting of 205 cm^{-1}. The difference essentially results from one order of magnitude difference in oscillator strength per unit volume: 0.0085 $Å^{-3}$ for NPO as compared to 0.0008 $Å^{-3}$ for anthracene. The polarization ratio of 1.93 is significantly higher than the value of 1.29 expected on the basis of an oriented gas model. Such deviations had already been noted in the Raman spectra of NPO single crystals [134]. A likely interpretation of these data calls upon the contribution of intermolecular charge transfer states to crystalline excited states, as confirmed by room temperature data that evidence a small contribution to the oscillator strength (0.04) in a direction perpendicular to the molecular plane. The most likely pathway for efficient deactivation points towards a $n\pi^*$ state at 23281 cm^{-1}. Similar features have been observed in POM. Because of its rapid decay, the ICT state cannot delocalize into an extended exciton, thus limiting the enhancement of nonlinear susceptibilities beyond the oriented gas value. The improvement of nonlinear materials for both $\chi^{(2)}$ and $\chi^{(3)}$ should therefore be targeted towards molecules where dark states (such as the $n\pi^*$ state discussed here) are pushed upwards, so that no fast and efficient pathway is offered to the exciton. For both $\chi^{(2)}$ and $\chi^{(3)}$ materials, involvement of intermolecular configurations, as can be inferred from the study of NPO, may play a more important role than had been postulated in the past.

As far as extension of the transparency window is concerned, similar considerations have been tentatively applied to DMACB (see Secs. 3–4) to account for its transparency up to 420 nm. It is expected that excited state considerations will increasingly contribute to molecular engineering guidelines in the future.

6. NEW PERSPECTIVES IN MOLECULAR ENGINEERING: FROM DIPOLAR TO MULTIPOLAR SYSTEMS

6.1. Rotational Invariance and Nonlinear Susceptibilities: Octupolar Nonlinear Molecules

While most of the attention has traditionally concentrated on polar nonlinear molecules and their subsequent macroscopic assemblies, a number of drawbacks are attached to the implementation of such structures: polar molecules tend to set up more readily an antiparallel molecular arrangement and their crystallization may be disrupted by dipolar aggregation. Furthermore, the highly anisotropic structure of an optimized electrooptic material such as DMACB [32], however attractive, will limit its application to electro-optic configurations whereby modulating field and optical beam polarizations are both aligned along 3 (3 being the common crystalline and molecular polar axis). In more general terms, one may be wrongly induced, by this strategy, to concentrate exclusively on one-dimensional objects, thus ignoring the possibilities of three-dimensional chemistry. This approach has been recently questioned: it was shown, based on experimental evidence as well as on general tensorial and quantum-mechanical considerations, that investigations should indeed be widened to encompass a more diversified range of molecules with attached nonlinearities of so-called "octupolar" or more general "multipolar" origin [114,135–137]. Symmetry considerations have demonstrated the possibility of designing molecules with exact-symmetry-based cancellation of their dipole moments in the excited as well as ground states and allowing for a nonzero, eventually optimized, quadratic hyperpolarizability β tensor. This approach is initially more geometric than that which had led to POM whereby mutual quasi-cancellation of sub-molecular dipole moments is ensured only for the ground state. Furthermore, the maximum degree of isotropy compatible with the existence of a nonzero susceptibility tensor of order 3 (at both molecular and macroscopic levels) may be shown to correspond to octupolarlike systems. The intuitive relevance of these issues appears clearly when considering an atom as the ultimate example of a fully isotropic spheroidal system: owing to centrosymmetry, it cannot possess a nonzero β value. There must consequently exist a boundary separating highly anisotropic dipolar systems such as push-pull polyenes from atomlike spheroidal objects and hence molecular types with nonzero β (or $\chi^{(2)}$) values from those with cancelling values.

The proper framework to discuss this point naturally involves the irreducible representation of rotations over tensorial sets and an adequate definition of *multipolar groups* [136,137]. We assume in the following, for the sake of simplicity, nonresonant processes resulting in Kleinman symmetry and tensors invariant with respect to permutation of their cartesian indices to describe susceptibilities. The universal nature of irreducible representations permits the decomposition of any tensor $T^{(n)}$ of arbitrary rank n in a sum of so-called irreducible tensorial components $T_{J,m}^{(n),\tau_J}$ following:

$$T^{(n)} = \sum_{J=0}^{n} \sum_{m=-J}^{J} \sum_{\tau_J} T_{J,m}^{(n),\tau_J}. \tag{6.1.1}$$

The so-called seniority index τ_J designates different components of identical weight J. Most noteworthy is the fact that the summation over the weight J extends up to rank n only and will only involve, owing to index permutation symmetry, odd integers for $n = 2p + 1$ (odd-rank tensor or even order susceptibility such as β or $\chi^{(2)}$ according to current conventions) and even order integers for $n = 2p$ (even rank tensors or odd-order susceptibilities such as $\chi^{(1)}$ or $\chi^{(3)}$). Furthermore, $T^{(n),\tau_J}_{J,m}$ satisfies a closure condition with respect to the application of rotations, namely:

$$R(\theta, \phi, \psi)T^{(n),\tau_J}_{J,m} = \sum_{m'=-J}^{J} D^J_{m,m'}(\theta, \phi, \psi)T^{(n),\tau_J}_{J,m'}. \tag{6.1.2}$$

The $D^j_{m,m'}$ are the Wigner matrix elements depending on the three Euler angles that define the rotation [138]. These expressions are not surprisingly reminiscent of similar ones encountered in atomic physics where spherical harmonics $Y^l_m(\theta, \phi)$ are constantly referred to in view of their adequate rotational invariance properties. This decomposition has been in fact developed and widely documented in the framework of atomic angular momentum composition, while its transposition to the tensorial description of NLO phenomena has been worked out later [139–142], but some molecular and material engineering insights and implications, such as reviewed here, may not have been recognized at the time such as the identification of electronic and structural parameters underlying the irreducible components of nonlinear susceptibilities and the molecular engineering possibilities therefrom. We concentrate hereafter on the main ideas and some of their specific consequences, while more technical details are deferred to the original articles. In turns out from the truncation of J in (6.1.1) that β (or $\chi^{(2)}$) will only have two components of order 1 and 3, respectively, referred to as the vectorial and octupolar irreducible components:

$$\beta = \beta_{j=1} + \beta_{j=3}. \tag{6.1.3}$$

The vectorial component ($J = 1$) allows for three independent coefficients ($-1 \leq m \leq 1$), while the octupolar component accommodates seven components ($-3 \leq m \leq 3$). In a multipolar group of order J, as defined in [136,137], based on character table manipulations and application of Schur's lemma, all tensorial properties of order strictly lower than J strictly cancels as from intrensic symmetry requirements. If a molecule or a material then belongs to an octupolar group ($J = 3$), all dipolarlike quantities (e.g., dipole moment, vector part of β) vanish, but there still remains a nonzero β tensor as from the symmetry allowed octupolar component contribution.

Octupolar groups, such as the D_{3h} group, the tetrahedral cubic group T, or quadrupolar groups ($J = 2$) such as the orthorhombic 222 Group (e.g.

POM crystalline structure) will lead to more isotropic objects than those belonging to lower-order (polar) groups, but are still compatible with β or $\chi^{(2)}$ properties. Such would not be the case for hexadecapolar groups ($J = 4$) as well as multipolar groups of order J strictly larger than 3.

A slight restriction, however, obviously comes in in the case of odd-order susceptibilities, which always contain a nonzero scalar ($J = 0$) component corresponding to the so-called isotropic or fully symmetrical component. The D_{4h} hexadecapolar group of phthalocyanines can then be viewed as the highest n-fold symmetry group compatible with the existence of a nonzero nonisotropic (i.e., $J \neq 0$) γ cubic hyperpolarizability contribution (namely $\chi^{(3)}_{J=4}$). Above $n = 4$, $\chi^{(3)}$ will reduce to its isotropic scalar component.

Finally, an interesting nonintuitive consequence of these considerations can be derived as to the issue of a possible nonzero β tensor for any noncentrosymmetric molecule: if the noncentrosymmetric molecule belongs to a multipolar group of order strictly higher than 3 (i.e., 5 or more) then, owing to the absence of irreducible components of order higher than 3 in the rotational spectrum of a rank 3 tensor such as β, one may conclude that β will strictly vanish. For example, an hypothetical planar five-fold symmetry molecule, such as ferrocene, has no β although it is noncentrosymmetric.

6.2. A General Classification Scheme for Nonlinear Molecules: Towards the Engineering of Octupolar Nonlinear Molecules

Traditional comparison and classification schemes of nonlinear molecules have been biased by the prevailing EFISH experimental technique, thus ignoring other contributions to β than the dipolar one. A more adequate scheme that encompasses previous ones while fully taking into account anisotropic contributions has been proposed [137] and is recalled in Fig. 11(a). In this vectorial representation, the β or $\chi^{(2)}$ tensorial space is decomposed into $J = 1$ (dipolar) and $J = 3$ (octupolar) orthogonal subspaces, the projections of β onto these subspaces standing respectively for $\beta_{J=1}$ and $\beta_{J=3}$. The corresponding third-rank *field tensor* [114,139,149] defined as

$$F^{(2)} = (E^{2\omega})^* \otimes E^{\omega} \otimes E^{\omega} \tag{6.2.1}$$

can be similarly decomposed. Its relative angular position with respect to β (point M in Fig. 11(a)) will reflect the efficiency of the field–matter interaction for a given polarization combination. Similar consideration may be applied to the problem of *multipoling* [114,137] where the multipolar charge

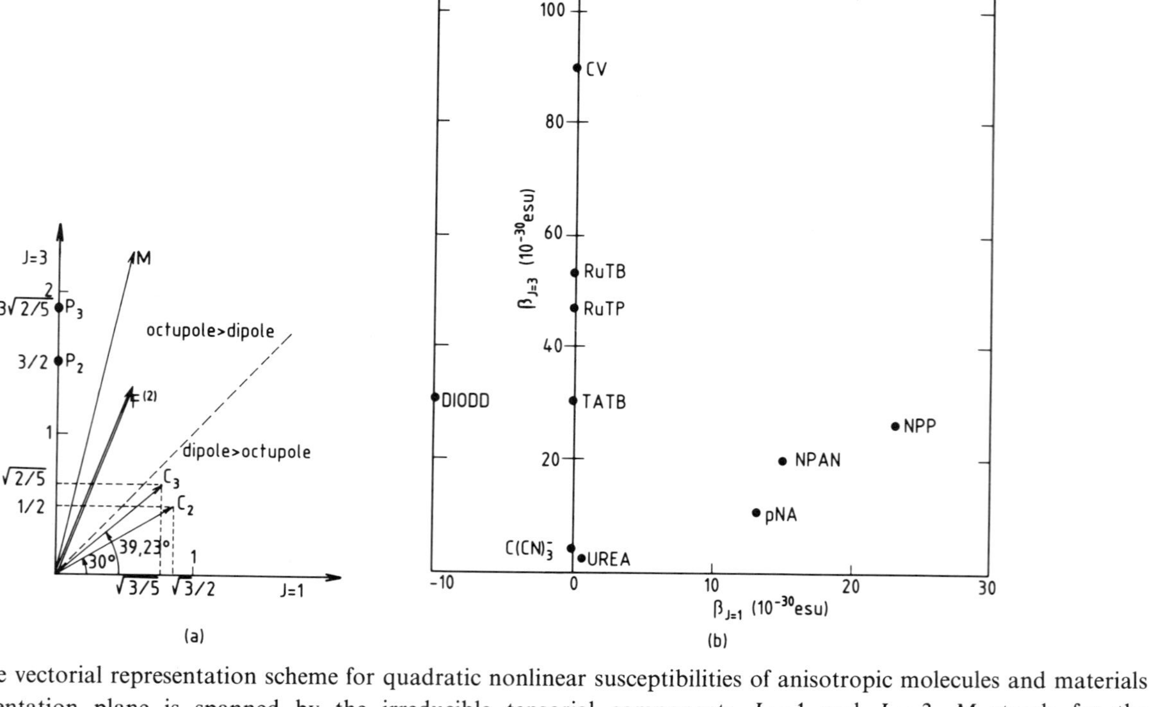

Fig. 11. Reference vectorial representation scheme for quadratic nonlinear susceptibilities of anisotropic molecules and materials in (a): the representation plane is spanned by the irreducible tensorial components $J = 1$ and $J = 3$. M stands for the quadratic susceptibility of a given molecule or material and $F^{(2)}$ for the corresponding field tensor, $C_{2,3}$ stands for a purely one-dimensional β tensor with $P_{2,3}$ deduced by simple tensorial addition. Utilization of a 2D or 3D formalism is reflected by the suffix 2 or 3 [137].

In (b), various examples of molecular systems of mixed or purely octupolar nature with data originating from either experiments or theory are plotted, namely: NPP [143]; NPAN [143]; *p*NA [157]; TATB [157,158]; urea [144]; crystal violet [145]; 3-amino-4,6-dinitroaniline (DIODD) [146]; ruthenium^{II}-tris(2,2′-bipyridil) (RuTB) and ruthenium^{II}-tris(1,10-phenanthroline (RuTP) [147]; and tricyanomethanide anion ($C(CN)^{3-}$) [148].

distribution and the multipoling field distribution are represented in a similar framework.

We concentrate hereafter on so-called "planar" systems [150] whereby the β tensor is four-dimensional: whether the molecule is actually planar in the geometric sense is not crucial inasmuch as the two-level model applies. The β tensor may then be shown to be confined in a plane spanned by $\boldsymbol{\mu}_{01}$, the transition dipole, and $\Delta\boldsymbol{\mu}$, the difference between excited and ground-state dipole moments, making the actual position of nuclei inessential. Two sets of expressions have been worked out for the irreducible components of β, depending on the molecular environment: one may want to consider a planar molecule either as a truly 2-D confined system with interactions also confined to the plane or, more realistically, as a 2D system embedded in 3D spaces, that is, allowing for out-of-plane Coulomb interactions between in-plane particles or out-of-plane intermolecular interactions. Detailed expressions for both cases and different molecular symmetries are to be found elsewhere [137]. A simple and widely used reference model is that of a one-dimensional system whereby β is limited to a single tensorial term, namely

$$\beta = \beta_{XXX}\mathbf{X} \otimes \mathbf{X} \otimes \mathbf{X},$$

and $\mathbf{X}$ stands for the charge-transfer axis in paranitroanilinelike chromophores. One may then derive the β tensor of the simplest corresponding octupolar molecule by mere tensorial addition of the βs of the individual one-dimensional component moieties arranged in a three-fold planar arrangement of D_{3h} symmetry (see Fig. 12). Such purely geometric procedure is based on the assumption that interactions between one-dimensional sub-systems are negligible. Following this definition, P_2 (resp. P_3) will then correspond to C_2 (resp. C_3) in the framework of Fig. 11a, where the C's stand for one dimensional systems and the P's to their "additive" octupolar D_{3h} counterparts, suffixes 2 and 3 reflecting the choice of a 2D or 3D representation formalism. In most interesting cases, and in particular for the paradigmatic paranitroaniline (pNA)-TATB couple, the additive assumption is not valid as mutual interactions between the NO_2 and NH_2 groups in TATB will completely scramble individual pNA contributions. The P point is then meant to serve as a reference position, its distance to the measured or computationally estimated $\beta_{J=3}$ component of the actual system being a measure of the amount of *octupolar interaction* in this system. Part (b) of Fig. 11 depicts various molecular systems with data of either theoretical or experimental origin. Of particular interest is the inspection of the $J = 3$ axis (purely octupolar systems), which evidences that high β values can indeed be achieved with trigonal molecular systems (see also Fig. 13) as compared

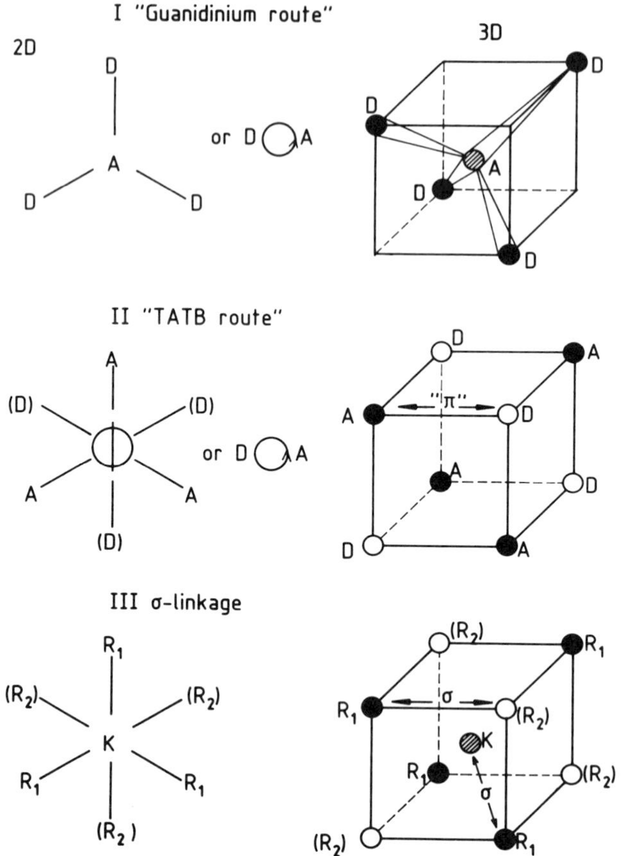

Fig. 12. Various molecular engineering routes towards the optimization of the β susceptibilities of octupolar nonlinear molecules in both planar and nonplanar cases. Note that in all cases, the donors (D) can be permuted with the acceptors (A). In case II, either D or A can be omitted leading to trisubstituted TATB analogs. In case III, R_1 (resp. R_2) corresponds to a molecule of the same type I (resp. II), or to a dipolar molecule such as belongs to the paranitroaniline family. There again, R_1 or R_2 can be omitted.

to the more classical *p*NA-like systems. Particularly noteworthy is crystal violet [145], which can be viewed as a more extended and conjugated version of the guanidinium cation with the central carbon playing the role of an electron donor. While the $\beta_{J=1}$ component is available from the now well documented EFISH experiment, the $\beta_{J=3}$ component has remained somewhat more elusive in this context owing to the absence of a permanently orientable permanent dipole. The so-called *Elastic Second-Harmonic Light*

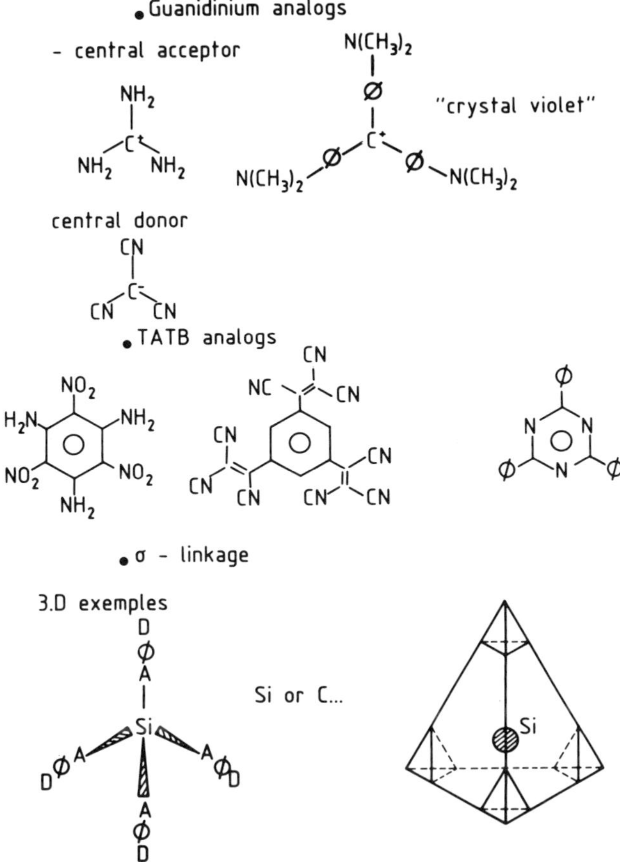

Fig. 13. Specific examples of implementation of the engineering routes depicted in Fig. 12.

Scattering (labelled EHLS) experiment demonstrated and analyzed in a series of two pioneering papers [139,151] provides the clue: carbon tetrachloride, a typical octupolar tetrahedral molecule, had been fully characterized by Terhune and Maker, using this technique as early as 1965, with a full spectral and temperature analysis reported in the 1970 *Physical Review* paper by Maker [139]. This experiment provides a combination of squared β coefficients weighed by coefficients that reflect intermolecular-correlation-induced statistical orientation. The noncoherent nature of the observed harmonic signal makes it proportional to the number of nonlinear molecules rather than to its square, as would have resulted from coherent SHG, thus requiring a sensitive off-axis, however tractable, detection system. Combination of

EFISH and EHLS may provide, in some cases, additional relevant data such as the angle between the dipole moment and $\beta_{J=1}$ in systems where the octupolar contribution of $\beta_{J=3}$ can be legitimately neglected, such as long chain polyenes. Note that this technique has been recently revived under the name *Hyper-Rayleigh Scattering* [152].

Various general 2D and 3D strategies towards the optimization of $\beta_{J=3}$ are displayed in Fig. 12 and illustrated by more specific examples in Fig. 13. The first one in Fig. 12, based on a central acceptor (resp. donor) atom interacting with surrounding donor (resp. acceptor) moieties in a trigonal arrangement, is inferred from the guanidinium cation and has naturally led to its more conjugated version as the crystal violet cation. Similarly, the fully planar tricyanomethanide anion can be viewed as an inverted guanidinium, whereby the central atom and peripheral groups are now respectively donor and acceptor. Whether the molecule is actually planar or not does not really matter provided its symmetry cancels the net dipole moment. Comparable 3D systems can be thought of with a heavier central atom so as to allow for conjugation via higher order orbitals. The TATB route generalizes over the initial observation of a strong signal from TATB, whereby the interaction between acceptor and donor groups is mediated by a conjugated system (an aromatic ring in the case of TATB). The conformation of TATB-like derivatives has been extensively studied [153,154] and shown to exhibit distorted boat structures. The highly connected intermolecular H-bonding network of TATB crystals keeps molecules in a perfectly planar structure as from the earlier structure determination by Cady and Larson [155]. However, a different, interesting case is that where the central atom or moiety is just meant to orient type I and type II molecules, as well as *p*NA-like ones, in an adequate octupolar assembly. This strategy can be illustrated by four identical *p*NA-type molecules aligned along the direction of sp^3 tetragonally hybridized orbitals originating from a central binding silicon or carbon atom. Alternatively, four tetrahedral octupolar molecules, as also shown at the bottom of Fig. 13, can be accommodated at the apex of a regular tetrahedron in a sort of "self-similar" octupolar arrangement held together by a central sp^3 binding atom or tetragonal moiety.

6.3. Quantum Mechanical Considerations: From Two-Level to Three-Level Systems

Modeling of the optically driven polarization properties of octupolar systems based on quantum mechanical approaches provides interesting insights into the underlying excited states, transparency, polarized charge redistribution and β magnitudes. It had been pointed out [114] that, regardless of the actual level of quantum chemical description of the octupolar system, the two-level quantum model all-pervadingly invoked by *p*NA analogs is irrelevant in the context of octupoles owing to the cancellation of all vectorial quantities including $\Delta\mu$, the difference between excited and ground-state dipole moments. The simplest framework capable of accounting for octupolar quadratic nonlinearities must be a three-level system whereby a nonzero β will originate from the product of dipolar transition moments coupling the three states, e.g. $\mu_{12}\mu_{23}\mu_{31}$. Indeed, it turns out from mere inspection of character tables that trigonal groups, such as the D_{34} group exhibit sets of *E*-labelled degenerate irreducible subspaces of dimension 2 that are reportedly playing a pivotal role in octupolar nonlinearities [156]: three-level systems may then be evidenced as from a triangular pathway with circularly polarized light connecting the ground state up to one of the degenerate *E* states, from there on to the other and back to the ground state.

Calculations have been performed at different levels of approximation using different schemes:

- finite-field differences within the semi-empirical Hartree-Fock Austin model (AM1);
- *ab initio* coupled perturbated Hartree-Fock (CPHF) whereby tensor coefficients are obtained analytically by derivation of the total energy with respect to the electric field;
- the sum-over-state method based on the INDO Hamiltonian with singly and doubly excited π-state configuration interactions (all three schemes in [157]); and
- partially up to fully correlated Pariser-Parr-Pople π-electron calculation [156] offering the possibility of continuous tuning of side group donor or acceptor characters via modulation of on-site energy terms and finite-field as well as sum-over-state derivations of β.

Results of all three methods consistently point out similar values for β_{xxx} of paranitroaniline (*p*NA) and TATB (*x* being along the para amino to nitro axis for either molecule). In general, the ratio of β magnitudes for two arbitrary polar or nonpolar molecules of C2v symmetry or sub-symmetry labelled by 1 and 2 is given by:

$$\frac{\|\beta^1\|}{\|\beta^2\|} = \frac{\beta^1_{XXX}(1 + 3u_1^2)^{1/2}}{\beta^2_{XXX}(1 + 3u_2^2)^{1/2}} \tag{6.3.1}$$

with $u_i = \beta^i_{XYY}/\beta^i_{XXX}$ and $i = 1, 2$. The main difference between *p*NA and

TATB lies in the ratios u_i: while the anisotropy of polarizability is strong in the case of *p*NA, making the parameter $3u^2$ almost negligible with respect to 1 in Eq. (6.3.1), u is exactly equal to -1, as from D_{3h} symmetry, in the case of TATB. This gives a strong advantage, at almost equal β_{XXX} value, to octupolar over corresponding dipolar species, the origin of the factor of 2 almost gained thereby becoming clear from Eq. (6.3.1). Furthermore, the excited-state charge distribution and dipolar transition moments of TATB resulting from the sum-over-state approach evidence a more intricate mixture of states, blurring any clear assignment to charge transfer states of para and ortho nature [157]. In strong contrast with the two-level model, a set of three doubly degenerate excited states, the S_a, S_b, and S_c states, respectively located at 4.64, 5.26, and 6.64 eV above the ground state, plays a predominant part.

Among the noteworthy indications in the PPP-Hamiltonian-based model by Joffre *et al.* [156] is the indication of an improved transparency-efficiency tradeoff in octupolar systems as compared to corresponding dipolar ones. Hyperpolarizabilities as well as absorption edge energies are indeed increased, the latter because of selection rule type of considerations: while essentially all states are accessible in the absence of stringent symmetry constraints for dipolar systems, *octupolar* symmetry will introduce selection rules that will preclude access to some states, such as low-lying *B*-labelled ones, located below the main optically relevant *E* states, and uselessly responsible for absorption. From calculations by Joffre *et al.*, β magnitude (resp. absorption onset energy) increases from 4.26 (resp. 4.6) to 9.10^{-30} esu (resp. 5.2 eV) as one moves from *p*NA to TATB.

6.4. Macroscopic Ordering and Optical Interactions

Lamellar 2D octupolar macroscopic ordering has been discussed for TATB [157,158] in terms of "heteroaffinity" between side-groups of different origins: a H-bonding network locks the nitro group of a given molecule to the amino group of one of its six neighbors, thus inducing perfect trigonal arrangement. Conversely, homoaffinity would be prone to induce a detrimental centrosymmetric hexagonal packing. The metaborate mineral ionic crystal (BBO) exemplifies octupolar symmetry at macroscopic as well as microscopic levels as can be recognized in the cyclic $(B_3O_6)^{3-}$ anionic structure [159,160]. The active nonlinear moiety in LAP [161] and dLAP [113] crystals can be ascribed to a quasi-trigonal guanidiniumlike fragment. Cocrystallization of guanidinium with tartaric acid has been recently reported [118] and shown to lead to a noncentrosymmetric, however far from optimum, crystalline structure: its 222 orthorhombic structure is composed of strongly interlocked strands of aggregated *L*-monohydrogentartrate polyanion chains hosting guanidinium ions at their intersections. Its SHG powder efficiency compares to that of urea and could be improved by one order of magnitude at least by increasing the angle between corresponding octupolar guanidinium planes from an actual 16° to the optimal 90°.

It can be shown from simple group theoretical considerations (inspection of the character tables of the abelian planar rotation group), that the relevant order parameters for lamellar or planar *statistical* octupolar ordering are $\langle \cos 3\theta \rangle$ and $\langle \sin 3\theta \rangle$, defined as averaged values over the statistical distribution of molecular orientation angles away from a fixed laboratory framework parallel to the octupolar planes.

Such a disordered statistical octupolar phase may either result from self-assembly and reflect an intricate mixture of shape, affinity, and charge-related features (possible "octupolar" mesophases) or be obtained by an adaptation of the well-known poling procedure traditionally used in the context of amorphous polymers embedding dipolar nonlinear moieties. The classical poling procedure [8] will not operate for octupoles in the absence of dipole moment. A possible solution, at least in principle, is to resort to a "multipoling" procedure [114] by way of extension of the classical poling technique, whereby a symmetry-adapted multipolar field distribution is generated in a properly electroded cell. Such a multipolar potential would couple to the multipolar charge distribution of the same multipolar order and ensure the poling energy. However the magnitude of the multipoling coupling term is expected to decrease exponentially with the multipolar order n, following an r^n/R^{n+1} law where r scales with the molecular dimension and R with that of the electroded cell, while n refers to the 2^n order of the multipole. Multipoling could then be ideally achieved in guest-host "supramolecular" compounds whereby the multipoling host size is scaled down to the molecular scale (i.e., $r \approx R$). Octupoling, for example, could then be ensured by hosting a trigonal nonlinear microscopic molecule in an octupoling trigonal or tetragonal lattice.

Finally, the polarization behavior of light in nonlinear interaction with an octupolar medium is unusual. The averaged harmonic energy conversion rate is given by:

$$\frac{dW}{dt} \propto - \sum_{J=1,3} \chi_J^{(2)} \cdot F_J^{(2)} . \tag{6.4.1}$$

In this expression, summation is performed over tensorial contractions of the irreducible components of identical weight J of $\chi^{(2)}$ and of the tensor product of interacting fields $F^{(2)}$ as defined in Eq. (6.2.1).

In the case of an octupolar nonlinear medium, the summation in Eq. (6.4.1) reduces to $J = 3$: a purely $J = 3$ $F^{(2)}$ tensor would then represent the symmetry matched combination of polarizations. Such octupolar $(2\omega, \omega, \omega)$ third-rank field tensor can be shown to correspond to right-handed (resp. left-handed) circular polarization for the incoming (resp. outgoing) fundamental (resp. harmonic) beam or vice-versa for left or right.

By way of conclusion to Sec. 6, general types of molecular and material structures that do not suffer from the drawbacks linked to more traditional designs have been described. The definition of such structures results from a deliberate attempt to optimize the octupolar part of β and $\chi^{(2)}$ while strictly cancelling their vectorial components. As for the main drawbacks associated to dipolar structures, they may now be faced with more favorable assets based on the specific features of octupolar structures:

- The more rounded-off shape of octupolar molecules should ease their packing in a single crystalline lattice, as opposed to less favorable elongated, dipolar, sometimes rodlike molecules.
- The absence of dipole moments should improve the statistics for favorable noncentrosymmetric structures as had been initially argued, however in a less systematic way, for POM crystals.
- A trigonal lamellar structure ensures $r_{13} = -r_{33}$, which should allow the increase of the capacity of electro-optic modulators; equally efficient modulation of TE and TM modes are then possible, contrary to traditional dipolar structures favoring r_{33} over r_{13} and hence favouring a polarization mode along 3.
- The tendency towards aggregation, as for polar dyes, should be strongly influenced by the absence of dipole, and in some cases decreased.
- Furthermore, transparency and efficiency of adequately tailored octupolar molecules have been shown in different instances to significantly improve over those of their dipolar counterparts.

Development of a new, fairly general family of both molecules and materials is now being undertaken following the molecular engineering guidelines proposed here. Whereas current research strategies may be narrowing the synthetic exploration domain to simpler quasi-one-dimensional molecular systems, more advanced molecular engineering guidelines aiming at the optimization or adequate balancing of irreducible tensorial components of nonlinear susceptibilties are to be developed; this approach has been exemplified here in the case of octupolar nonlinear systems, consistently so at both molecular and macroscopic levels. Further work currently under way and expanding over the concepts exposed here should permit us to fully benefit, within the context of nonlinear optics, from the hitherto poorly exploited three-dimensional nature of organic systems.

7. CONCLUSION

At the end of this chapter, the reader, possibly initially unfamiliar with the domain, may feel overwhelmed by the apparently unlimited diversity of molecules and assemblies therefrom of present or potential interest for nonlinear optics. We would like to stress herein, by way of general conclusion, that such a wealth of possibilities is not out of scale with the strategic importance of the field. A historical metaphore may be found useful to illustrate this point of view: tailor-made synthesis of artificial organic dyes has been at the root of the development, in the middle of the nineteenth century, of a modern chemical industry to which we owe, among other benefits, a colorful daily environment previously restricted to a few privileged individuals. In modern terms, one could define the scientific approach at the root of this still ongoing revolution in terms of molecular engineering for linear optics. By way of extension, one may subsequently apply the strategy outlined here the label of molecular engineering at the nonlinear level. Although it is premature to forecast an equally bright future for the latter direction, going from passive to active interactions of molecules with light updates and considerably broadens the ensuing application potential of organics while enriching our insight on optical processes in molecular media.

This last point, regardless of the payoff in applications, is of great importance: complex features of different natures such as composition, shape, anisotropy, and polarizability at both microscopic and macroscopic scales are not fully amenable to classical linear optics. Multiphoton processes, involving different wavelength and polarization of the interacting waves, provide a more accurate probe of molecular properties, capable of reflecting and unveiling, to a higher degree of intimacy, their fascinating spectral and geometric intricacy.

REFERENCES

1. J. Zyss and D. S. Chemla, "Quadratic Nonlinear Optics and Optimization of the Second-Order Nonlinear Optical Response of Molecular Crystals", in *Nonlinear*

Optical Properties of Organic Molecules and Crystals, D. S. Chemla and J. Zyss, eds. (Academic Press, New York, 1987).
2. J-F. Nicoud and R. J. Twieg, "Design and Synthesis of Organic Molecular Compounds for Efficient Second-Harmonic-Generation," in *Nonlinear Optical Properties of Organic Molecules and Crystals*, D. S. Chemla and J. Zyss, eds. (Academic Press, New York, 1987).
3. J-L. Oudar, *J. Chem. Phys.* **67**, 446 (1977).
4. D. Josse, S. X. Dou, J. Zyss, P. Andreazza, and A. Périgaud, *Applied Physics Letters* **61**, 121 (1992).
5. J. Zyss, J-F. Nicoud, and M. Coquillay, *J. Chem. Phys.* **81**, 4160 (1984).
6. J. Zyss, "Organic Crystals and Quadratic Nonlinear Optics: the Transparency-Efficiency Trade-off," in *Conjugated Polymeric Materials: Opportunities in Electronics, Optoelectronics and Microelectronics*, J-L. Bredas and R. R. Chance, eds. (Kluwer, Dordrecht, 1990), p. 545.
7. G. R. Meredith in *Nonlinear Optical Properties of Organic and Polymeric Materials*, D. J. Williams, ed., ACS Symposium Series, **223**, 30 (1983).
8. K. Singer, M. Kuzyk, and J. Sohn, *J. Opt. Soc. Am. B* **4**, (6), 968 (1987).
9. W. Tam, B. Buerin, J. C. Calabrese, and S. H. Stevenson, *Chem. Phys. Lett.* **154**, 93 (1989).
10. J. D. Bierlein, L. K. Cheng, Y. Wang, and W. Tam, *Appl. Phys. Lett.* **56**, 423 (1990).
11. M. Okazaki, S. Kibodera, H. Funkunaga, N. Uchino, and M. Ishihara, *Fujifilm Res. & Dev.* **36**, 55 (1991).
12. A. Harada, Y. Okazaki, K. Kamiyama, and S. Umegaki, *Appl. Phys. Lett.* **59**, 1535 (1991).
13. N. V. Kondratenko, V. I. Popov, G. N. Timofeeva, N. V. Ignat'ev, and L. M. Yagupolskii, *J. Org. Chem. USSR (Engl. Trans.)* **21**, 2367 (1985).
14. W. Tam, L. T. Cheng, J. D. Bierlein, L. K. Cheng, Y. Wang, A. E. Feiring, G. R. Meredith, D. F. Eaton, J. C. Calabrese, and G. L. J. A. Rikken, in *Materials for Nonlinear Optics—Chemical Perspectives*, Marder *et al.*, eds., ACS Symposium Series, **455**, 158 (1991).
15. J. O. Morley, V. J. Docherty, and D. Pugh, *J. Chem. Soc. Perkin Trans. II*, 1361 (1987).
16. D. W. Robinson, H. Abdel-Halim, S. Inoue, M. Kimura, and D. O. Cowan, *J. Chem. Phys.* **90**, 3427 (1989).
17. A. Ulman, C. S. Willand, W. Köhler, D. R. Robello, D. J. Williams, and L. Handley, *J. Am. Chem. Soc.* **112**, 7083 (1990).
18. M. Barzoukas, D. Josse, J. Zyss, P. Gordon, and J. O. Morley, *Chem. Phys.* **139**, 359 (1989).
19. G. Mignani, M. Barzoukas, J. Zyss, G. Soula, F. Balegroune, D. Grandjean, and D. Josse, *Organometallics* **10**, 3660 (1991).
20. G. Mignani, A. Krämer, G. Puccetti, I. Ledoux, G. Soula, and J. Zyss, *Mol. Engineer.* **1**, 11 (1991).

21. G. Mignani, A. Krämer, G. Puccetti, I. Ledoux, J. Zyss, and G. Soula, *Organometallics* **10**, 3656 (1991).
22. H. J. Berwin, *J. Chem. Soc. Chem. Comm.* 237 (1972).
23. C. Eaborn, *J. Chem. Soc. Chem. Comm.* 1255 (1972).
24. G. Mignani, Krämer, G. Puccetti, I. Ledoux, G. Soula, J. Zyss, and R. Meyrueix, *Organometallics* **9**, 2640 (1990).
25. C. C. Teng and A. F. Garito, *Phys. Rev. Lett.* **50**, 350 (1989).
26. D. N. Beratan, "Electronic Polarizability and Chemical Structure," in *Materials for Nonlinear Optics*, S. E. Marder, J. E. Sohn, and G. D. Stucky, eds., ACS Symposium Series **455**, American Chemical Society, Washington DC (1991).
27. M. M. Choy, S. Ciraci, and R. K. Byer, *IEEE J. Quant. Electron.* **QE-11**, 40 (1975).
28. G. Berkovic, T. Rasing, and Y. R. Shen, *J. Opt. Soc. Am. B* **4**, 945 (1987).
29. G. Berkovic, Y. R. Shen, and M. Shadt, *Mol. Cryst. Liquid Cryst.* **150b**, 607 (1987).
30. C. Combellas, H. Gauthier, J. Simon, A. Thiébault, F. Tournilhac, M. Barzoukas, D. Josse, I. Ledoux, C. Amatore, J-N. Verpeaux, *J. Chem. Soc. Chem. Comm.* 203 (1988).
31. I. Ledoux, J. Zyss, A. Jutand, and C. Amatore, *Chem. Phys.* **150**, 117 (1991).
32. J. Zyss, I. Ledoux, M. Bertault, and E. Toupet, *Chem. Phys.* **150**, 125 (1991).
33. G. Puccetti, I. Ledoux, J. Zyss, A. Jutand, and C. Amatore, *Chem. Phys.* **160**, 467 (1992).
34. J. O. Morley, V. L. Docherty, and D. Pugh, *J. Chem. Soc. Perkin Trans. II*, 1351 (1987).
35. M. Kasha, H. R. Rawls, and M. Ashraf El-Bayoumi, *Pure Applied Chem* **11**, 371 (1965).
36. N. F. Phelan and M. Orchin, *J. Chem. Ed.* **45**, 633 (1968).
37. Y. Itho, K. Oono, M. Isogai, and A. Kakuta, *Mol. Cryst. Liq. Cryst.* **170**, 259 (1989).
38. S. Guha, C. C. Frazier, and W. Chen in *Nonlinear Optical Properties of Organic Materials*, G. Khanarian, ed., Proceed. SPIE, **971**, 89 (1988).
39. R. J. Twieg and K. Jain in *Nonlinear Optical Properties of Organic Materials*, D. J. Williams, ed., ACS Symposium Series, **233**, 57 (1983).
40. J. F. Nicoud in *Organic Materials for Nonlinear Optics*, R. A. Hann and D. Bloor, eds., Royal Society of Chemistry, special publication **69**, 157 (1989).
41. C. Serbutoviez, J. F. Nicoud, I. Ledoux, and J. Zyss, unpublished results (1993).
42. D. Fichou, T. Watanabe, T. Takeda, S. Miyata, Y. Goto, and M. Nakayama, *Jpn. J. Appl. Phys.* **27**, L249 (1988).
43. G. Zhang, T. Kinoshita, K. Sasaki, Y. Goto, and M. Nakayama, *J. Cryst. Growth* **100**, 411 (1990).
44. Y. Goto, A. Hayashi, G. J. Zhang, M. Nakayama, Y. Kitaoka, T. Sasaki, T. Watanabe, S. Miyata, K. Honda, and M. Goto in *Nonlinear Optical Properties of Organic Materials III*, G. Khanarian, ed., Proceed. SPIE **1337**, 297 (1990).
45. Y. Kitaoka, T. Sasaki, S. Nakai, and Y. Goto, *Appl. Phys. Lett.* **59**, 19 (1991).

46. L. S. Pu in *Materials for Nonlinear Optics—Chemical Perspectives*, S. R. Marder *et al.*, eds., ACS Symposium Series, **455**, 331 (1991).
47. L. S. Pu, *J. Chem. Soc. Chem. Comm.*, 249 (1991).
48. T. Tsunekawa, T. Gotho, and M. Iwamoto, *Chem. Phys. Lett.* **166**, 353 (1990).
49. T. Tsunekawa, T. Gotho, H. Matakin, T. Kondoh, S. Fukuda, and M. Iwamoto in *Nonlinear Optical Properties of Organic Materials III*, G. Khanarian, ed., Proc. SPIE, **1337**, 272 (1990).
50. P. Skrabal, J. Steiger, and H. Zollinger, *Helv. Chim. Acta* **58**, 800 (1975).
51. D. Lupo, W. Prass, U. Scheunemann, A. Laschewsky, H. Ringsdorf, and I. Ledoux, *J. Opt. Soc. Am. B* **5**, 300 (1988).
52. I. Ledoux, W. Paulus, A. Laschewsky, and H. Ringsdorf, unpublished results (1990).
53. C. Bubeck, A. Laschewsky, D. Lupo, D. Neher, P. Ottenbreit, W. Paulus, W. Prass, H. Ringsdorf, and G. Wegner, *Adv. Mater.* **3**, 54 (1991).
54. C. W. Spangler, K. O. Havelka, M. W. Becker, T. A. Kelleher, and L-T. Cheng, *Proceedings of the SPIE* **1560**, 139 (1991).
55. T. Kurihara, H. Tabei, and T. Kaino, *J. Chem. Soc. Chem. Comm.*, 959 (1987).
56. H. Tabei, T. Kurihara, and T. Kaino, *Appl. Phys. Lett.* **50**, 1855 (1987).
57. A. E. Stiegman, V. M. Miskowski, J. W. Perry, and D. R. Coulter, *J. Am. Chem. Soc.* **109**, 5884 (1987).
58. A. E. Stiegman, E. Graham, K. J. Perry, L. R. Khundkar, L. T. Cheng, and J. W. Perry, *J. Am. Chem. Soc.* **113**, 7658 (1991).
59. M. Barzoukas, A. Fort, G. Klein, A. Boeglin, C. Serbutoviez, L. Oswald, and J. F. Nicoud, *Chem. Phys.* **153**, 457 (1991).
60. G. R. Desiraju and T. S. R. Krishna, *J. Chem. Soc. Chem. Comm.* 192 (1988).
61. J. F. Nicoud in *Nonlinear Optical Properties of Organic Materials*, G. Khanarian, ed., Proc. SPIE, **971**, 68 (1988).
62. J. F. Nicoud, R. J. Twieg, C. Moylan, and D. Burland, unpublished results (1992).
63. D. Jungbauer, I. Teraoka, D. Y. Yoon, B. Reck, J. D. Swalen, and R. J. Twieg, *J. Appl. Phys* **69**, 8011 (1991).
64. H. Sasabe, T. Wada, M. Hosoda, H. Ohkawa, A. Yamada, and A. F. Garito, *Mol. Cryst. Liq. Cryst.* **189**, 155 (1990).
65. K. C. Rustagi and J. Ducuing, *Opt. Comm.* **10**, 258 (1974).
66. M. Blanchard-Desce, I. Ledoux, J-M. Lehn, J. Malthête, and J. Zyss, *J. Chem. Soc. Chem. Comm.*, 237 (1988).
67. M. Barzoukas, M. Blanchard-Desce, D. Josse, J-M. Lehn, and J. Zyss, *Chem. Phys.* **133**, 323 (1989).
68. M. Barzoukas, M. Blanchard-Desce, D. Josse, J-M. Lehn, and J. Zyss, *Inst. Phys. Conf. Ser.* **103**, 239 (1989).
69. R. A. Huijts and G. L. J. Hesselink, *Chem. Phys. Lett.* **156**, 209 (1989).
70. H. Ikeda, Y. Kawabe, T. Sakai, and K. Kawasaki, *Chem. Phys. Lett.* **179**, 551 (1991).

71. B. M. Pierce, "A Theoretical Analysis of the Third-Order Nonlinear Optical Properties of Linear Cyanines and Polyenes," in *Nonlinear Optical Organic Materials IV*, SPIE Proceedings, **1560**, 148 (1991).
72. J. Messier, F. Kajzar, C. Sentein, M. Barzoukas, J. Zyss, M. Blanchard-Desce, and J.-M. Lehn, *Nonlinear Optics*, **2**, 53 (1992).
73. G. P. Agrawal, C. Cojan, and C. Flytzanis, *Phys. Rev. B* **17**, 776 (1978).
74. F. Meyers and J-L. Brédas, *Electronic Structure and Nonlinear Optical Properties of Push-Pull Polyenes: Theoretical Investigation of Benzodithiapolyenals and Dithiolenepolyenals, Organic Materials for Nonlinear Optics*, R. A. Hann and D. Bloor, Eds., Royal Soc. Chem. London (1992).
75. H. M. Hutchinson and L. E. Sutton, *J. Chem. Soc.* 4382 (1958).
76. A. Dulcic, C. Flytzanis, C. L. Tang, D. Pépin, M. Fétizon, and Y. Hopilliard, *J. Chem. Phys.* **74**, 1559 (1981).
77. F. Krausz, E. Winter, and G. Leising, *Phys. Rev. B* **39**, 3701 (1989).
78. G. Puccetti, I. Ledoux, J. Zyss, M. Blanchard-Desce, and J-M. Lehn, *J. Phys. Chem.* (1992).
79. H. H. Jaffe and M. Orchin, *Theory and Applications of Ultraviolet Spectroscopy*, Wiley, New York, chap. 11 (1962).
80. K. Knoll and R. R. Schrock, *J. Am. Chem. Soc.* **111**, 7989 (1989).
81. G. Craig, R. R. Schrock, R. Silbey, G. Puccetti, I. Ledoux, and J. Zyss, *J. Am. Chem. Soc.* **115**, 860 (1993).
82. J. Oudar and R. Hierle, *J. Appl. Phys.* **48**, 2669 (1977).
83. J. Zyss, D. S. Chemla, and J-F. Nicoud, *J. Chem. Phys.* **74**, (9), 4800 (1981).
84. J. K. Whitesell, R. E. Davies, L. L. Saunders, R. J. Wilson, and J. P. Feagins, *J. Am. Chem. Soc.* **113**, 3267 (1991).
85. P. Günter, C. Bosshard, K. Sutter, H. Arend, G. Chapuis, R. J. Twieg, and D. Dobrowolski, *Appl. Phys. Lett.* **50**, 486 (1987).
86. C. Bosshard, K. Sutter, P. Günter, and G. Chapuis, *J. Opt. Soc. Am. B* **6**, 721 (1989).
87. T. Uemiya, N. Uenishi, N. Shiomizu, T. Yoneyama, and K. Nakatsu, *Mol. Cryst. Liq. Cryst.* **182**, 51 (1990).
88. K. Sutter, G. Knöpfle, N. Saupper, J. Hulliger, P. Günter, and W. Petter, *J. Opt. Soc. Am. B* **7**, 1483 (1991).
89. S. Tomaru, S. Matsumoto, T. Kutihara, H. Suzuki, N. Ooba, and T. Kaino, *Appl. Phys. Lett.* **58**, 2583 (1991).
90. T. Watanabe, H. Yamamoto, T. Hosomin, and S. Miyata in *Organic Molecules for Nonlinear Optics and Photonics*, J. Messier *et al.*, eds. NATO ASI Series, **194**, 151 (1991).
91. T. Tsunekawa, T. Gotho, H. Mataki, and M. Iwamoto in *Nonlinear Optical Properties of Organic Materials III*, G. Khanarian, ed., Proc. SPIE, **1337**, 285 (1990).
92. H. E. Katz, M. L. Schilling, W. R. Holland, and T. Fang in *Materials for Nonlinear Optics—Chemical Perspectives*, S. R. Marder *et al.*, eds. ACS Symposium Series **455**, 267 (1991).

93. S. M. Arakelyan, G. A. Lyakhov, and Yu. S. Chilingaryan, *Sov. Phys. Usp.* (*Engl. Transl.*) **23**, 245 (1980).
94. C. Fouquey, J. M. Lehn, and J. Malthête, *J. Chem. Soc. Chem. Comm.*, 1424 (1987).
95. F. Tournilhac, J. F. Nicoud, J. Simon, S. Weber, D. Guillon, and A. Skoulios, *Liq. Cryst.* **2**, 55 (1987).
96. A. Taguchi, K. Kajikawa, Y. Ouchi, H. Takezoe, and A. Fukuda in *Nonlinear Optics of Organics and Semiconductors*, T. Kobayashi, ed., Springer Proceedings in Physics, 36, 250 (1989).
97. M. Ozaki, M. Utsumi, T. Goto, Y. Morita, K. Daido, Y. Sadohara, and K. Yoshino, *Ferroelectrics* **121**, 259 (1991).
98. D. M. Walba, M. B. Ros, N. A. Clark, R. Shao, M. G. Robinson, L. Y. Liu, and K. M. Johnson, *J. Am. Chem. Soc.* **113**, 5471 (1991).
99. A. Samoc, M. Samoc, J. Fünfschilling, and I. Zschokke-Gränacher, *Chem. Phys. Lett.* **114**, 423 (1985)
100. A. Miniewicz, A. Samoc, and J. Sworakowski, *J. Molec. Electr.* **4**, 25 (1988).
101. D. Kholer, M. Staehelin, T. Enderle, J. Fünfschilling, and I. Zschokke-Gränacher, *Helv. Phys. Acta* **62**, 912 (1989).
102. N. Okamoto, T. Abe, D. Chen, H. Fujimura, and R. Matsushima, *Opt. Commun.* **4**, 421 (1990).
103. R. Matsushima, H. Takeshita, and N. Okamoto, *J. Mater. Chem.* **1**, 591 (1991).
104. T. W. Panunto, Z. Urbanczyk-Lipowska, R. Johnson, and M. C. Etter, *J. Am. Chem. Soc. Am.* **109**, 7786 (1987).
105. M. C. Etter, *Acc. Chem. Res.* **23**, 120 (1987).
106. M. C. Etter, G. M. Frankenbach, and D. A. Adsmond, *Mol. Cryst. Liq. Cryst.* **187**, 25 (1990).
107. M. C. Etter and G. M. Frankenbach, *Chem. Mater.* **1**, 10 (1989).
108. S. J. Geib, S. C. Hirst, C. Vicent, and A. D. Hamilton, *J. Chem. Soc. Chem. Comm.* 1283 (1991).
109. G. Puccetti, I. Ledoux, A. Périgaud, J. Badan, and J. Zyss., *J. Opt. Soc. Am. B* **10** (in press) (1993).
110. R. Masse and J. Zyss, *Mol. Eng.* **1**, (2), 141 (1991).
111. R. Masse and A. Durif, *Zeit. für Kristallogr.* **190**, 19 (1990).
112. K. Aoki, K. Agano, and Y. Iitaka, *Acta Cryst.* **B27**, 11 (1971).
113. D. Eimerl, S. Velsko, L. Davis, F. Wang, G. Loiacono, and G. Kennedy, *IEEE J. Quant. Electron.* **QE25**, 179 (1989).
114. J. Zyss, *Nonlinear Optics* **1**, 3 (1991).
115. Z. Kotler, R. Hierle, D. Josse, J. Zyss, and R. Masse, *J. Opt. Soc. Am. B*, **9**, 534 (1992).
116. R. T. Bailey, F. R. Cruickshank, S. M. G. Guthrie, B. J. McArdle, H. Morrison, D. Pugh, E. A. Shepherd, J. N. Sherwood, C. S. Yoon, R. Kashyap, B. K. Nayar, and K. I. White, *Opt. Commun.* **65** (3), 229 (1988).

117. C. Bosshard, M. Küpfer, P. Günter, C. Pasquier, S. Zahir, and M. Seifert, *Appl. Phys. Lett.* **56**, 1204 (1990).
118. J. Zyss, J. Pécaud, J-P. Lévy, and R. Masse, submitted to *Acta Cryst. B* (1992).
119. A. Samoc, M. Samoc, P. Prasad, and A. Krajewska-Cizio, *J. Opt. Soc. Am B*, **9**, 1819 (1992).
120. A. Samoc, M. Samoc, D. Kohler, M. Stähelin, J. Fünfschilling, and I. Zschokke-Gränacher, *Nonlinear Optics*, **2**, 13 (1992).
121. D. F. Eaton, *Tetrahedron* **43**, 1551 (1987).
122. S. Tomaru, S. Zembutsu, M. Kawachi, and M. Kobayashi, *J. Chem. Soc. Chem. Comm.* 1207 (1984).
123. S. Tomaru, S. Zembutsu, M. Kawachi, and M. Kobayashi, *J. Inclusion Phenom.* **2**, 885 (1984).
124. Y. Wang and D. S. Eaton, *Chem. Phys. Lett.* **120**, 441 (1985).
125. D. F. Eaton, A. G. Anderson, W. Tam, and Y. J. Wang, *J. Am. Chem. Soc.* **109**, 1886 (1987).
126. W. Tam, D. F. Eaton, J. C. Calabrese, I. D. Williams, Y. Wang, and A. G. Anderson, *Chem. Mater.* **1**, 128 (1989).
127. S. D. Cox, T. E. Gier, and G. D. Stucky, *Chem. Mater.* **2**, 609 (1990).
128. G. D. Stucky and J. E. MacDougall, *Science* **247**, 4943, 669 (1990).
129. S. Palacin, M. Blanchard-Desce, J-M. Lehn, and A. Barraud, *Thin Sol. Films* **178**, 387 (1989).
130. V. Dentan, M. Blanchard-Desce, S. Palacin, I. Ledoux, A. Barraud, J-M. Lehn, and J. Zyss. *Thin Sol. Films* **210–211**, 221 (1992).
131. M. Kasha, M. Ashraf El-Bayoumi, and W. Rhodes, *J. Chem. Phys.* **58**, 916 (1961).
132. D. P. Craig and S. H. Walmsley, *Exciton in Molecular Crystals* (Benjamin, Amsterdam, 1968).
133. M. Pierre, P. L. Baldeck, D. Block, R. Georges, and P. Trommsdorf, *Chem. Phys.* **156**, 103 (1991).
134. M. Joyeux and N. Q. Dao, *J. Raman Spectr.* **19**, 441 (1988).
135. J. Zyss, in *Hommage to Galileo*, P. Mazzoldi, Ed., p. 399 (Cleup, Padova, 1992).
136. J. Zyss in *Nonlinear Optics: Fundamentals, Materials and Devices*, S. Miyata, ed., Elsevier, Amsterdam, p. 33 (1992).
137. J. Zyss, *J. Chem. Phys.* **98**, 6583 (1993).
138. E. P. Wigner, *Group Theory*, Academic Press, New York (1959).
139. P. D. Maker, *Phys. Rev.* **A1**, 923 (1970).
140. J. Jerphagnon, *Phys. Rev.* **B2**, 1091 (1970).
141. J. Jerphagnon, C. S. Chemla, and R. Bonneville, *Adv. Phys.* **27**, 609 (1978).
142. D. S. Chemla, J. L. Oudar, and J. Jerphagnon, *Phys. Rev.* **B12**, 4534 (1975).
143. M. Barzoukas, D. Josse, P. Fremaux, J. Zyss, J. F. Nicoud, and J. O. Morley, *J. Opt. Soc. Am. B* **4**, 967 (1987).
144. J. Zyss and G. Berthier, *J. Chem. Phys.* **77**, 3635 (1982).
145. J. Zyss, T. Chau Van, C. Dhenaut and I. Ledoux, submitted to *Chem. Phys.* and *CLEO '93 Proceedings* paper QTuA2.

146. H. S. Nalwa, K. Nakajima, T. Watanabe, K. Nakamura, A. Yamada, and S. Miyata, *Jpn. J. Appl. Phys.* **30**, 983 (1991).
147. J. Zyss, C. Dhenaut, T. Chau Van, and I. Ledoux, to be published in *Chem. Phys. Lett.*
148. A. Persoons, T. Verbiest, K. Clays, F. Meyers, and J. L. Brédas, to be published (1993).
149. B. Boulanger and G. Marnier, *J. Phys.: Condens. Matter* **3**, 8327 (1991).
150. J. L. Oudar and J. Zyss, *Phys. Rev.* **A26**, 2016 (1982).
151. R. W. Terhune, P. D. Maker, and C. M. Savage, *Phys. Rev. Lett.* **14**, 681 (1965).
152. K. Clays and A. Persoons, *Phys. Rev. Lett.* **66**, 2980 (1991).
153. J. J. Wolff, S. F. Nelsen, P. A. Petillo, and D. R. Powell, *Chem. Ber.* **124**, 1719 (1991).
154. J. J. Wolff, S. F. Nelsen, and D. R. Powell, *J. Org. Chem.* **56**, 5908 (1991).
155. H. H. Cady and A. C. Larson, *Acta Crystallogr.* **18**, 485 (1965).
156. M. Joffre, D. Yaron, R. Silbey, and J. Zyss, *J. Chem. Phys.* **97**, 5607 (1992).
157. I. Ledoux, J. Zyss, J. S. Siegel, J. Brienne, and J-M. Lehn, *Chem. Phys. Lett.* **172**, 440 (1990).
158. J-L. Brédas, F. Meyers, B. Pierce, and J. Zyss, submitted to *J. Am. Chem. Soc.* **114**, 4928 (1992).
159. C. J. Chen, B. Wu, A. Jiang, and G. You, *Sci. Sinica* **28**, 235 (1985).
160. C. S. Willand and A. C. Albrecht, *Opt. Commun.* **57**, 146 (1985).
161. D. Xu, M. Jiang, and Z. Tan, *Acta Chim. Sinica* **2**, 230 (1983)

Chapter 5

QUADRATIC NONLINEAR OPTICAL EFFECTS IN CRYSTALLINE WAVEGUIDING STRUCTURES

Takashi Kondo

and

Ryoichi Ito

Department of Applied Physics, Faculty of Engineering, The University of Tokyo, Bunkyo-ku, Tokyo, Japan

1. INTRODUCTION

Nonlinear optics is attracting increasing attention not only from the physics community but also from industry because of the remarkable growth of

ISBN 0-12-784450-3

photonics technologies such as optical data storage, laser printing, and fiber-optic communications. Most of these new technologies involve laser diodes as their light sources. The laser diode is one of the most important key devices in photonics technologies. It is only natural that researchers have begun searching for further ways to expand and enhance the performance of laser diodes with the aim of improving the performance of the existing technologies and, more importantly, developing novel photonics technologies such as photonic switching and optical computing. Nonlinear optics is expected to play a key role in these new developments.

In particular, frequency doubling of laser diodes via optical second-harmonic generation (SHG), a typical example of quadratic nonlinear optical effects, has been a topic of intense research efforts in recent years. A major reason for this is the strong demand for compact short-wavelength coherent light sources in the optical data storage technology, the largest market for lasers. Combination of laser diodes and frequency conversion devices is obviously a highly attractive means to achieve high-density optical data storage, because it will offer compact light sources with coherent emission down to the blue or ultraviolet (UV) portion of the spectrum. Although optical data storage is a most visible application area for the frequency-doubled laser diodes, there are many other important application areas for them, such as color xerography, high-resolution laser printing, spectroscopy and even photo-chemistry. Furthermore, frequency-doubled laser diodes that emit coherent light covering the near-infrared to UV wavelength regime will no doubt open up a new area of both scientific and technological importance.

Nonlinear optics using laser diodes as light sources presents a special challenge to materials and device scientists. The largest problem arises because of relatively low powers available from laser diodes. First of all, the choice of materials with very large nonlinear optical coefficients is, of course, mandatory in frequency conversion devices compatible with laser diodes. Moreover, appropriate designs of the devices are required to enhance optical power densities, because the efficiency of SHG is proportional to the fundamental optical power density in the nonlinear medium. Nonlinear integrated optics or nonlinear waveguiding optics offers the possibility of enhancing nonlinear optical interactions even at relatively low power levels [1,2]. Since the high concentration of optical power can be kept over extended interaction lengths—the diffraction-free characteristic of waveguides—highly efficient nonlinear optical interactions can take place in appropriately designed optical waveguides.

Efficient frequency conversion also requires phase matching. Phase matching in optical waveguides can be achieved using several configurations as follows:

(a) Phase matching between guided modes [3,4]: Frequency doubling takes place by coupling the fundamental guided mode into the second-harmonic guided mode. Phase matching is achieved by adjusting dimensional parameters of the waveguide so that the propagation constant of the second-harmonic guided mode coincides with that of the fundamental guided mode.
(b) Quasi phase matching [5,6]: Phase matching is achieved, in effect, by compensating the phase mismatch between the fundamental and the second-harmonic guided modes taking advantage of the periodic modulation of the nonlinearity.
(c) Balanced phase matching [7]: Phase matching is achieved in segmented structures, each segment consisting of sections, in which the phase mismatch of each section is "balanced" with that of other sections so that the overall mismatch is canceled out at the end of each segment.
(d) Phase matching in the Cerenkov-radiation scheme [8]: Frequency doubling takes place by coupling guided mode into the second-harmonic radiation mode. In this case, phase matching can be achieved automatically, because of the continuous distribution of the propagation constant of the radiation mode, which is free from the constraint of the standing-wave formation in the guiding layer.

These phase-matching schemes have been demonstrated in optical waveguides of the well-known inorganic nonlinear optical crystals with relatively large nonlinear optical coefficients, such as $LiNbO_3$ [9–11], $LiTaO_3$ [12] and $KTiOPO_4$ (KTP) [7,13]. Among the configurations mentioned above, SHG in the Cerenkov-radiation scheme (Cerenkov-type SHG) is probably most useful in practical devices at present, though conversion efficiencies obtainable in Cerenkov-type SHG are intrinsically not so high as those in other configurations. Taniuchi and Yamamoto demonstrated the usefulness of the Cerenkov-type SHG using a proton-exchanged $LiNbO_3$ channel waveguide [10]. They obtained a conversion efficiency as high as one percent for a fundamental power of 40 mW at $\lambda = 0.84\ \mu m$.

As to nonlinear optical materials, organic crystals have received considerable attention because some of them exhibit very large optical nonlinearities [14]. Despite their high potentiality, organic crystals have been far behind inorganics in the application to practical devices. However, remarkable progress in the past few years in the research on organic-crystalline waveguiding structures is changing such a situation. Reasonably high conversion efficiencies comparable to that of a $LiNbO_3$ waveguide have been demonstrated in some organic-crystalline waveguides using the Cerenkov-radiation scheme [15,16].

In this chapter, we present a detailed description of SHG in organic-crystalline waveguiding structures, placing emphasis on the Cerenkov-type SHG. In Sec. 2, typical features of organic nonlinear optical crystals are summarized from the standpoint of the applicability to practical devices. Section 3 is devoted to an introductory description of SHG in bulk and waveguiding configurations based on theoretical analyses, emphasizing phase-matching characteristics and conversion efficiencies. We review the recent progress in experimental work on SHG in organic-crystalline waveguiding structures, and then present the guidelines for efficient SHG in Sec. 4. Finally, conclusions and future outlook are given in Sec. 5.

2. ORGANIC NONLINEAR OPTICAL CRYSTALS

It is now widely recognized that some organic crystals are far superior to the conventional inorganic nonlinear optical crystals in respect of quadratic optical nonlinearities. Many researchers, including chemists, physicists, and optical and device scientists, have come to be interested in organic nonlinear optical crystals stimulated by the systematic work, dating back to the early 1970s, done chiefly by researchers of CNET in France and Bell Laboratories in the United States. In particular, the discovery of 2-methyl-4-nitroaniline (MNA) that possesses an extremely large nonlinear optical coefficient, first reported by Levine *et al.* in 1979 [17], directed researchers' attention to this newly developing field.

From the early stage of the studies on organic nonlinear optical crystals, it has frequently been pointed out that organic crystals with high nonlinearities are suitable materials for the frequency doubling of laser diode light and that organic-crystalline waveguides are the most favorable device structure to that purpose. However, the usefulness of a nonlinear optical crystal in practical devices cannot be judged only by its nonlinear optical coefficient but also by other factors such as crystal quality, processability, and stability. Moreover, the performance of a device fabricated using a nonlinear optical crystal must be judged again on the basis of a number of parameters, such as conversion efficiencies, phase-matching tolerances to size, temperature, angle and wavelength deviations, compactness, durability, and even production costs. Therefore, the usefulness of organic crystals in practical devices is not so obvious. In the following, let us summarize typical characteristics of organic nonlinear optical crystals investigated so far, considering their applicablity to practical devices.

2.1. Nonlinear Optical Properties

Some organic crystals exhibit very large quadratic optical nonlinearities. For example, an organic crystal 4′-nitrobenzylidene-3-acetamino-4-methoxyaniline (MNBA) has an extremely large nonlinear optical coefficient; d_{11} of MNBA was measured to be 13 times larger than d_{33} of $LiNbO_3$, the largest coefficient among the inorganics that are transparent in the visible range, and the corresponding figure of merit for SHG, d^2/n^3, where d is the second-order nonlinear optical coefficient and n is the refractive index, is larger than that of $LiNbO_3$ (d_{33}) by more than two orders of magnitude [18]. The figures of merit for SHG at a fundamental wavelength of 1.064 μm for some organic crystals are shown in Fig. 1 along with those for some conventional inorganic crystals. Their extraordinary large figures of merit are indeed highly attractive because the conversion efficiency is proportional to the figure of merit of the material used in the device, as far as the other factors—e.g., the overlap integral in the waveguiding structure—are the same.

Their large nonlinearities originate from the large hyperpolarizabilities of the individual molecules, the building blocks of the organic molecular crystals. Intramolecular charge transfer (CT) between conjugated donor and acceptor substituents dominates the fundamental excitation of the most nonlinear optical molecules and is recognized to be responsible for their large hyperpolarizabilities. The strong CT absorption band commonly appears in the UV region, centered around $0.3 \leq \lambda \leq 0.4$ μm. Nonlinear optical coefficients of the organic crystals exhibit large dispersion showing strong resonance to the CT excited states. Figure 2(a) shows the dispersion of d_{11} of MNA [17,25]. The enhancement of the values of d_{11} due to the CT transition resonance is clearly seen even at the fundamental wavelength of 1.064 μm, although the corresponding second-harmonic wavelength, 0.532 μm, is longer than the wavelength resonant to the CT state, 0.37 μm in an ethanol solution (see Fig. 2(c)). The remarkable resonance enhancement is due to the large oscillator strength of the CT transition, and is common to all the organic nonlinear optical materials based on the CT excitation. We may utilize their resonance-enhanced nonlinearities for efficient frequency conversion.

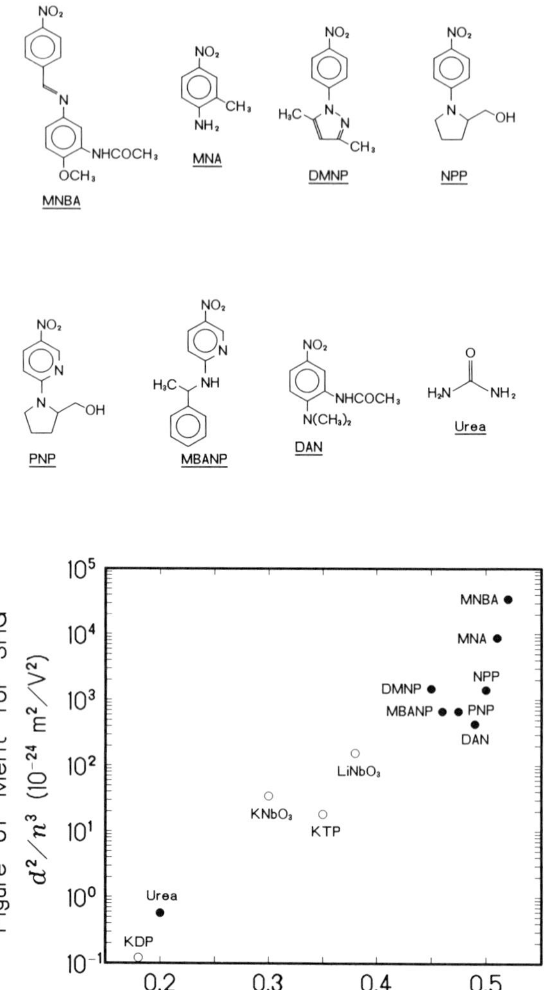

Fig. 1. Figures of merit for SHG at the fundamental wavelength of 1.064 μm and cutoff wavelengths for some organic and inorganic crystals. Molecular structures of organic compounds are also shown. The following components are chosen to calculate figures of merit: MNBA (4′-nitrobenzylidene-3-acetamino-4-methoxyaniline) [18], d_{11}; MNA (2-methyl-4-nitroaniline) [17], d_{11}; DMNP (3,5-dimethyl-1-(4-nitrophenyl)pyrazole) [19], d_{32}; NPP (*N*-(4-nitrophenyl)-(*L*)-prolinol) [20], d_{21}; PNP (2-(*N*-prolinol)-5-nitropyridine) [21], d_{21}; MBANP ((−)2-(α-methylbenzyl-amino)-5-nitropyridine) [22], d_{22}; DAN (4-(*N*,*N*-dimethylamino)-3-acetamidonitro-benzene) [23], d_{23}; $LiNbO_3$, d_{33}; $KNbO_3$, d_{33}; KTP ($KTiOPO_4$), d_{33}; urea [24], d_{14}; KDP (KH_2PO_4), d_{36}.

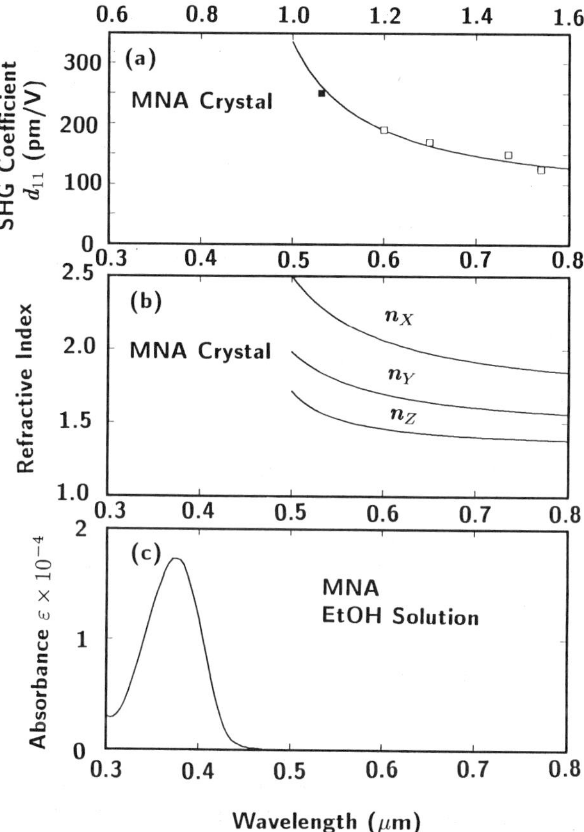

Fig. 2. Nonlinear and linear optical properties of MNA. (a) Dispersion of the nonlinear optical coefficient d_{11} of MNA crystal. Upper abscissa represents the fundamental wavelength. Closed square denotes the experimental value measured using an Nd:YAG laser [17], and the open squares denote the experimental values measured using laser diodes [25]. The solid line represents a fit to the two-level model. (b) Dispersion of the principal refractive indices of MNA crystal according to the Sellmeir formula given by Morita *et al.* [26]. (c) Absorption spectrum of MNA measured in an ethanol solution.

2.2. Other Optical Properties

Extremely large nonlinearities of organic crystals are partly offset by their relatively long cutoff wavelengths. The abscissa in Fig. 1 gives the absorption cutoff wavelengths of the nonlinear optical crystals. All the organic compounds plotted in this figure, excepting urea, contain a similar basic element as the reservoir of the nonlinearly responding electrons—the nitro-substituted aromatic ring. As a result, their absorption cutoff wavelengths, which are dominated by the tail of the strong CT absorption band, are longer than about 0.45 μm; they are absorbing in the wavelength range of the second-harmonics of typical GaAs laser diodes. Although the molecular engineering for nonlinear optics in organic molecules has made remarkable progress in the last decade, we cannot overcome the barrier of the efficiency-transparency tradeoff yet. In order to invent highly nonlinear compounds that are transparent down to $\lambda \simeq 0.4$ μm, or to still shorter wavelengths, new strategies, probably involving new molecular structures other than nitro-aromatic derivatives, will be needed.

As to the refractive index, the most important parameter closely connected with the phase-matching characteristics, organic crystals exhibit significant features—large dispersion and birefringence. Figure 2(b) shows the dispersion of the principal refractive indices of MNA [26]. The large index dispersion is caused by the resonance to the CT transition with large oscillator strength, as in the case of the large dispersion of the nonlinear optical coefficients. The birefringence of MNA is extremely large; $n_X - n_Z \simeq 0.8$ at $\lambda = 0.5$ μm. The very large optical anisotropy is attributable to the orientation of the molecular dipoles in the crystal lattice; the dipole moment of an MNA molecule, which is approximately parallel to the CT axis, lies in the XY plane and nearly parallel to the X axis [27]. In general, nonlinear CT molecules with quasi-one-dimensional characters align in a highly asymmetric and anisotropic manner in crystals exhibiting large quadratic optical nonlinearities. Therefore, most organic nonlinear optical crystals exhibit strong optical anisotropies. It is difficult to conclude generally whether such a feature, large dispersion and birefringence, of a crystal is a merit for its phase matching characteristics in the bulk configuration. It may result in small angle- and wavelength-tolerances and large walk-off angles. These are, however, not necessarily serious limitations on the usefulness of organic crystals if one takes into account their very large figures of merit which enable us to make crystals shorter. On the other hand, the situation is partly disappointing in the case of SHG in waveguiding structures. The large dispersion of the refractive index drastically reduces the conversion efficiencies of the guided-guided and the Cerenkov-type SHG when the diagonal element of the d tensor, d_{ii}, is utilized in the nonlinear interaction, i.e. the polarizations of the fundamental and the second-harmonic waves are parallel. This means that the extremely large nonlinear optical coefficient of, for example, MNBA or MNA cannot be utilized effectively in the guided-guided or the Cerenkov-type interaction, even when the phase matching is achieved. This is, of course, not the case in SHG via the off-diagonal d-tensor elements where the dispersion can be compensated by the birefringence. The theoretical background of the situation mentioned here is described in detail in Sec. 3.2.

We only have quite limited information on the temperature dependence of the refractive indices of organic crystals. The temperature acceptances in organic crystals reported so far are of the order of 2 to 20°C·cm [19,28], comparable to those of inorganic crystals.

Another important property of organic nonlinear crystals is their high optical damage thresholds, which make them attractive in high-power applications. Some measurements using pulsed high-power lasers show that optical damage thresholds of organic crystals are extremely high ($\gg 1$ GW/cm^2). The mechanism of the damage is very complicated, and the question still remains as to why organic crystals are more resistant to the irradiation of laser light than inorganics. It should be noted that the situation may be different in the waveguiding devices for frequency doubling of cw laser diodes, where the optical power densities are of the order of 10 MW/cm^2. It may be necessary to eliminate residual absorptions in order to avoid possible thermal effects due to their low thermal conductivities.

2.3. Processability

In order to make organic nonlinear optical crystals truly usable in practical devices, suitable technologies for device fabrication must be developed. The first important step in the device fabrication process is, of course, purification and crystal growth. Some efforts have been directed at the purification of materials and the growth of single crystals of good optical quality [29,30], including crystal-cored fibers and thin-film crystals as well as bulk crystals, and devices of several configurations have been demonstrated using organic crystals. Technologies required in the fabrication of bulk-crystal-based devices, where organic crystals are often placed in a cavity, include cutting, polishing, and coating techniques. In the fabrication of waveguiding devices, especially of the channel-waveguide type, another important technique—lithography—is needed in addition to the techniques listed above.

Conventional fabrication technologies used for inorganics—dielectrics and semiconductors—can hardly be applied to organic crystals because of their troublesome intrinsic properties as follows.

1. Organic nonlinear optical crystals tend to be mechanically weak, because, in molecular crystals, molecules in the crystal lattice are bound

Table 1.

Properties of Organic and Inorganic Crystals.

Crystal	Bonding	Vickers hardness	Melting point (°C)
DAN[a]	van der Waals[e]	7	168
(−)MBANP[b]	van der Waals[e]	9	83
(±)MBANP[c]	van der Waals[e]	27	124
Urea	van der Waals[e]	9	135
KTP[d]	ion	570	1180
$LiNbO_3$	ion	450	1250
Si	covalent	940	1410

[a] 4-(*N*,*N*-Dimethylamino)-3-acetamidonitrobenzene.
[b] (−)2-(α-Methylbenzylamino)-5-nitropyridine.
[c] (±)2-(α-Methylbenzylamino)-5-nitropyridine, racemic form.
[d] $KTiOPO_4$.
[e] Hydrogen bondings also play a significant role in these crystals.

to each other by relatively weak bondings, such as van der Waals, multipolar, and hydrogen bondings. Table 1 shows Vickers hardness, a measure of the material hardness, of some organic crystals along with those of two ionic inorganic crystals, KTP and $LiNbO_3$, and covalent-bonding Si. One can see that the Vickers hardness of the organic molecular crystals is extremely low compared with that of inorganics. In addition, most organic crystals tend to be cleaved easily. These mechanical properties make it difficult to cut and polish organic crystals. Recent progress in the applications of organic crystals to bulk-crystal-based devices is indeed encouraging [31–34], but further improvement is badly needed. Although waveguide structures can provide mechanical protection by tough cladding or substrate materials, the polishing of input and output faces still remains a serious problem.

2. Organic crystals melt at relatively low temperatures. As shown in Table 1, the melting temperatures of the organic crystals are quite low (~100°C). High-reflection or antireflection coatings, which are necessary in cavity-enhanced nonlinear processes, cannot be formed on organic crystals using the conventional technique because of this drawback. This also makes it impossible to bake conventional photoresists used in the lithographic process.
3. Organic crystals are soluble in polar organic solvents. Therefore, organic photoresists widely used in the conventional lithography

process of semiconductors and other inorganics cannot be used for the lithography of organic crystals, because those resists are dissolved in polar organic solvents.

As to the fabrication of organic-crystalline channel waveguides, we can produce several types of channel waveguides, even if no lithographic technique is available; those fabrication methods are described in Sec. 4.1. However, it is very important to develop new lithographic techniques suitable for organic crystals, because they will enable precise control of device dimensions and provide more flexibility and variety of device configurations; in Sec. 4.1, we present a lithographic technique newly developed for organic crystals.

3. WAVEGUIDING STRUCTURES AND SECOND-HARMONIC GENERATION

3.1. Fundamentals of Second-Harmonic Generation

Quadratic nonlinear optical effects are characterized by the relation,

$$\boldsymbol{P}_{\mathrm{NL}}^{\omega_1+\omega_2} = \varepsilon_0 g \chi^{(2)}(-\omega_1 - \omega_2; \omega_1, \omega_2)\boldsymbol{E}^{\omega_1}\boldsymbol{E}^{\omega_2}, \tag{3.1.1}$$

where $\boldsymbol{P}_{\mathrm{NL}}^{\omega_1+\omega_2}$ is the Fourier complex amplitude at angular frequency $\omega_1 + \omega_2$ of the induced second-order nonlinear polarization, $\boldsymbol{E}^{\omega}$ is the Fourier amplitude at ω of the optical electric field, ε_0 is the permittivity of free space, g is the degeneracy factor arising from the intrinsic permutation symmetry, and $\chi^{(2)}(-\omega_1 - \omega_2; \omega_1, \omega_2)$ is the second-order nonlinear susceptibility tensor representing a variety of quadratic nonlinear optical effects. For the specific case of SHG, $\omega_1 = \omega_2 = \omega$ and $g = \frac{1}{2}$, and the second-harmonic polarization at angular frequency 2ω is given by

$$\begin{aligned} P_i^{2\omega} &= \varepsilon_0 \sum_{j,k} \tfrac{1}{2}\chi_{ijk}^{(2)}(-2\omega; \omega, \omega)E_j^{\omega}E_k^{\omega} \\ &= \varepsilon_0 \sum_{j,k} d_{ijk}E_j^{\omega}E_k^{\omega}, \end{aligned} \tag{3.1.2}$$

where d_{ijk}, defined as

$$\chi_{ijk}^{(2)}(-2\omega; \omega, \omega) = 2d_{ijk}, \tag{3.1.3}$$

is the second-order nonlinear optical coefficient or the SHG coefficient. The second-order nonlinear optical coefficient d_{ijk} is a third-rank tensor. Taking into account its permutation symmetry, it can be contracted to the 3×6

matrix form through the relation,

$$d_{il} = d_{ijk}; \qquad l = \begin{cases} j & \text{if } j = k \\ 9 - j - k & \text{if } j \neq k. \end{cases} \tag{3.1.4}$$

In the contracted form, the second-harmonic polarization is expressed in the following familiar format,

$$\begin{bmatrix} P_x^{2\omega} \\ P_y^{2\omega} \\ P_z^{2\omega} \end{bmatrix} = \varepsilon_0 \begin{bmatrix} d_{11} & d_{12} & d_{13} & d_{14} & d_{15} & d_{16} \\ d_{21} & d_{22} & d_{23} & d_{24} & d_{25} & d_{26} \\ d_{31} & d_{32} & d_{33} & d_{34} & d_{35} & d_{36} \end{bmatrix} \begin{bmatrix} (E_x^{\omega})^2 \\ (E_y^{\omega})^2 \\ (E_z^{\omega})^2 \\ 2E_y^{\omega}E_z^{\omega} \\ 2E_z^{\omega}E_x^{\omega} \\ 2E_x^{\omega}E_y^{\omega} \end{bmatrix}. \tag{3.1.5}$$

The nonlinear polarization $\boldsymbol{P}^{2\omega}$ generates a frequency-doubled optical electric field $\boldsymbol{E}^{2\omega}$ at angular frequency 2ω, which is the second-harmonic wave. The second-harmonic power generated by a plane-wave fundamental beam incident on a plane parallel crystal of length L (with negligible walk-off) is given by

$$\mathbf{P}^{2\omega} = \frac{2\omega^2}{\varepsilon_0 c^3} \frac{d_{\text{eff}}^2}{n_{2\omega} n_{\omega}^2} \frac{(\mathbf{P}^{\omega})^2}{A} L^2 \frac{\sin^2(\Delta k L/2)}{(\Delta k L/2)^2}, \tag{3.1.6}$$

where $\mathbf{P}^{2\omega}$ and $\mathbf{P}^{\omega}$ are the optical powers internal to the crystal of the second-harmonic and fundamental waves, respectively, d_{eff} the effective nonlinear optical coefficient, $n_{2\omega}$ and n_{ω} the refractive indices at 2ω and ω, respectively, and A the beam cross-sectional area of the fundamental wave. d_{eff}^2/n^3 is called a figure of merit for SHG and is a measure of a material capacity for efficient SHG. Δk is the wave vector mismatch between the second-harmonic and fundamental waves given by

$$\Delta k = k_{2\omega} - 2k_{\omega} = \frac{2\omega}{c}(n_{2\omega} - n_{\omega}). \tag{3.1.7}$$

If phase matching is achieved, i.e., $\Delta k = 0$, the conversion efficiency is given by

$$\eta = \frac{\mathbf{P}^{2\omega}}{\mathbf{P}^{\omega}} = \frac{2\omega^2}{\varepsilon_0 c^3} \frac{d_{\text{eff}}^2}{n^3} \frac{\mathbf{P}^{\omega}}{A} L^2. \tag{3.1.8}$$

Equations (3.1.7) and (3.1.8) show that efficient SHG requires (i) materials with large figures of merit d^2/n^3, (ii) phase matching $\Delta k = 0$, (iii) high fundamental power density $\mathbf{P}^{\omega}/A$, and (iv) long interaction length L.

Unfortunately, the last two conflict in the bulk configuration owing to diffraction, an intrinsic property of light. The high fundamental power density requires tight focusing, but the stronger the focusing, the smaller the confocal parameter beyond which the fundamental beam cross-sectional area is increased resulting in the reduction of the conversion efficiency. Therefore, an optimum focusing condition exists for a given crystal length L. For Gaussian beams, the confocal parameter b, the distance over which the beam cross-sectional area remains within twice that at the beam waist and thus the beam may be approximated as a plane wave, is given by

$$b = \frac{2\pi n_{\omega}}{\lambda} W_0^2, \tag{3.1.9}$$

where W_0 is the beam radius at the beam waist, and λ is the wavelength in the free space. A rigorous theoretical analysis [35] shows that the second-harmonic power reaches, in the case of noncritical phase matching, its maximum value,

$$\mathbf{P}^{2\omega} = 1.068 \frac{2\omega^3}{\pi\varepsilon_0 c^4} \frac{d_{\text{eff}}^2}{n^2} (\mathbf{P}^{\omega})^2 L, \tag{3.1.10}$$

when the focusing condition is optimized so that $L = 2.84b$. In the bulk configuration, the optimized second-harmonic power is proportional to the crystal length L rather than L^2.

Let us make a rough estimate of the nonlinear optical coefficient required for efficient SHG on the basis of Eq. (3.1.10). In the case where $\lambda = 1$ μm, $n = 1.8$, $L = 10$ mm and $\mathbf{P}^{\omega} = 100$ mW, an optimized second-harmonic power of 5 mW (5 percent for a 100 mW input) can be obtained through $d_{\text{eff}} = 50$ pm/V, provided that $b = 3.5$ mm and $W_0 = 18$ μm. Recent progress in the material development revealed that organic crystals even have potentially larger phase-matchable effective nonlinear optical coefficients (e.g., $d_{\text{eff}} = 80$ pm/V for 3,5-dimethyl-1-(4-nitrophenyl)pyrazole (DMNP) [19]). However, it should be noted that Eq. (3.1.10) is derived assuming an ideal situation, where reflection loss at the interfaces and propagation loss due to residual absorption and scattering are negligible, and noncritical phase matching is achieved. Further progress in the crystal growth technique and the fabrication technology for bulk organic crystals, such as cutting, polishing, and coating, is needed in order to achieve the theoretical efficiencies. More importantly, the wavelength, at which noncritical phase matching is possible, still remains an uncontrollable parameter. "Molecular and crystal engineering" does not offer us any means to produce a crystal that is noncritically phase-matchable at a given wavelength. This difficulty in bulk-configuration devices is a reason for our interests in waveguiding structures in which some parameters relating to the phase-matching characteristics can be controlled as described in the following.

3.2. Second-Harmonic Generation in Waveguiding Structures

3.2.1. Optical Waveguides and Phase Matching

The optical waveguide is composed of a guiding portion embedded in a cladding material with a refractive index smaller than that of the guiding portion. In such a structure, optical waves are totally reflected at the interfaces between the guiding and cladding portions, and thus are confined in the guiding portion. Advantages of waveguiding structures for efficient SHG are briefly summarized:

1. The optical waveguide can hold the optical field within the guiding portion of dimensions comparable to the wavelength of light over extended propagation lengths without any diffraction spreading. This ensures very high fundamental optical power densities of the order of 10 MW/cm^2 even for relatively low power levels (~ 100 mW) available from laser diodes and enhances nonlinear optical interactions.
2. Phase matching may be achieved using the model dispersion of optical modes propagating in waveguides. The propagation constant and the effective index of a propagating mode of a waveguide depends on dimensions of the guiding portion, as well as on the refractive indices of the guiding and cladding materials. Therefore, the propagation constant can be controlled by tailoring waveguide dimensions so that phase matching is achieved at a given fundamental wavelength. At the same time, this makes it possible to utilize otherwise unusable nonlinear optical coefficients including the diagonal elements of the d tensor, d_{11}, d_{22}, and d_{33}, and the other elements that are non-phase-matchable owing to their unfavorable combinations of the dispersion and birefringences of the refractive indices.
3. Optical waves in waveguides are free from the walk-off effect. Optical anisotropy, of course, affects the propagation characteristics of the waveguide, but it does not result in the walk-off of the second-harmonic beam from the fundamental beam that reduces the effective interaction lengths and conversion efficiencies in bulk crystals.

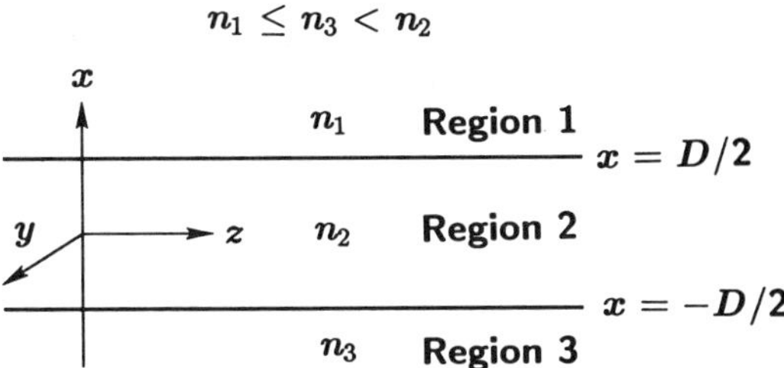

Fig. 3. Schema of a planar waveguide structure infinitely extending along the y and z directions. The propagation direction is parallel to the z axis.

4. Organic crystals, which tend to be chemically unstable and mechanically weak, may be protected by tough cladding materials.

In order to illustrate the basic features of optical modes in waveguiding structures, let us recall the fundamentals of guided-wave optics [36]. In practice, we have to deal with waveguides of two-dimensional confinement structures composed of, in general, anisotropic materials. However, we will here discuss a planar waveguide, which confines optical fields only in one direction and lacks the lateral confinement, composed of isotropic materials, because it makes mathematical treatments easier and yet involves essential features of waveguide modes needed for an understanding of SHG and phase matching in waveguiding structures.

Let us consider an isotropic planar waveguide structure shown in Fig. 3. Region 2 represents the guiding layer of thickness D, and regions 1 and 3 represent cladding layers. The refractive indices of regions 1, 2, and 3 are designated by n_1, n_2, and n_3, respectively, and we assume that $n_1 < n_3 < n_2$. Assuming z- and t-dependence of the form $\exp\{i(\beta z - \omega t)\}$ for the electromagnetic field, with β the longitudinal propagation constant, the wave equation for an angular frequency ω is written as

$$\frac{\partial^2}{\partial x^2} \boldsymbol{E}^{\omega}(x) + [k_{\omega}^2\{n^{\omega}(x)\}^2 - \beta_{\omega}^2]\boldsymbol{E}^{\omega}(x) = 0, \tag{3.2.1}$$

where $k_{\omega} = \omega/c = \omega\sqrt{\varepsilon_0\mu_0}$. There exist two independent solutions for Eq. (3.2.1)—the transverse electric (TE) mode with nonzero components E_y, H_x, and H_z, and the transverse magnetic (TM) mode with nonzero components H_y, E_x, and E_z. Physically realizable solutions exist for the eigenvalues β_{ω} in the range of $\beta_{\omega} < k_{\omega}n_2^{\omega}$.

For $k_{\omega}n_3^{\omega} < \beta_{\omega} < k_{\omega}n_2^{\omega}$, the solution is an oscillating sinusoidal function in the guiding region 2 with evanescent exponential tails in the cladding regions 1 and 3. This corresponds to a guided mode. The mode distribution of the guided wave is written as follows:

For TE mode,
$$E_y^\omega = \begin{cases} A_1 e^{-\delta_\omega x} & \text{Region 1} \\ A_2 e^{i\kappa_\omega x} + B_2 e^{-i\kappa_\omega x} & \text{Region 2} \\ B_3 e^{\gamma_\omega x} & \text{Region 3} \end{cases} \tag{3.2.2}$$

$$H_x^\omega = -\frac{\beta_\omega}{\omega\mu_0} E_y^\omega \tag{3.2.3}$$

$$H_z^\omega = -\frac{i}{\omega\mu_0}\frac{\partial E_y^\omega}{\partial x}; \tag{3.2.4}$$

for TM mode,
$$H_y^\omega = \begin{cases} a_1 e^{-\delta_\omega x} & \text{Region 1} \\ a_2 e^{i\kappa_\omega x} + b_2 e^{-i\kappa_\omega x} & \text{Region 2} \\ b_3 e^{\gamma_\omega x} & \text{Region 3} \end{cases} \tag{3.2.5}$$

$$E_x^\omega = \frac{\beta_\omega}{\omega\varepsilon_0\{n^\omega(x)\}^2} H_y^\omega \tag{3.2.6}$$

$$E_z^\omega = \frac{i}{\omega\varepsilon_0\{n^\omega(x)\}^2}\frac{\partial H_y^\omega}{\partial x}, \tag{3.2.7}$$

where
$$\delta_\omega = \sqrt{\beta_\omega^2 - k_\omega^2(n_1^\omega)^2} \tag{3.2.8}$$

$$\kappa_\omega = \sqrt{k_\omega^2(n_2^\omega)^2 - \beta_\omega^2} \tag{3.2.9}$$

$$\gamma_\omega = \sqrt{\beta_\omega^2 - k_\omega^2(n_3^\omega)^2}. \tag{3.2.10}$$

From the boundary condition at the interfaces, we get the eigenvalue equation

$$\kappa_\omega D = \tan^{-1}\left(\frac{\zeta_1\delta_\omega}{\kappa_\omega}\right) + \tan^{-1}\left(\frac{\zeta_3\gamma_\omega}{\kappa_\omega}\right) + m\pi \qquad (m = 0, 1, 2, \ldots), \tag{3.2.11}$$

with
$$\zeta_j = \begin{cases} 1 & \text{(TE)} \\ (n_2^\omega/n_j^\omega)^2 & \text{(TM)}, \end{cases} \tag{3.2.12}$$

where m is the mode number of the guided mode. The eigenvalue equation, Eq. (3.2.11), implies that the optical wave is totally reflected at the interfaces and that a standing wave is constructed in the guiding layer. This gives rise to the discrete distribution of the propagation contant β_ω and the corresponding effective index defined as

$$n_{\text{eff}}^{\omega} = \frac{\beta_{\omega}}{k_{\omega}} = \frac{c\beta_{\omega}}{\omega}. \tag{3.2.13}$$

From Eq. (3.2.11), the D- and m-dependence (modal dispersion) of the effective index n_{eff}^{ω} can be obtained.

For $k_{\omega}n_1^{\omega} < \beta_{\omega} < k_{\omega}n_3^{\omega}$, the solution in the region 3 becomes sinusoidal; this solution corresponds to the optical field radiated into the region 3. It is sometimes called a substrate radiation mode; we will refer to this hereafter as a one-side radiation mode. In this case, since the optical wave is not totally reflected at the 2–3 interface and thus free from the constraint of the standing-wave formation, the propagation constant and the effective index take arbitrary values in the ranges of $k_{\omega}n_1^{\omega} < \beta_{\omega} < k_{\omega}n_3^{\omega}$ and $n_1^{\omega} < n_{\text{eff}}^{\omega} < n_3^{\omega}$, respectively.

For $\beta_{\omega} < k_{\omega}n_1^{\omega}$, the optical field is radiated into regions 1 and 3; this is called a substrate-cladding radiation mode. We will refer to this as a two-side radiation mode. The values of the effective indices of the two-side radiation modes are in the range of $n_{\text{eff}}^{\omega} < n_1^{\omega}$.

Modal dispersion curves for these modes in an asymmtric planar waveguide are shown in Fig. 4, where we assume $n_1 = 1.65$, $n_2 = 1.82$, $n_3 = 1.70$ and $\lambda = 1.0$ µm. The values of the effective indices of the guided modes are indicated by discrete lines in the range of $n_3 < n_{\text{eff}} < n_2$, while the values of

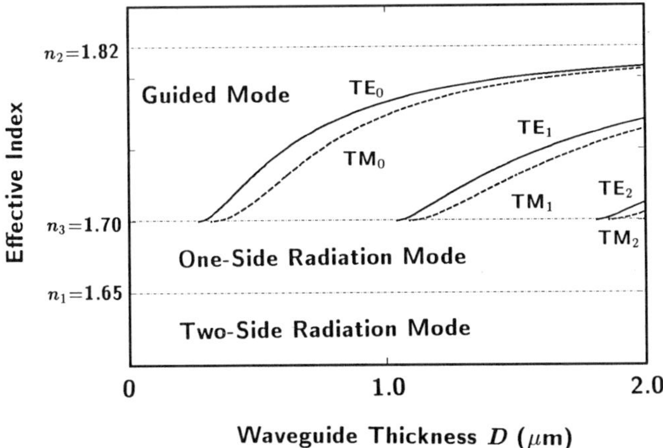

Fig. 4. Modal dispersion curves for a planar waveguide; $n_1 = 1.65$, $n_2 = 1.82$, $n_3 = 1.70$ and $\lambda = 1.0$ µm.

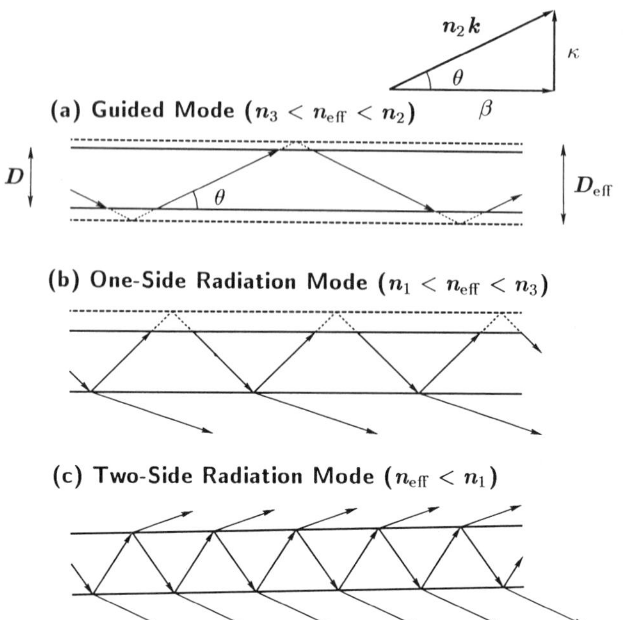

Fig. 5. Zig-zag paths of optical rays for a guided mode (a), a one-side radiation mode (b), and a two-side radiation mode (c).

the effective indices of the one-side and the two-side radiation modes form continuum in the range of $n_1 < n_{\text{eff}} < n_3$ and $n_{\text{eff}} < n_1$, respectively.

Figure 5 shows the illustrations of the waveguide modes based on the ray-optical approach, a description using zig-zag paths of optical rays in the waveguide. The dotted lines in the paths depicted in Figs. 5(a) and 5(b) correspond to Goos-Haenchen shifts reflecting the phase shifts the optical waves suffer on total internal reflection at interfaces. The effective thickness for the guided mode, D_{eff}, designated in Fig. 5(a) is given by

$$D_{\text{eff}} = D + \frac{\kappa_\omega^2 + \delta_\omega^2}{\kappa_\omega^2 + \zeta_1^2\delta_\omega^2}\frac{\zeta_1}{\delta_\omega} + \frac{\kappa_\omega^2 + \gamma_\omega^2}{\kappa_\omega^2 + \zeta_3^2\gamma_\omega^2}\frac{\zeta_3}{\gamma_\omega}, \tag{3.2.14}$$

which gives a measure of the thickness of the electromagnetic field spreading in the waveguide. Since D_{eff} is of the order of the wavelength of light in the suitably designed waveguides, the fundamental wave propagating as a guided mode provides very high power densities, the first advantage of the guided-wave SHG.

The second advantage is, as already mentioned, its flexibility in phase matching. Phase matching in guided-wave SHG is achieved by matching the longitudinal propagation constant of the second-harmonic mode with that of the fundamental guided mode;

$$\Delta\beta = \beta_{2\omega} - 2\beta_{\omega} = 0. \tag{3.2.15}$$

This phase-matching condition is equivalent to the effective-index-matching condition,

$$n_{\text{eff}}^{2\omega} = n_{\text{eff}}^{\omega}. \tag{3.2.16}$$

The effective indices are controllable parameters in the waveguiding structure, in sharp contrast to the case of the bulk configuration where the refractive indices are inherent built-in parameters of a crystal. The phase-matching condition, Eq. (3.2.15) or (3.2.16), can be satisfied by adjusting the waveguide dimensions to the phase-matching size at which the dispersion curves for the second-harmonic and the fundamental modes intersect each other.

Figure 6 shows the dispersion of the effective indices n_{eff} for the fundamental ($\lambda = 1.0$ μm) and the second-harmonic ($\lambda/2 = 0.5$ μm) TE modes in a "model" isotropic asymmetric planar waveguide. The parameters chosen to draw Fig. 6 are such that $n_1^{\omega} < n_3^{\omega} < n_1^{2\omega} < n_3^{2\omega} < n_2^{\omega} < n_2^{2\omega}$. This specific

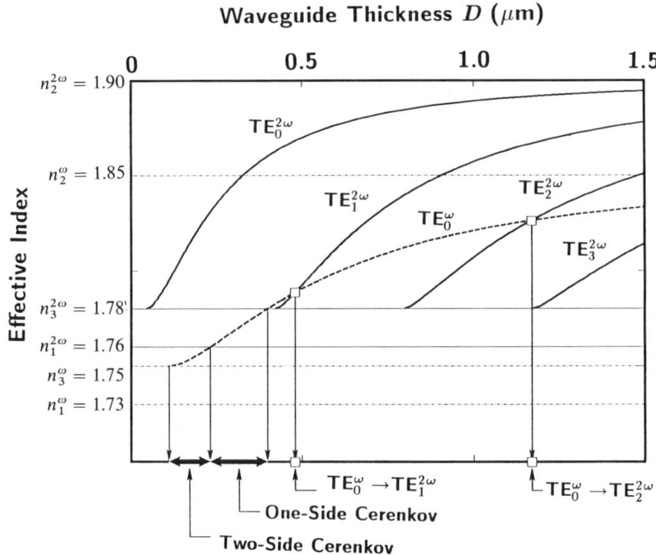

Fig. 6. Phase-matching scheme in a planar waveguide; $n_1^{\omega} = 1.73$, $n_2^{\omega} = 1.85$, $n_3^{\omega} = 1.75$, $n_1^{2\omega} = 1.76$, $n_2^{2\omega} = 1.90$, $n_3^{\omega} = 1.78$ and $\lambda^{\omega} = 1.0$ μm. Higher-order modes at ω are omitted for simplicity.

case involves almost all the features needed to understand the principles in the phase matching in the guided-wave SHG.

The phase-matching condition for the conversion from a fundamental guided mode to a second-harmonic guided mode (guided-guided-type SHG) is satisfied at a crossing point of the two dispersion curves corresponding to the modes. Phase matching between guided modes can be achieved only at a specific thickness for a given fundamental wavelength.

The second-harmonic modes phase-matchable to a fundamental guided mode include not only guided modes but also radiation modes; second-harmonic radiation modes also take part in phase-matched SHG. The phase-matching condition for the conversion from a fundamental guide mode to second-harmonic one-side radiation modes are satisfied when $n_1^{2\omega} < n_{\text{eff}}^{\omega} < n_3^{2\omega}$. For the conversion to two-side radiation modes, the phase-matching condition is $n_{\text{eff}}^{\omega} < n_1^{2\omega}$. Phase-matched conversions to the radiation modes occur when the effective index of the fundamental guided mode is smaller than the refractive index at 2ω of the cladding material (region 3 or 1), i.e., when the phase veocity of the nonlinear second-harmonic polarization wave excited in the nonlinear medium ($2\omega/2\beta_{\omega} = c/n_{\text{eff}}^{\omega}$) exceeds that of the free second-harmonic wave radiated into the cladding region ($c/n_{1,3}^{2\omega}$). This type of SHG is called SHG in the Cerenkov-radiation scheme or Cerenkov-type SHG, because the situation resembles the original Cerenkov radiation generated by a charged particle moving faster than the light propagating in the medium. The most significant advantage of the Cerenkov-type SHG as compared to the guided-guided SHG is its larger flexibility in phase matching. As is readily seen in Fig. 6, the phase-matching thickness for the Cerenkov-type SHG is not indicated by a point as in the case of the guided-guided SHG, but by a line owing to the continuous distribution of the effective indices of the radiation modes. As a result, stringent design and process requirements are greatly eased in the Cerenkov-type SHG devices. In addition, phase matching is achieved over a wide wavelength range in a single device; for the guided-guided conversion, we have to design and fabricate a specific device for a specific wavelength. Despite these fascinating features, the Cerenkov-type SHG is often regarded as less favorable than the guided-guided conversion, because the conversion efficiencies obtainable in the Cerenkov-type SHG are intrinsically not so high compared to the guided-guided conversion. However, it should be emphasized that some organic crystals with very large figures of merit make it possible to realize reasonably efficient Cerenkov-type SHG devices.

After these introductory remarks, we proceed to discuss the conversion efficiency of each type of SHG.

3.2.2. *Guided-Guided-Type Second-Harmonic Generation*

Conversion efficiencies obtainable in the guided-guided-type SHG has already been analyzed by several researchers [3,4]. The starting point of the analyses is, of course, the nonlinear wave equation which contains the nonlinear polarization term:

$$\nabla^2 \boldsymbol{E}^{2\omega} - \mu_0 \varepsilon \frac{\partial^2 \boldsymbol{E}^{2\omega}}{\partial t^2} = \mu_0 \frac{\partial^2 \boldsymbol{P}_{\mathrm{NL}}^{2\omega}}{\partial t^2}, \tag{3.2.17}$$

or its equivalence

$$\nabla^2 \boldsymbol{H}^{2\omega} - \mu_0 \varepsilon \frac{\partial^2 \boldsymbol{H}^{2\omega}}{\partial t^2} = -\frac{\partial}{\partial t}(\nabla \times \boldsymbol{P}_{\mathrm{NL}}^{2\omega}). \tag{3.2.18}$$

We will deal with a planar waveguide shown in Fig. 3 where the guiding layer consists of nonlinear medium and may be anisotropic. For simplicity, we take one of the principal dielectric axes parallel to the waveguide coordinate axis y so that the waveguide modes can be separated into TE and TM modes. The generated second-harmonic power from a portion of a width W in a planar waveguide uniformly excited over an infinite width can be easily calculated on the basis of the coupled-mode approach. For the conversion from a fundamental TE_n^{ω} guided mode to a second-harmonic $\mathrm{TE}_m^{2\omega}$ guided mode via d_{yyy}, the generated second-harmonic power is given by

$$\mathbf{P}^{2\omega} = \frac{2\omega^2}{\varepsilon_0 c^3} \frac{d_{yyy}^2}{n_{\mathrm{eff}}^{2\omega}(n_{\mathrm{eff}}^{\omega})^2} \frac{(\mathbf{P}^{\omega})^2}{W} \frac{S_{nm}^2}{(D_{\mathrm{eff}}^{\omega})^2 D_{\mathrm{eff}}^{2\omega}} L^2 \frac{\sin^2(\Delta\beta L/2)}{(\Delta\beta L/2)^2}, \tag{3.2.19}$$

with

$$S_{nm} = \int_{-D/2}^{D/2} \{f_n^{\omega}(x)\}^2 f_m^{2\omega}(x)dx, \tag{3.2.20}$$

where $D_{\mathrm{eff}}^{\omega}$ and $D_{\mathrm{eff}}^{2\omega}$ are the effective thicknesses for the fundamental and the second-harmonic guided modes, respectively, S_{nm} is the overlap integral defined as the overlap in the nonlinear region of the normalized amplitude distribution function of the second-harmonic mode $f_m^{2\omega}(x)$ and the square of that of the fundamental mode $f_n^{\omega}(x)$, and L is the waveguide length. When the phase-matching condition, $\Delta\beta = 0$, is satisfied, the second-harmonic power is proportional to the square of the waveguide length L^2 owing to the diffraction-free characteristics of the waveguide.

An important parameter that must be optimized to attain high conversion efficiencies is the overlap integral S_{nm}. It is easily shown that $S_{nm} = 0$ for odd-*m*th second-harmonic modes (odd modes) in symmetric waveguides and that $S_{nm} \simeq 0$ for those in asymmetric waveguides. Let us consider the physical interpretation of the overlap integral taking an example of a symmetric waveguide. In a symmetric planar waveguide, the overlap integral containing a second-harmonic even mode, $f_m^{2\omega} = \sqrt{2}\cos(\kappa_{2\omega}x)$, is given by

$$S_{nm} = \frac{D}{\sqrt{2}}\left[\frac{\sin\{(\kappa_{2\omega} - 2\kappa_{\omega})D/2\}}{(\kappa_{2\omega} - 2\kappa_{\omega})D/2} + \frac{\sin\{(\kappa_{2\omega} + 2\kappa_{\omega})D/2\}}{(\kappa_{2\omega} + 2\kappa_{\omega})D/2} \pm 2\frac{\sin(\kappa_{2\omega}D/2)}{\kappa_{2\omega}D/2}\right], \tag{3.2.21}$$

with

$$\kappa_{2\omega} = \sqrt{k_{2\omega}^2(n_2^{2\omega})^2 - \beta_{2\omega}^2}, \tag{3.2.22}$$

where the plus (minus) sign applies to the case of the even fundamental mode, $f_n^{\omega} = \sqrt{2}\cos(\kappa_{\omega}x)$ (odd mode, $f_n^{\omega} = \sqrt{2}\sin(\kappa_{\omega}x)$). Three $\sin(x)/x$ terms in Eq. (3.2.21) originate in "transverse phase mismatch". The phase-matching condition requires only to match the longitudinal wave vector, β, while the mismatch of the transverse wave vectors, κ, remains and lessens the conversion efficiencies. The nonlinear second-harmonic polarization wave induced in the nonlinear guiding layer consists of three plane waves with transverse wave vectors, $2\kappa_{\omega}$, $-2\kappa_{\omega}$ and 0, as shown in Fig. 7. Each term in overlap integral represents the phase mismatch between each component of the nonlinear polarization wave and the generated free second-harmonic wave with a transverse wave vector $\kappa_{2\omega}$. From Eq. (3.2.21), it is clear that the most favorable situation is realized when both the fundamental and the second-harmonic modes are the lowest-order modes ($m = n = 0$). It is impossible, however, to achieve phase matching between the lowest-order modes in an isotropic material because of the dispersion of the material. In anisotropic waveguides, phase matching between the lowest-order modes can be realized by taking advantage of the birefringence of the

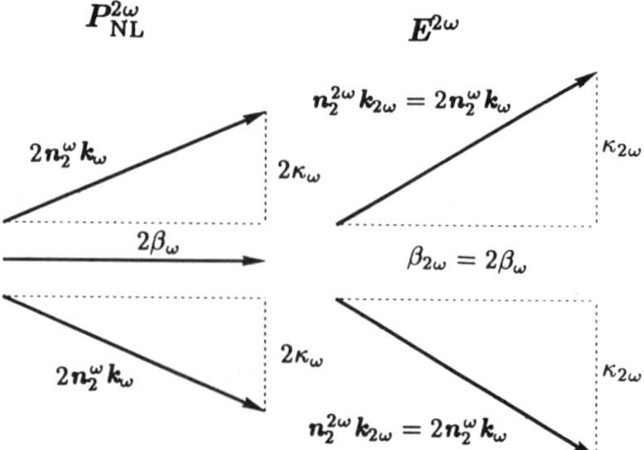

Fig. 7. Wave vectors of the nonlinear second-harmonic polarization wave $P_{NL}^{2\omega}$ and the generated free second-harmonic wave $E^{2\omega}$.

anisotropic crystal to compensate for the dispersion. This requires the perpendicular-polarization configurations; i.e., $TE_0^\omega + TE_0^\omega \to TM_0^{2\omega}$, $TE_0^\omega + TM_0^\omega \to TE_0^{2\omega}$, $TE_0^\omega + TM_0^\omega \to TM_0^{2\omega}$ and $TM_0^\omega + TM_0^\omega \to TE_0^{2\omega}$ conversions are most favorable. Therefore, the diagonal elements of d tensors, d_{11}, d_{22}, and d_{33}, are not useful in the waveguides of usual geometries.

In the case of an optimized geometry where $S_{nm} \sim D/\sqrt{2}$, the conversion efficiency is given by

$$\eta = \frac{\mathbf{P}^{2\omega}}{\mathbf{P}^{\omega}} \sim \frac{\omega^2}{\varepsilon_0 c^3} \frac{d_{\text{eff}}^2}{n_{\text{eff}}^3} \frac{\mathbf{P}^{\omega}}{WD_{\text{eff}}} L^2. \tag{3.2.23}$$

A comparison with Eq. (3.1.10) shows that the conversion efficiencies in the appropriately designed waveguide are larger than those in the bulk configurations by $\sim L/\lambda > 10^3$. A rough estimation shows that an effective nonlinear optical coefficient of $d_{\text{eff}} \simeq 10$ pm/V is sufficient for a 5-percent-efficiency conversion from a 100 mW fundamental input of $\lambda = 1$ μm in a waveguide of $L = 1$ mm, $W = 5$ μm, $D = 1$ μm, provided that $n_{\text{eff}} = 1.8$. Organic crystals with larger nonlinear optical coefficients will make it possible to realize an almost 100 percent conversion from a 100 mW input even in the waveguide of $L = 1$ mm, if the waveguide geometry is optimized.

However, this rather optimistic expectation is defied by another important parameter—the dimensional tolerances in phase matching. This phase-matching tolerance can be evaluated from the term $\{\sin(\Delta\beta L/2)/(\Delta\beta L/2)\}^2$ as

$$\left|\frac{\Delta\beta L}{2}\right| = \left|(n_{\text{eff}}^{2\omega} - n_{\text{eff}}^{\omega})\frac{\pi L}{\lambda}\right| < \frac{\pi}{2}. \tag{3.2.24}$$

Provided that $\lambda = 1$ μm and $L = 1$ mm, the acceptable effective index mismatch, $\Delta n_{\text{eff}} = n_{\text{eff}}^{2\omega} - n_{\text{eff}}^{\omega}$, is only 5×10^{-4}. This quite severe condition usually requires extremely stringent control and uniformity of the waveguide thickness. As an example, in the asymmetric planar waveguide adopted to draw Fig. 6, in order to ensure the phase matching between the TE_0^ω mode and the $TE_2^{2\omega}$ mode, the thickness deviation needs to be smaller than ± 14 Å for the waveguide length of 1 mm. The uniformity and the stability of the waveguide dimensions must usually be better than ~ 50 Å/mm, in order to ensure coherent superposition of the generated second-harmonic wave over the full length of the waveguide. Although organic crystals with very large figures of merit offer the possibilities to shorten the waveguide lengths in appropriately designed waveguides, design and process requirements still remain extremely stringent. Unfortunately, present technologies of crystal growth and fabrication of organic crystals are not compatible with such stringent requirements. As a matter of course, efficient frequency conversion using the guided-guided interactions has not been realized yet.

3.2.3. Cerenkov-Type Second-Harmonic Generation

As already mentioned, the stringent design and process requirements are greatly eased in the Cerenkov-type SHG owing to the continuous distribution of the second-harmonic radiation modes, at the expense of the conversion efficiencies. In order to discuss important characteristics of the Cerenkov-type SHG, let us consider only even TE modes ($TE_n^{\omega} \rightarrow TE^{2\omega}$ conversion via d_{yyy} of the guiding layer) in a symmetric waveguide where the symmtric two-side Cerenkov-type SHG is possible. For planar waveguides, the Cerenkov-type SHG can be analysed by directly solving the nonlinear wave equation; the detailed procedure is described by Li *et al.* [37] and Onda and Ito [38]. Neglecting the pump depletion, the fundamental TE guided mode is expressed as follows:

$$E_y^{\omega} = \begin{cases} A' e^{-\rho_{\omega} x} & \text{Region 1} \\ A \cos(\kappa_{\omega} x) & \text{Region 2} \\ A' e^{\rho_{\omega} x} & \text{Region 3,} \end{cases} \tag{3.2.25}$$

where

$$\rho_{\omega} = \sqrt{\beta_{\omega}^2 - k_{\omega}^2 (n_{1,3}^{\omega})^2}. \tag{3.2.26}$$

The phase-matching condition, Eq. (3.2.16), is automatically satisfied when the second-harmonic wave is radiated into the cladding regions at the Cerenkov angle α:

$$\alpha = \cos^{-1}\left(\frac{\beta_{2\omega}}{k_{2\omega} n_{1,3}^{2\omega}}\right) = \cos^{-1}\left(\frac{n_{\text{eff}}^{\omega}}{n_{1,3}^{2\omega}}\right). \tag{3.2.27}$$

The electric field distribution of the $TE^{2\omega}$ radiation mode is obtained by solving the nonlinear wave equation, Eq. (3.2.17), as follows:

$$E_y^{2\omega} = \begin{cases} F' e^{i\rho_{2\omega} x} & \text{Region 1} \\ F \cos(\kappa_{2\omega} x) - 2A^2 \omega^2 \varepsilon_0 \mu_0 d_{yyy} \left[\dfrac{\cos(2\kappa_{\omega} x)}{\kappa_{2\omega}^2 - 4\kappa_{\omega}^2} + \dfrac{1}{\kappa_{2\omega}^2}\right] & \text{Region 2} \\ F' e^{-i\rho_{2\omega} x} & \text{Region 3,} \end{cases} \tag{3.2.28}$$

where

$$\rho_{2\omega} = \sqrt{k_{2\omega}^2 (n_{1,3}^{2\omega})^2 - \beta_{2\omega}^2}. \tag{3.2.29}$$

The second-harmonic power generated from a portion of a length L and a width W in an infinitely long planar waveguide uniformly excited over an infinite width is given by

$$\mathbf{P}^{2\omega} = \frac{2\omega^2}{\varepsilon_0 c^3} \frac{d_{yyy}^2}{n_{1,3}^{2\omega}(n_{\text{eff}}^{\omega})^2} \frac{(\mathbf{P}^{\omega})^2}{W} \frac{T_1 T_2 S_C^2}{(D_{\text{eff}}^{\omega})^2 \sin\alpha} L, \tag{3.2.30}$$

with

$$S_C = \frac{D}{\sqrt{2}}\left[\frac{\sin\{(\kappa_{2\omega} - 2\kappa_{\omega})D/2\}}{(\kappa_{2\omega} - 2\kappa_{\omega})D/2} + \frac{\sin\{(\kappa_{2\omega} + 2\kappa_{\omega})D/2\}}{(\kappa_{2\omega} + 2\kappa_{\omega})D/2} + 2\frac{\sin(\kappa_{2\omega}D/2)}{\kappa_{2\omega}D/2}\right] \tag{3.2.31}$$

$$T_1 = \frac{1}{1 + \dfrac{(\kappa_{2\omega}^2 - \rho_{2\omega}^2)^2}{4\kappa_{2\omega}^2\rho_{2\omega}^2}\sin^2(\kappa_{2\omega}D)} \tag{3.2.32}$$

$$T_2 = 1 - \frac{\kappa_{2\omega}^2 - \rho_{2\omega}^2}{\kappa_{2\omega}^2}\sin^2(\kappa_{2\omega}D/2). \tag{3.2.33}$$

The most remarkable feature of the Cerenkov-type SHG compared to the guided-guided-type SHG is that the generated second-harmonic power is proportional to the waveguide length L instead of L^2, resulting in lower conversion efficiencies.

The expression for the overlap integral S_C reflecting the transverse phase mismatch is formally identical with that for the guided-guided interactions. The conversion efficiencies in the Cerenkov-type SHG also critically depend on this term as in the guided-guided-type SHG. The optimization of the overlap integral requires that the transverse wave vector mismatch, $\Delta\kappa = \kappa_{2\omega} - 2\kappa_{\omega}$, should be minimized; this is equivalent to the requirement that $n_2^{2\omega} - n_2^{\omega}$ should be sufficiently small. Since the diagonal elements of the d tensors of many organic crystals are inevitably accompanied by very large index dispersions, they cannot effectively be used in efficient SHG even in the Cerenkov-type SHG. Again, the birefringence of the anisotropic crystals plays an important role in compensating for the dispersion in order to obtain higher conversion efficiencies. In order to take advantage of the birefringence, we have to adopt the perpendicular-polarization configurations, i.e., $\text{TE}^{\omega} + \text{TE}^{\omega} \to \text{TM}^{2\omega}$, $\text{TE}^{\omega} + \text{TM}^{\omega} \to \text{TE}^{2\omega}$, $\text{TE}^{\omega} + \text{TM}^{\omega} \to \text{TM}^{2\omega}$ and $\text{TM}^{\omega} + \text{TM}^{\omega} \to \text{TE}^{2\omega}$ conversions. We will not go into detail on the expressions for the conversion efficiencies in these configurations, since a general expression for the Cerenkov-type SHG in planar waveguides is described in [39] in detail. It should be mentioned that the theoretical analysis of these configurations has revealed that reasonably high conversion efficiencies will indeed be obtained by appropriately utilizing the birefringence of anisotropic crystals [40]. Some numerical examples are given in Sec. 4.2.

Other important parameters that are connected with the device optimization are the terms T_1 and T_2 appearing in Eq. (3.2.30). They represent the effects of multiple reflections of the second-harmonic wave in the guiding layer. The second-harmonic waves radiated upward (with the transverse wave vector $+\kappa_{2\omega}$) and downward (with $-\kappa_{2\omega}$) propagate in the guiding layer, being partially reflected at the interfaces and partially transmitted to the cladding layers. T_1 originates in the interference of the second-harmonic waves that are initially radiated upward and transmitted to a cladding layer; T_1 is a waveguide analogue to the well-known interference formula of a Fabry-Perot etalon. On the other hand, T_2 represents the effects of the interference of the "upward beam" and the "downward beam." In order to optimize these parameters, the refractive index step for the second-harmonic wave between the guiding and the cladding material, $n_2^{2\omega} - n_{1,3}^{2\omega}$, should be sufficiently small.

A rough examination of Eq. (3.2.30) shows that, assuming

$$T_1 T_2 S_C^2/(D_{\mathrm{eff}})^2 \sin\alpha \sim 1,$$

the conversion efficiencies expected for the Cerenkov-type SHG are reduced by $\sim D/L \sim 10^{-3}$ compared with those for the guided-guided-type SHG. Therefore, the conversion efficiencies of the Cerenkov-type SHG are not so high and comparable with those of the bulk configurations; nevertheless, some organic crystals make it possible to attain reasonably high conversion efficiencies even in the Cerenkov-type SHG from the waveguide of the practical size ($L \simeq 10$ mm).

The most significant features of the Cerenkov-type SHG is its large waveguide dimensional tolerances. The constraint on the waveguide dimensions in the Cerenkov-type SHG does not come from the phase-matching condition—the longitudinal-wave-vector matching—as in the case of the guided-guided interaction, but comes from the transverse-wave-vector matching, because the longitudinal phase matching is automatically achieved. The acceptable thickness deviation evaluated from the function $\sin(\Delta\kappa D/2)/(\Delta\kappa D/2)$ is usually larger than 0.1 μm, in sharp contrast to the case of the guided-guided SHG (~ 50 Å). The dimensional tolerances are drastically relaxed, because the second-harmonic waves are radiated out of the guiding layer, and thus they need not be superimposed coherently over the whole interaction length.

Let us consider more realistic cases—one-side Cerenkov-type SHG from an asymmetric planar waveguide. In the case that $n_1^{2\omega} < n_{\mathrm{eff}}^{\omega} < n_3^{2\omega}$, the second-harmonic light is radiated only into region 3. Although the situation becomes somewhat complicated compared with the symmetric case, the generated second-harmonic power can be expressed in a similar form to the symmetric case. The result for the $TE^{\omega} \rightarrow TE^{2\omega}$ conversion in the waveguide in which the nonlinear medium occupies the guiding layer (region 2) is

$$\mathbf{P}^{2\omega} = \frac{4\omega^2}{\varepsilon_0 c^3} \frac{d_{yyy}^2}{n_3^{2\omega}(n_{\text{eff}}^{\omega})^2} \frac{(\mathbf{P}^{\omega})^2}{W} \frac{T' S_C'^2}{(D_{\text{eff}}^{\omega})^2 \sin\alpha} L, \tag{3.2.34}$$

where
$$S_C' = \frac{D}{\sqrt{2}} \left[\frac{\sin\{(\kappa_{2\omega} - 2\kappa_{\omega})D/2\}}{(\kappa_{2\omega} - 2\kappa_{\omega})D/2} \cos\left(\frac{\kappa_{2\omega} D}{2} - \phi - \psi \right) + \frac{\sin\{(\kappa_{2\omega} + 2\kappa_{\omega})D/2\}}{(\kappa_{2\omega} + 2\kappa_{\omega})D/2} \cos\left(\frac{\kappa_{2\omega} D}{2} - \phi + \psi \right) + 2 \frac{\sin(\kappa_{2\omega} D/2)}{\kappa_{2\omega} D/2} \cos\left(\frac{\kappa_{2\omega} D}{2} - \phi \right) \right] \tag{3.2.35}$$

$$T' = \frac{1}{1 + \dfrac{\kappa_{2\omega}^2 - \gamma_{2\omega}^2}{\gamma_{2\omega}^2} \sin^2(\kappa_{2\omega} D - \phi)} \tag{3.2.36}$$

$$\phi = \tan^{-1} \frac{\delta_{2\omega}}{\kappa_{2\omega}} \tag{3.2.37}$$

$$\psi = \tan^{-1} \frac{\kappa_{\omega}(\gamma_{\omega} - \delta_{\omega})}{\kappa_{\omega}^2 + \gamma_{\omega}\delta_{\omega}} \tag{3.2.38}$$

$$\gamma_{2\omega} = \sqrt{k_{2\omega}^2 (n_1^{2\omega})^2 - \beta_{2\omega}^2} \tag{3.2.39}$$

$$\delta_{2\omega} = \sqrt{\beta_{2\omega}^2 - k_{2\omega}^2 (n_3^{2\omega})^2}. \tag{3.2.40}$$

Again, S_C' and T' represent the overlap integral and the multiple reflection effects of the second-harmonic wave, respectively.

For the case where the substrate (region 3) is made of the nonlinear medium, another realistic device configuration, the essence remains the same, except that the expression of the overlap integral becomes somewhat different from Eq. (3.2.35).

For the Cerenkov-type SHG in a channel waveguide, theoretical analysis becomes somewhat difficult. Tamada has analyzed the Cerenkov-type SHG in $LiNbO_3$ planar and channel waveguides using the coupled-mode approach [41,42]. He pointed out that the conversion efficiency in a channel waveguide of a width W becomes about 10 times larger than that from a portion of a width W in a planar waveguide. This large difference is caused by the three-dimensional cone-shaped radiation of the second-harmonic wave from the actual channel waveguide, which is neglected in the planar-waveguide analyses, which only take into account the two-dimensional radiation.

As to fiber waveguides, theoretical analyses of the Cerenkov-type SHG have been made by White and Nayar [43] and, in more detail, by Chikuma and Umegaki [44,45].

We will not go into details of the analyses on the channel and the fiber waveguides here; nevertheless, it should be pointed out that the guidelines for efficient SHG are common to all the waveguide structures.

4. ORGANIC-CRYSTALLINE WAVEGUIDING STRUCTURES

In this section, we will describe the fabrication of the organic-crystalline waveguides and their application to guided-wave SHG, including our own work. Section 4.1 is devoted to the description of the experimental studies made to date; and the favorable waveguide parameters, including the material choice, for efficient SHG are discussed on the basis of the theoretical analysis in Sec. 4.2.

4.1. Second-Harmonic Generation in Organic-Crystalline Waveguides

4.1.1. Planar Waveguides

SHG in an organic-crystalline planar waveguide was first demonstrated by Hewig and Jain [46]. They formed a thin film of *p*-chlorophenylurea (PCPU), 0.9 μm thick, on a glass substrate by the conventional vacuum-deposition technique. The deposited films were polycrystalline. Phase matching was achieved by scanning the fundamental wavelength. Phase-matched SHG via the $TM_0^{\omega} \rightarrow TM_2^{2\omega}$ conversion process was observed at a fundamental wavelength of ~0.9 μm.

The first attempt at SHG in a single-crystalline planar waveguide was reported by Sasaki *et al.* [47]. They placed a thick MNA single crystal (50 μm thick) prepared by vapor phase growth on a tapered planar glass waveguide as a high-index top layer. Phase matching in the guided-guided interaction was achieved by adjusting the waveguide thickness with lateral movement of the tapered waveguide. Phase-matched SHG for a fundamental wavelength of 1.064 μm was observed at a waveguide thickness of 1.68 μm. Later, they reported another planar waveguide structure made of an MNA single crystal [48]. A tapered MNA single-crystalline thin film ($\sim 3\ \mu m$) was prepared from melt. The conversion efficiency in the demonstrated phase-matched SHG was 3×10^{-4} for an input power of 600 W.

A similar preparation method was applied to grow single-crystalline films of *N*-(4-nitrophenyl)-(*L*)-prolinol (NPP) by Ledoux *et al.* [20]. However, no guided-wave SHG experiment of the NPP planar waveguides was reported.

We have investigated the Cerenkov-type SHG in organic-crystalline waveguiding structures. An organic crystal (−)2-(α-methylbenzylamino)-5-nitropyridine (MBANP) was used as a nonlinear medium. MBANP was first studied as a nonlinear optical material by Twieg *et al.* [49] and then, in more detail, by Bailey *et al.* [50,51] and ourselves [22,52]. The largest nonlinear optical coefficient of this crystal is $d_{bbb} = d_{22} = 60$ pm/V. The absorption cutoff in the visible range is around 0.45 μm.

Planar waveguides were fabricated by growing thin MBANP single-crystal films between glass substrates with appropriate indices [53]. For the growth of single-crystal films we used the solvent evaporation method; a saturated solution of MBANP was carried into the narrow gap between two glass substrates by the capillary action, and then the solvent was left to evaporate slowly at a constant temperature. The typical dimensions of a homogeneous single crystal grown from an acetone solution are 2 to 4 mm long (along *c*) and 1 mm wide (along *b*). The thickness (approximately along *a*) can be controlled between 0.5 and 2.0 μm using spacers with appropriate thicknesses. A photograph of a grown MBANP single-crystal film is shown in Fig. 8. The orientation of the MBANP single-crystal film grown from the acetone solution is the most suitable for our experiments, because the d_{22} coefficient can be fully utilized in the nonlinear interaction when an end-fire-coupled fundamental light propagates along the *c* axis as a TE wave.

For the SHG experiment, we used a planar waveguide shown in Fig. 9(a). The MBANP single-crystalline guiding layer was 2 mm long and 1 μm thick, and the substrate material was SF10 glass (Schott). The fundamental light sources used in this experiment were a laser diode oscillating at $\lambda = 0.87\ \mu m$, a Nd:YAG laser oscillating at $\lambda = 1.064\ \mu m$ and a laser diode oscillating at $\lambda = 1.30\ \mu m$. The focused fundamental beam was fed into the waveguide and guided as a TE wave (polarized along the *b* axis). The planar waveguide shown in Fig. 9(a) is phase-matchable in the Cerenkov-radiation scheme for all the fundamental light sources used; three primary colors, blue (0.435 μm), green (0.532 μm), and red (0.65 μm), can be generated from a single device. The obtained output powers were 0.14 nW at $\lambda = 0.435\ \mu m$ for a fundamental

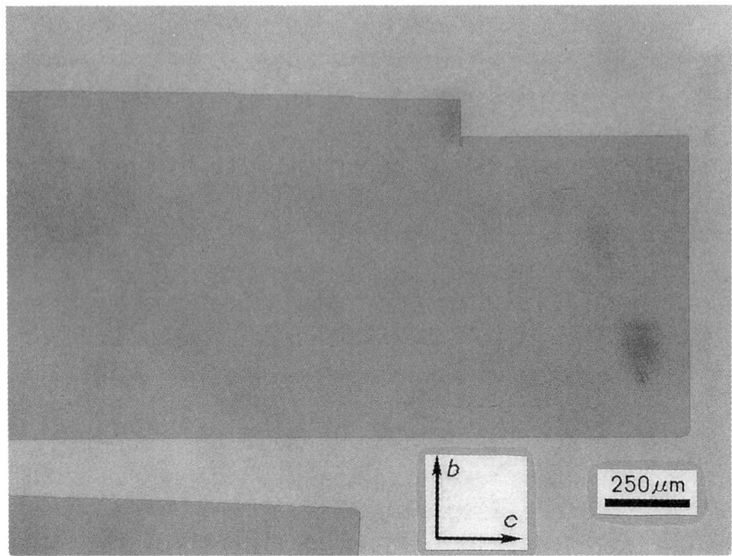

Fig. 8. A photograph of a grown MBANP single-crystal film.

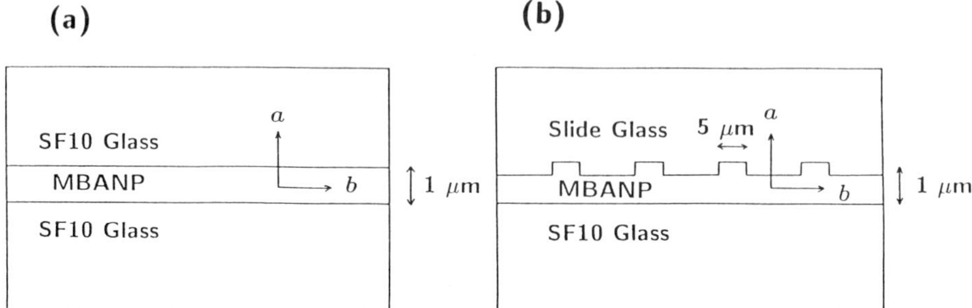

Fig. 9. Schematic cross-sections of the MBANP waveguides used in the SHG experiments; (a) planar waveguide, (b) rib waveguide.

incident power of 16 mW, and 8 nW at $\lambda = 0.65$ μm for a fundamental incident power of 60 mW. These powers are quite low; a possible reason is the diffraction of the fundamental beam due to the lack of the lateral confinement in the planar structure.

4.1.2. Channel Waveguides

The number of attempts at fabrication of organic-crystalline channel waveguides is quite limited. Tomaru *et al.* succeeded in the growth of a single crystal of *m*-nitroaniline (*m*NA) in a narrow grove (~100 μm wide and ~100 μm thick) formed on a glass substrate [54]. The *m*NA-single-crystalline embedded-stripe waveguide was fabricated by a zone-melting process, where the CO_2 laser beam scanning along the groove changed the polycrystalline state into a single crystal. They found that *m*NA crystal tends to grow with its polar (*c*) axis parallel to the channel direction. The two largest nonlinear optical coefficient, d_{13} and d_{33}, of *m*NA cannot then be utilized in guided-wave nonlinear interactions owing to the unfavorable crystal orientation. Vidakovic *et al.* also used a similar process to fabricate NPP embedded-strip waveguides [55].

We have sought ways to fabricate channel waveguides using organic crystals. Fig. 10 shows cross-sections of possible structures of organic-crystalline channel waveguides. The fabrication of rib waveguide (a), diffused-strip-loaded waveguide (b), raised-strip waveguide (c) and strip-loaded waveguide (d) is based on planar processes in which the first step is the growth of thin-film single crystals. These are highly promising because planar processes may provide more flexibility in the control of crystal orientations and the device architecture.

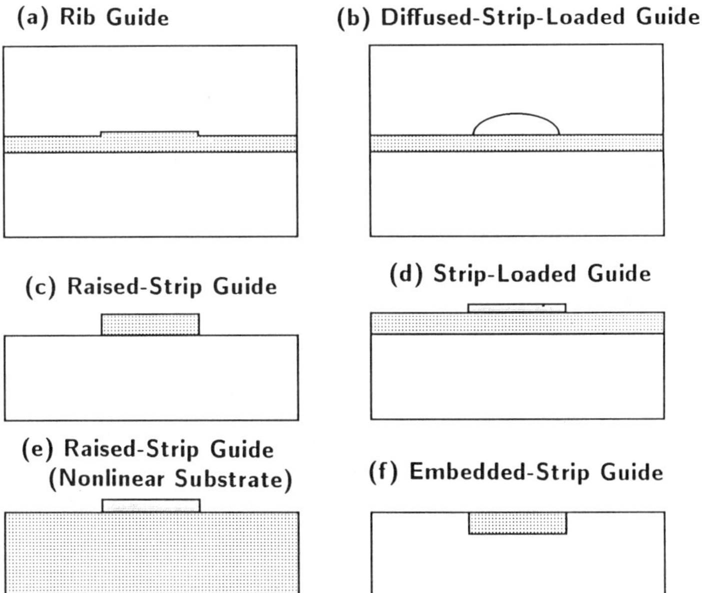

Fig. 10. Schematic cross-sections of possible structures of organic-crystalline channel waveguides.

We have fabricated several types of channel waveguides using MBANP single crystals. The fabrication technique developed to produce those waveguides will be briefly described.

(a) **Rib Waveguide.** Rib waveguides were produced by growing MBANP crystal between glass substrates, one of which had shallow grooves on its surface. The grooves, typically 0.3 μm deep and 5 μm wide, were defined by the usual photolithographic technique. MBANP crystal grown from a solution filled the grooves, thus forming rib waveguides [53].

(b) **Diffused-Strip-Loaded Waveguide.** Diffused-strip-loaded waveguides were produced by growing MBANP crystal between glass substrates, one of which had high-index channels on its surface. The high-index channels, typically 2 μm deep and 5 μm wide, were formed on the glass substrate by the usual ion-exchange technique. When the ion-exchanged channel has a refractive index larger than that of the substrate and smaller than that of MBANP, it acts as a diffused loading strip.

(c) **Raised-Strip Waveguide.** In order to produce raised-strip waveguides, we have to develop a new lithographic technique that enables us to fabricate thin-film crystals into fine channels. The largest problem is that organic photoresists widely used in lithographic processes for semiconductors and dielectrics are not applicable to our purpose. Recently, Gotoh *et al.* reported that MNBA-crystal waveguides were successfully fabricated using a popular organic photoresist [56]. This was made possible because the solubility of MNBA in organic solvents was exceptionally low. However, since most organic molecular crystals are, in general, soluble in polar organic solvents, conventional organic resists dissolved in organic solvents cannot be used in the lithography of organic crystals.

On the basis of many considerations, we selected an inorganic resist based on peroxopolyniobotungstic acid (HPA) for the fabrication of MBANP crystals [40]. HPA was originally developed for the two-layer resist process in silicon fabrication technologies [57]. HPA is an amorphous material of typical empirical formula $CO_2 \cdot 12(W, Nb)O_3 \cdot 7H_2O_2 \cdot 25H_2O$. HPA has some attractive properties as follows:

1. HPA is soluble in water, and uniform thin films can be formed from its water solution using the conventional spin-coating technique.

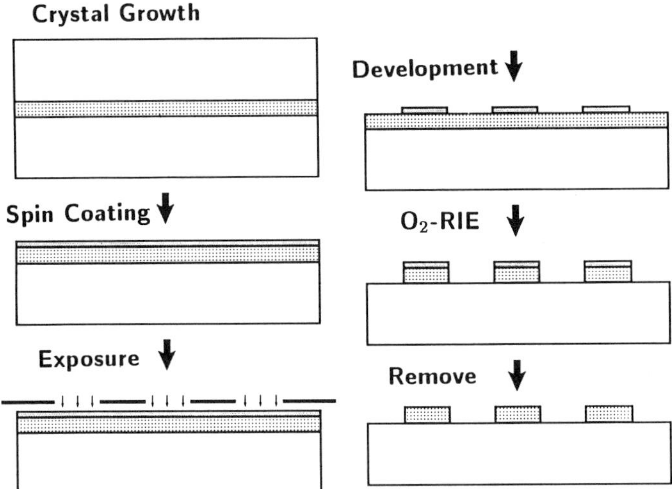

Fig. 11. Fabrication process using the water-soluble inorganic resist HPA and O_2-RIE. Exposure is carried out using a Xe-Hg lamp, followed by development using an aqueous H_2SO_4 solution. The remaining HPA films are removed using an aqueous KOH solution.

2. The film becomes insoluble in water when irradiated when the deep UV light, electron beam, or X-ray; HPA is a negative resist.
3. HPA has a high O_2-RIE (oxygen reactive ion etching) durability.

These properties make it possible to produce MBANP-crystal strip waveguides.

Figure 11 shows schematically the fabrication process using HPA and O_2-RIE [58]. A photograph of MBANP-crystal strip waveguides fabricated in this way is shown in Fig. 12. This fabrication technique is expected to be applicable to almost any organic crystal, because the whole process is organic-solvent-free and, moreover, does not require high temperature baking.

(d) **Strip-Loaded Waveguide.** HPA can be used as a material of loading strips. If the process described above is stopped after the development, the structure shown in Fig. 10(d) is realized. Since the refractive index of HPA is about 1.75 in the near-infrared, slightly smaller than n_Y of MBANP, and HPA is transparent down to 0.32 μm, HPA strips on MBANP films act as loading strips.

We have demonstrated the frequency doubling of a laser diode oscillating

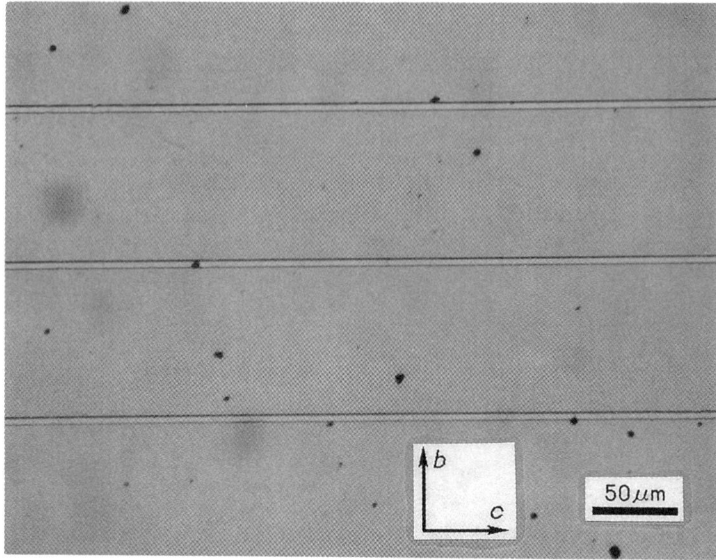

Fig. 12. A microscope photograph of the fabricated MBANP raised-strip waveguides. Guide width is 5 μm.

at $\lambda = 0.87\ \mu m$ in an MBANP rib waveguide [53]. The structure of the rib waveguide used in the SHG experiment is shown in Fig. 9(b). The waveguide dimensions were 2 mm long, 5 μm wide, and 1 μm thick. With a fundamental power of 15 mW, a second-harmonic power of 6 nW at $\lambda = 0.435$ μm was obtained. The conversion efficiency was increased by a factor of about 50 compared with the case of the planar waveguide, demonstrating the effect of lateral confinement of the fundamental beam. The second-harmonic power obtained in this experiment, 6 nW, is smaller than the value theoretically calculated taking into account the "channel effect," 40 nW. This discrepancy is mainly attributed to the coupling loss of the fundamental light. In addition, since the wavelength of the second-harmonic light, $\lambda = 0.435$ μm, is slightly in the absorption band of MBANP, some absorption of second-harmonic light has possibly taken place. In order to attain the theoretical conversion efficiency, the improvement of fabrication processes and polishing of the input face of the waveguide are imperative.

Unfortunately, MBANP is not suitable for efficient SHG in such a simple device architecture owing to a large index dispersion; nevertheless, we believe that the fabrication techniques developed for MBANP will no doubt contribute to the future development of organic-crystalline waveguiding devices.

4.1.3. *Organic-Crystal-Cored Fibers*

A waveguide structure that has been a subject of the most intense research efforts is the organic-crystal-cored fiber.

The first attempt at the growth of organic nonlinear optical crystals was made by Stevenson [59,60]. He succeeded in growing *m*NA crystal in hollow glass capillaries with core diameters of 8 to 40 μm. However, the polar axis of *m*NA was found to align the fiber axis, hindering the effective nonlinear optical interaction.

The first successful SHG was reported by Nayer [61]. Benzil crystal was grown in a 30 mm long capillary with a 3.75 μm diameter. He observed SHG of Nd:YAG laser oscillating at $\lambda = 1.064$ μm phase-matched by coupling the fundamental HE_{11} mode into the second-harmonic radiation mode, demonstrating the usefulness of the Cerenkov-type SHG in organic-crystal-cored fibers.

Umegaki *et al.* fabricated a MNA-crystal-cored fiber with a 2.3 μm core diameter [62]. They found that MNA tends to grow with its X axis aligned parallel to the fiber axis, and the nonlinear optical coefficients, d_{11}, the largest element of d tensor of this material, cannot be effectively utilized in SHG.

Vidakovic *et al.* have demonstrated the fabrication of crystal-cored fibers using relatively new materials, such as NPP and *N*-(4-nitrophenyl)-*N*-methylaminoacetonitrile (NPAN) [63].

The most popular organic crystal that is suitable for guided-wave SHG is 4-(*N*,*N*-dimethylamino)-3-acetamidonitrobenzene (DAN). Many researchers have undertaken the fabrication of crystal-cored fibers for efficient SHG using DAN crystals [16,64–66], stimulated by the first report on the DAN-cored fiber fabricated by Tomaru and Zembutsu [67]. DAN crystal grows with its *a* axis parallel to the fiber axis [65]. This orientation makes it possible to utilize the nonlinear optical coefficients $d_{23} = 50$ pm/V, the largest coefficient of this material [23].

Recently, reasonably efficient coherent blue light generation was demonstrated by Kawaharada *et al.* using the Cerenkov-radiation scheme [16]. They obtained 6.2 μW second-harmonic light at $\lambda = 0.493$ μm from a 4.7-mm-long DAN-crystal-cored fiber for an input fundamental power of 6 mW from a laser diode oscillating at $\lambda = 0.987$ μm. They also observed that the conversion efficiencies rapidly decreased for shorter fundamental wavelengths, probably owing to the absorption cutoff around $\lambda = 0.49$ μm.

Uemiya *et al.* investigated the focusing characteristics of the second-harmonic light generated in the form of Cerenkov radiation using a

DAN-crystal-cored fiber [66]. They collimated conically radiated second-harmonic light at $\lambda = 0.532$ μm using an axicon lens and focused it into a spot of a diameter of 1.8 μm using an objective lens of N.A. = 0.3. This spot size is 0.8 times smaller than the usual diffraction-limited spot for a circular aperture, owing to the ring-shaped radiation pattern. This clearly demonstrates the usefulness of organic-crystal-cored fibers in the application to the optical data storage.

Another significant achievement in the Cerenkov-type SHG using an organic-crystal-cored fiber was reported by researchers of the Fuji Photo Film in Japan. Harada *et al.* reported that a second-harmonic blue light at $\lambda = 0.442$ μm of 64.5 μW was obtained for a fundamental power of 16.6 mW from a laser diode oscillating at $\lambda = 0.884$ μm in a 5-mm-long DMNP-crystal-cored fiber [15]. DMNP is an excellent material developed by these authors. The value of its largest nonlinear optical coefficient d_{32} reaches 90 pm/V, and the absorption cutoff wavelength is around 0.45 μm [19]. The crystal orientation in the capillary is most favorable for efficient SHG; since c (X) axis aligns parallel to the fiber axis, one can make full use of d_{32}. They also investigated the temperature dependence of the Cerenkov angle and found that temperature needed to be controlled within ~0.4°C in order to ensure diffraction-limited focusing [68].

4.2. Guidelines for Efficient Second-Harmonic Generation

Conversion efficiencies in the Cerenkov-type SHG depends on a number of parameters. However, we can summarize guidelines for attaining high conversion efficiencies in the Cerenkov-radiation scheme as follows:

1. The most important parameter is the dispersion of the refractive index of the nonlinear medium. $n_{NL}^{2\omega} - n_{NL}^{\omega}$ should be minimized in order to ensure a large overlap integral, which allows a high conversion efficiency. In general, it is preferable to compensate for the dispersion by taking advantage of the birefringence of the optically anisotropic nonlinear medium; crystals that are phase-matchable by the angular tuning in the bulk configuration are desirable.
2. The refractive index step at the interface between the nonlinear guiding layer and the linear cladding region at the second-harmonic wavelength should be small in order to minimize the efficiency reduction due to the multi-reflection effect.
3. The refractive index step at the interface between the nonlinear guiding layer and the linear cladding region at the fundamental wavelength

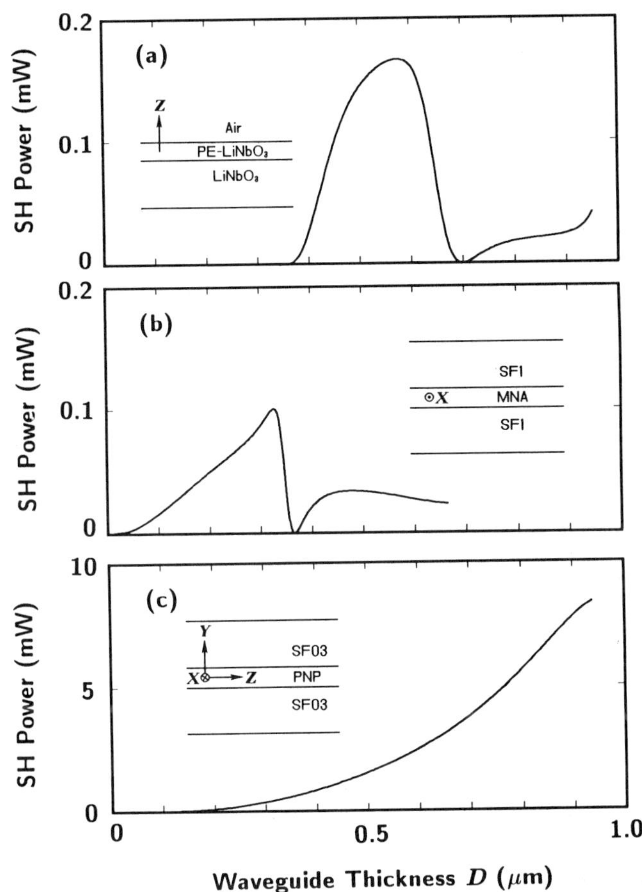

Fig. 13. Calculated second-harmonic powers for $LiNbO_3$, (a), MNA (b), and PNP (c) planar waveguides. Parameters used in the calculation are $\lambda^{\omega} = 1.0\ \mu m$, $L = 10$ mm, $W = 2.0\ \mu m$ and $\mathbf{P}^{\omega} = 100$ mW. For the $LiNbO_3$ waveguide, the nonlinear optical coefficient in the proton-exchanged (PE) guiding layer is assumed to be zero.

should be large in order to realize strong confinement of the fundamental optical field in the nonlinear guiding region.

To make the first point clearer, we will take some numerical examples. Figure 13 shows the guide-thickness dependent conversion efficiencies of planar waveguides of $LiNbO_3$, MNA and 2-(N-prolinol)-5-nitropyridine (PNP). These are calculated using the following parameters; $\lambda = 1.0\ \mu m$, $\mathbf{P}^{\omega} = 100$ mW, $L = 10$ mm, and $W = 2\ \mu m$. We have assumed crystal orientations as shown in the insets in Fig. 13, so that the largest element of the d

tensor of each material (d_{33} for $LiNbO_3$, d_{11} for MNA, and d_{21} for PNP) can be fully utilized in the nonlinear interaction. Note that the "channel effect" corrections are not considered in this calculation.

As shown in Fig. 13(b), conversion efficiency obtainable in the MNA waveguide is smaller than that in the $LiNbO_3$ waveguide, despite its extremely large figure of merit. The high nonlinearity of MNA is offset by the large transverse mismatch that originates in the large dispersion; compare $n_X^{2\omega} - n_X^{\omega} = 0.53$ (MNA) with $n_e^{2\omega} - n_e^{\omega} = 0.15$ ($LiNbO_3$). On the other hand, the conversion efficiency calculated for the PNP waveguide is 50 times larger than that for the $LiNbO_3$ waveguide owing to the small transverse phase mismatch; in PNP, the dispersion is compensated for by the birefringence so that $n_Y^{2\omega} - n_X^{\omega} = -0.03$.

Reasonably high conversion efficiencies experimentally obtained in DAN- and DMNP-fibers are naturally understandable in the same context. At the same time, we can conclude that the other crystals that possess large phase-matchable effective nonlinear optical coefficients, such as PNP and NPP, are also good candidates for efficient guided-wave SHG devices if they are fabricated with favorable crystal orientations.

5. CONCLUSIONS

At present, there are several inorganic crystals with sufficiently large optical nonlinearities to allow efficient frequency doubling of laser diodes or diode-pumped solid state lasers. Undoubtedly, some of them will be put into practical use in high-density optical data storage and other applications. Our research efforts have been based on the expectation that organic materials with nonlinearities that are orders of magnitude larger than those of the best inorganics will allow us to obtain even higher efficiencies with very low optical powers available from laser diodes in configurations that are much simpler than those currently employed with inorganic crystals. Remarkable progress in the past few years in the research on organic-crystalline waveguiding structures has clearly proved that the high potential of organic crystals is indeed coming into full play in practical device structures.

There are, however, many problems that must be solved before organic nonlinear optical crystals become truly usable.

As to the material development, the efficiency-transparency tradeoff still remains to be overcome to realize shorter-cutoff-wavelength materials. Extension of the transparent window toward shorter wavelength (<0.4 μm) is mandatory in order to make organic crystals compatible with highly reliable AlGaAs/GaAs laser diodes oscillating around 0.8 μm. On the other hand, rapid progress of InGaAs/GaAs strained quantum well lasers oscillating around 1 μm [69], developed as pumping sources for Er-doped fiber amplifiers, may provide an alternative answer, because they may make currently available organic crystals with high nonlinearities usable in the application to the compact coherent blue/green light sources.

Another problem is the device processing technology. First of all, purification of materials and crystal growth techniques need to be further improved in order to ensure the optical and the dimensional homogeneity of the grown crystals. Techniques for the precise crystal orientation control also need to be developed. In addition, waveguiding structures necessitate further improvement on the lithographic technique to define waveguides with sufficient accuracy and on the polishing technique to produce good input and output faces.

Reliability and stability of the device performance also must be investigated before organic-crystalline waveguiding devices become loaded in commercial products. Long-term stability, including optical as well as chemical stability, of organic crystals when irradiated by cw intense light has not been studied to date. Considering the application to optical data storage, focusing characteristics become very important when the Cerenkov-type SHG is employed. Tolerances to temperature and wavelength deviations need to be thoroughly investigated.

As to the device architecture, we claim that we should direct our attention to channel-type waveguides. They are inherently compatible with laser diodes, because the laser diode itself is a typical channel waveguide device. Moreover, channel waveguide structures will provide flexibility of the device architecture. For example, extremely large nonlinear optical coefficients d_{ii} of organic crystals may be effectively utilized in nonlinear interactions with some modifications of the device structure. In order to make use of the diagonal element d_{ii}, which is otherwise not usable in efficient SHG, we have to employ multi-layer structures that enable us to optimize overlap integrals or quasi-phase-matching that leads to the efficient coupling between appropriate modes; such artificial manipulation on waveguide modes is possible only in channel waveguides but almost impossible in fiber waveguides. Furthermore, channel structures are compatible with guided-wave electro-optic light modulators. This may push up organic crystals to a great competitor against the conventional inorganic ferroelectrics and the newly developed poled polymers in the field of electro-optic devices.

Future researches on the application of organic nonlinear optical crystals to practical devices will demand that more device-oriented researchers, in addition to material scientists, should take part in this interesting research area. When research results from different disciplines are successfully combined, organic crystals will acquire positions in photonics that may be compared to those occupied by semiconductors and inorganic dielectrics in electronics.

ACKNOWLEDGMENTS

We gratefully acknowledge the support of Prof. S. Umegaki of the Department of Material Science of Keio University and Prof. N. Ogasawara of the Department of Electronics Engineering of the University of Electro-Communications, who provided us an opportunity for starting the research on organic waveguides and have continuously cooperated with us. We appreciate valuable contributions to the work on lithographic process from Prof. T. Kudo of the Institute of Industrial Science of the University of Tokyo and Mrs. F. Saito of Nitto Denko. We are also very grateful to Drs. R. Morita and T. Onda, and Messrs. K. Tsuda, N. Nashizume, S. Miyoshi, and Y. Taito for their enthusiastic collaborations. Thanks are also due to Prof. K. Onabe for his continuous support and encouragement. Part of this work was supported by the Grant-in-Aid for Scientific Research from the Ministry of Education, Science and Culture, Japan.

REFERENCES

1. G. I. Stegeman and C. T. Seaton, *J. Appl. Phys.* **58**, R57 (1985).
2. J. Zyss, *J. Mol. Electron.* **1**, 25 (1985).
3. E. M. Conwell, *IEEE J. Quantum Electron.* **QE-9**, 867 (1973).
4. A. Yariv, *IEEE J. Quantum Electron.* **QE-9**, 919 (1973).
5. J. A. Armstrong, N. Bloembergen, J. Ducuing, and P. S. Pershan, *Phys. Rev.* **127**, 1918 (1962).
6. S. Somekh and A. Yariv, *Opt. Commun.* **6**, 301 (1972).
7. J. D. Bierlein, D. B. Laubacher, J. B. Brown, and C. J. van der Poel, *Appl. Phys. Lett.* **56**, 1725 (1990).
8. P. K. Tien, R. Ulrich, and R. J. Martin, *Appl. Phys. Lett.* **17**, 447 (1970).
9. R. Regener and W. Sohler, *J. Opt. Soc. Am. B* **5**, 267 (1988).
10. T. Taniuchi and K. Yamamoto, *Dig. CLEO'87*, WP6 (1987).
11. G. A. Magel, M. M. Fejer, and R. L. Byer, *Appl. Phys. Lett.* **56**, 108 (1990).
12. K. Mizuuchi, K. Yamamoto, and T. Taniuchi, *Dig. CLEO'91*, CTuV3 (1991).

13. C. J. van der Poel, J. D. Bierlein, J. B. Brown, and S. Colak, *Appl. Phys. Lett.* **57**, 2074 (1990).
14. See, for example, *Nonlinear Optical Properties of Organic Molecules and Crystals*, D. S. Chemla and J. Zyss, eds., Vol. 1 (Academic Press, Orlando, 1987).
15. A. Harada, Y. Okazaki, K. Kamiyama, and S. Umegaki, *Appl. Phys. Lett.* **59**, 1535 (1991).
16. M. Kawaharada, Y. Yamanaka, H. Endoh, S. Ishikawa, S. Takano, and N. Ookubo, *Dig. CLEO'91*, CTuP3 (1991).
17. B. F. Levine, C. G. Bethea, C. D. Thurmond, R. T. Lynch, and J. L. Bernstein, *J. Appl. Phys.* **50**, 2523 (1979).
18. T. Tsunekawa, T. Gotoh, H. Mataki, T. Kondoh, S. Fukuda, and M. Iwamoto, *SPIE Proc.* **1337**, 272 (1990).
19. Y. Okazaki, K. Kamiyama, and S. Umegaki, *Dig. Cryst. Growth Org. Mater.* 256 (1989).
20. I. Ledoux, D. Josse, P. Vidakovic, and J. Zyss, *Opt. Eng.* **25**, 202 (1986).
21. K. Sutter, Ch. Bosshard, and P. Günter, *Ferroelectrics* **92**, 395 (1989).
22. T. Kondo, R. Morita, N. Ogasawara, S. Umegaki, and R. Ito, *Jpn. J. Appl. Phys.* **28**, 1622 (1989).
23. P. Kerkoc, M. Zgonik, K. Sutter, Ch. Bosshard, and P. Günter, *J. Opt. Soc. Am. B* **7**, 313 (1990).
24. J.-M. Halbout, S. Blit, W. Donaldson, and C. L. Tang, *IEEE Quantum Electron.* **QE-15**, 1176 (1979).
25. R. Morita, T. Kondo, Y. Kaneda, A. Sugihashi, N. Ogasawara, S. Umegaki, and R. Ito, *Jpn. J. Appl. Phys.* **27**, L1131 (1988).
26. R. Morita, N. Ogasawara, S. Umegaki, and R. Ito, *Jpn. J. Appl. Phys.* **26**, L1711 (1987).
27. G. F. Lipscomb, A. F. Garito, and R. S. Narang, *J. Chem. Phys.* **75**, 1509 (1981).
28. T. Sasaki, A. Yokotani, K. Fujioka, Y. Kitaoka, and S. Nakai, *Dig. Cryst. Growth Org. Mater.* 246 (1989).
29. B. J. McArdle and J. N. Sherwood, in *Advanced Crystal Growth*, P. M. Dryburgh, B. Cockayne, and K. G. Barraclough, eds., (Prentice-Hall, New York, 1987) p. 179.
30. J. Badan, R. Hierle, A. Perigaud, and P. Vidakovic, in *Nonlinear Optical Properties of Organic Molecules and Crystals*, D. S. Chemla and J. Zyss, eds., Vol. 1 (Academic Press, Orlando, 1987), p. 297.
31. J.-M. Halbout and C. L. Tang, in *Nonlinear Optical Properties of Organic Molecules and Crystals*, D. S. Chemla and J. Zyss, eds., Vol. 1 (Academic Press, Orlando, 1987), p. 385.
32. I. Ledoux, J. Badan, J. Zyss, A. Migus, D. Hulin, J. Etchepare, G. Grillon, and A. Antonetti, *J. Opt. Soc. Am. B* **4**, 987 (1987).
33. S. Bucharme, W. P. Risk, W. E. Moerner, V. Y. Lee, R. J. Twieg, and G. C. Bjoklund, *Appl. Phys. Lett.* **57**, 537 (1990).
34. Y. Goto, A. Hayashi, G. J. Zhang, M. Nakayama, Y. Kitaoka, T. Sasaki, T. Watanabe, S. Miyata, K. Honda, and M. Goto, *SPIE Proc.* **1337**, 297 (1990).

35. G. D. Boyd and D. A. Kleinman, *J. Appl. Phys.* **39**, 3597 (1968).
36. See, for example, D. Marcuse, *Theory of Dielectric Optical Waveguides* (Academic Press, New York, 1974).
37. M. J. Li, M. de Micheli, Q. He, and D. B. Ostrowsky, *IEEE J. Quantum Electron.* **26**, 1384 (1990).
38. T. Onda and R. Ito, *Jpn. J. Appl. Phys.* **30**, 957 (1991).
39. N. Hashizume, T. Kondo, T. Onda, N. Ogasawara, S. Umegaki, and R. Ito, *IEEE J. Quantum Electron.* **28**, 1798 (1992).
40. T. Kondo, N. Hashizume, K. Tsuda, R. Morita, N. Ogasawara, S. Umegaki, and R. Ito, *Int. J. Nonlin. Opt. Phys.* **1**, 367 (1992).
41. H. Tamada, *IEEE J. Quantum Electron.* **26**, 1821 (1990).
42. H. Tamada, *IEEE J. Quantum Electron.* **27**, 502 (1991).
43. K. I. White and B. K. Nayar, *J. Opt. Soc. Am. B* **5**, 317 (1988).
44. K. Chikuma and S. Umegaki, *J. Opt. Soc. Am. B* **7**, 768 (1990).
45. K. Chikuma and S. Umegaki, *J. Opt. Soc. Am. B.* **9**, 1083 (1992).
46. G. H. Hewig and K. Jain, *Opt. Commun.* **47**, 347 (1983).
47. K. Sasaki, T. Kinoshita, and N. Karasawa, *Appl. Phys. Lett.* **45**, 333 (1984).
48. H. Ito, K. Hotta, H. Takara, and K. Sasaki, *Appl. Opt.* **25**, 1491 (1986).
49. R. Twieg, A. Azema, K. Jain, and Y. Y. Cheng, *Chem. Phys. Lett.* **92**, 208 (1982).
50. R. T. Bailey, F. R. Cruickshank, S. M. G. Guthrie, B. J. McArdle, H. Morrison, D. Pugh, E. A. Shepherd, J. N. Sherwood, C. S. Yoon, R. Kashyap, B. K. Nayar, and K. I. White, *Opt. Commun.* **65**, 229 (1988).
51. R. T. Bailey, F. R. Cruickshank, S. M. G. Guthrie, B. J. McArdle, H. Morrison, D. Pugh, E. A. Shepherd, J. N. Sherwood, C. S. Yoon, R. Kashyap, B. K. Nayar, and K. I. White, *J. Mod. Opt.* **35**, 511 (1988).
52. T. Kondo, N. Ogasawara, R. Ito, K. Ishida, T. Tanase, T. Murata, and M. Hidai, *Acta Cryst. C* **44**, 102 (1988).
53. T. Kondo, N. Hashizume, S. Miyoshi, R. Morita, N. Ogasawara, S. Umegaki, and R. Ito, *SPIE Proc.* **1337**, 53 (1990).
54. S. Tomaru, M. Kawachi, and M. Kobayashi, *Opt. Commun.* **50**, 154 (1984).
55. P. Vidakovic, J. Badan, R. Hierle, and J. Zyss, *Proc. IQEC*, PD-C5-1 (1984).
56. T. Gotoh, T. Tsunekawa, T. Kondoh, S. Fukuda, H. Mataki, M. Iwamoto, and Y. Maeda, *Dig. Cryst. Growth Org. Mater.* 234 (1989).
57. T. Kudo, A. Ishikawa, H. Okamoto, K. Miyauchi, F. Murai, K. Mochiji, and H. Umezaki, *J. Electrochem. Soc.* **134**, 2607 (1987).
58. K. Tsuda, T. Kondo, F. Saito, T. Kudo, and R. Ito, *Jpn. J. Appl. Phys.* **31**, 134 (1992).
59. J. L. Stevenson and R. B. Dyott, *Electron. Lett.* **10**, 449 (1974).
60. J. L. Stevenson, *J. Cryst. Growth* **37**, 116 (1977).
61. B. K. Nayar, *ACS Symp. Ser.* **233**, 153 (1983).
62. S. Umegaki, Y. Takahashi, A. Manabe, and S. Tanaka, *Proc. MRS Nonlin. Opt. Mater.* 97 (1985).
63. P. V. Vidakovic, M. Coquillay, and F. Salin, *J. Opt. Soc. Am. B* **4**, 998 (1987).

64. J. D. Rush, G. E. Holdcroft, and P. L. Dunn, *SPIE Proc.* **1017**, 135 (1988).
65. P. Kerkoc, Ch. Bosshard, H. Arend, and P. Günter, *Appl. Phys. Lett.* **54**, 487 (1989).
66. T. Uemiya, N. Uenishi, S. Okamoto, K. Chikuma, K. Kumata, T. Kondo, R. Ito, and S. Umegaki, *Appl. Opt.* **20**, 7581 (1992).
67. S. Tomaru and S. Zembutsu, *Preprints SPSJ Int. Polym. Conf.* **2**, 145 (1986).
68. C. Goto, K. Kamiyama, and S. Umegaki, *Ext. Abst. Jpn. Soc. Appl. Phys.* (*Spring Meeting*), 28p-PA-4 (1991) (in Japanese).
69. D. F. Welch, W. Streifer, C. F. Schaus, S. Sun, and P. L. Gourley, *Appl. Phys. Lett.* **56**, 10 (1990).

PART III

NONLINEAR OPTICS IN POLYMERIC MEDIA

Chapter 6

QUADRATIC NONLINEAR OPTICS IN POLED POLYMER FILMS: FROM PHYSICS TO DEVICES

K. D. Singer

and

J. H. Andrews

Case Western Reserve University, Department of Physics, Cleveland, Ohio

ISBN 0-12-784450-3

1. INTRODUCTION

The study of second-order nonlinear optical properties in polymeric materials is a field of growing and evolving interest [1–4]. Originally, interest centered on crystalline diacetylene systems [5,6] but later moved on to glass-forming materials [7–10]. In both cases, the origin of the optical nonlinearity is molecular in nature, and the role of the polymer is largely structural. Glass-forming materials have become the favored system for applications due to the inherent optical quality of glasses and their high degree of processability [11]. In these materials, however, intrinsic polar order is absent, making the materials centrosymmetric and thus incapable of exhibiting second-order nonlinear optical effects.

To overcome this important shortcoming, electric field poling is employed to impart an overall polar orientation. This process is depicted in Fig. 1. In its original manifestation, dipolar nonlinear optical molecules are dissolved in a glass-forming material [8]. At room temperature, the molecules are relatively immobile with respect to orientational motion. As the temperature is raised, orientational motion increases until, above the glass-rubber transition temperature, the molecules rotate freely. In this state, an electric

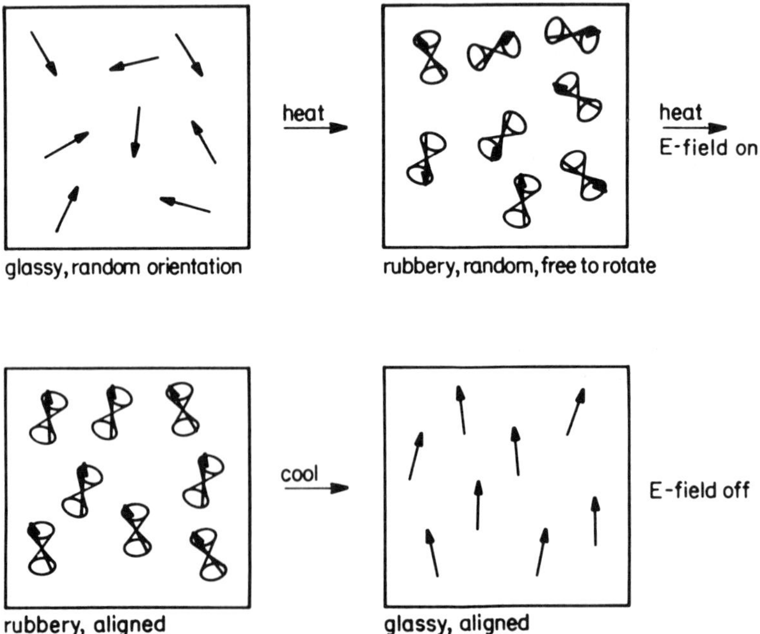

Fig. 1. Electric field poling process.

field is applied, which orients the molecules through coupling to the molecular dipole moment. The glass is then cooled with the field applied, thus "freezing-in" the orientation. The original experiments were carried out in liquid crystal polymer hosts, and the orientation was found to be extremely unstable [7]. Later work on polymethylmethacrylate (PMMA) hosts produced orientations stable enough for systematic study, but not stable enough for applicability to photonic devices [8]. Recent work to develop more stable polymers has led to extensions of the poling process [12,13]. For example, the mobile state might consist of an uncured polymer state, and the "freezing-in" would consist of an immobilizing chemical reaction. These new processes allow for more technologically interesting materials.

Since the potential advantages of glass-forming second-order nonlinear optical materials were first identified, their study and development have continually grown. These advantages include the intrinsic optical quality, the favorable dielectric properties, and the inherent processability of glasses. Further, glassy materials have been found to be compatible with the organic molecules exhibiting exceptional optical nonlinearities. Research interest has been directed in three principal ways: understanding of the physical processes at work in the materials [14–26], the study of potential applications of these materials [27–35], and the development of new and higher performance materials [9,36–42]. In the study of physical processes, most work has centered on the relation between the poling process, poled order and the nonlinear optical properties. In addition, studies of molecular motion and its relationship to the decay of the optical nonlinearity have been carried out. More recently, attention has also turned to photorefractive processes in glass-forming polymers.

The study of the nonlinear optical properties of glass-forming polymers is mainly driven by the potential applications of these polymers. Electro-optic applications have always appeared most promising, and, indeed, much of this promise has been experimentally demonstrated. It is anticipated that commercial realization of these applications is not far off. The potential of polymeric materials for frequency conversion is also a subject of study, but efficient conversion has not yet been demonstrated. Issues regarding the development of new materials appear elsewhere in this volume.

We begin the chapter with a discussion of the physical mechanisms responsible for the second-order nonlinearity in polymeric materials starting with a short review of molecular mechanisms in Sec. 2.1 followed by a discussion of the relationship between orientational order and the nonlinear optical susceptibility in Sec. 2.2. In Sec. 3 we describe the poling process, with particular attention to the electronic processes involved and their relationship to materials and structure. That section ends with a description of orientational relaxation phenomena in poled polymers. Sec. 4 describes potential applications of poled polymers with particular attention to materials issues. Finally, the theory and experiment of photorefraction is described in Sec. 5, again with attention to polymeric materials.

2. PHYSICAL MECHANISMS

2.1. Molecular Response

In the usual description of nonlinear optics, one expresses the electric polarization per unit volume as a series expansion in powers of the electric field [43],

$$P_i(\omega) = P_i^0 + \chi_{ij}^{(1)}(-\omega)E_j(\omega) + \chi_{ijk}^{(2)}(-\omega;\omega_1,\omega_2)E_j(\omega_1)E_k(\omega_2) + \chi_{ijkl}^{(3)}(-\omega;\omega_1,\omega_2,\omega_3)E_j(\omega_1)E_k(\omega_2)E_l(\omega_3) + \cdots, \tag{2.1.1}$$

where the χs are the linear and nonlinear optical susceptibility tensors that describe the interaction between the electric fields, E_i, and the materials. Polymeric second-order nonlinear optical materials are molecular materials. The underlying physical mechanisms in molecular materials are, by definition, traced to the nonlinear optical molecular constituents. In describing the interaction between light and molecules, one can identify a polarization due to the molecular units included in the polymer. In this case, one can write the molecular polarization as

$$p_I = \mu_I + \alpha_{IJ}F_J + \beta_{IJK}F_JF_K + \gamma_{IJKL}F_JF_KF_L + \cdots \tag{2.1.2}$$

where the molecular susceptibilities appear in analogy to the bulk ones in Eq. (2.2.1) and the local fields F apply at analogous frequencies. The molecular susceptibilities must be related to $\chi_{ijk}^{(2)}$ carefully with particular attention to the permutation symmetries so that, for instance, when comparing studies of molecular susceptibilities, β_{IJK} is not the quantity usually measured using electric-field-induced second harmonic generation (EFISH), but is actually twice the value measured. See, for example, [44] for a discussion of the various degeneracy and symmetry factors needed to compare different expansions for the susceptibilities. (The β usually measured in EFISH is analogous to the second harmonic, d, coefficient).

In molecular materials such as guest-host or side-chain functionalized polymers, the intermolecular interactions are weak; the nonlinear optical moieties interact with their environment only through the dipolar interaction.

In this van der Waals limit, the bulk polarization can be expressed as a sum over the molecular units. The origin of the nonlinear optical properties, in this limit, can be traced to the molecular constituents. Much of the underlying physics responsible for the second-order nonlinearity was studied some time ago using Hartree-Fock methods and perturbation theory directly to calculate the molecular nonlinear optical susceptibility [45,46]. Though, in general, the perturbation calculation of β involves a full set of quantum mechanical states of the molecule, it is found that in certain organic molecules, the susceptibility is dominated by a limited set of the quantum states.

These quantum calculations have been shown to describe qualitatively and quantitatively the second-order nonlinear optical susceptibility in a wide range of donor/acceptor substituted conjugated organic molecules. The results of these calculations have confirmed the two-level model for β in some of the smaller conjugated systems, such as substituted benzene molecules. The two-level model for second harmonic generation in these one-dimensional molecules arises from the experimental and calculational fact that, in the sum-over-states, the magnitude and sign of β is largely determined by the interaction between the electric field and the ground and lowest-lying excited molecular electronic states. In this model, the susceptibility is given by [47,48]

$$\beta_{xxx}(-2\omega;\omega;\omega) = \frac{e^3|\mu_{01}|^2(\mu_{11}-\mu_{00})}{\hbar^2}F(\omega), \tag{2.1.3}$$

where
$$F(\omega) = \frac{3\omega_0^2}{(\omega_0^2-\omega^2)(\omega_0^2-4\omega^2)}, \tag{2.1.4}$$

μ_{01} is the transition moment between the two states, μ_{00} and μ_{11} are ground- and excited-state dipole moment, respectively, and ω_0 gives the excited state energy, $\hbar\omega_0$.

Though more extended π-electron systems do not, in general, conform to the two-level behavior observed in substituted benzene systems, the essential features of charge rearrangement in the excited state can be understood qualitatively by examining Fig. 2 where electron density contour diagrams are shown for a representative aromatic extended donor-acceptor substituted molecule, Disperse Orange [48,49]. This molecule consists of an NH_2 electron-donating group and NO_2 electron-withdrawing group located on the termini of an azo (N=N) bridged double benzene ring conjugated electron system. The figures are shown for the highest occcupied and lowest unoccupied one-electron state belonging to the ground and excited state,

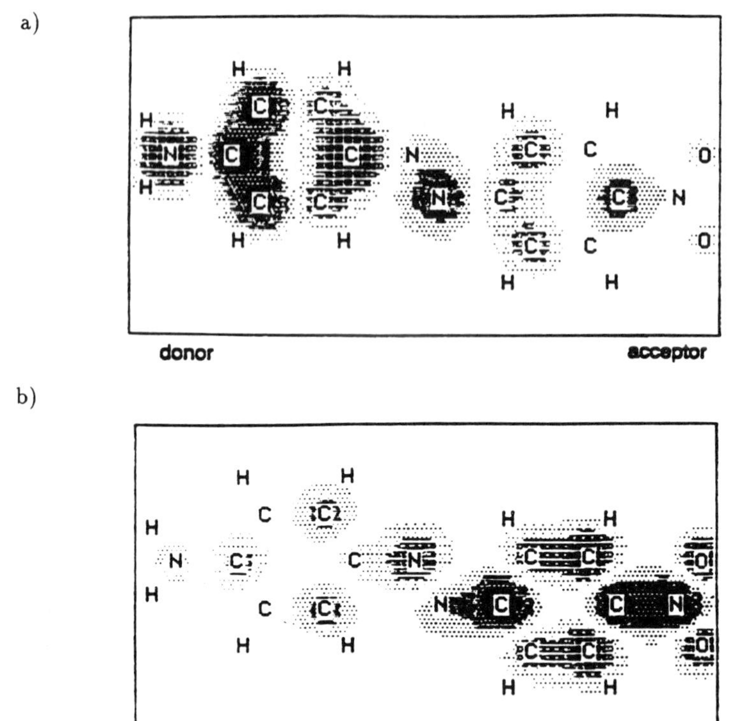

Fig. 2. Electron-density maps of (a) highest occupied and (b) lowest unoccupied one-electron states for the molecule Disperse Orange. Darker colors indicate higher density.

respectively. The contours represent the electron density slightly above the molecular plane. It is evident that the highest occupied one-electronic state exhibits enhanced charge density near the electron donating NH_2 group. This state, then, makes a negative contribution to the ground-state dipole moment of the molecule. In the lowest unoccupied state, enhanced charge density is observed near the electron-withdrawing group. This charge rearrangement across the entire molecular length leads to a large value of $\mu_{11} - \mu_{00}$ and, along with the large transition moments observed in such transitions, leads to a large value of β as given by Eqs. 2.1.3 and 2.1.4. If other states are involved in the susceptibility, such as would be expected in the extended Disperse Orange molecule, they would contribute in a similar manner. This behavior has been observed in a number of aromatic molecules [48].

In quinoid molecules, where one of the benzene ring π bonds is absent in the ground state, similar results are obtained except that the charge density in the highest occupied one-electron state is located at the electron withdrawing group and the lowest unoccupied one-electron state at the electron donating group. Thus, the sign of $\mu_{11} - \mu_{00}$ and β_{xxx} is opposite that of the aromatics. This has been observed in measurements of β [50]. The two-level model has been further confirmed by comparing measurements where $\omega < \omega_0$ and $2\omega < \omega_0$ with those where $\omega < \omega_0$ and $2\omega > \omega_0$. The sign change required by the two-level model was confirmed, and the measured magnitude was nearly constant [51].

The physical understanding gained through calculation and measurement has led to a great deal of "molecular engineering" aimed at optimizing molecular properties such as β_{xxx}, the ground state dipole moment, and molecular electronic absorption for various applications and material systems. For example, a study on the effect of various donors and acceptors chosen for their strength based on chemical principles has led to the development of new molecular systems. A plot of the nonlinear susceptibility measured by electric-field induced second harmonic generation versus the electron-withdrawing strength as measured by the Hammett constant σ_R^- for substituted benzene molecules is shown in Fig. 3 [48,52]. This study led to the development of exceptional nonlinear optical molecules such as those given in Table 1.

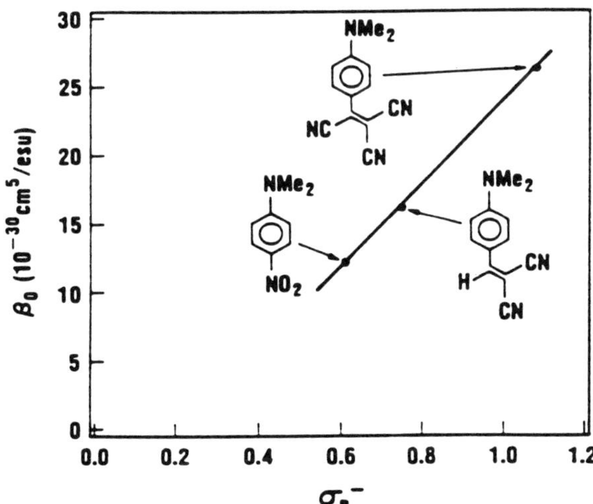

Fig. 3. Plot of the second-order nonlinear optical susceptibility as a function of electron withdrawing strength, σ_R^-.

Table 1.

Results of Measurements of the Molecular Nonlinear Optical Susceptibility of Several Molecules (β in units of 10^{-30} cm^5/esu).

Compound	Structure	β_0 (10^{-30} cm^5/esu)
DMNA	$(H_3C)_2N-C_6H_4-NO_2$	12
DMA-NS	$(H_3C)_2N-C_6H_4-CH{=}CH-C_6H_4-NO_2$	52
Disperse Red 1	$(HOH_2CH_2C)(H_3CH_2C)N-C_6H_4-N{=}N-C_6H_4-NO_2$	47
NB-DMAA	$(H_3C)_2CH-C_6H_4-N{=}CH-C_6H_4-NO_2$	37
DMA-DCVS	$(H_3C)_2N-C_6H_4-CH{=}CH-C_6H_4-CH{=}C(CN)_2$	133
DWA-TCVAB	$(H_2CH_2C)_2N-C_6H_4-N{=}N-C_6H_4-C(CN){=}C(CN)_2$	154

2.2. Bulk Response

In molecular materials, the responses of the molecular components are additive, and the determination of the bulk properties mostly involves the arrangement of the molecules within the material taking into account the effect of local fields. In this case, the relationship between the molecular and bulk response is a sum,

$$P_i(t) = \frac{1}{V} (\sum p_I(t))_i, \tag{2.2.1}$$

where V is the molecular volume, the indices i refer to the bulk laboratory coordinate system, and the indices I refer to the molecular coordinate system. The molecular ordering determines the methods used to evaluate the sum in Eq. (2.2.1). The ordering can be highly structured, as in a crystal, in which case x-ray diffraction can be used to locate the molecules within the unit cell.

In polymeric materials, the location and orientation of each molecule within the polymer is not known. In calculating the bulk nonlinear response of such a system, we must consider an ensemble average over the orientational distribution and the local field correction factors [8,14,53]. The summation over individual molecules for the second-order bulk susceptibility as required in crystals, is replaced by a thermodynamic average [14,54],

$$\chi^{(2)}_{ijk} = N \langle \beta^*_{IJK} \rangle_{ijk}, \tag{2.2.2}$$

where N is the molecular density, $\langle \beta^*_{IJK} \rangle_{ijk}$ the ijkth component of the orientational average of the molecular second-order susceptibility, and where * denotes that the local field effects have been included within the susceptibility. The problem of local fields is discussed generally in texts on nonlinear optics. See, for example, [43]. The orientational ensemble average for the tensor β^*_{IJK} is calculated using the following integral,

$$\langle \beta^*_{IJK} \rangle_{ijk} = \int_0^{2\pi} d\phi \int_0^{\pi} \sin\theta \, d\theta \int_0^{2\pi} d\psi \beta^*_{IJK} a_{iI} a_{jJ} a_{kK} G(\phi, \theta, \psi), \tag{2.2.3}$$

where $G(\phi, \theta, \psi)$ is the normalized orientational distribution function, which is a function of the Euler angles ϕ, θ, and ψ, and where a_{mM} are components of the rotation matrix. The appropriate Euler angles are given in Fig. 4. If the 3-axis is denoted as the direction of the poling field, then the axis of rotation ψ will denote the major axis z of the molecule, which will be the dipole direction in the case of dipolar molecules. The molecules will, on average, by symmetric for rotations about ψ, and we assume that the films will be symmetrical about any rotation in ϕ. The distribution then becomes one

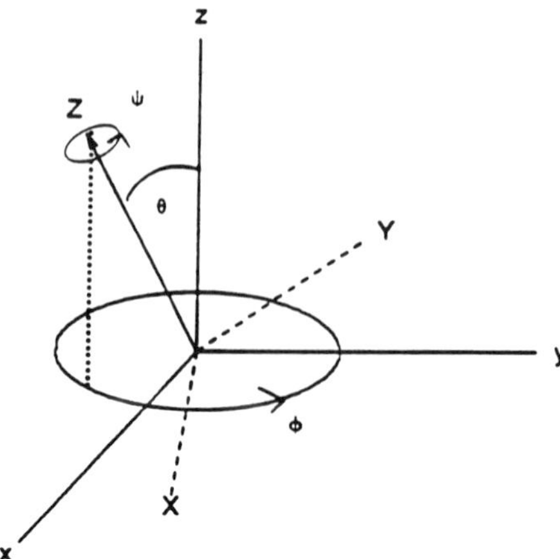

Fig. 4. Euler angles describing molecular orientation in laboratory frame. The direction z is the poling direction, and Z' the direction of the molecular dipole moment.

over θ, the angle between the poling and the dipolar direction. The symmetry group we have just described is ∞mm and consists of a unique axis 3 about which there is an infinite-fold rotation, and an infinity of mirror planes containing the axis. In this symmetry group, the nonzero tensor components corresponding to the third rank second-order nonlinear optical tensor are $\chi^{(2)}_{333}$, $\chi^{(2)}_{311}$, and $\chi^{(2)}_{113}$. Further, far from resonance (when Kleinman symmetry holds [55]), $\chi^{(2)}_{311} = \chi^{(2)}_{113}$, so that there are but two independent tensor components. The thermal averaging integrals also involve the transformation matrix,

$$\mathbf{a} = \begin{bmatrix} \cos\theta\cos\phi\cos\psi - \sin\phi\sin\psi & \cos\theta\sin\phi\cos\psi + \cos\phi\sin\psi & -\sin\theta\cos\psi \\ \cos\theta\cos\phi\sin\psi - \sin\phi\cos\psi & -\cos\theta\sin\phi\sin\psi + \cos\phi\cos\psi & \sin\theta\sin\psi \\ \sin\theta\cos\phi & \sin\theta\sin\phi & \cos\theta \end{bmatrix}, \quad (2.2.4)$$

where now the distribution function has been made independent of ψ and ϕ, so that the integrand in Eq. (2.2.3) depends on ψ and ϕ only through $\mathbf{a}$.

Two different distribution functions have been applied to Eq. (2.2.3) to evaluate the polarization density. In the first case, one assumes that only the poling field acts on the nonlinear optical molecules during poling so that a true oriented dipolar gas results [8,56,57]. A Gibbs distribution is usually assumed, in which the dipolar energy contributes in the usual way and the distribution function is given by

$$G(\Omega, E_p) = \frac{\exp\left[\frac{1}{kT}(\mathbf{m}^* \cdot \mathbf{E_p})\right]}{\int d\Omega \exp\left[\frac{1}{kT}(\mathbf{m}^* \cdot \mathbf{E_p})\right]}, \tag{2.2.5}$$

where $\mathbf{m}^*$ is the molecular dipole moment including local fields arising from the poling field $\mathbf{E_p}$. Contributions from the linear polarizability are ignored.

The susceptibilities can then be calculated by evaluating the integrals of Eq. (2.2.3) using the distribution function of Eq. (2.2.5). Kielich has reported these as [58]

$$\langle \beta^*_{IJK} \rangle_{333} = 3\beta^*_{zxx} L_1(p) + (\beta^*_{zzz} - 3\beta^*_{zxx}) L_3(p) \tag{2.2.6}$$

and
$$\langle \beta^*_{IJK} \rangle_{311} = (\beta^*_{zzz} - \beta^*_{zxx}) L_1(p) + (3\beta^*_{zxx} - \beta^*_{zzz}) L_3(p), \tag{2.2.7}$$

where $p = \mathbf{m}^* \cdot \mathbf{E_p}/kT$, the reduced energy of a dipole in an electrostatic field, and L_n is the generalized Langevin function,

$$L_n(p) = \frac{\int_0^\pi \cos^n \theta \exp(p \cos \theta) \sin \theta \, d\theta}{\int_0^\pi \exp(p \cos \theta) \sin \theta \, d\theta}. \tag{2.2.8}$$

The functions of interest here are approximated by

$$L_1(p) = \coth p - \frac{1}{p} \tag{2.2.9}$$

and
$$L_3(p) = \left(1 + \frac{6}{p^2}\right) L_1(p) - \frac{2}{p} \cdots \tag{2.2.10}$$

The molecules have been assumed symmetric in x and y, leading to two nonzero tensor components. When one considers the extended linear dye molecules such as azo and stilbene dyes, it is also reasonable to assume that the molecules are one-dimensional, i.e., exhibit a nonzero β^*_{zzz} only. In addition, for solid solutions of amorphous polymers, it is possible to approximate the local fields as Lorentz-Lorenz type for β^* and Onsager type for the $\mathbf{m}^*$ [8]. In this case, Eq. (2.2.2) yields

$$\chi^{(2)}_{333} = N f^{2\omega} f^{\omega} f^{\omega} f^{0} \beta^*_{zzz} L_3(p), \tag{2.2.11}$$

and
$$\chi^{(2)}_{311} = N f^{2\omega} f^{\omega} f^{\omega} f^{0} \beta^*_{zzz} (L_1(p) - L_3(p)), \tag{2.2.12}$$

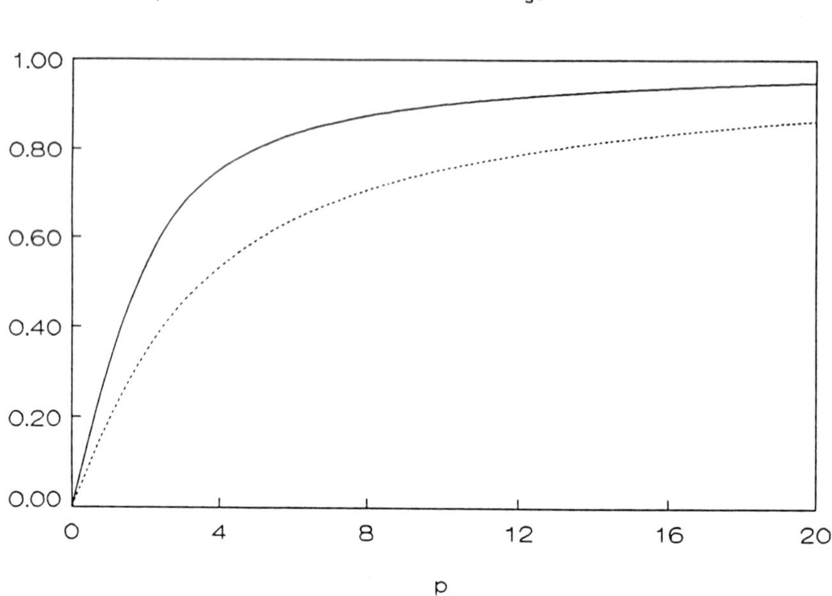

Fig. 5. Langevin functions.

where
$$f^{\omega} = \frac{n_{\omega}^2 + 2}{3} \tag{2.2.13}$$

and
$$f^{0} = \left(\frac{\varepsilon(n^2 + 3)}{n^2 + 2\varepsilon}\right). \tag{2.2.14}$$

ε is the static dielectric constant, and n_{ω} is the index of refraction at frequency ω. The Langevin functions $L_1(p)$ and $L_3(p)$ are plotted in Fig. 5. At very high fields, the functions tend toward unity as the molecules attain total alignment in the applied field. This state of total alignment is, in practice, unattainable due to the finite dielectric strength of polymeric materials. Close examination of the Langevin functions indicates that they are closely approximated by a line up to $p \simeq 1$. Dielectric strengths are generally around this value and generally do not exceed 3. Thus, in real materials, the alignment is described by the rapidly rising portion of the Langevin functions and does not really approach the saturation regime. In the limit of small field, i.e., $p \simeq 1$, the susceptibilities for a one-dimensional molecule are approximated by

$$\chi_{333} = N \frac{\mu E_p}{5kT} f^{0} f^{2\omega} f^{\omega} f^{\omega} \beta_{zzz} \tag{2.2.15}$$

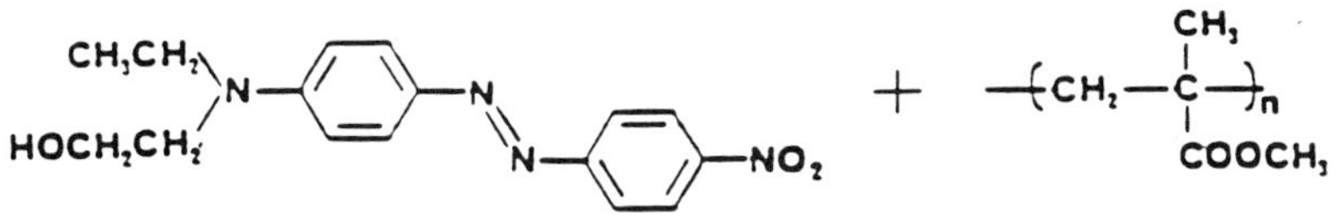

CH_3CH_2 N— —N=N— —NO_2 + —(CH_2—C(CH_3)($COOCH_3$))$_n$—
$HOCH_2CH_2$

Fig. 6. Disperse Red 1 (DR1) and polymethylmethacrylate (PMMA).

and

$$\chi_{311} = N \frac{\mu E_p}{15kT} f^0 f^{2\omega} f^{\omega} f^{\omega} \beta_{zzz}, \qquad (2.2.16)$$

where μ is the molecular dipole moment. These forms have been shown to apply very well to experimental data on doped and functionalized poled amorphous polymeric materials, an early example of which is shown in Fig. 6, consisting of an azo dye Disperse Red 1 (DR1) dissolved in PMMA. In this case, the susceptibility is seen to agree with the model with no adjustable parameters over a significant density range as seen in Fig. 7.

A second distribution function that has been used to evaluate the thermodynamic average integrals includes the effects of ordering forces other than the poling field [14,20]. These forces may be internal and include forces such as those leading to liquid crystalline order, or they may be externally applied. In any case, the effect is to impose additional *axial* ordering, which

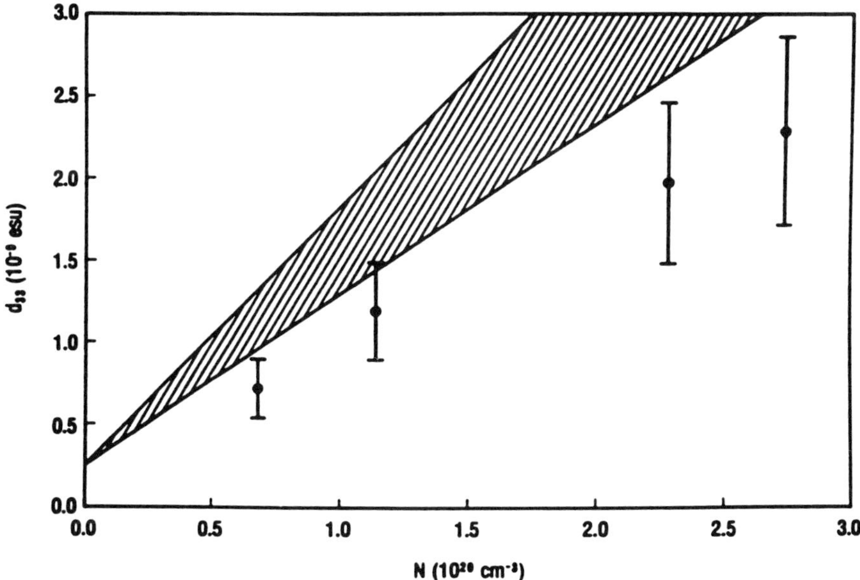

Fig. 7. Second harmonic coefficient as a function of number density of DR1 in PMMA films.

in turn affects the polar order due to the poling field. For a multispecies material, the appropriate Gibbs distribution is given by

$$G_u(\Omega, E_p) = \sum_v \frac{\exp\left[-\frac{1}{kT}(U_{uv} - \mathbf{m}_\mathbf{u}^* \cdot \mathbf{E}_\mathbf{p})\right]}{\int d\Omega \exp\left[-\frac{1}{kT}(U_{uv} - \mathbf{m}_\mathbf{u}^* \cdot \mathbf{E}_\mathbf{p})\right]}, \tag{2.2.17}$$

where $\mathbf{m}_\mathbf{u}^*$ is the molecular dipole moment including local fields arising from the poling field $\mathbf{E}_\mathbf{p}$. The term U_{uv} is the mean field potential between species u and v, which we will hereafter call the uniaxial potential. If $\mathbf{m}_\mathbf{u}^* \cdot \mathbf{E}_\mathbf{p} < kT$, then the dipolar energy can be expanded in a Taylor series up to first order and a semi-empirical analytical solution can be obtained. (This is referred to as the linear model [14].) The symmetry assumptions noted above lead to the distribution function being independent of the Euler angles, ϕ and ψ. To carry out the integration, we expand the distribution function in terms of Legendre polynomials,

$$G(\theta) = \sum_0^\infty \frac{(2l+1)}{2} A_l P_l(\cos\theta), \tag{2.2.18}$$

with the coefficients A_l determined from Eq. (2.2.18) as

$$\langle P_l \rangle \equiv A_l = \int_0^\pi \sin\theta \, d\theta G(\theta) P_l(\cos\theta). \tag{2.2.19}$$

The A_l are the ensemble averages of the P_l and are known in the liquid crystal literature as microscopic order parameters. The even-order polynomial terms describe axial ordering, while the odd order describe polar ordering. The expansions and integrations for the second-order susceptibility can be carried out to relate the susceptibility to the order parameters. These can be carried out most easily in the weak field limit where the exponential in the Gibbs distribution can be expanded to first order in the parameter $m_z^* E_p/kT$. In this case, one obtains

$$\chi_{333} = N \frac{\beta_{zzz}^* m_z^* E_p}{kT}\left(\frac{1}{5} + \frac{4}{7}\langle P_2 \rangle + \frac{8}{35}\langle P_4 \rangle\right) \tag{2.2.20}$$

and $$\chi_{311} = \chi_{113} = \chi_{131} = N \frac{\beta_{zzz}^* m_z^* E_p}{kT}\left(\frac{1}{15} + \frac{1}{21}\langle P_2 \rangle - \frac{8}{70}\langle P_4 \rangle\right). \tag{2.2.21}$$

This is a general solution applying to polymers and liquid crystals, where axial ordering existing independently of the poling order is specified by the

Table 2.

The Effect of Axial Order Parameters on Nonlinear Optical Susceptibilities.

Material	$\langle P_2 \rangle$	$\langle P_4 \rangle$	$\chi^{(2)}_{333}/Z$ [b]	$\chi^{(2)}_{311}/Z$ [b]
Nematic 40.8 [a]	0.65	0.25	0.63	0.07
Smectic B 40.8 [a]	0.91	0.79	0.90	0.02
Polymer glass	0	0	0.20	0.07

[a] Raman scattering data from S. Jen, N. A. Clark, P. S. Pershan, and E. B. Priestly, *J. Chem. Phys.* **66**, 4635 (1977).
[b] $Z = N\beta^* m^* E_p/kT$.

order parameters $\langle P_2 \rangle$ and $\langle P_4 \rangle$. One may apply Eqs. (2.2.20) and (2.2.21) to the case of liquid crystals, where the order parameters, $\langle P_2 \rangle$ and $\langle P_4 \rangle$, can be independently determined in order to examine the effect of independently occurring axial order on the polar order induced by the electric poling field. Table 2 shows this effect of axial order on the susceptibility. It is seen that typical nematic ordering increases $\chi^{(2)}_{333}$ by a factor of 2.5 or so, while the other component is hardly affected. A typical smectic A sees nearly a 4.5-fold increase with a substantial decrease in the off-diagonal tensor component. It should be noted that the limiting case of the Ising model [53] where $\langle P_2 \rangle = \langle P_4 \rangle = 1$ leads to a 5-fold increase in $\chi^{(2)}_{333}$ with $\chi^{(2)}_{311} = 0$. In actual liquid crystals, the susceptibilities lie between the Ising model case and the amorphous case where $\langle P_2 \rangle = \langle P_4 \rangle = 0$.

The development leading to Eqs. (2.2.20) and (2.2.21) are applicable only in the low-field limit, since the integrals are easily soluble only in this limit. However, van der Vorst and Picken have solved the problem numerically in the case of arbitrarily large fields [20]. They have also taken the linear polarizability into account, which makes a small contribution. In the high-field limit it is found that the axial order parameters are dependent on the poling field leading to nonlinear behavior requiring iterative solutions. These calculations determined that the poling field can induce liquid crystallinity when above the clearing temperature (the isotropic liquid crystal transition temperature) due to this nonlinear behavior. This behavior is depicted in Fig. 8 where the large change in order parameter corresponds to a temperature above the clearing temperature where liquid crystalline order is induced by the poling field. In materials that are in the liquid crystalline phase (mesophase), the change in order parameter due to the field

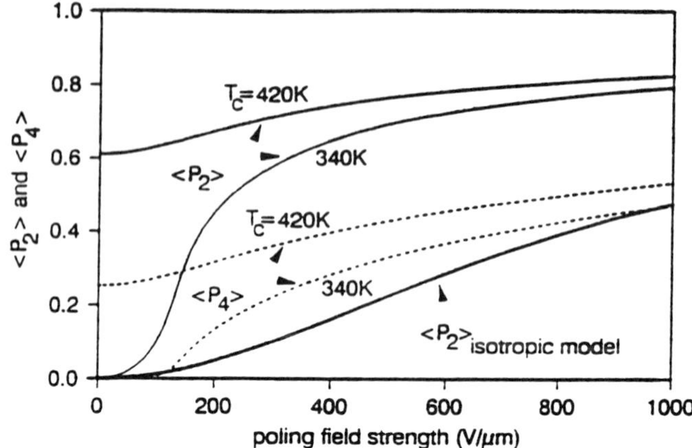

Fig. 8. Axial order parameters, $\langle P_2 \rangle$ and $\langle P_4 \rangle$ as a function of the local poling field for $m^{(0)} = 7D$, $\Delta\alpha = 47$ Å^3, $T_g = 380$ °K and T_c is as indicated. From [20].

is much less significant even at high fields. The saturation effect, however, is significant at high fields. Fig. 9 depicts $\langle \cos^3(\theta) \rangle$, which is proportional to $\langle \beta_{zzz} \rangle_{333}$ as a function of the poling field for an isotropic model as described above, for the Ising model, and for the liquid crystal models both in the linear model and the van der Vorst and Picken model for a realistic uniaxial

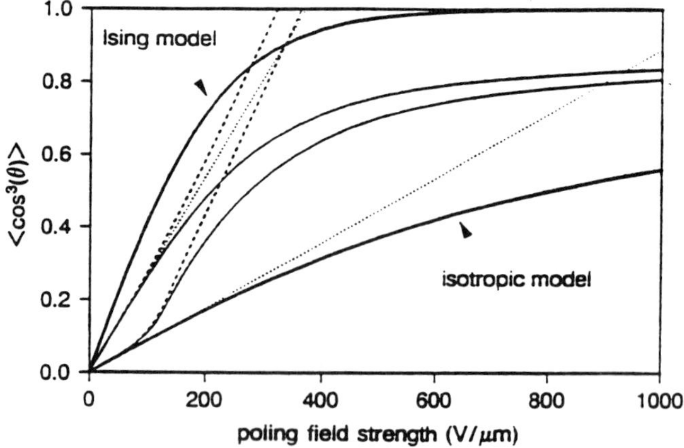

Fig. 9. Plot of $\langle \cos^3(\theta) \rangle$ as a function of the electric field with the same parameters as Fig. 8. The thick solid curves are from the exact solution of the Ising and isotropic model. The dotted and dashed curves are from the linear model. The thin solid curves are from the van der Vorst and Picken work. From [20].

potential. It is seen that the Ising model does not approximate a realistic system in any limit. The linear model is applicable up to approximately 1 MV/cm (depending on the value of the uniaxial potential). Above that level saturation effects set in making the full model necessary.

The models, including the uniaxial potential, were developed to describe the effect of liquid crystalline order on the nonlinear optical susceptibility. However, the applicability of these models is more general. In one application, an amorphous polymer was poled while a large uniaxial stress was applied. The stress induces a flow in the polymer, which in turn causes the included molecules to flow [59]. The effect of the stress energy on the orientation of the molecules takes the same form as the liquid crystal mean-field potential (except with opposite sign), so that with a moderate poling field, Eqs. (2.2.20) and (2.2.21) can be applied to determine polar and axial order. With this model, measurements of the magnitude of the two nonzero tensor components determined $\langle P_2 \rangle$ and $\langle P_4 \rangle$, while independent measurements of the poling and stress energies determined $\langle P_1 \rangle$ and $\langle P_3 \rangle$. Thus, the first four terms of the molecular orientation distribution were measured. The molecular orientation distribution so derived is shown in Fig. 10. The distribution indicates the flow of the polymer (and the molecules) under stress. The effect of the stress is to reduce the anisotropy of the optical nonlinear susceptibility tensor and to reduce the diagonal component of that

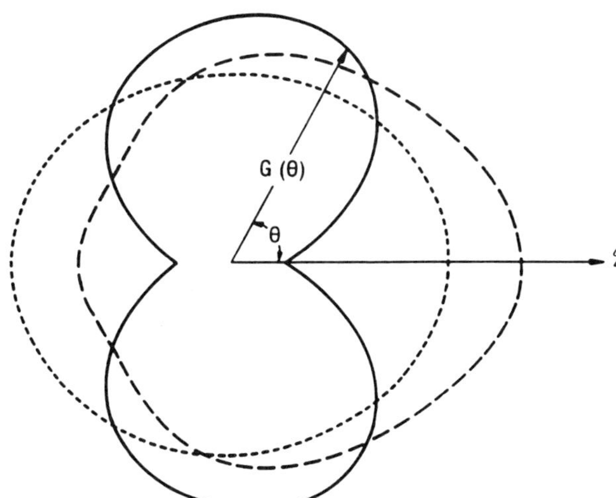

Fig. 10. Molecular orientational distribution $\langle P_1 \rangle$ through $\langle P_4 \rangle$ from second harmonic generation in poled stressed films for three cases: dotted is unpoled, dashed is poled with no stress flow, and solid is poled with stress flow.

tensor. In addition to this application, it is presumed that the models described above could be applied to determine polar and axial order in anisotropic polymers used as hosts in poled polymer materials.

3. POLING AND RELAXATION

The poling and relaxation behavior of poled nonlinear optical polymers depends on the electrical and dielectric properties of the polymers involved. The poling behavior is dependent on the choice of process, electrode materials, and cladding layers, and the relaxation behavior on the choice of materials and process. We describe here some of the issues related to poling and studies of relaxation phenomena in doped polymer films.

3.1. Poling of Electro-optic Polymers

Early experiments on poling utilized contact electrode poling of a single layer of nonlinear optical polymer. For example, a film of PMMA with the dye Disperse Red 1 was spin cast on indium tin oxide coated glass. Following this, a gold electrode was deposited on top of the film. A poling field was then applied as depicted in Fig. 1 [8]. For deposited electrodes, the dielectric strength of the film limits the maximum field to approximately 1 MV/cm. This maximum field can be increased through careful processing including film formation in a clean room environment. Film defects and inclusions limit the applied poling field.

It is desirable to attain higher poling fields which are less critically dependent on processing. To this end, one needs a large resistor in series with the power supply to limit the currents. This has been accomplished in two ways. One method is to use corona poling [10,36]. A typical corona poling setup is shown in Fig. 11. Here a needle or wire is placed at a high voltage (on the order of 10 kV), creating a corona near the electrode. A current begins to flow toward the ground leading to the deposition of charge on the surface of the electro-optic polymer. The large surface impedance as well as the high resistance of the corona acts to limit the currents that can be developed in film defects, thus catastrophic breakdown does not occur. In high resistance films, very large fields (>3 MV/cm) can be developed. While this can cause pinholes to open up at film defects, the undamaged part of the film is indeed poled at this large field. Further, the pinhole damage can be eliminated by controlling the amount of charge deposited with an intermediate grid. When the grid is present, the surface

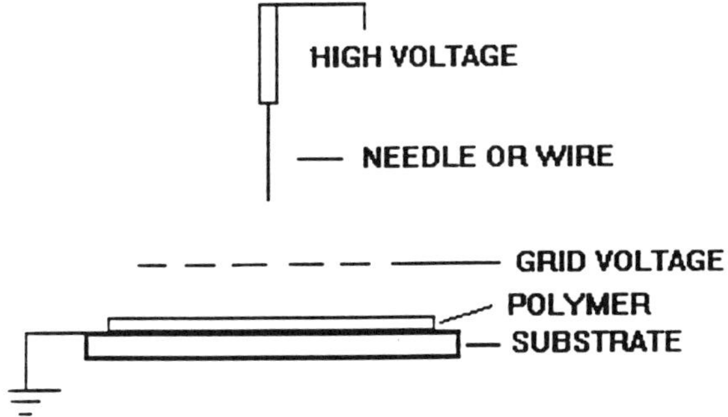

Fig. 11. Corona poling setup.

potential is limited to the grid potential, and thus, the charge deposited can be also limited.

The maximum surface charge is related to the film resistivity. A mechanism for the limiting of deposited positive charge can be seen in Fig. 12. On the left is the semiconductor-like band structure often used for polymeric materials. One has a filled band, a mobility gap ΔE, and a conduction band. One also has a number of trap states in the gap. In a low resistivity material, either the gap is smaller or is filled with a large number of trap states. The traps lead to conduction, generally, through hopping mechanisms. When a positive ion approaches the surface, an electron at the surface can recombine with the ion, thus neutralizing it. If U_{rec} is large enough, then the difference in energy between the electronic state in the polymer and the recombination energy, U_{rec}, will appear in Auger-like processes where the excess energy is imparted as kinetic energy to other electrons in the valence band. As the resistivity is lowered, the probability that one of these Auger-like electrons will be promoted to a trap level and, through the hopping mobility, will be

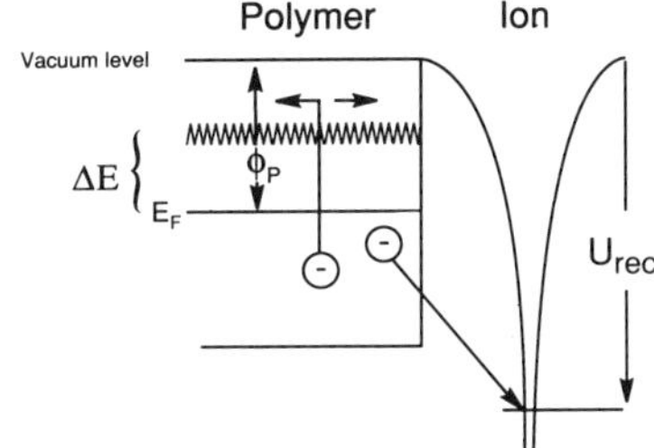

Fig. 12. Energies at a polymer film surface. ΔE is the polymer mobility gap, ϕ_P the polymer work function, E_F the Fermi energy, and U_{rec} the recombination energy.

transported to recombine with the hole left behind at the surface from the electron-ion recombination process becomes significant. This process leads to a reduction of surface charge and a concomitant increase in space charge thus reducing the poling field. The large fields developed can also lead to significant electron drifts so that electronic charge injected from the grounding electrode neutralizes holes within the polymer and at the polymer surface.

Another, more technologically significant, method for attaining high poling fields with a large effective resistance in series with the electro-optic layer is observed in multilayer structures. In this case, one may be dealing with optical cladding layers associated with optical waveguides. One can make a resistive network that optimizes poling through a careful choice of polymer and electrode materials. If the cladding polymers are chosen properly, the poling voltage is optimized across the nonlinear optical layer. A typical device geometry consists of three layers, the electro-optic polymer located between the two optical cladding layers. The cladding layers are chosen for a lower refractive index than the electro-optic layer so that optical waveguiding can occur. If the material of the cladding layer is selected so that the resistance of these layers *at the poling temperature* is large but is at least an order of magnitude lower than that of the electro-optic layer at the poling temperature, then most of the poling voltage will appear across the electro-optic layer. Since, in a multilayer structure, it is unlikely that defects will occur in the same location in all three layers, problems with dielectric breakdown are minimized in this structure. Thus, the cladding layers also behave as large resistors in series with the power supply. With the appropriate voltage division, then, fields of the order of 3 MV/cm are attainable in the electro-optic layer. This voltage division effect has been studied in a dicyanovinyl/methacrylate copolymer clad with an ultraviolet curable polymer [60]. It was found, over a broad temperature range, that the layers behaved as resistors in series.

Critical to this resistive behavior is the nature of the electrode/polymer interface. To understand this we need to consider the energetics of the interface [61]. The important energies are depicted in Fig. 13. Here, ϕ_m is the work function of the metal, ϕ_p the work function of the polymer, E_F the Fermi energy of the metal, and ψ the electron affinity of the polymer. The nature of the contact is determined by the relative work functions of the polymer and metal. If the work function of the metal is less than or equal to that of the polymer, then one has Ohmic contact, where charge can be injected or neutralized. In either case, the structure is resistive [60]. If the work function of the metal is greater than that of the polymer, then a blocking electrode results. In this case, the energy barrier prevents charge injection

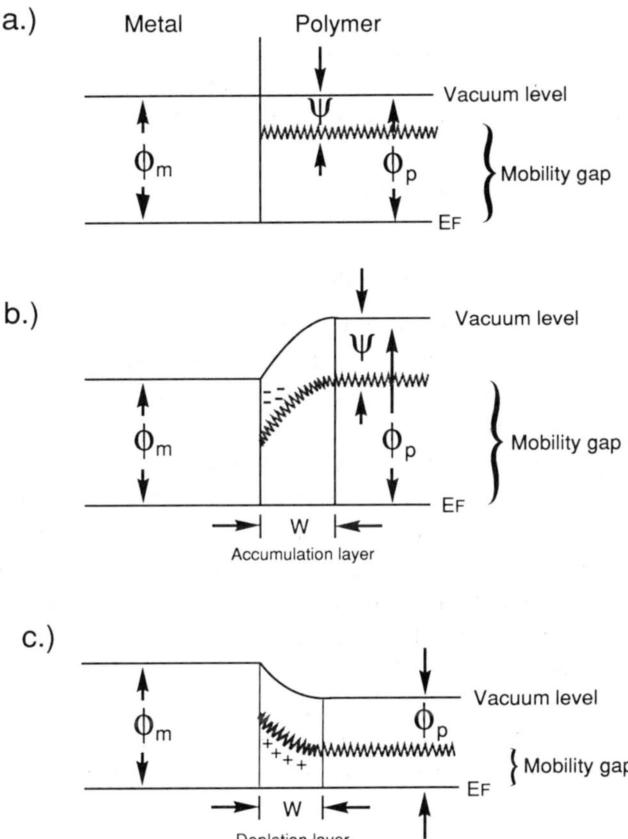

Fig. 13. The metal polymer interface for (a) neutral electrodes, (b) Ohmic electrodes, (c) blocking electrodes. ϕ_m is the metal work function, ϕ_p the polymer work function, E_F the Fermi energy of the metal, and ψ the electron affinity of the polymer.

and neutralization. For blocking electrodes, then, the sample will behave more like a capacitor than a resistor, and capacitive voltage division may appear to dominate the poling process. Another issue arises with blocking contacts which can be understood by examining the depletion layer that is created at the interface. When the sample is poled the depletion layer grows or shrinks considerably depending on the direction of the applied field. As the temperature is lowered, the depletion layer gets frozen-in as the mobility decreases. Because the electrodes are blocking, the change in depletion layer cannot be neutralized, and this polarization becomes stored. A similar process occurs with the accumulation layer in Ohmic contacts, but the layer is easily neutralized by carrier transport across the metal/polymer interface. Thus, the polarization is not stored, although space-charge is certainly injected. Other processes complicate the poling process, such as inhomogeneous space charge which can mimic thermoelectric effects. In any case, considerably more work is necessary in order to fully understand the role of real charge in the poling process.

3.2. Relaxation Phenomena

It is found that the nonlinear optical susceptibility decays with time [9,24,36,62,63,64]. This decay is due both to thermally activated rotational motion of the chromophore within the polymer [24,25,63] and to relaxation motion of the polymer [17,22,23]. In polymer relaxation studies, Eich and coworkers [17] found that the relaxation behavior of a side-chain functionalized polymer followed Williams-Landel-Ferry (WLF) behavior [65]. These studies involving corona poling indicated that the relaxation rates depended on the value of the poling field. This behavior is not unexpected with corona poling as a great deal of charge is injected (and probably inhomogeneously) so that space charge development and relaxation may complicate the observed relaxation behavior. The same group observed similar behavior in main-chain systems [23]. In this study, the decays were seen to fit to Williams-Watt stretched exponentials [66] with WLF temperature dependence. The WLF behavior would be expected in main-chain polymers and side-chain polymers near the glass transition since the motions in both cases are highly coupled to the polymer chain.

Systems which are less tightly coupled to the gross polymer motions were also studied [15,24]. In these cases, the temperatures were far below the glass-transition temperature, so that WLF behavior would not be expected to be observed, and side-chain or dissolved chromophore materials were studied. In the earlier study, Williams-Watt behavior was not always apparent [15]. The later study found that Williams-Watt behavior would be observed when care was taken to eliminate effects of trapped charge, and when short enough time scales were used [24]. Short times are necessary to avoid mixing gross polymer motions with the molecular motions.

These time-temperature effects can be understood for side chain and dissolved chromophores by considering the Williams-Watt behavior to describe a broad distribution of local exponential relaxations. The distribution of relaxation times arises from the orientational trapping of the chromophores in a distribution of free volumes [15]. However, when polymer aging becomes important at long times or at temperatures near the glass transition, the distribution becomes highly time-dependent, and one must consider the time-temperature coupling. At short times and low temperature, the distribution can be considered constant in time, and fits a single Williams-Watt stretched exponential. One expects to find that, at short times, the relaxation will fit a stretched exponential. As time goes on the polymer ages around the chromophores, usually leading to a narrower and higher energy distribution [67] so that the decay of the nonlinearity slows and deviates in this way from the original Williams-Watt distribution. Depending on the size and shape of a dissolved chromophore, its motions will be more or less coupled to the polymer motion. Thus, at short times, the behavior may resemble side-chain relaxation in a polymer and at longer times reflect the gross polymer relaxation. This behavior was confirmed in dielectric studies [68]. One then expects the activation energies one measures in side-chain and dissolved chromophore systems to be intermediate between typical main-chain (α) and small side-chain (β) energies. This is consistent with the fact that nonlinear optical chromophores are much bulkier than typical polymer side groups but less bulky than the main chain itself. For main-chain chromophores, one expects the relaxation behavior always to be described by main-chain dynamics. Bulky chromophores or side-chain systems may be incompletely coupled to the polymer motion. One may or may not observe WLF behavior depending on the details of the particular chromophore/ polymer system. To illustrate these concepts we now discuss in detail the studies we carried out on dissolved chromophore materials. These studies relate the orientation relaxation to a nonisothermal poling process.

Samples were prepared by dissolving purified Disperse Red 1 dye in PMMA to 10 percent by weight in a common solvent and spin coating thin films on indium tin oxide coated glass slides. A double electrode structure was formed by placing two of these films face to face and fusing them. The films were placed in a temperature-controlled cell in a second harmonic generation apparatus. The second harmonic intensity from a 1.58 μm fundamental wavelength was monitored while the sample temperature was heated at a constant rate with a 30-V potential applied. The sample was then cooled to a fixed temperature, the potential removed, and the electrodes shorted. The decay of the second harmonic intensity was monitored over time. This measurement was repeated for several final temperatures.

The data during the heat ramp are shown in Fig. 14 where the heating rate, h, was 1.7°C/min. The data for the isothermal decays are shown in Fig. 15. In both figures, the vertical scale is the square root of the intensity, which is proportional to the nonlinear optical polarization and thus the second

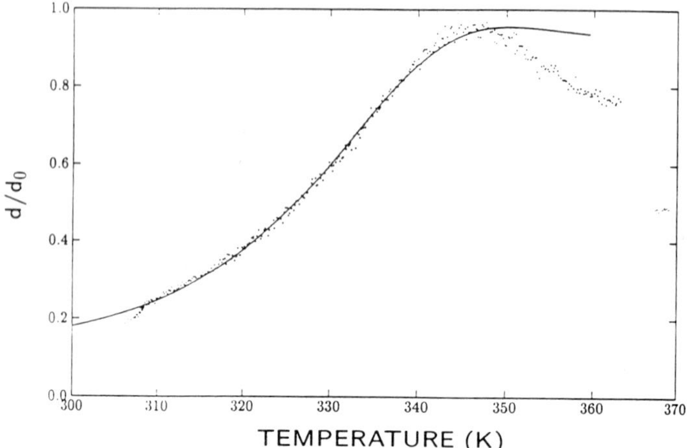

Fig. 14. Electric-field induced second-harmonic generation coefficient as a function of temperature for a constant heating rate of 1.7 °C/min for the DR1/PMMA system. The line is the fit to Eq. (3.2.3) with the parameters of Table 3.

harmonic coefficient. The data of Fig. 15 cannot be fitted to a single-exponential, a biexponential, or a simple single-diffusion model. The data suggest behavior consistent with a distribution of local relaxation rates. These relaxation rates would be determined by the distribution of free volume. In this type of model, the relaxation would follow the set of differential equations [61],

$$\frac{dP_i(t)}{dt} + \alpha_i(T)P_i(t) = 0. \tag{3.2.1}$$

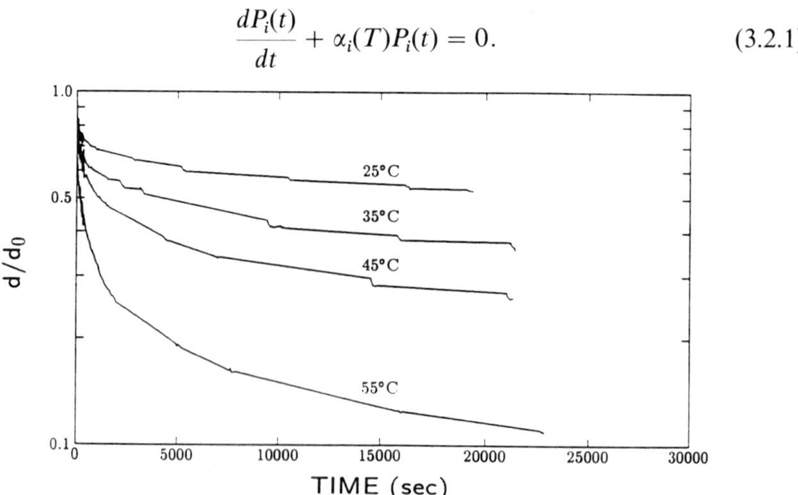

Fig. 15. Isothermal decay of the second-harmonic coefficient for a series of temperatures.

where $P_i(t)$ is the nonlinear optical polarization of the ith relaxing dipole with relaxation rate $\alpha_i(T)$ in the distribution of temperature-dependent relaxation rates. The relaxation for each dipole follows a simple Debye relaxation for a local relaxer. A continuous distribution of relaxation rates can be described in a variety of ways. We have found it convenient to describe it by a Williams-Watt-Kohlrausch stretched exponential so [66]

$$P(t) = P_0 \exp[-(\alpha(T)t)^{\beta}], \tag{3.2.2}$$

where the parameter β describes the width of the distribution.

We fit the data to the relaxation time, $\tau = 1/a$, and β by plotting $\ln[-\ln(d_{\text{eff}}/d_0)]$ versus $\ln(t)$ for each data set. An example is shown in Fig. 16. The width parameter was found to be relatively temperature insensitive and describes a rather broad distribution for this material system. A linear temperature dependence fit the data for β with the regression parameters given in Table 3. The relaxation time τ fit well to an Arrhenius plot with the fit parameters also given in Table 3.

The nonisothermal temperature ramp data would follow a related differential equation for each relaxation, which leads to the following expression for the temperature dependent poling process [61],

$$P(t) = \frac{P_f}{T}\{1 - \exp[-(\alpha(T)t)^{\beta}]\}, \tag{3.2.3}$$

where P_f is an amplitude constant proportional to the applied electric field.

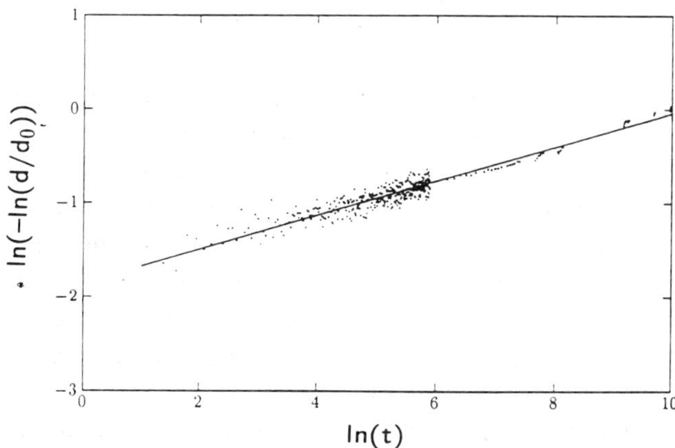

Fig. 16. Linearized plot of the isothermal decay of second harmonic coefficient at 35°C. The line is a least squares fit.

Table 3.

Activation Energies, A/k, Exponential Prefactors, $-\ln \tau_0$, and Slopes, m, and Intercepts, b, of β Versus Temperature Derived from Isothermal Decay Data and from a Fit of Temperature Ramped EFISH.

	A/k (1/K)	$-\ln \tau_0$	m	b
Isothermal decay	2.16×10^4	-59.4	0.00317	-0.795
Ramped EFISH	2.22×10^4	-57	0.0033	-0.73

With a constant heating rate, h (leading to $T = ht$), and again with the relaxation time obeying Arrhenius behavior, and assuming the width parameter β linear in temperature in this range, we fitted the data of Fig. 14 to Eq. (3.2.3) with the resulting parameters given in Table 3. The fitted line shown in Fig. 14 describes the data well except at high temperatures, where the increased space charge due to the rise in polymer conductivity screens the poling field.

Examination of Table 3 reveals that both experiments result in the same phenomenological parameters indicating that the dynamical behavior is locked in during the poling process. Thus, an experimental program varying the poling process while monitoring the susceptibility will directly measure the expected lifetime. For example, a poling rate leading to a high temperature peak, a higher glass transition temperature, or a steeper temperature dependence will result in a longer-lived polarization. This was directly verified for a side-chain copolymer [24]. The poling and relaxation processes involve a rather broad distribution of local Debye relaxation rates which may follow Arrhenius behavior. Measurements over a broader temperature range would determine how tightly coupled the chromophore motion is to the polymer motion. The activation energy deduced for the side-chain polymer is comparable to that of the main chain motion, so that this system may indeed exhibit WLF behavior near the glass transition. The dissolved system has a lower activation energy, so that it is more decoupled from the polymer. In addition, the exact form of the distribution that governs these processes cannot be discerned from the data obtained here due to a limited dynamic range. Typical distributions differ most in the distribution tails. The rates are related to the rotational binding forces in the polymer, such as the steric forces between the molecule and the local polymer environment. Understanding of the underlying physical mechanisms requires further study.

4. DEVICES

Poled polymers possess many important properties that are key to the fabrication and operation of guided-wave devices. Foremost among these properties is their inherent processability, which derives from their similarity to materials used in the photolithography of electronic integrated circuits. Polymer glasses are easily dissolved and formed into thin films using spin-coating techniques. The resulting waveguides possess the requisite optical clarity for centimeter scale waveguides. The large nonlinear susceptibilities and excellent dielectric and refractive properties also contribute to their applicability to guided wave nonlinear optical devices. In addition, the multicomponent nature allows for material tailoring for a particular application. Of particular importance are frequency conversion devices [69] and electro-optic devices [30,32,70].

4.1. Frequency Conversion Devices

For frequency conversion devices, the phase-matching property along with the optical nonlinearity mainly determines the device efficiency [71]. Phase-matching refers to the fact that the nonlinear polarization and the light radiated by the polarization have different spatial frequencies due to the dispersion of the refractive index. These different spatial frequencies interfere over the propagation length, thus limiting the nonlinear optical conversion. In the phase-matched condition, the dispersion is canceled by one of several means, so that no interference occurs.

In integrated optics, the light propagates within a waveguide. Since the light in a waveguide can be confined to a very small cross-sectional area over long lengths, more efficient frequency conversion interaction can take place. The advantage due to increased energy density in waveguided devices over bulk devices can be as high as the ratio of the length of the waveguide to the wavelength of light, which can, in principle, be 10^4 [72]. The quantized properties of waveguide modes complicate the light propagation, which allows for more phase-matching schemes but adds practical difficulties in achieving phase matching in a real system. In the case of waveguides, it is not the bulk refractive indices that must be phase-matched, but rather the mode indices. Because guided waves must satisfy the mode conditions, the two polarizations propagate with different refractive indices even in an isotropic material, and the dispersion in the effective mode index of refraction is greater than that of the bulk index of refraction.

In addition to phase matching, efficient operation requires low-loss waveguides where the guided-wave energy is highly confined to the nonlinear optical waveguide so that a large fraction of energy is converted, and the fundamental and second-harmonic waves need to be confined in such a way that much of their energy interferes constructively. All the modes exhibit confinement not only in the central guiding region but also in the cladding region as an evanescent field. The fraction of energy confined in each depends on the thickness of the guiding region, the relative bulk indices in the regions, the wavelength, and the order of the mode. For nonlinear waveguides, the efficiency is optimized with the lowest order mode since it yields the most energy in the waveguide. For nonlinear cladding, a higher order mode will yield more energy in the cladding. In most schemes, both the fundamental and second harmonic waves are confined to the waveguide, and the efficiency is related to the overlap integral between the fundamental and harmonic guided modes.

Phase matching can be achieved in some cases in a waveguide fabricated from isotropic materials by matching, for instance, two modes of orthogonal polarization. However, when the overlap is calculated, it is found that efficiency would be rather low. Therefore, it is still desirable to employ phase-matching of the lowest-order modes. Several schemes for phase-matching appear most appropriate for molecular polymeric films. The first is birefringence phase-matching. In this case, the waveguide composition, birefringence (through poling), and dimensions are adjusted so that the lowest order mode curves for two polarizations intersect [72]. In trying to phase-match using birefringence, the critical thickness condition may make specific frequency operation difficult due to device-processing tolerances.

Another method of phase-matching consists of having the second harmonic light not confined in the waveguide [73]. In this case, the lowest-order mode cutoff occurs at some wavelength between the fundamental and the second harmonic. The phase-matching condition is attained in a noncritical way for certain wave vectors in the radiated mode. The efficiency is reduced, but using the large nonlinearity in organic materials it may be possible to attain useful devices. This type of interaction has been observed in crystal-cored waveguides containing organic materials [74].

A phase-matching approach that would take advantage of the fabrication flexibility of poled polymers involves a periodic modulation of the guided-wave interaction using a grating overlying the waveguide [69]. This geometry can facilitate phase matching in two different ways. In one way, the effective index is periodically modulated by the grating, which adds a component of wave vector in the guide that can be used to attain phase matching. In addition, a grating in the nonlinearity can be induced that periodically changes the phase of the nonlinear interaction. If the grating is chosen properly, a quasi-phase-matched condition is obtained. For a channel waveguide, the grating fabrication is critical to efficiency.

Another method of phase matching that is particularly suited to guest-host polymer materials and may yield efficient devices is to use anomalous dispersion to achieve phase matching [75–77]. In this case, the pump and generated wavelengths lie on either side of the energy of an excited state of the nonlinear optical guest. The linear polarizability of the guest exhibits anomalous dispersion, which, when added to the normal dispersion of the host, can lead to zero dispersion at the appropriate concentration.

This method of phase matching is attractive for several reasons. First, it allows the use of the large optical nonlinearities found in visible absorbing dyes for second harmonic generation of semiconductor laser light. This can mean a 100- to 10,000-fold increase in second harmonic generation efficiency. This can be seen graphically in Fig. 17, where we have plotted the effective nonlinearity as a function of the second harmonic wavelength. The conventional dyes would be utilized in other phase-matching schemes and thus

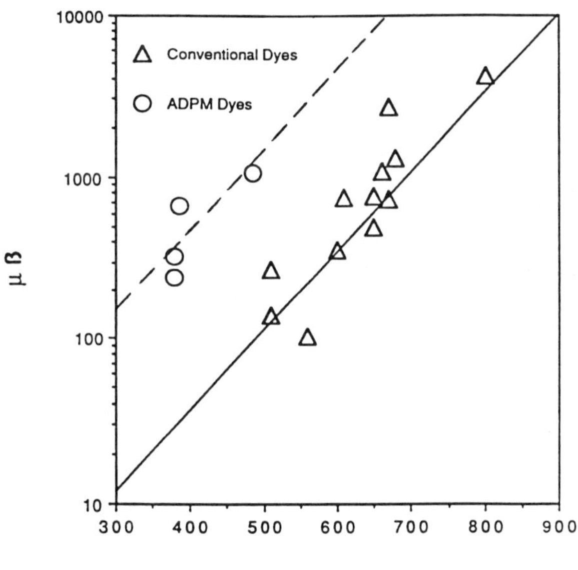

Fig. 17. Optical nonlinearities $\mu\beta$ for various chromophores versus effective transparency.

require transparency at all wavelengths longer than the second harmonic wavelength. This constrains them to the lower line, since the value of β is tied closely to the wavelength of the lowest energy absorption. The anomalous dispersion dyes possess an absorption at a wavelength *longer* than the second harmonic and thus possess much larger values of β. Useful frequency conversion devices require that the nonlinearities occur in the upper left quadrant, as the anomalous dispersion phase-matching (ADMP) materials do. An additional advantage of ADPM is that any nonlinear optical tensor component can be utilized, especially those with their electric polarization along principal dielectric directions. In the bulk, this implies the absence of walk-off, which is the deviation and subsequent reduction in overlap of the fundamental and harmonic beams due to birefringence as the beams traverse a thick crystal.

Phase-matched second harmonic generation using anomalous dispersion was demonstrated using electric-field-induced second harmonic generation in liquid solution of Foron Brilliant Blue (FBB-SR) dye in acetonitrile. Although the phase-matched condition was attained, the conversion efficiency was limited by residual absorption of the second harmonic by the dye. A figure of merit for absorption and anomalous dispersion can be defined as the ratio of the molar absorption at the absorption peak wavelength divided by the molar absorption at the second harmonic wavelength. For FBB-SR, this figure of merit is 133. It was found that the barbituric acid (BA) molecule shown in Fig. 18 would be more appropriate for second harmonic generation. It possesses a figure of merit of about 540.

Films of BA in PMMA were spin-deposited onto silicon wafers on which were grown several-micron-thick oxide layers. This geometry was appropriate for waveguiding purposes. The bulk indices of refraction were determined from the transverse electric (TE) mode coupling angles. The concentration dependence of the bulk refractive index dispersion is shown in Fig. 19. It is apparent that the phase-matched condition is achieved at a weight fraction of about 6 percent. The absorption coefficient in this concentration range is 20 to 25 cm^{-1}.

In order to assess the potential second harmonic efficiency one may expect in anomalous dispersion phase-matching, we follow the calculation of Oudar [78]. The expected second harmonic generation efficiency defined by the ratio of the power of the second harmonic to the power of the fundamental, $\eta = P^{2\omega}/P^{\omega}$, is given by

$$\eta = \frac{16\pi}{c} I_{\omega} T_Q Q_b^2 K, \tag{4.1.1}$$

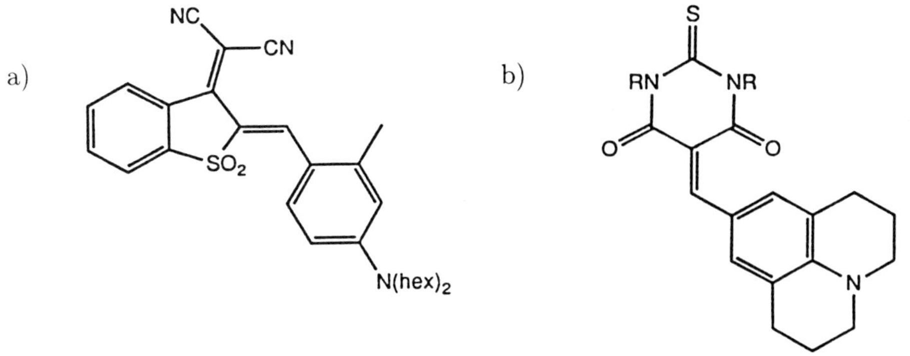

Fig. 18. Anomalous dispersion phase matched second harmonic generation dyes: (a) Foron Brilliant Blue (FBB) and (b) barbituric acid (BA).

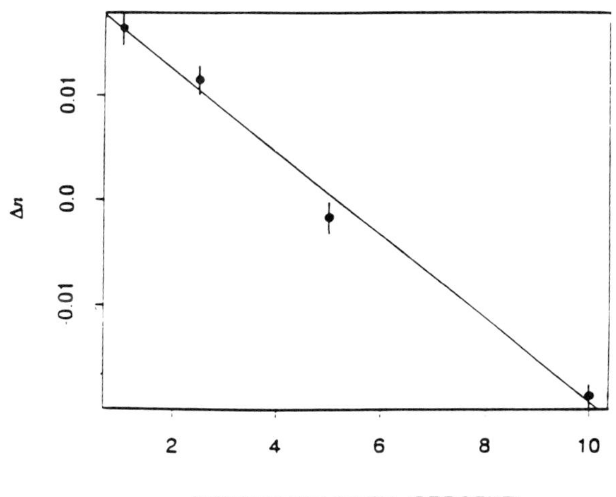

Fig. 19. Refractive index dispersion versus concentration in BA films.

where
$$T_Q = 2\frac{n}{1+n},$$

$$Q_b = \frac{32\pi^2 d}{\lambda(1+n)\sqrt{\left[\frac{2\pi}{l_c}\right]^2 + \alpha_{2\omega}^2}}, \tag{4.1.2}$$

$$\mathrm{K} = \exp\left[-\left(\alpha_\omega + \frac{\alpha_{2\omega}}{2}\right)l\right]\left[\cosh\left(\alpha_\omega - \frac{\alpha_{2\omega}}{2}\right)l - \cos\left(\frac{\pi l}{l_c}\right)\right],$$

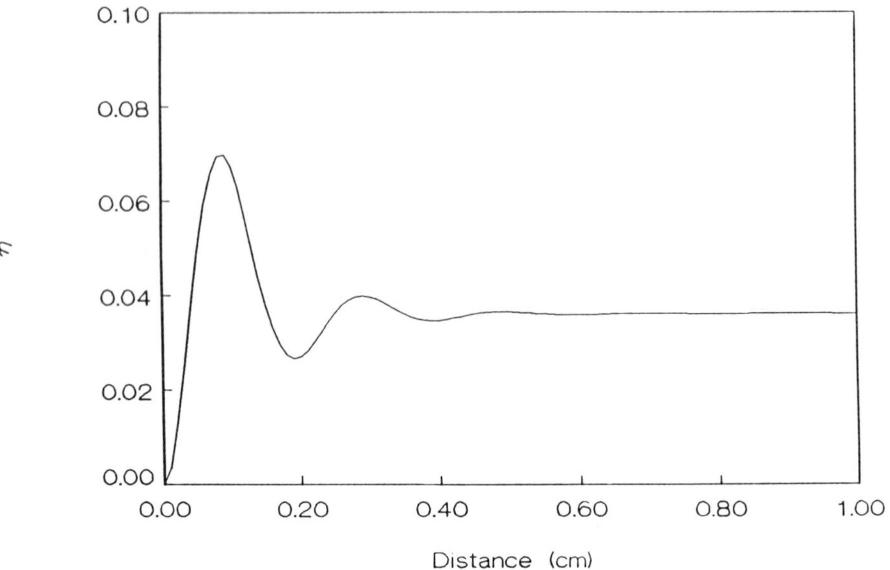

Fig. 20. Second harmonic efficiency for guided-wave anomalous dispersion phase-matched second harmonic generation from Eq. (4.1.1).

and where l_c is the coherence length, l the optical path length, ω the fundamental frequency, α_ω the absorption at frequency ω, d the second harmonic coefficient, and n the refractive index. Based on the experimental results above, the second harmonic efficiency was estimated for a 2 μm square waveguide with a $100mW$ incident intensity at a fundamental wavelength of $800nm$, $l_c = 0.1$ cm. $\alpha_\omega = 0\ \text{cm}^{-1}$, $\alpha_{2\omega} = 20\ \text{cm}^{-1}$, $d = 5 \times 10^{-8}$ esu, and $n = 1.5$. The results as a function of optical path length l are shown in Fig. 20. It is clear that even with a 0.1 cm coherence length, useful conversion efficiencies can be attained. It should be possible to obtain more efficient conversion by shifting to higher concentrations with zero dispersion.

Material design considerations for anomalous dispersion phase-matching also center on the nonlinear optical moiety and the polymer host. For the nonlinear moiety, the susceptibility and the optical loss spectrum must be considered, while for the polymer, the optical quality, stability, and dispersion are of prime importance in design.

Although many potential schemes for waveguide phase matching exist, all involve critical device-fabrication steps mostly involving thickness which, due to processing control, may limit the efficiency of the devices. The schemes that do not require a critical fabrication step (radiated modes and multiple layers) produce energy profiles that may represent a limitation of the device in systems. The processing flexibility of molecular polymeric films may be useful in realizing phase-matching schemes.

4.2. Electro-optic Devices

While phase-matching is of prime importance in frequency conversion devices, it is not required for electro-optic devices. For these devices, the dielectric properties of the material at the modulating frequency are important. In addition, electro-optic devices involve only one optical frequency so transparency requirements are generally less stringent than in all-optical devices. In fact, the first devices made from guest-host polymers have been fabricated using selective poling techniques. The function of these devices with respect to both their optical and electro-optic behavior has been as expected from basic material properties. Based on the progress in the development of these materials, it is anticipated that truly high performance devices are not far off.

Guided-wave channel electro-optic devices include modulators and switches that can perform a variety of system functions such as modulation, switching, deflection, multiplexing, and analog-to-digital conversion [27,32]. The properties of guided-wave electro-optic devices are best illustrated by considering a transverse phase modulator for which two electrode configurations are shown in Fig. 21. The behavior of this device is particularly simple and is suitable for discussing the material properties and device considerations that impact on all electro-optic devices. The phase change induced by the electric field is given by [79]

$$\Delta\phi = \frac{\pi}{\lambda} n^3 r E l, \tag{4.2.1}$$

where E is the applied electric field and l the length of the electrode. For the electrode geometry in Fig. 21(b), the electric field is V/t, where V is the applied voltage and t the thickness. The tensor nature of the electro-optic coefficient has been ignored for simplicity. The amount of modulation for a given voltage is determined by the geometry, the index of refraction, and electro-optic coefficient. In general, the electro-optic coefficients of organic and polymeric materials compare favorably with inorganic materials. The index of refraction is generally somewhat smaller in organics, but a larger electro-optic coefficient will produce comparable phase shifts and index-match more closely to optical fibers. Further, the trends in poled-polymer films suggest that the electro-optic coefficients may increase substantially in the future. An important aspect of an electro-optic film technology that

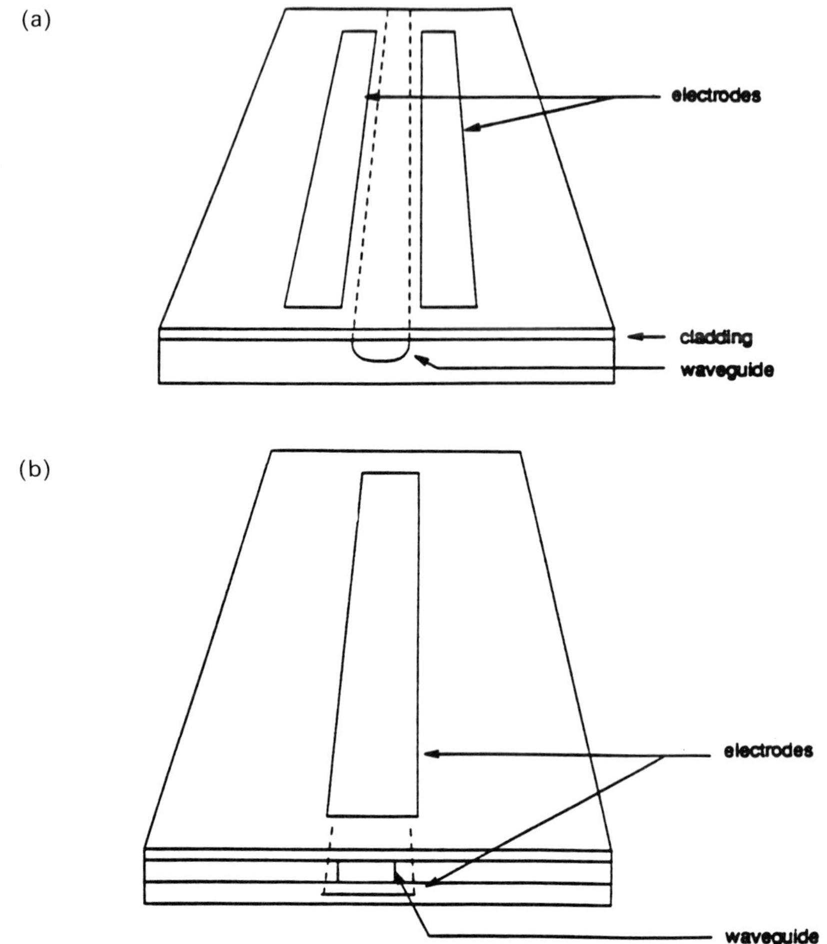

Fig. 21. Schematic of transverse phase modulator: (a) bulk crystal electrode geometry, and (b) thin-film electrode geometry.

cannot be attained in a bulk crystal approach is the ability to form layered structures that allow modulators in the configuration of Fig. 21(b) to be formed as opposed to Fig. 21(a). It has been shown that a several-fold increase in the phase shift can be obtained in the Fig. 21(b) geometry over the Fig. 21(a) geometry [80]. Thus, the geometric fabrication capability of polymers will improve device performance.

The true advantage of polymeric electro-optic materials is expected to come in high-speed performance. It has been shown previously that the electro-optic effect in organic crystals and polymers is mainly due to the electronic structure [57]. Thus, the nonlinear optical response is expected to be in the subpicosecond regime. The speed of electro-optic devices will be limited, then, by the modulating field. In the simplest case, the electro-optic device appears as a lumped *RC* element to the modulating circuit. In the case of Fig. 21(b), one can show that the modulation power to attain a peak retardation Γ_m is given by [79]

$$P = \frac{\Gamma_m^2 \lambda^2 A t \varepsilon \Delta\nu}{4\pi l^2 n_0^6 r^2}, \tag{4.2.2}$$

where A is the cross-sectional area of the electrodes, l the length of the device, and $\Delta\nu$ the modulating bandwidth. Thus, the modulating power per unit bandwidth is proportional to the dielectric constant, ε, and inversely proportional to $(n_0^3 r)^2$. Since one wishes to minimize the modulating power, a figure of merit for high speed modulation of a lumped modulator is $n_0^3 r/\sqrt{\varepsilon}$. Since the dielectric constant of many polymers is an order of magnitude smaller than ferroelectrics, such as $LiNbO_3$, polymers would be appropriately faster when considered as a lumped electrical element. The reduction in bandwidth due to the higher capacitance of the geometry of Fig. 21(b) does increase the modulating power, but this is more than offset by the increased field of this geometry. In fact, the power is linear in the thickness t. One must take care to minimize the electrode area A, though, in designing such a device.

True high-speed devices do not occur in the lumped element limit, and one must consider the transit time of both the optical and modulating wave [79]. If the modulating field is assumed to be uniform across the device during the transit of the optical wave, the modulating field will change during the transit of the light according to the time dependence of the modulating field. If the transit time is τ_d and the modulating field is sinusoidal with angular frequency ω_m, then the peak modulation is reduced by a factor $r = (1 - e^{-i\omega_m \tau_d})/i\omega_m \tau_d$. This can limit useful retardation to the gigahertz regime, but can be overcome by designing a modulator where both the modulating and optical field are traveling waves, so that the optical wave can, in principle, keep up with the modulating wave. In this case, both waves are integrated over their propagation across the crystal. The retardation is then given by [79]

$$\Gamma(t) = \Gamma_0 r = \Gamma_0 e^{i\omega_m t}\left[\frac{e^{i\omega_m \tau_d(1 - c/nc_m)} - 1}{i\omega_m \tau_d(1 - c/nc_m)}\right], \tag{4.2.3}$$

where c/n is phase velocity of the optical wave and $c_m = c/\sqrt{\varepsilon}$ is the phase velocity of the modulating wave (c is the speed of light in vacuum). If the maximum modulating frequency is taken as that for which $\omega_m \tau_d(1 - c/nc_m) = \pi/2$, then

$$(v_m)_m\text{ax} = \frac{c}{4nl(1 - c/nc_m)}. \tag{4.2.4}$$

Based on Eq. (4.2.4), a figure of merit for the bandwidth or useful device length is $1/(\sqrt{\text{e}} - n)$. This figure can be an order of magnitude or larger for polymers than for $LiNbO_3$ or GaAs. When this figure of merit is coupled with the ability to deposit and process these materials on many substrates (eg., silicon and GaAs), molecular polymeric materials may be capable of forming truly high-performance guide-wave electro-optic devices.

5. PHOTOREFRACTION IN POLED POLYMERS

When light is transmitted through certain materials, it induces the generation, migration, and separation of photogenerated charge carriers (electrons and/or holes). If the incident light is not uniform in intensity, these charge carriers may become trapped in the less-illuminated regions and give rise to internal "space-charge" electric fields within the material. These fields produce measurable changes in the refractive index through the linear electro-optic (Pockels) effect in noncentrosymmetric materials (or the Kerr effect in the case of centrosymmetric materials).[1] This index change replicates the phase of the optical intensity distribution as in a hologram and is called the photorefractive effect [81,82]. It is well-known that many inorganic electro-optic materials will display photorefractive properties owing to space-charge mediated index of refraction changes.

It is important to characterize the photorefractive properties of molecular-polymeric material for two reasons. First, photorefractive effects, if present, can lead to light-induced changes that can be problematic in the operation of electro-optic devices at certain wavelengths. This "optical damage" has been the subject of considerable study for over 20 years in ferroelectric crystals that are commonly used in electro-optic devices, such as $LiNbO_3$ [82–84]. Second, it is probable that molecular-polymeric films will themselves be useful as photorefractive materials. Photorefractive materials are prime

[1] Because the focus of this chapter is poled polymers, we restrict the discussion here to the applications of second-order nonlinear effects. Thus it is to be understood in this section that we are considering changes in the index of refraction of a material arising from the linear electro-optic effect.

candidates for all-optical data-processing applications. They are expected to be used for high-density optical data storage; associative-image-processing techniques, including dynamic holography and image amplification; spatial light modulation, such as beam-fanning limiters, phase conjugators, and self-induced optical resonators; programmable interconnections in integrated optics and simulations of neural networks; and associative memories with parallel signal processing [82,85].

Application of inorganic ferroelectric crystals to these devices has been hindered by at least two factors. First, the electro-optic effects from space charge in inorganics appear to be limited, in that any increase in a component of the electro-optic coefficient of the material is effectively counterbalanced by an increase in the corresponding dielectric constant. Second, the growth and preparation of crystals is generally an expensive and difficult task. This is even more of a factor when attempting to engineer the properties of the crystal by the modification of the crystal basis to include materials with desired functionalities. Growth of doped organic crystals is difficult because crystal preparation causes most dopants to be expelled [86]. On the other hand, the photorefractive properties of polymers may be improved by either doping or covalently incorporating different functional groups into the polymer once the appropriate combinations are identified, provided the different functionalities are compatible with each other and with the poling process. Also, polymers are generally more amenable to being processed into a device structure by spin coating or other methods to make large-area thin-film waveguides and other useful geometries [87–90]. We offer a simplified description of the theory of photorefraction in subsection 5.1 and a discussion of some of the other processes that lead to refractive index changes and how they can be distinguished from photorefraction in Sec. 5.2. In Sec. 5.3 we discuss some important aspects of the formulation of photorefractive polymers in terms of the functional properties which they must incorporate and the results of some of the recent experiments at that formulation.

5.1. A Simple Model

The useful properties of the photorefractive effect involve the writing of a phase grating through the interference of two or more beams in the material. To write such a phase grating requires that the material be able to generate sufficient photoionizable charge, provide for the transport of the charge over distances comparable to the inverse of the grating wave vector, trap the mobile charges at suitable photoionizable trapping sites, and modulate the index of refraction of the material through the space-charge field arising from

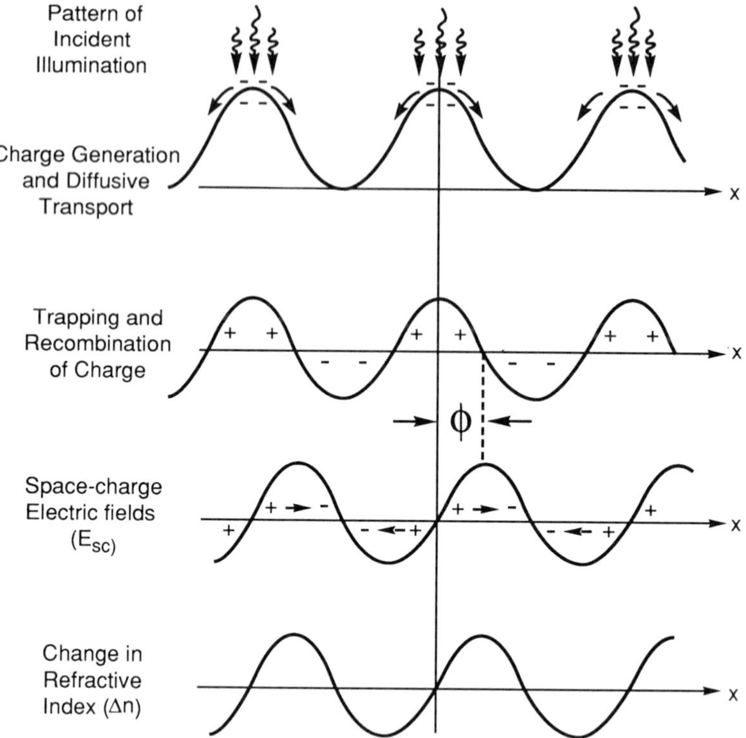

Fig. 22. Schematic of the photorefractive process. Note the phase shift, ϕ, between the incident intensity grating and the resulting photorefraction grating in the material.

the trapped charge acting via the electro-optic effect. Attempts to synthesize polymers that simultaneously satisfy these requirements have met significant obstacles. Photorefraction in polymers was only recently reported for the first time [85,87]. In this section we explore reasons that poled polymers are well suited for the multifunctional requirements of photorefraction while exploring some of the problems that must be overcome in the formulation of useful multifunctionality in multicomponent polymers. To begin we offer a brief introduction to a simple model for photorefraction.

In Fig. 22 the steps in the process of writing a photorefractive grating are shown schematically. Photorefractive behavior can be roughly modeled by assuming a one-dimensional index of refraction change caused by a space-charge field [82,91],

$$\Delta n(z, t) \sim -\frac{n^{3\,4}r}{2} E_{sc}(z, t), \tag{5.1.1}$$

where n is the index of refraction, r is the electro-optic coefficient, and $E_{sc}(z, t)$ is the time- and space-dependent space-charge field. As discussed further below, $E_{sc}(z, t)$ is inversely proportional to the relative permittivity ε. Thus, the factor n^3r/ε represents an important figure of merit for the sensitivity and indicates a significant advantage of poled polymers over conventional inorganic photorefractives. With inorganic crystals, increases in a component of the electro-optic coefficient r are generally accompanied by comparable increases in a corresponding component of the dielectric constant ε. The electro-optic mechanism in polymers is intrinsically different from that in inorganic crystals. In ferroelectric crystals, acoustic and optical phonons contribute significantly to the electro-optic effect. As discussed in Sec. 3.2, the electro-optic response in organic polymers is primarily due to the electronic excitations of individual molecules and/or molecular units, thus yielding low dielectric constants and minimizing dispersion over DC to optical frequencies [92]. The factor n^3r/ε has already been shown to be ten times greater in available poled polymers than in known inorganic materials.

For a photorefractively written diffraction grating, the space-charge field can be described by

$$E_{sc} \sim \left(\frac{\Phi}{h\nu}\right)(eL_{\text{eff}})\left(\frac{mI_0\alpha}{\varepsilon_0\varepsilon}\right) \tag{5.1.2}$$

where Φ is the photocarrier generation efficiency, $h\nu$ the photon energy, e the electronic charge, L_{eff} the effective transport length, m the modulation index, I_0 the incident energy density, α the absorption coefficient, and ε_0 the vacuum permittivity. The modulation index is a function of the grating geometry. For an elementary hologram made by crossed beams in a bulk sample, the light intensity distribution is $I(z, t) = I_0(t)[1 + m\cos(k_g z)]$, where k_g is the wave vector defined by the spacing Λ_g between adjacent intensity maxima in the incident grating (the grating "wavelength"). Thus $k_g = 2\pi/\Lambda_g = 2\pi \sin\theta/\lambda$, where λ and θ are the incident wavelength (assuming the interfering beams are the same wavelength) and the crossing angle between those beams. E_{sc} is due to the trapped charge within the material. Conduction in electro-optic polymers is generally modeled using hopping mechanisms instead of the band transport model used for inorganic ferroelectric crystals. The hopping model assumes that conductivity is controlled not by the size and accessibility of the conduction band but by the presence and location of suitable trapping and recombination centers between which charges move via electric-field-dependent or thermally activated jumps. (Traps are distinguished from recombination centers by the relative likelihood of the carrier either being released from a *trap* to participate further in conduction or being neutralized through *recombination* with a carrier of the opposite sign.) Photoconductivity is then an increasing function of incident frequency. The complexities of trap-controlled transport, especially the dispersion of the time dependence of release of carriers from traps, have been modeled by a continuous time random walk analysis by Scher and Montroll [93]. In photoconduction, the charge hopping is between photo-ionizable traps that are sensitized by the presence of a drift (DC) electric field across the sample as in the Poole-Frankel model [87,88,94,95].

The quantity L_{eff} describes the transport length of the generated photo-carriers and depends on the presence of applied fields, the grating geometry, and the intrinsic mechanisms of carrier transport. These transport mechanisms are parameterized by the electron and hole mobilities, the trap and recombination center densities, and energies and photovoltaic constants. It is important to maximize L_{eff}, which, as described below, is defined so as to be always less than the inverse grating wavevector (of the order of 0.1 to 1.0 μm) [82].

In its simplest form for one carrier and only drift transport due to an applied electric field E_0 or the photovoltaic effect and without significant diffusion, we can represent L_{eff} in terms of the grating wavevector k_g and the average drift length L^{ph} or L_E

$$L_{\text{eff}} = \frac{L_E^{ph}}{[1 + (k_g L_E^{ph})^2]^{1/2}}, \tag{5.1.3}$$

where L^{ph} is the characteristic photovoltaic transport length (probably small in polymers) and $L_E = \mu\tau(E_0 + E_{sc})$, the transport length due to an applied field E_0. Here, μ is the effective carrier mobility, and τ is the effective free lifetime of the carrier over large distances compared to the distances between photoionizable traps (τ in polymers would be practically limited by the presence of "deep traps" from which further photoionization is inhibited). For transport involving significant diffusion and a drift field but no photovoltaic effect, the transport length becomes

$$L_{\text{eff}} = \frac{[L_E^2 + k_g^2 L_D^4]^{1/2}}{[[1 + (k_g L_D)^2]^2 + (k_g L_E)^2]^{1/2}}, \tag{5.1.4}$$

where $L_D = (\mu\tau(k_B T)/e)^{1/2}$ is the diffusion transport length.

We see from these equations that to maximize L_{eff} for a given grating spacing and either diffusive or drift transport requires maximization of the μ–τ product. A major difficulty in the development of poled polymers for photorefraction arises due to the need both to maximize the mobility for photoconductive transport and yet to minimize the dark conductivity to allow poling and for storage of the photorefractive grating. For many applications, a photorefractive material will need to produce clear gratings quickly. Because poling introduces charge carriers into the sample, if the mobility is appreciable in the presence of the applied poling field, the sample will short-circuit before the molecules have adequately aligned. (Cf. Sec. 3.1.)

One way to characterize the relationship between the mobility and the free-carrier lifetime between traps is to determine the density of trapping centers in the material. This density highlights another criterion for the usefulness of a doped polymer for photorefraction. The space charge field E_{sc} is limited by the number of trapping centers available, reaching a maximum value at

$$E_{sc}^{\max} = \frac{eN_A}{k_g \varepsilon \varepsilon_0}, \tag{5.1.5}$$

where N_A is the trap density [82]. For most inorganics the trap density is sufficiently high that this limit is inconsequential in light of the limitations on the other parameters involved. In doped polymers, trapping centers are present due to the defects and conformational disorder inherent in the amorphous structure, but these traps may not be useful for photorefraction if they are not photoionizable. In the recent work of Moerner *et al.* [85], a trap density about a factor of ten smaller than that necessary to avoid a space-charge limit in ferroelectric crystals was measured, but the low dielectric constant found in the polymer avoided the limitation as can be seen in Eq. (5.1.5). The space-charge fields attainable are thus not limited by the trap density, so that large efficiencies could be realized by the application of large external fields. Thin-film polymers are amenable to the application of large fields, which can lead to large efficiencies by compensating for small values of the μ–τ product.

An important material property associated with photorefraction is the time constant associated with the writing and erasure of the grating. In the case of relatively large grating spacings and small applied fields, the response time is given by [90]

$$\tau_R = \frac{\varepsilon \varepsilon_0 h\nu}{\alpha I e \mu \tau}. \tag{5.1.6}$$

Here we see that a small μ–τ product leads to a long response time, which has been observed [85]. Unlike the efficiency, however, the application of a strong external field does not compensate this. In fact, the correction to Eq. (5.1.6) in the presence of a strong field is proportional to E^2 [90];

thus, the application of the strong field does not tend to decrease the response time.

Similarly, the photocarrier generation efficiency Φ depends upon the applied electric field and the density of nonionized donors. While the initial quantum yield may approach unity in some polymeric photoconductors, the photogeneration efficiency is limited by geminate recombination or back electron transfer due to the small thermalization radius of many polymers, i.e., freed carriers cannot separate sufficiently so that the energy of their coulombic attraction is less than $k_B T$ [96]. A strong applied field will help to separate the charge carriers and so improve the effective carrier generation efficiency. Typically, the field dependence of the charge generation and the effects of geminate recombination are understood using the Onsager model [85,97]. To determine the quantum efficiency experimentally, we calculate the number of absorbed photons (N_p) for the sample from the intensity of the pulse, the wavelength and the duration and shape of the pulse and the absorption coefficient. We then calculate the number of charge carriers created in the sample by integrating over the characteristic photocurrent decay curve to find the total photocharge Q in the sample using $Q = Ne$. The quantum efficiency is then N/N_p, a function of the frequency and the intensity of the applied field [94].

One figure of merit for photorefraction is given by the energy required to achieve a specified diffraction efficiency (usually 0.1 percent) in a given thickness of material. The diffraction efficiency is the ratio of the diffracted to the incident reading beam power for a grating. This criterion may be difficult to compare for different device structures such as waveguide versus bulk materials. The theoretical diffraction efficiency can be expressed as [82]

$$\eta = e^{-\alpha d/\cos\theta} \sin^2\left(\frac{\pi \Delta n d G}{\lambda}\right), \tag{5.1.7}$$

where α is the absorption coefficient, d is the depth of the grating, and G is a function of the geometry and polarization of the incident and reading beams on the material. $G = 1/\cos\theta$ in our simplified model, where θ is the half angle between the intersecting beams [85,91]. For thin films or materials where the absorption is small, it is usually sufficiently accurate to drop the exponential decay factor and to approximate the sine function by its argument [82,85]. Using the measured diffraction efficiency and Eq. (5.1.1), one can determine the magnitude of the space-charge field E_{sc} in the material.

Using Eqs. (5.1.1) and (5.1.2), another figure of merit for photorefractive sensitivity is given by the change in the refractive index per absorbed incident intensity,

$$S_n = \frac{dn_i}{d(\alpha I_0)} \sim \left(\frac{n^3}{2}\frac{r}{\varepsilon\varepsilon_0}\right)\left(\frac{\Phi}{h\nu}\right)(eL_{\mathrm{eff}})m, \tag{5.1.8}$$

where αI_0 is the absorbed energy per volume. This sensitivity is also referred to as the *internal* sensitivity [82]. The external sensitivity is the change in the refractive index with change in incident energy intensity. The internal sensitivity is a better figure of merit for an optical device when the absorption of the beam is to be minimized. To be useful, sensitivity must be determined over relevant dynamic ranges of intensity. A material may be sensitive for very low intensities yet not diffract sufficient power for device needs. Thus, both the sensitivity and the diffraction efficiency of a material are important in that the change in the index of refraction may saturate at a specific laser power density. Finally, for many applications the speed with which a useful grating can be written is an important parameter and, as discussed further below, may present a special challenge for the design of photorefractive poled polymers.

While the simplified model of photorefraction discussed here includes its best understood aspects, the model is far from complete. Researchers have found it necessary to alter the model to consider more complicated forms for the intensity dependence of the photoconductivity and absorption [98], including nonlinear and geminate recombination of charge carriers, persistent photoconductivity [96,98], and other complicated phenomena such as secondary photorefraction centers [99]. Because of the dispersive nature of the transport, μ and τ are less well defined quantities in polymers. In polymers, the mobilities are field dependent, often in a highly nonlinear way, and the presence of significant trapping in a polymer complicates their determination [94]. The literature on photoconductivity in polymers is voluminous, and the variety of effects to be accounted for mirrors the variety of applications to which polymers may be adapted. Helpful overviews can be found in [94], [95], and [100].

5.2. Identifying Characteristics of Photorefraction

Illumination may lead to changes in the refractive indices of a polymer in ways other than photorefraction [86]. Some of these other processes are photochromism or photobleaching [101] including photodimerization, photopolymerization [102], and photoisomerization [103]; thermo-optic effects [104], such as the thermochromic effect and simple heating, causing photophysical changes in the molecular structure; transient combined pyroelectric and electro-optic effects [105]; the piezoelectric effect [106]; excited states in the constituent molecules; and conventional χ_3 processes [85]. Though not as useful for the devices we have associated with photorefraction, photochromic and photochemical effects have been the subject of significant recent studies and may have important applications [107,108,109] These effects may be useful for the fabrication of channel waveguides or in processes requiring more permanent refractive index changes over small regions. The polymeric material can be masked to form channel waveguides by photochemically lowering the index of refraction in selected areas of thin highly photosensitive films. Two-photon photochromic effects have been proposed for use in optical memory devices but lack the robustness necessary for repeated writing and erasure. While somewhat reversible, these effects become noticeably degraded after repeated writings and have low photosensitivities, since they require changes in the electronic structure [86].

Though not always experimentally simple, there are many ways to distinguish photorefraction from competing mechanisms that lead to refractive index changes. We list some of the most important of these identifying characteristics of photorefraction:

1. *Phase shift of the index grating.* The most unmistakable sign of photorefraction (and one of its most promising features) is asymmetric two-beam coupling [110]. A critical element in this coupling is the nonzero phase shift between the phase grating and the incident interference pattern. When the grating is produced solely by the migration of charge by diffusion away from the illuminated regions, the phase of the index grating is shifted by $\pi/2$ with respect to the intensity grating. See Fig. 22. For absorption gratings from photochemical or photochromic effects, the index and intensity gratings are in phase. Photorefractive effects involving transport by the photovoltaic effect alone also do not show a $\pi/2$ phase shift. Techniques for measuring grating phase are discussed in [110].

   Because photorefractive effects can be considerably strengthened in many materials by the application of a drift field, techniques have been developed to cause beam coupling under such circumstances [85,86]. These techniques, such as nearly degenerate two-wave mixing, involve the translation of the material along the grating wavevector or the introduction of a phase delay in one of the incident beams over a time much shorter than that needed to write a new grating in the material. Generally, the phase shift of the grating under a drift field is a decreasing function of time and of grating spacing and an increasing function of the effective transport length [82].

Two-beam coupling in the absence of a moving grating or an a.c. electric field is the strongest evidence of photorefraction. Its absence does not alone signify the absence of photorefraction [87]. Other measurements will be required to identify strong candidate materials for photorefractive application.

2. *Hologram erasability*. The traps that lead to the space-charge fields in photorefractive materials must themelves be photoionizable. If so, the photorefractive diffraction grating is erased simply by bathing the material in a uniform field of the proper frequency. The photorefractive index changes can be written and erased over and over again during months or even years of use without noticeable change in the material [81]. For photochemical effects, reversibility generally requires heating or other chemical treatment. Most photochromic gratings tend to be only partially erasable under uniform illumination.
3. *Correlation with photoconductivity*. The photoconductivity of a material should be measurable separately at the same wavelengths as used for photorefraction. Comparisons between the absorption spectrum and the wavelength dependence of the photoconductivity of a material help identify the underlying photoconduction mechanism and resultant charge density gradients [94]. These are not simple measurements. Some of the difficulties in quantifying relevant photoconductivity parameters in polymeric photorefractives are discussed in [111]. For the material considered there, the mobility and photocharge generation efficiency were found to be orders of magnitude below the levels of inorganic photorefractives or of other polymeric materials developed solely for their photoconducting properties.
4. *Electro-optic response*. Most photochromic and photochemical effects do not depend upon the absence of a center of symmetry. If a candidate photorefractive-poled polymer shows refractive index changes without poling or some mechanism for introducing asymmetry, the changes are not due to the linear electro-optic effect.
5. *Enhancement by external fields*. Photoconductivity generally scales at least with the square root of the applied drift field. Application of an external field enhances photocarrier generation both by reducing the barrier to charge release and by assisting the separation of photogenerated charges to lessen the effects of geminate recombination.
6. *Polarization anisotropy of the grating readout*. The photorefractive response is dependent upon the polarization of the incident beam. If pure *s*- or *p*-polarized light is used in writing the grating, the diffraction efficiency will be different for the *s*- and *p*-polarized components of an

oblique reading beam. The geometrical and polarization G factors in Eq. (5.1.7) are modified in the three-dimensional case so that the diffraction efficiencies for s and p reading beams become

$$\eta_s = \sin^2\left(\frac{\pi n^3 r_{13} E_{sc}\, d \sin\theta_g}{2\lambda(\cos\theta_2 \cos\theta_1)^{1/2}}\right) \tag{5.2.1}$$

and

$$\eta_p = \sin^2\left(\frac{\pi n^3 r_{\mathrm{eff}}^p E_{sc}\, d \cos 2\theta_0}{2\lambda(\cos\theta_2 \cos\theta_1)^{1/2}}\right), \tag{5.2.2}$$

respectively, where θ_g is the angle between the grating wavevector and the film plane, θ_1 and θ_2 are the angles of incidence of each of the writing beams, and $2\theta_0$ is the full angle between the writing beams [85].

5.3. Composition of Photorefractive Polymers

Formulation of the ideal photorefractive polymer requires finding materials that either already combine necessary photogeneration and transport with nonlinear optical properties or can be functionalized to include these attributes in the polymeric network architecture by attaching to a polymer backbone groups that combine the necessary characteristics. The largest nonlinearities in organics have been in polymers that combine an extended system of highly mobile pi-electrons with electron donor and acceptor groups at para positions [86]. It is usually necessary to include one or more additional functionalities, however, by doping after polymerization. Charge transport and trapping properties can be tailored by attaching molecules to the polymer structure with the appropriate ionization energy or electron affinity [88]. For instance, charge transfer complexing is accomplished by adding donor sensitizers to acceptor polymers or vice versa so that the electron released by the donor is taken by the accepter to create an absorption band corresponding to the energy level difference between the donor and acceptor energies. Also because many otherwise appropriate polymers tend to absorb primarily in the ultraviolet spectrum, dye sensitization of these polymers leads to greater absorption of visible light through the release of photocarriers in the dye [94]. Yet another approach is to use a $\chi^{(3)}$ polymer in an electric field to produce the nonlinear optical response [85,87].

In the photorefractive polymer composition reported by Moerner and coworkers [85,87,110,111], an electro-optic material (partially crosslinked epoxy, bis-phenol-A-diglycidylether acid, with a nonlinear chromophore, 4-nitro 1,2-phenylenediamine [NPDA]) provided both the nonlinear optical properties as well as the optical absorption necessary for charge generation. To become photorefractive, however, this polymer had to be doped with a hole transporting agent (diethylamino benzaldehyde diphenylhydrazone [DEH]). The photorefractive diffraction efficiency of this polymer was found to be significantly lower ($\Delta n \sim 4.5 \times 10^6$ [110]) than is available in inorganic crystals of the same thickness, in part because of the relatively small electro-optic coefficient of the material. Also, the low mobilities present resulted in a relatively slow response.

In Schildkraut's and coworkers' [88,112] formulation of photoconducting electro-optic polymer film, a nonlinear chromophore was attached as a side chain to an acrylic backbone, and both a photocharge-generating sensitizer and a hole-transport molecule were added. When adding a photogeneration sensitizer it is important that it absorbs at a longer wavelength than the nonlinear optical chromophore so that attenuation of the beam by the sensitizer is minimized. Obviously, this requires careful balancing of functions that compete for absorbance within the polymer. Sensitizers for photogeneration are available throughout the visible and near-infrared spectra, which may permit tuning of the photorefractive effect for an otherwise identical polymer [95,100]. Measurements on a grating from two crossed beams with these doped materials yielded a maximum diffraction efficiency of only 3.2×10^{-5}. While not reported, the mobility and thus the speed of response of these polymers are expected to be slow. Photorefraction, as opposed to another mechanism for photoinduced refractive index change, was inferred from the strong field dependence of the diffraction efficiency [112].

Optimizing the nonlinear response of the polymer usually requires saturating the polymer with the appropriate chromophore. When combining functions by combining sensitizers and transport agents, it is necessary to get by with less than the optimum amount of the nonlinear response to accommodate other functionalities. Also, doping of materials often leads to an increase in phase segregation, which causes undesirable light scattering and can severely limit the useful lifetime of the polymer and/or its processability [87]. Plasticization due to added sensitizers and charge transport agents may lead to a lower glass transition temperature [89].

Perhaps a more promising approach is found in the methods of Li *et al.*, who have reported the development of a dye-doped photocrosslinkable nonlinear optical polymer (polyvinylcinnamate 3-cinnamoyloxy-4-[4-(*N*,*N*, dimethylamino)-2-cinnamoyloxy phenylate] in a guest [CNNBR]-host [PVLN] structure), which is strongly photoconducting without the introduction of any photosensitizer or photo-charge transport agents [89].

Photocrosslinking involves using photosensitive chromophores which are irradiated with ultraviolet light during the late stages of poling to "cure" the composition [113]. The estimated value for n^3r/ε is expected to be $12pm/V$ for this polymer.

Utility in waveguides generally imposes additional restrictions on the polymer composition. All dopants should be soluble in an appropriate spin-coating solvent and of sufficiently high molecular weight so as not to evaporate with the solvent. Problems with the use of waveguides in the study of photorefractive effects also include the presence of thermo-optic changes in the guide and/or coupling mechanism, whereby small temperature changes may significantly change the coupling angles. Thus, temperature control may be imperative in practical applications.

In summary, the benefits from the development of photorefractive polymers are many but there are numerous problems to overcome before these benefits can be realized. Foremost among the problems may be balancing the competing need for high mobility polymers (large μ–τ product) to attain fast and efficient response as found in some highly conjugated systems with the need to be able to pole the device to optimize its nonlinear optical properties. In poly(*p*-phenylene vinylene), it is found that in certain doping situations it may be possible to achieve ratios of photo to dark conductivity of nearly 1000; however, the observed dark conductivity is still too high to support large poling fields [114]. It may be necessary to find applications where the tradeoff between dark and photoconductivity is advantageous, or to seek new alignment processes.

6. CONCLUSIONS

Molecular polymeric materials can be formed into efficient second-order nonlinear optical materials. These materials simultaneously take advantage of the large optical nonlinearities of conjugated organic molecules and the favorable optical and dielectric properties of polymer glasses. The origin of the optical nonlinearity is traceable to the molecular constituents, whose susceptiblities can be calculated using molecular Hartree-Fock theory. These calculations reveal that the susceptibility of donor- and acceptor-terminated charge-conjugated systems arises from the large difference in dipole moment between the ground and first excited states. Once the molecular properties are known, the bulk material's susceptibility is calculated as a thermodynamic sum over the molecules. These sums have been calculated for both amorphous and anisotropic materials as well as for moderate and strong poling fields. It is found that existing anisotropy in the host affects the nonlinear susceptibility in a definite manner.

The poling process involves both the orientation of the dipoles of the nonlinear optical chromophores and the injection and transport of real charges. It is found that material selection can grossly affect the poling behavior, but this selection can also lead to optimized structures and processes. In closely related phenomena, both the orientation and disorientation of the nonlinear chromophores proceed from a distribution of relaxation rates. We describe our own experiments, which have deduced some of the underlying parameters responsible for the orientation in doped polymer glasses and its subsequent decay.

The potential applications of polymeric second-order nonlinear optical material include electro-optic switches and modulators as well as frequency conversion devices. We have seen that polymeric materials are uniquely suited to electro-optic applications, and the flexibility of fabrication may lead to novel phase-matching schemes for frequency conversion devices. We also see that multifunctional polymeric materials may exhibit interesting photorefractive properties that certainly deserve further study.

Since the promise of polymeric materials for second-order nonlinear optics was first described, *all* of their favorable properties for application have been realized. No unanticipated problems concerning their application have appeared, and the anticipated problem of thermal stability is well on the way to being solved. One must remain extremely optimistic about their potential application and especially concerning their key role in the emerging technologies of optical interconnection and optical data processing and switching. This promise will no doubt stimulate further work and workers and lead to a deeper understanding of physical properties, the development of new devices and applications, and the discovery of new materials.

REFERENCES

1. D. S. Chemla and J. Zyss, eds., *Nonlinear Optical Properties of Organic Molecules and Crystals* (Academic Press, New York, 1987).
2. A. J. Heeger, J. Orenstein, and D. R. Ulrich, eds., *Nonlinear Optical Properties of Polymers*, vol. 109 (Materials Research Society, Pittsburgh, 1988).
3. P. N. Prasad and D. J. Williams, *Nonlinear Optical Effects in Molecules and Polymers* (John Wiley, New York, 1991).
4. J. Messier, F. Kajzar, and P. Prasad, eds. *Organic Molecules for Nonlinear Optics and Photonics* (Kluwer, Dordrecht, 1991).

5. A. F. Garito and K. D. Singer, *Laser Focus* **80**, 59 (1982).
6. A. F. Garito, K. D. Singer, K. Hayes, G. F. Lipscomb, S. J. Lalama, and K. N. Desai, *J. Opt. Soc. Am.* **70**, 1399 (1980).
7. G. R. Meredith, J. G. VanDusen, and D. J. Williams, *Macromol.* **15**, 1385 (1982).
8. K. D. Singer, J. E. Sohn, and S. J. Lalama, *Appl. Phys. Lett.* **49**, 248 (1986).
9. C. Ye, T. J. Marks, J. Yang, and G. K. Wong, *Macromol.* **20**, 2322 (1987).
10. M. A. Mortazavi, A. Knoesen, S. T. Kowel, B. G. Higgins, and A. Dienes, *J. Opt. Soc. Am. B* **6**, 733 (1987).
11. K. D. Singer, J. E. Sohn, S. J. Lalama, M. G. Kuzyk, and R. D. Small, *Proc. SPIE* **704**, 240 (1986).
12. D. Jungbauer, B. Reck, R. Twieg, D. Y. Yoon, C. G. Willson, and J. D. Swalen, *Appl. Phys. Lett.* **56**, 2610 (1990).
13. J. W. Wu, E. S. Binkley, J. T. Kenney, R. Lytel, and A. F. Garito, *J. Appl. Phys.* **69**, 7366 (1991).
14. K. D. Singer, M. G. Kuzyk, and J. E. Sohn, *J. Opt. Soc. Am. B* **4**, 968 (1987).
15. H. L. Hampsch, J. Yang, G. K. Wong, and J. M. Torkelson, *Macromol.* **21**, 526 (1988).
16. G. T. Boyd, *Thin Solid Films* **152**, 295 (1987).
17. M. Eich, A. Sen, H. Looser, G. C. Bjorklund, J. D. Swalen, R. Twieg, and D. Y. Yoon, *J. Appl. Phys.* **66**, 2559 (1989).
18. D. Broussoux, E. Chastaing, S. Esselin, P. LeBarny, P. Robin, Y. Bourbin, J. P. Pocholle, and J. Raffy, *Rev. Tech. Thomson-CSF* **20-21**, 151 (1989).
19. M. G. Kuzyk, R. C. Moore, and L. A. King, *J. Opt. Soc. Am. B* **7**, 64 (1990).
20. C. P. J. M. van der Vorst and S. J. Picken, *J. Opt. Soc. Am. B* **7**, 320 (1990).
21. J. F. Valley, J. W. Wu, and C. L. Valencia, *Appl. Phys. Lett.* **57**, 1084 (1990).
22. W. Köhler, D. R. Robello, P. T. Dao, C. S. Willand, and D. J. Williams, *J. Chem. Phys.* **93**, 9157 (1990).
23. I. Teraoka, D. Jungbauer, B. Reck, D. Y. Yoon, R. Twieg, and C. G. Willson, *J. Appl. Phys.* **69**, 2569 (1991).
24. K. D. Singer and L. A. King, *J. Appl. Phys.* **70**, 3251 (1991).
25. J. W. Wu, *J. Opt. Soc. Am. B* **8**, 142 (1991).
26. G. T. Boyd, C. V. Francis, J. E. Trend, D. A. Ender, *J. Opt. Soc. Am. B* **8**, 887 (1991).
27. J. I. Thackara, G. F. Lipscomb, M. A. Stiller, A. J. Ticknor, and R. Lytel, *Appl. Phys. Lett.* **52**, 1031 (1988).
28. S. R. Marder, J. E. Sohn, and G. D. Stucky, *Materials for Nonlinear Optics: Chemical Perspectives* (ACS Symp. Ser. 455, Washington, D.C., 1991).
29. K. D. Singer, W. R. Holland, M. G. Kuzyk, G. L. Wolk, H. E. Katz, M. L. Schilling, and P. A. Cahill, *Proc. SPIE* **1147**, 233 (1989).
30. E. Van Tomme, P. Van Daele, R. Baets, G. R. Möhlmann, and M. B. J. Diemeer, *J. Appl. Phys* **69**, 6273 (1991).
31. J. I. Thackara, D. M. Bloom, and B. A. Auld, *Appl. Phys. Lett.* **59**, 1159 (1991).

32. G. F. Lipscomb, R. S. Lytel, A. J. Ticknor, T. E. Van Eck, S. L. Kwiatkowski, and D. G. Girton, *Proc. SPIE* **1337**, 23 (1990).
33. D. R. Yankelevich, A. Dienes, A. Knoesen, R. W. Schoenlein, C. V. Shank, and G. A. Lindsay, *Proc. CLEO*, 602 (1990).
34. C. B. Rider, J. S. Schildkraut, and M. Scozzafava, *J. Appl. Phys.* **70**, 29 (1991).
35. T. E. Van Eck, A. J. Ticknor, R. S. Lytel, and G. F. Lipscomb, *Appl. Phys. Lett.* **58**, 1588 (1991).
36. K. D. Singer, M. G. Kuzyk, W. R. Holland, J. E. Sohn, S. J. Lalama, R. B. Comizzoli, H. E. Katz, and M. L. Schilling, *Appl. Phys. Lett.* **53**, 1800 (1988).
37. R. DeMartino, D. Haas, G. Khanarian, T. Leslie, H. T. Man, J. Riggs, M. Sansone, J. Stamatoff, C. Teng, and H. Yoon, "Nonlinear Optical Polymers for Electro-Optic Devices," in *Nonlinear Optical Properties of Polymers*, A. J. Heeger, J. Orenstein, and D. R. Ulrich, eds. (Materials Research Society, Pittsburgh, 1988).
38. L. M. Hayden, G. F. Sauter, F. R. Ore, P. L. Pasillas, J. M. Hoover, G. A. Lindsay, and R. A. Henry, *J. Appl. Phys.* **68**, 456 (1990).
39. L. R. Dalton, L. Yu, and L. Sapochak, *Mol. Cryst. Liq. Cryst., Nonl. Opt.* **189**, 49 (1990).
40. S. K. Tripathy, *Chemtracts-Macromol. Chem.* **1**, 349 (1990).
41. J. D. Stenger-Smith, J. W. Fischer, R. A. Henry, J. M. Hoover, and G. A. Lindsay, *Makromol. Chem., Rapid Commun.* **11**, 141 (1990).
42. K. D. Singer and J. E. Sohn, "Organic Materials for Second-Order Nonlinear Optics" in *Electroresponsive Molecular and Polymeric Systems*, T. Skotheim, ed. (Marcel-Dekker, New York, 1991).
43. M. Schubert and B. Wilhelmi, *Nonlinear Optics and Quantum Electronics* (Wiley, New York, 1986).
44. D. M. Burland, C. A. Walsh, F. Kajzar, and C. Sentein, *J. Opt. Soc. Am B* **8**, 2269 (1991).
45. S. J. Lalama and A. F. Garito, *Phys. Rev. A* **20**, 1179 (1979).
46. D. Pugh and J. O. Morley, "Molecular Hyperpolarizabilities of Organic Materials," in *Nonlinear Optical Properties of Organic Molecules and Crystals*, D. S. Chemla and J. Zyss, eds. (Academic Press, New York, 1987).
47. J. L. Oudar and D. S. Chemla, *J. Chem. Phys.* **66**, 2664 (1977).
48. K. D. Singer, J. E. Sohn, L. A. King, H. M. Gordon, H. E. Katz, and C. W. Dirk, *J. Opt. Soc. Am. B* **6**, 1339 (1989).
49. S. J. Lalama and J. E. Sohn, unpublished.
50. S. J. Lalama, K. D. Singer, A. F. Garito, and K. N. Desai, *Appl. Phys. Lett.* **39**, 940 (1981).
51. P. A. Cahill, K. D. Singer, and L. A. King, *Opt. Lett.* **14**, 1137 (1989).
52. H. E. Katz, K. D. Singer, J. E. Sohn, C. W. Dirk, L. A. King, and H. M. Gordon, *J. Am. Chem. Soc.* **109**, 6561 (1987).
53. G. R. Meredith, J. G. Van Dusen, and D. J. Williams, "Characterization of Liquid Crystalline Polymers for Electro-Optic Applications" in *Nonlinear Optical Prop-*

erties of Organic and Polymeric Materials, D. J. Williams, ed. (Amer. Chem. Soc., 233, Washington, DC, 1982).
54. J. D. LeGrange, M. G. Kuzyk, and K. D. Singer, *Mol. Cryst. Liq. Cryst.* **150b**, 567 (1987).
55. D. A. Kleinman, *Phys. Rev.* **126**, 1977 (1962).
56. D. J. Williams, "Nonlinear Optical Properties of Guest-Host Polymer Structures," in *Nonlinear Optical Properties of Organic Molecules and Crystals*, D. S. Chemla and J. Zyss, eds. (Academic Press, New York, 1987).
57. K. D. Singer, S. L. Lalama, J. E. Sohn, and R. D. Small, "Electro-Optic Organic Materials" in *Nonlinear Optical Properties of Organic Molecules and Crystals*, D. S. Chemla and J. Zyss, eds. (Academic Press, New York, 1987).
58. S. Kielich, *IEEE J. Quant. Electron.* **QE-5**, 562 (1969).
59. M. G. Kuzyk, K. D. Singer, L. A. King, and H. E. Zahn, *J. Opt. Soc. Am. B* **6**, 742 (1989).
60. H. C. Ling, W. R. Holland, and H. M. Gordon, *J. Appl. Phys.* **70**, 6669 (1991).
61. J. van Turnhout, "Thermally Stimulated Discharge of Electrets," in *Electrets*, G. M. Sessler, ed. (Springer-Verlag, Berlin, 1980).
62. B. Reck, M. Eich, D. Jungbauer, R. J. Twieg, C. G. Willson, D. Y. Yoon, and G. C. Bjorklung, *Proc. SPIE* **1147**, 74 (1989).
63. H. L. Hampsch, J. Yang, G. K. Wong, and J. M. Torkelson, *Polym. Commun.* **30**, 40 (1989).
64. M. Eich, H. Looser, D. Y. Yoon, R. Twieg, G. Bjorklund, and J. C. Baumert, *J. Opt. Soc. Am. B* **6**, 1590 (1989).
65. M. L. Williams, R. F. Landel, and J. P. Ferry, *J. Am. Chem. Soc.* **77**, 3701 (1955).
66. F. Kohlrausch, *Pogg. Ann. Physk.* **119**, 352 (1863); G. Williams and D. C. Watts, *Trans. Faraday. Soc.* **66**, 80 (1970).
67. L. C. E. Struik, *Physical Aging in Amorphous Polymers and Other Materials* (Elsevier, Amsterdam, 1978).
68. M. A. Schen and F. I. Mopsik, *Proc. SPIE* **1560**, 315 (1991).
69. G. Khanarian and R. A. Norwood, *Proc. SPIE* **1337**, 44 (1990).
70. G. R. Mohlmann, W. H. Horsthuis, A. M. Donach, J. M. Copeland, C. Duchet, P. Fabre, M. B. Diemeer, E. S. Trommel, F. M. Suyten, E. Van Tomme, P. Baquero, and P. Van Daele, *Proc. SPIE* **1337**, 215 (1990).
71. K. D. Singer and J. E. Sohn, "Organic Materials for Second-Order Nonlinear Optical Devices" in *Electroresponsive Molecular and Polymeric Systems*, T. A. Skotheim, ed. (Marcel Dekker, New York, 1991).
72. J. Zyss and D. S. Chemla, "Quadratic Nonlinear Optics and Optimization of the Second-Order Nonlinear Optical Response of Molecular Crystals", in *Nonlinear Optical Properties of Organic Molecules and Crystals*, D. S. Chemla and J. Zyss, eds. (Academic Press, New York, 1987).
73. K. I. White and B. K. Nayar, *J. Opt. Soc. Am. B* **5**, 317 (1988).
74. T. Kondo, N. Hashizume, S. Miyoshi, R. Morita, N. Ogasawara, S. Umegaki, and R. Ito, *Proc. SPIE* **1337**, 53 (1990).

75. P. A. Cahill, "Nonlinear Optical Waveguides Containing a Transition Metal-Based Dye Molecule," in *Nonlinear Optical Properties of Polymers*, A. J. Heeger, J. Orenstein, and D. R. Ulrich, eds. (Materials Research Society, vol. 109, Pittsburgh, 1988).
76. P. A. Cahill and K. D. Singer, "Chemistry of Anomalous-Dispersion Phase-Matched Second Harmonic Generation," in *Materials for Nonlinear Optics: Chemical Perspectives*, S. R. Marder, J. E. Sohn, and G. D. Stuckey, eds. (ACS Symp. Ser. 455, 1991).
77. K. D. Singer, M. G. Kuzyk, T. Fang, W. R. Holland, and P. A. Cahill, "Design Considerations for Multi-Component Molecular-Polymeric Nonlinear Optical Materials," in *Organic Molecules for Nonlinear Optics and Photonics*, J. Messier, *et al.*, eds. (Kluwer, Dordrecht, 1991).
78. J. L. Oudar, *J. Chem. Phys.* **67**, 446 (1977).
79. A. Yariv, *Quantum Electronics* (Wiley, New York, 1989).
80. R. V. Mustacich, *Appl. Opt.* **27**, 3732 (1988).
81. J. Feinberg, D. Heiman, A. R. Tanguay, Jr., and R. W. Hellwarth, *J. Appl. Phys.* **51**, 1297 (1980); *J. Appl. Phys.* **52**, 537 (1980).
82. P. Günter and J.-P. Huignard, "Photorefractive Effects and Materials," in *Photorefractive Materials and Their Applications I*, P. Günter and J.-P. Huignard, eds. (Springer-Verlag, Berlin, 1988).
83. A. Ashkin, G. D. Boyd, J. M. Dziedzic, R. G. Smith, A. A. Ballman, and K. Nassau, *Appl. Phys. Lett.* **9**, 72 (1966).
84. F. S. Chen, *J. Appl. Phys.* **38**, 3418 (1967).
85. W. E. Moerner, C. Walsh, J. C. Scott, S. Ducharme, D. M. Burland, G. C. Bjorklund, and R. J. Twieg, *Proc. SPIE* **1560**, 278 (1991).
86. In 1990, K. Sutter *et al.* reported observation of photorefractive gratings for the first time in an organic crystal. See K. Sutter and P. Günter, *J. Opt. Soc. Am. B* **7**, 2274 (1990).
87. S. Ducharme, J. C. Scott, R. J. Twieg, and W. E. Moerner, *Phys. Rev. Lett.* **66**, 1846 (1991).
88. J. S. Schildkraut, *Appl. Phys. Lett.* **58**, 340 (1991).
89. L. Li, R. J. Jeng, J. Y. Lee, J. Kumar, and S. K. Tripathy, *Proc. SPIE* **1560**, 243 (1991).
90. A. M. Glass and J. Strait, "The Photorefractive Effect in Semiconductors," in *Photorefractive Materials and Their Applications I*, P. Günter and J.-P. Huignard, eds. (Springer-Verlag, Berlin, 1988).
91. A. M. Glass, *Opt. Eng.* **17**, 470 (1978).
92. Y. Shuto, M. Amano, and T. Kaino, *Jap. J. Appl. Phys.* **30**, 320 (1991).
93. H. Scher and E. W. Montroll, *Phys. Rev. B* **12**, 2455 (1975).
94. P. J. Reucroft, "Photoconduction Mechanisms in Polymeric Solids," "Photocarrier Generation Mechanisms in Polymers," and "Sensitization of Photoconduction," in *Photoconductivity in Polymers: An Interdisciplinary Approach*, A. V. Patsis and D. A. Seanor, eds. (Technomic, Westport, 1976).

95. W. D. Gill, "Polymeric Photoconductors," in *Photoconductivity and Related Phenomena*, J. Mort and D. Pai, eds. (Elsevier Scientific, Amsterdam, 1976).
96. A. Twarowski, *J. Appl. Phys.* **65**, 2833 (1989).
97. S. C. Freilich, *Macromolecules* **20**, 973 (1987).
98. D. Fluck, P. Amrhein, and P. Günter, *J. Opt. Soc. Am. B* **8**, 2196 (1991).
99. G. A. Brost, R. A. Motes, and J. R. Rotge, *J. Opt. Soc. Am. B* **5**, 1879 (1988).
100. J. Mort and G. Pfister, "Photoelectronic Properties of Photoconducting Polymers," in *Electronic Properties of Polymers*, J. Mort and G. Pfister, eds. (John Wiley and Sons, New York, 1982).
101. S. E. Kanellopoulos, V. A. Handerek, H. Jamshidi, and A. J. Rogers, *IEEE Phot. Tech. Lett.* **3**, 244 (1991).
102. W. J. Tomlinson, E. A. Chandross, R. L. Fork, C. A. Pryde, and A. A. Lamola, *Appl. Opt.* **11**, 533 (1972).
103. M. Ottelenghi and D. S. McClure, *J. Chem. Phys.* **46**, 4613 (1967).
104. H. J. Eichler, P. Günter, and D. Phohl, *Laser Induced Dynamic Gratings* (Springer-Verlag, Heidelberg, 1986).
105. S. Ducharme, *Opt. Lett.* **16**, 1791 (1991).
106. G. Pauliat, P. Mathey, and G. Roosen, *J. Opt. Soc. Am. B* **8**, 1942 (1991).
107. D. A. Parthenopoulos and P. M. Rentzepis, *Science* **245**, 843 (1989).
108. Y. Shi, W. H. Steier, L. Yu, M. Chen, and L. R. Dalton, *Appl. Phys. Lett.* **58**, 1131 (1991).
109. K. W. Beeson, K. A. Horn, M. McFarland, and J. T. Yardley, *Appl. Phys. Lett.* **58**, 1955 (1991).
110. C. A. Walsh and W. E. Moerner, *J. Opt. Soc. Am. B* **9**, 1642 (1992).
111. J. C. Scott, L. T. Pautmeier, and W. E. Moerner, *J. Opt. Soc. Am. B* **9**, 2059 (1992).
112. Y. Cui, Y. Zhang, P. W. Prasad, J. S. Schildkraut, and D. J. Williams, *Appl. Phys. Lett.* **61**, 2132 (1992).
113. B. K. Mandel, Y. M. Chen, J. Y. Lee, J. Kumar, and S. Tripathy, *Appl. Phys. Lett.* **58**, 2459 (1991).
114. J. Obrzut, M. J. Obrzut, and F. E. Karasz, *Synth. Met.* **29**, E103 (1989).

Chapter 7

QUADRATIC ELECTRO-OPTICS OF GUEST-HOST POLYMERS

Mark G. Kuzyk

and

Constantina Poga

Department of Physics, Washington State University, Pullman, Washington

1. INTRODUCTION

Guest-host polymers have been widely studied owing to their potential applications in optical devices [1,2,3]. These materials combine the good

ISBN 0-12-784450-3

optical and dielectric properties of the host polymer with the large nonlinear-optical response of the guest molecules. (We note here that by guest-host polymer, we use the broader definition that includes guest molecules that are chemically attached to the polymer.) Aside from their good optical properties, such polymeric materials are easy to process into large-area thin films [4] on a wide variety of substrates as well as free-standing fibers [5,6,7,8] that define a waveguiding region. Furthermore, standard photolithography can be applied to define channel waveguides of the polymeric material on semiconductor substrates [9]. This added flexibility allows for the fabrication of thin-film optical devices that are integrable with electronics [10,11]. The possibility of synthetically tailoring both the guest and host system to suit a given application makes this material class both technologically important and scientifically challenging.

While device demonstrations have shown that these materials can be built into working modulators and switches, certain material reliability issues need to be addressed. Electro-optic devices, for example, rely on second-order nonlinear-optical susceptibilities that require polar-ordered materials; that is, materials in which the guest molecules' vector dipole moments are aligned. A stable material thus requires that the polar order should not relax with time. Recent studies of orientational decay [12,13] and methods for increasing stability have centered on understanding the mechanisms of molecular reorientation as well as the role of the host polymer.

All-optical devices, on the other hand, are based on the third-order nonlinear-optical susceptibility and are nonzero even in isotropic media. Such devices operate on the principle that one beam of light affects the propagation of another beam through the material nonlinearity: an effect that can be used to perform a logic or switching operation. While guest-host polymers have a large and fast electronic response, the slower guest reorientation also contributes [14]. Such slow effects can be a nuisance to designing fast devices and to building experiments that measure the fast response.

In common to both electro-optical and all-optical devices, reorientational effects need to be well understood to be avoided. On the other hand, guest chromophore reorientation can be used as a helpful tool for studying properties of the polymeric system [16]. In this chapter, we concentrate on the mechanisms of the reorientational contribution to the nonlinear-optical response and how such a response depends on the microscopic properties of the host polymer matrix. Particularly, we derive a model of the third-order nonlinear response of a guest-host polymer in the presence of both an optical field and a quasi-static field through a generalized dielectric-response formalism that accounts for both the electronic and reorientational response. The model is applied to the quadratic electro-optic response of the material and the experimental technique for measuring such a response is described.

2. THEORY

2.1. Background

In the late 1800s, Kerr showed that the polarization of a beam of light rotates when it passes through a liquid in the presence of a static field [17]. Electrochromism, defined as the color change of a material in the presence of an applied field has been studied by Liptay [18] and generalized in the strong-field limit by Yamaoka [19]. These effects are examples of electro-optic effects where the linear optical properties are affected by a static field. Electrochromism studies of molecules in liquids have been used to determine molecular dipole moments and transition moments [18].

Havinga and Van Pelt first applied these techniques to solid solutions of dye chromophores dispersed in polymers [20–22]. These studies are carried out in three steps. First, the linear absorption spectrum is measured with and without an applied field at room temperature—well below the polymer's glass transition temperature. The guest molecules are assumed to be immobile, and the electrochromic response is attributed to the polarization of charge within the stationary guest chromophores (Fig. 1a). Next, the polymer system is heated above the glass transition temperature where the molecules are free to rotate. (If the absorption spectrum is again measured in the presence of an applied static field, the change in the spectrum relative to the zero-field value can be attributed to a combination of charge polarization within the guests and guest reorientation.) The system is then cooled in the presence of the field, which is removed at room temperature. Assuming that the molecules' orientation is locked in place below the glass transition temperature until the measurement is performed, the difference in the absorption spectrum in this poled system relative to the isotropic system can be assumed to give the purely reorientational response (Fig. 1b). When an electric field is applied to the poled system, the absorption peak shifts either to the left or to the right, depending on the sign of the applied field relative to the poled guest molecules (Fig. 1c). These three measurements together are used to determine the transition and excited state dipolar moments of a guest molecule.

Havinga's method of poling a polymer and locking in the polarization

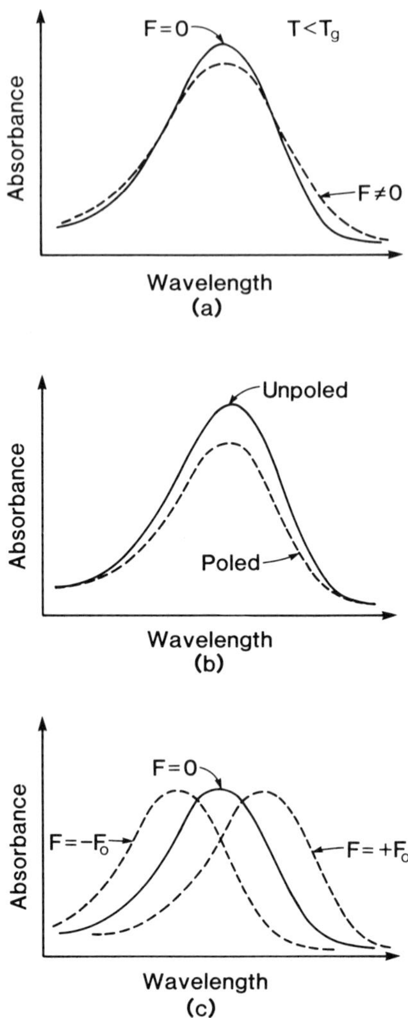

Fig. 1. Absorption spectra of a doped polymer: (a) zero-field spectrum and field-induced spectrum; (b) spectrum before and after electric field poling; (c) electric-field-induced spectrum of poled polymer. (Reprinted from Ref. 22 with permission from the publisher, Plenum Publishing Corp.)

was used by Singer and coworkers to make a guest-host material with a large second-order susceptibility as measured with second harmonic generation [4]. A thermodynamic model of the poling process that predicted the nonlinearity was developed and later refined to include ordered materials such as liquid crystals [23]. This second-order response was subsequently measured with electro-optic techniques that showed a response consistent with the second harmonic results [11]. The polarization of these materials, however, was shown to decay over time [13]. Further efforts succeeded in increasing stability by attaching the chromophore to the polymer matrix [11]. In later studies, Mortazavi and coworkers [24] and Page and coworkers [25] determined the polar order of a poled polymer with electrochromism.

Later, Kuzyk and coworkers developed a second-harmonic-generation method to measure the motion of the dye chromophores in polymers at room temperature in response to a static field, and the degree of reorientation was found to be substantial [16]. These reorientational effects were modeled as elastically impeded rotors that showed weak coupling to the polymer matrix. Later work of Boyd and coworkers confirmed this weak coupling and described their findings with a free-volume model [26].

More recently, Kuzyk and coworkers showed that reorientational effects of the guest dye chromophores in the polymer matrix contribute substantially to the third-order nonlinear-optical susceptibility as measured with quadratic electro-optic modulation [14,15]. In this contribution, we develop a generalized theory of such a response and show that in its limiting cases reduces to either the elastic response as calculated by Kuzyk and coworkers or to the free volume model of Boyd and coworkers.

2.2. Nonlinear Response Functions

Electric fields, either as components of an electromagnetic wave or as an applied static field, are often used to probe a material's structure and its elementary excitations. Historically, the response of a material to an applied static field, for example, was related to its dielectric function while the optical response was related to its refractive index or linear absorption. In both cases, the electric field induces a polarization (induced dipole moment per unit volume) that follows the applied field. The refractive index and dielectric functions are simply related to the high ($\approx 10^{15}$ Hz) and low frequency components ($< 10^{9}$ Hz), respectively, of the general response function.

Each underlying elementary excitation dominates the response function within a range of frequencies. The electronic excitation energies of most organic molecules, for example, are of the order of 1 eV to 5 eV and therefore dominate the optical response. The motion of polymer chains, on the other hand, takes place in the gigahertz range and is excited by microwaves. The full frequency dependence of the response function thus incorporates all underlying mechanisms that result in excitations.

A response function that is often used to describe the interaction of light with a material is the susceptibility, χ. The susceptiblity is a tensor that is a function of the applied electric field, $\vec{E}$, and relates the polarization, $\vec{P}$, to this applied field. In tensor notation, the relationship is

$$P_i = \chi_{ij}(\vec{E})E_j, \tag{2.2.1}$$

where repeated indices are summed. (The rest of this chapter assumes the summation convention.) Both the refractive index and the dielectric function are related to the susceptibility. When the susceptibility is independent of the electric field, the response is said to be linear, and the polarization frequency is equal to the frequency of the applied field.

When the susceptibility is a function of the electric field, the polarization will no longer be linear in the applied field. An applied field at one frequency can then result in a polarization at another frequency. More generally, if several fields of distinct frequencies are applied, a polarization with many new frequency components results [27]. In frequency space, the nth-order polarization at a given frequency, $P_i^{(n)}(\omega)$, is related to the field amplitudes, $E_i^{\omega_\sigma}$ at other frequencies, ω_σ,

$$P_i^{(n)}(\omega) = K(-\omega; \omega_1, \omega_2, \omega_3, \ldots, \omega_n)\chi^{(n)}_{ijkl\ldots\xi}(-\omega; \omega_1, \omega_2, \omega_3, \ldots, \omega_n) \times E_j^{\omega_1}E_k^{\omega_2}E_l^{\omega_3}\ldots E_\xi^{\omega_n}, \tag{2.2.2}$$

where $\chi^{(n)}_{ijkl\ldots\xi}(-\omega; \omega_1, \omega_2, \omega_3, \ldots, \omega_n)$ is the nth-order susceptibility tensor of rank $n+1$, and relates the n-applied fields to the nth-order polarization at frequency ω. $K(-\omega; \omega_1, \omega_2, \omega_3, \ldots, \omega_n)$ is a degeneracy factor that arises from the definition of the susceptibility. A detailed description of its meaning can be found in Butcher and Cotter [28]. It is defined as

$$K(-\omega; \omega_1, \omega_2, \omega_3, \ldots, \omega_n) = 2^{l+m-n}\pi, \tag{2.2.3}$$

where π is the number of distinct permutations of the frequencies $(\omega_1, \omega_2, \omega_3, \ldots, \omega_n)$, n is the order of the nonlinearity, m is the number of zero frequencies from the set $(\omega_1, \omega_2, \omega_3, \ldots, \omega_n)$ and where $l = 1 - \delta_{\omega,0}$ with $\delta_{\omega,0}$ the Kronecker delta. The total polarization of order n thus depends on the product of n electric field amplitudes.

The total response at a given frequency is then the sum over all orders, i.e., $\vec{P}(\omega) = \sum_{n=0}^{\infty} \vec{P}^n(\omega)$ and the total polarization is the Fourier transform of $\vec{P}(\omega)$. Unlike the linear response where the applied field and response are both at the same frequency, the nonlinear susceptibility mixes several frequencies. Typically, the nonlinear optical susceptibility refers to the limit where all of the applied and resultant fields are at optical frequencies while the nonlinear dielectric susceptibility refers to the limit where all the fields are at low frequency. This chapter deals with the mixed case where both optical and static fields are applied. In the language of nonlinear optics, such effects are often referred to as electro-optic effects. We will, in particular, focus on the third-order response where one optical field mixes with two low-frequency quasi-static fields, resulting in an optical response. Owing to the quadratic dependence of the polarization on the applied static field, such a response is called the quadratic electro-optic effect.

2.3. The Model

Based on the observations, it is clear that the guest chromophores interact with the polymer matrix and that the strength of the interaction depends on the temperature. A polymer matrix, though, is known to have a distribution of sites; that is, guest molecules at different sites interact differently with the matrix. Modeling this distribution of sites, however, is beyond the scope of this chapter. Instead, we model only the average values of the parameters of interest. When, for example, the strength of interaction between a molecule and the polymer is calculated—the term *interaction potential* implies the *averaged* interaction potential over the distribution of sites. While such an average response model may limit the range of its applicability, generalization of our formalism is straightforward albeit messy.

We begin by considering a single molecule that is imbedded in the polymer. We assume that this molecule sees a uniformly smeared-out polymer, that is, an average site. A coordinate frame denoted by double primes is fixed to the principle axes of the bare molecule so that the molecular dipole moment, $\vec{p}$, is related to the local field at the molecule, $\vec{F}$, through the molecular susceptibility tensor ζ:

$$p_{j''}^{(n)} = K\zeta_{j''k''l''\ldots}^{(n)} F_{k''} F_{l''} \ldots, \tag{2.3.1}$$

where K is again the degeneracy factor that depends on the field frequencies. The local field has two sources: the applied field and the polarization field. The polarization field includes contributions from both other guest molecules and the polymer host. In general—the local field, $\vec{F}$, and the applied field, $\vec{E}$—are not collinear and are related by a local field tensor$L$:

$$F_{k''} = L_{k''p'} E_{p'}, \tag{2.3.2}$$

where the dressed susceptibilities (defined below) are defined to have as their principle axes the primed coordinate frame.

The induced dipole moment can be related to the applied electric field by substituting Eq. (2.3.2) into Eq. (2.3.1) and taking the tensor product with L, which leads to:

$$p_{j''}^{(n)} L_{j''r'} = K \zeta_{j''k''l''\ldots}^{(n)} L_{j''r'} L_{k''p'} L_{l''q'\ldots} \times E_{p'} E_{q'\ldots}. \tag{2.3.4}$$

Defining the dressed susceptibilities, ζ^*, as

$$\zeta_{r'p'q'\ldots}^{*(n)} = \zeta_{j''k''l''\ldots}^{(n)} L_{j''r'} L_{k''p'} L_{l''q'\ldots}, \tag{2.3.5}$$

the nonlinear polarization can be expressed in the dressed molecular frame as:

$$p_{r'}^{*(n)} = \zeta_{r'p'q'\ldots}^{*(n)} E_{p'} E_{q'\ldots}. \tag{2.3.6}$$

While simple local field models for L such as that of Lorentz-Lorenz or Onsager have enjoyed some success in relating the local fields to the applied fields, more complicated systems may not be so easily treated. Because most experiments measure the dressed susceptibility tensor directly, it is convenient to sweep away our ignorance of the local fields by using Eq. (2.3.6) as the constitutive relationship.

Models of the nonlinear response must include all possible sources of induced dipole moment, such as the electronic polarization due to the electrons within the guest molecule and the polarization from molecular reorientation. The degree of induced polarization depends on the dressed electronic structure of the molecule and the physical interaction of the molecule with the polymer matrix. Our treatment, however, is limited to guest molecules that are fixed with respect to translational motion, thereby effectively excluding such mechanisms as phonons and electrostriction. These restrictions are obeyed for electrically neutral molecules in the presence of small field gradients. We also limit our treatment to linear-optical polymer hosts whose spatial coordinates are fixed so that the polymer does not contribute to the nonlinear polarization.

Under the above assumptions, we can determine the equilibrium orientation of a molecule in an electric field as follows. The guest molecule is influenced by the torque of the applied static field and the restraining force of the polymer. The magnitude of the electric-field-induced torque depends on the orientation of the molecule relative to the field while the internal force due to the polymer generally depends on the molecule's orientation relative to its equilibrium orientation within the polymer. We assume that the two forces are independent, that is, the electric field does not induce changes in the polymer that alters its action on the guest molecule. Such assumptions are generally true when the interaction between host polymer units is weak so that spontaneous orientation is not favored (unlike liquid crystalline systems, for example, where spontaneous ordering can be induced with applied fields and/or surface treatment) [29].

To simplify matters, we assume that the bulk material has azimuthal symmetry and that the molecules are randomly ordered along the axial angular coordinate in the dressed molecular frame. Under this assumption, the potential energy of a molecule depends only on its polar angle relative to the applied field, θ, and its polar angle relative to its equilibrium orientation. Defining the molecule's zero-field equilibrium polar angle as θ_0, the potential becomes:

$$U(\theta, \theta_0) = U_E(\theta) + U_{\text{int}}(\theta, \theta_0), \tag{2.3.7}$$

where $U_{\text{int}}(\theta, \theta_0)$ is the internal restraining potential and $U_E(\theta)$ is the electric potential:

$$U_E = -\int_0^{\vec{E}} p^*_{q'}(\vec{E})dE_{q'}. \tag{2.3.8}$$

The electric field potential, then, depends only on the dressed molecule's polar coordinate in the laboratory frame while the interaction energy of the molecule with the polymer matrix is independent of the field. It is this separation of the potential that leads to a tractable solution of the nonlinear response function.

The applied electric field, then, has two effects. First, it induces a dipole moment in the molecule; and secondly, it causes the molecule to reorient through U_E. It is this combination of polarization and reorientation that leads to the nonlinear response of the molecule. The electronic dipole moment in the dressed frame, p^*, is specifically written as:

$$p^*_{i'} = \mu^*_{i'} + K'_1\alpha^*_{i'j'}E_{j'} + K'_2\beta^*_{i'j'k'}E_{j'}E_{k'} + K'_3\gamma^*_{i'j'k'l'}E_{j'}E_{k'}E_{l'} + \cdots \tag{2.3.9}$$

where μ^* is the dressed ground state dipole moment, α^* the dressed polarizability, β^* the dressed second-order polarizability, and so on. The K's are the degeneracy factors.

Assuming that the material is in thermal equilibrium, the angle θ_0 is just the average orientation of all molecules that experience the same zero-field potential. For noninteracting guest molecules, the bulk polarization in the laboratory frame (denoted by unprimed subscripts), $\vec{P}$, is related to the induced dipole moment through the vector sum of the individual induced molecular dipole moments. The orientational distribution function of the dipoles is determined by a Gibbs distribution and takes the form:

$$P_i = N\langle p^*_{i'}\rangle_i. \tag{2.3.10}$$

The brackets denote the statistical average. For a continuous distribution of guest molecules, this average will be expressed as an integral over both the zero-field equilibrium orientation and the laboratory coordinate:

$$N\langle p_{i'}^* \rangle_i = N \int d\Omega_0 \int d\Omega [G(\theta, \theta_0) p_{i'}^*] a_{i'i}(\Omega) \Big/ \int d\Omega_0, \qquad (2.3.11)$$

where $G(\theta, \theta_0)$ is the orientational distribution function that gives the probability density of finding a molecule orientated at θ given that its zero-field orientation is θ_0. a is the rotation matrix and is a function of the three Euler angles $\vec{\Omega} = (\theta, \phi, \psi)$.

The bulk polarization, on the other hand, is expressed by a power series in the applied electric field, and in the laboratory frame, is given by:

$$P_i = \chi_i^{(0)} + K_1 \chi_{ij}^{(1)} E_j + K_2 \chi_{ijk}^{(2)} E_j E_k + K_3 \chi_{ijkl}^{(3)} E_j E_k E_l + \cdots. \qquad (2.3.12)$$

Note here that while we have suppressed the frequency dependence of the susceptibilities and the electric fields in Eqs. (2.3.9) through (2.3.11), they are implied. The bulk nth-order nonlinear susceptibility is then calculated from Eq. (2.3.11) by differentiating the polarization with respect to the electric field n times:

$$\chi_{ij\ldots}^{(n)}(-\omega; \omega_1, \omega_2, \ldots) = \frac{1}{K_n(-\omega; \omega_1, \omega_2, \ldots) D_n(E_j^{\omega_1}, E_k^{\omega_2}, \ldots)} \times \left. \frac{\partial^n P_i^{(\omega)}}{\partial E_j^{\omega_1} \partial E_k^{\omega_2}, \ldots} \right|_{\vec{E}=0}, \qquad (2.3.13)$$

where D_n is the degeneracy factor that accounts for the number of distinct fields (fields are distinct if they have different frequency or different vector component or both) and is given by $n_1! n_2! n_3 \ldots !$. $n_1, n_2, n_3, \ldots$ are the number of distinct fields at frequency $\omega_1, \omega_2, \omega_3, \ldots$ with $n_1 + n_2 + n_3 + \cdots = n$.

The bulk susceptibility, then, is related to the molecular susceptibility and the polymer-induced molecular restraining force through Eq. (2.3.13). Substituting Eqs. (2.3.10) and (2.3.11) into Eq. (2.3.13) yields

$$\chi_{ijkl}^{(3)}(-\omega; \omega_1, \omega_2, \omega_3) = \frac{N}{K_3(-\omega; \omega_1, \omega_2, \omega_3) D_3(E_j^{\omega_1}, E_k^{\omega_2}, E_l^{\omega_3})} \times \left. \frac{\int d\Omega_0\, \partial^3 (\int d\Omega\, [G(\theta, \theta_0) p_{i'}^{*\omega}] a_{i'i}(\Omega))}{\partial E_j^{\omega_1} \partial E_k^{\omega_2} \partial E_l^{\omega_3}} \right|_{\vec{E}=0} \Big/ \int d\Omega_0. \qquad (2.3.14)$$

Equation (2.3.14) is the general model-independent relationship between the molecular and bulk response. Models of the polymer's influence, then, depend on the choice of the orientational distribution function. The electric-field derivative in Eq. (2.3.14) operates on two terms: the dressed dipole moment and the orientational distribution function. The electronic response is a result of the former, while the reorientational response results from a combination of field-induced changes in the distribution function and lower-order electronic polarizations.

In our model, we calculate G from the Gibbs distribution by using the potential given by Eq. (2.3.7),

$$G(\theta, \theta_0) = \frac{\exp[-(-\int_0^{\vec{E}} p_{i'}^*(\vec{E})dE_{i'} + U_{\text{int}})/kT]}{\int d\Omega \exp[-(-p_{i'}^* E_{i'} + U_{\text{int}})/kT]}. \qquad (2.3.15)$$

The electric-field dependence in Eq. (2.3.14) is contained in the orientational distribution function and in the dipole moment $p_{i'}^*$ (the product in square brackets in Eq. (2.3.14)). The action of the third derivative with respect to an electric field can be brought under the integral and is of the form

$$\left.\frac{\partial^3[G(\theta, \theta_0)\, p_{i'}^{*\omega}]}{\partial E_j^{\omega_1} \partial E_k^{\omega_2} \partial E_l^{\omega_3}}\right|_{\vec{E}=0}. \qquad (2.3.16)$$

After the fields are set to zero, the integral over the orientational distribution function reduces to an integral over the zero-field orientational distribution function. We define this zero-field distribution function as $G_0(\theta, \theta_0)$:

$$G_0(\theta, \theta_0) = \frac{\exp[-U_{\text{int}}/kT]}{\int d\Omega \exp[-U_{\text{int}}/kT]}. \qquad (2.3.17)$$

In the following subsection, we evaluate the third-order susceptibility for the zero-frequency response and the quadratic electro-optic response. The zero-frequency response so evaluated is a generalization of the low-frequency dielectric function where molecule/polymer interactions are accounted for.

2.4. Static Response

In the static limit, all of the electric fields are degenerate in frequency. In this case $D = 1$ if all the field components are different; otherwise, D is determined by considering the number of fields with the same vector component. The derivative of Eq. (2.3.16) becomes:

$$\left.\frac{\partial^3[G(\theta, \theta_0)\, p_{i'}^{*\omega}]}{\partial E_j^{\omega_1} \partial E_k^{\omega_2} \partial E_l^{\omega_3}}\right|_{\vec{E}=0} = \left.\frac{\partial^3[G(\theta, \theta_0)\, p_{i'}^*]}{\partial E_j \partial E_k \partial E_l}\right|_{\vec{E}=0}. \qquad (2.4.1)$$

The third derivative of Eq. (2.3.16) is then:

$$\begin{aligned}
\left.\frac{\partial^3[G(\theta,\theta_0)p_{i'}^*]}{\partial E_j\partial E_k\partial E_l}\right|_{\vec{E}=0} = G_0(\theta,\theta_0)\Bigg[&\frac{\partial^3 p_{i'}^*}{\partial E_j\partial E_k\partial E_l} - \frac{1}{kT}\frac{\partial^3 U_E}{\partial E_j\partial E_k\partial E_l}p_{i'}^* \\
&-\frac{1}{kT}\mathscr{P}_{jkl}\left\{\frac{\partial^2 p_{i'}^*}{\partial E_j\partial E_k}\frac{\partial U_E}{\partial E_l} + \frac{\partial^2 U_E}{\partial E_j\partial E_k}\frac{\partial p_{i'}^*}{\partial E_l}\right\} \\
&+\left(\frac{1}{kT}\right)^2\mathscr{P}_{jkl}\left\{\frac{\partial U_E}{\partial E_j}\left(\frac{\partial U_E}{\partial E_k}\frac{\partial p_{i'}^*}{\partial E_l} + \frac{\partial^2 U_E}{\partial E_k\partial E_l}p_{i'}^*\right)\right\} \\
&-\left(\frac{1}{kT}\right)^3\frac{\partial U_E}{\partial E_j}\frac{\partial U_E}{\partial E_k}\frac{\partial U_E}{\partial E_l}p_{i'}^* \\
&+\left(\frac{1}{kT}\right)\mathscr{P}_{jkl}\left\{\frac{\partial p_{i'}^*}{\partial E_j}\left\langle\frac{\partial^2 U_E}{\partial E_k\partial E_l}\right\rangle_U\right\} \\
&-\left(\frac{1}{kT}\right)^2\mathscr{P}_{jkl}\left\{p_{i'}^*\frac{\partial U_E}{\partial E_j}\left\langle\frac{\partial^2 U_E}{\partial E_k\partial E_l}\right\rangle_U + \frac{\partial p_{i'}^*}{\partial E_j}\left\langle\frac{\partial U_E}{\partial E_k}\frac{\partial U_E}{\partial E_l}\right\rangle_U\right\} \\
&+\left(\frac{1}{kT}\right)^3\mathscr{P}_{jkl}\left\{p_{i'}^*\frac{\partial U_E}{\partial E_j}\left\langle\frac{\partial U_E}{\partial E_k}\frac{\partial U_E}{\partial E_l}\right\rangle_U\right\} \\
&-\left(\frac{1}{kT}\right)^2\mathscr{P}_{jkl}\Bigg\{p_{i'}^*\left\langle\frac{\partial U_E}{\partial E_j}\right\rangle_U\frac{\partial^2 U_E}{\partial E_k\partial E_l} \\
&\qquad + \mathscr{P}_{jk}\left\{\frac{\partial p_{i'}^*}{\partial E_j}\frac{\partial U_E}{\partial E_k}\right\}\left\langle\frac{\partial U_E}{\partial E_l}\right\rangle_U\Bigg\} \\
&+\left(\frac{1}{kT}\right)^3\mathscr{P}_{jkl}\left\{p_{i'}^*\frac{\partial U_E}{\partial E_j}\frac{\partial U_E}{\partial E_k}\left\langle\frac{\partial U_E}{\partial E_l}\right\rangle_U\right\} \\
&+\left(\frac{1}{kT}\right)\mathscr{P}_{jkl}\left\{\left\langle\frac{\partial U_E}{\partial E_j}\right\rangle_U\frac{\partial^2 p_{i'}^*}{\partial E_k\partial E_l}\right\} \\
&+\left(\frac{1}{kT}\right)p_{i'}^*\left\langle\frac{\partial^3 U_E}{\partial E_j\partial E_k\partial E_l}\right\rangle_U \\
&-\left(\frac{1}{kT}\right)^2\mathscr{P}_{jkl}\left\{p_{i'}^*\left\langle\frac{\partial^2 U_E}{\partial E_j\partial E_k}\frac{\partial U_E}{\partial E_l}\right\rangle_U\right\} \\
&+\left(\frac{1}{kT}\right)^3 p_{i'}^*\left\langle\frac{\partial U_E}{\partial E_j}\frac{\partial U_E}{\partial E_k}\frac{\partial U_E}{\partial E_l}\right\rangle_U
\end{aligned}$$

$$- 2\left(\frac{1}{kT}\right)^3 \mathscr{P}_{jkl}\left\{p_{i'}^* \frac{\partial U_E}{\partial E_j}\left\langle\frac{\partial U_E}{\partial E_k}\right\rangle_U \left\langle\frac{\partial U_E}{\partial E_l}\right\rangle_U\right\}$$

$$+ 2\left(\frac{1}{kT}\right)^2 \mathscr{P}_{jkl}\left\{\frac{\partial p_{i'}^*}{\partial E_j}\left\langle\frac{\partial U_E}{\partial E_k}\right\rangle_U \left\langle\frac{\partial U_E}{\partial E_l}\right\rangle_U\right\}$$

$$+ 2\left(\frac{1}{kT}\right)^2 \mathscr{P}_{jkl}\left\{p_{i'}^* \left\langle\frac{\partial^2 U_E}{\partial E_j \partial E_k}\right\rangle_U \left\langle\frac{\partial U_E}{\partial E_l}\right\rangle_U\right\}$$

$$- 2\left(\frac{1}{kT}\right)^3 \mathscr{P}_{jkl}\left\{p_{i'}^* \left\langle\frac{\partial U_E}{\partial E_j}\frac{\partial U_E}{\partial E_k}\right\rangle_U \left\langle\frac{\partial U_E}{\partial E_l}\right\rangle_U\right\}$$

$$+ 6\left(\frac{1}{kT}\right)^3 p_{i'}^* \left\langle\frac{\partial U_E}{\partial E_j}\right\rangle_U \left\langle\frac{\partial U_E}{\partial E_k}\right\rangle_U \left\langle\frac{\partial U_E}{\partial E_l}\right\rangle_U\Bigg], \tag{2.4.2}$$

where the permutation operation, $\mathscr{P}_{jkl}$, includes only even permutations, i.e., $\mathscr{P}_{jkl}\{f_{jkl}\} = f_{jkl} + f_{klj} + f_{ljk}$, and for two subscripts gives $\mathscr{P}_{jk}\{f_{jk}\} = f_{jk} + f_{kj}$. The averaging denoted by the angled brackets is over the zero-field distribution function and is an integral over θ only. Note that summation notation over repeated indices does not apply for the permutation operator.

Substituting Eqs. (2.4.1) and (2.4.2) into Eq. (2.3.14) with the help of Eqs. (2.3.8) and (2.3.9) yields:

$$\chi_{ijkl}^{(3)} = \frac{N}{K_3 D_3}\Bigg\langle 6K_3'\langle\gamma_{ijkl}^*\rangle$$

$$+ \left(\frac{2}{kT}\right)\mathscr{P}_{jkl}\{K_1'\eta_{jk}\langle\alpha_{il}^*\alpha_{jk}^*\rangle - \langle\alpha_{ij}^*\rangle\langle K_1'\eta_{kl}\alpha_{kl}^*\rangle\}$$

$$+ \left(\frac{6}{kT}\right)[K_2'\eta_{jkl}\langle\mu_i^*\beta_{jkl}^*\rangle - \langle\mu_i^*\rangle\langle K_2'\eta_{jkl}\beta_{jkl}^*\rangle]$$

$$+ \left(\frac{2}{kT}\right)\mathscr{P}_{jkl}\{K_2'\langle\mu_l^*\beta_{ijk}^*\rangle - K_2'\langle\mu_j^*\rangle\langle\beta_{ikl}^*\rangle\}$$

$$+ \left(\frac{1}{kT}\right)^2 \mathscr{P}_{jkl}\{\langle\mu_j^*\mu_k^*\alpha_{il}^*\rangle + 2K_1'\eta_{kl}\langle\mu_j^*\mu_i^*\alpha_{kl}^*\rangle + 2\langle\alpha_{ij}^*\rangle\langle\mu_k^*\rangle\langle\mu_l^*\rangle\}$$

$$+ \left(\frac{1}{kT}\right)^2 \mathscr{P}_{jkl}\{4\langle\mu_i^*\rangle\langle K_1'\eta_{jk}\alpha_{jk}^*\rangle\langle\mu_l^*\rangle - 2\langle\mu_j^*\rangle\langle K_1'\eta_{kl}\mu_i^*\alpha_{kl}^*\rangle\}$$

$$- \left(\frac{1}{kT}\right)^2 \mathscr{P}_{jkl}\{2\langle\mu_i^*\mu_j^*\rangle\langle K_1'\eta_{kl}\alpha_{kl}^*\rangle + 2\langle\mu_i^*\rangle\langle K_1'\eta_{jk}\alpha_{jk}^*\mu_l^*\rangle + \langle\mu_k^*\mu_l^*\rangle\langle\alpha_{ij}^*\rangle\}$$

$$- \left(\frac{1}{kT}\right)^2 \mathscr{P}_{jkl}\{\mathscr{P}_{jk}\{\langle\mu_i^*\rangle\langle\alpha_{ij}^*\mu_k^*\rangle\}\}$$

$$- \left(\frac{1}{kT}\right)^3 [\langle\mu_i^*\rangle\langle\mu_j^*\mu_k^*\mu_l^*\rangle + 6\langle\mu_i^*\rangle\langle\mu_j^*\rangle\langle\mu_k^*\rangle\langle\mu_l^*\rangle]$$

$$+ \left(\frac{1}{kT}\right)^3 \{\langle\mu_i^*\mu_j^*\mu_k^*\mu_l^*\rangle - \mathscr{P}_{jkl}\{\langle\mu_i^*\mu_j^*\rangle\langle\mu_k^*\mu_l^*\rangle - \langle\mu_i^*\mu_j^*\mu_k^*\rangle\langle\mu_l^*\rangle\}\}$$

$$+ 2\left(\frac{1}{kT}\right)^3 \mathscr{P}_{jkl}\{\langle\mu_i^*\mu_j^*\rangle\langle\mu_k^*\rangle\langle\mu_l^*\rangle + \langle\mu_i^*\rangle\langle\mu_j^*\mu_k^*\rangle\langle\mu_l^*\rangle\}\bigg\rangle_0, \quad (2.4.3)$$

where $\eta_{ijk\ldots}^{(n)}$ is a factor that arises from evaluating the integral of Eq. (2.3.8) with p^* given by Eq. (2.3.9) and is expressed as

$$\eta_{ijk\ldots}^{(n)} = \int_0^{\vec{E}} p_i^{*(n)} dE_i / \xi_{ijk\ldots}^{(n)} E_i E_j E_k \ldots .$$

Note that $\eta_{ijk\ldots}^{(n)}$ is a multiplicative factor that is not summed according to summation convention. Also, note that to conserve space, we have used shorthand notation for the averaging over the interaction potential, i.e., $\langle T_{ijk\ldots}\rangle = \langle T_{i'j'k'\ldots}\rangle_{ijk\ldots}$. The primed degeneracy factors refer to the dressed molecular frame. When the fields are indistinguishable in the laboratory frame (for example, when all the zero-frequency fields are in the laboratory 3-direction) they will in general be distinguishable in the dressed molecular frame when it does not coincide with the laboratory frame. In the static case, all η and K' factors are equal to unity. Note that the brackets again denote the average over the zero-field distribution function, i.e., with Eq. (2.3.17) as the distribution function. The brackets $\langle\ \rangle_0$ denote integration over the isotropic zero-field distribution of the guests, i.e., integration over θ_0.

Equation (2.4.3) is the static dielectric function. Any model of the polymer can be incorporated into this dielectric function by substituting the appropriate interaction potential between the guest molecule and the polymer into the distribution function followed by subsequent averaging. To model a gas, for example, the interaction potential can be set to zero and the results reduce to those of the standard dielectric response as given in Bottcher [30]. The polymer restraint, on the other hand, can be modeled as a harmonic potential if the polymer elasticity prevails; the free-volume model, where the molecules are allowed to rotate freely within a restricted range of angles; or any combination of these or other potentials. Such calculations, however, will not be presented here for the static case. We will, however, derive this generalized response for the quadratic effect, which is the central focus of this chapter.

2.5. Electro-optic Response

In quadratic electro-optic experiments, the propagation of an optical field is affected by the presence of a static field [31]. Formally, the response function at the optical frequency (ω) will depend on the product of the square of the static field and the input optical field, i.e., the response of interest is:

$$\chi^{(3)}_{ijkl}(-\omega; \omega, 0, 0). \tag{2.5.1}$$

Because an optical cycle is fast compared to thermal orientational relaxation times, the potential energy in the exponent of the Gibbs distribution function must be replaced by the average potential energy over one optical cycle:

$$\bar{U} = \frac{1}{\tau}\int_0^\tau U dt. \tag{2.5.2}$$

τ is the optical period. Because the interaction potential is explicitly time independent,

$$\bar{U} = \bar{U}_E + U_{\text{int}}. \tag{2.5.3}$$

The applied field has a zero-frequency component and an optical component ($\vec{E} = \vec{E}^\omega + \vec{E}^0$) so that the electric-field potential energy as given by Eq. (2.3.8) and Eq. (2.3.9) will have terms of the form $(\vec{E}^0)^n(\vec{E}^\omega)^m$ The time-averaged value of such a product vanishes for m-odd. The lowest-order nonvanishing optical field term is $m = 2$. Because the third-order polarization at the optical frequency is proportional to the square of magnitude of the static field and is directly proportional to the magnitude of the optical field, the $m = 2$ term will not contribute to the third-order response. The potential energy in the exponent of the Gibbs distribution function will thus be a function of only the static field:

$$\begin{aligned} -\bar{U}_E = \int_0^{\vec{E}^0} p^*_{i'}(\vec{E})dE^0_{i'} &= \mu^*_{i'}E^0_{i'} + \eta_{i'j'}\alpha^*_{i'j'}(0;0)E^0_{i'}E^0_{j'} \\ &+ \eta_{i'j'k'}\beta^*_{i'j'k'}(0;0,0)E^0_{i'}E^0_{j'}E^0_{k'} \\ &+ \eta_{i'j'k'l'}\gamma^*_{i'j'k'l'}(0;0,0,0)E^0_{i'}E^0_{j'}E^0_{k'}E^0_{l'} + \cdots, \end{aligned} \tag{2.5.4}$$

where according to Eq. (2.2.3) we have set $K'_1 = 1$, $K'_2 = 2$, $K'_3 = 3$.

To evaluate the third order susceptibility, we differentiate Eq. (2.3.11) as we had for the static case. The induced dressed dipole moment now has both a static and optical component, $\vec{p} = \bar{p}^0 + \bar{p}^\omega$. Because the distribution function is independent of optical frequency, the induced dipole that multiplies the distribution function will contribute only at the optical frequency:

$$p_{i'} = p_{i'}^{\omega} = \alpha^*_{i'j'}(-\omega; \omega)E_{j'}^{\omega} + 2\beta^*_{i'j'k'}(-\omega; \omega, 0)E_{j'}^{\omega}E_{k'}^{0} + 3\gamma^*_{i'j'k'l'}(-\omega; \omega, 0, 0)E_{j'}^{\omega}E_{k'}^{0}E_{l'}^{0} + \cdots \quad (2.5.5)$$

where we have set $K'_1 = 1$, $K'_2 = 2$, $K'_3 = 3$. Using Eq. (2.3.14) for the electro-optic susceptibility, we get:

$$\begin{aligned}\chi^{(3)}_{ijkl}&(-\omega; \omega, 0, 0)\\ &= \frac{N}{6}\Bigg\langle 3\mathscr{P}_{kl}\{\langle\gamma^*_{ijkl}(-\omega; \omega, 0, 0)\rangle\}\\ &\quad + \left(\frac{1}{kT}\right)\mathscr{P}_{kl}\{2(\langle\mu^*_l\beta^*_{ikj}(-\omega; \omega, 0)\rangle - \langle\mu^*_l\rangle\langle\beta^*_{ikj}(-\omega; \omega, 0)\rangle)\\ &\quad + (\langle\alpha^*_{ij}(0; 0)\alpha^*_{kl}(-\omega; \omega)\rangle - \langle\alpha^*_{lk}(0; 0)\rangle\langle\alpha^*_{ij}(-\omega; \omega)\rangle)\}\\ &\quad + \left(\frac{1}{kT}\right)^2[\langle\mu^*_l\mu^*_k\alpha^*_{ij}(-\omega; \omega)\rangle - \langle\mu^*_k\mu^*_l\rangle\langle\alpha^*_{ij}(-\omega; \omega)\rangle\\ &\quad + 2\langle\mu^*_l\rangle\langle\mu^*_k\rangle\langle\alpha^*_{ij}(-\omega; \omega)\rangle - \mathscr{P}_{kl}\{\langle\mu^*_l\rangle\langle\mu^*_k\alpha^*_{ij}(-\omega; \omega)\rangle\}]\Bigg\rangle_0,\end{aligned} \quad (2.5.6)$$

where we have set $D_3 = 2$ owing to the degeneracy of the two zero-frequency fields.

With this general result in hand, we can now evaluate the electro-optic response in terms of model interaction potentials. The following section considers two such model potentials.

3. MODEL POTENTIALS

In this section, we substitute model potentials into the more general expressions for the quadratic electro-optic susceptibility. First, for simplicity, we assume that the molecule is one-dimensional, that is, the only nonzero tensor components of the dressed susceptibilities are $\mu^*_{z'}$, $\alpha^*_{z'z'}$, $\beta^*_{z'z'z'}$, $\gamma^*_{z'z'z'z'}$, and so on. Many long-conjugated organic molecules have been shown to behave as nearly one-dimensional: thermodynamic models of their nonlinear-optical susceptibility agree with experiments (particularly with regard to the tensor ratios of the nonlinear susceptibility) [32]. Secondly, we assume that the external electric field is applied in the laboratory 3-direction so that the bulk material has azimuthal symmetry in the 1–2 plane. Typical quadratic electro-optic modulation experiments use a thin-film sample in which the modulating field is applied perpendicular to the film plane. In such samples, the laboratory 3-axis is defined to be the film normal.

In these electro-optic experiments, the optical field propagates through the sample at some arbitrary angle to the film normal. There are therefore two possible independent polarizations: *s*-polarization, where the optical field is polarized in the film plane; and *p*-polarized, where the optical field polarization has both a component in the film plane and perpendicular to it. There are therefore two tensor components that are probed under these conditions, that is, $\chi^{(3)}_{1133}(-\omega; \omega, 0, 0)$, $\chi^{(3)}_{3333}(-\omega; \omega, 0, 0)$ and any permutation of these tensor subscripts. We therefore specifically evaluate these two components for the model potentials.

The two model potentials that we choose to evaluate are the harmonic restraint:

$$U_{\text{int}} = \tfrac{1}{2}k_\theta(\theta - \theta_0)^2, \tag{3.0.1}$$

where k_θ is the microscopic polymer elasticity and the restricted free rotator,

$$U_{\text{int}} = \begin{cases} 0 & \text{if } \theta_0 - \dfrac{\Delta\theta}{2} < \theta < \theta_0 + \dfrac{\Delta\theta}{2}, \\ \infty & \text{otherwise} \end{cases} \tag{3.0.2}$$

where $\Delta\theta$ is the limit of angular motion and θ_0 the average zero-field orientation. In the following subsections, we evaluate the quadratic susceptibility for the elastic restraint and the impeded rotator potentials.

3.1. Elastic Potential

In this subsection, we evaluate the third-order susceptibility as given by Eq. (2.5.6) for the elastic potential (Eq. 3.0.1). Instead of presenting these calculations in detail, we summarize the results below. Note that the thermodynamic averages of Eq. (2.5.6) for the elastic potential give rise to integrals of the form:

$$\int_0^\pi d\theta \sin^n\theta \cos^m\theta \exp\left[-\frac{pk_\theta}{kT}(\theta - \theta_0)^2\right], \tag{3.1.1}$$

where n and m are integers and p a rational number. An integral of this form cannot be solved in closed form. For integration limits that span from $-\infty$ to ∞; however, the integral can be evaluated exactly [33]. Our integrals from 0 to π can be approximated with the integral of infinite range when the microscopic elasticity, k_θ, is much greater than thermal energies, kT; i.e., $kT/k_\theta \ll 1$.

In the opposite limit, where the elasticity is small relative to thermal energies, the exponential can be approximated by unity and the integrals become trivial to evaluate. In this section, we evaluate only the high-elasticity limit. The low-elasticity limit is treated in the next section.

In the high-elasticity limit, the $\chi^{(3)}_{1133}$-component as calculated from Eq. (2.5.6) and (3.0.1) is:

$$\begin{aligned}
&\chi^{(3)}_{1133}(-\omega;\omega,0,0)\\
&\quad=\frac{N}{6}\Bigg[\tfrac{3}{16}\gamma^*(-\omega;\omega,0,0)\left(2+\tfrac{1}{3}\exp\left[-4\frac{kT}{k_\theta}\right]-\tfrac{1}{5}\exp\left[-12\frac{kT}{k_\theta}\right]\right)\\
&\qquad+\frac{1}{8kT}\beta^*(-\omega;\omega,0)\mu^*\left(2-\tfrac{8}{3}\exp\left[-3\frac{kT}{k_\theta}\right]+\tfrac{1}{3}\exp\left[-4\frac{kT}{k_\theta}\right]\right.\\
&\qquad\left.+\tfrac{8}{15}\exp\left[-9\frac{kT}{k_\theta}\right]-\tfrac{1}{5}\exp\left[-12\frac{kT}{k_\theta}\right]\right)\\
&\qquad-\frac{1}{16kT}\alpha^*(-\omega;\omega)\alpha^*(0;0)\left(1+\tfrac{1}{3}\exp\left[-4\frac{kT}{k_\theta}\right]\right.\\
&\qquad\left.-\tfrac{23}{15}\exp\left[-8\frac{kT}{k_\theta}\right]+\tfrac{1}{5}\exp\left[-12\frac{kT}{k_\theta}\right]\right)\\
&\qquad-\frac{1}{32}\left(\frac{1}{kT}\right)^2\mu^{*2}\alpha^*(-\omega;\omega)\left(1-\tfrac{8}{3}\exp\left[-3\frac{kT}{k_\theta}\right]+\tfrac{1}{3}\exp\left[-4\frac{kT}{k_\theta}\right]\right.\\
&\qquad+\tfrac{56}{15}\exp\left[-7\frac{kT}{k_\theta}\right]-\tfrac{23}{15}\exp\left[-8\frac{kT}{k_\theta}\right]-\tfrac{16}{15}\exp\left[-9\frac{kT}{k_\theta}\right]\\
&\qquad\left.+\tfrac{1}{5}\exp\left[-12\frac{kT}{k_\theta}\right]\right)\Bigg],
\end{aligned}\tag{3.1.2}$$

where we have set $D_3 = 2$ owing to the degeneracy of the two zero-frequency fields and where, owing to the one-dimensional approximation, we have dropped the tensor indices of the molecular polarizabilities; i.e., $\mu^* = \mu^*_{z'}$, $\alpha^* = \alpha^*_{z'z'}$, etc.

Similarly, the $\chi^{(3)}_{3333}$-component is:

$$\begin{aligned}
&\chi^{(3)}_{3333}(-\omega;\omega,0,0)\\
&\quad=\frac{N}{6}\Bigg[\tfrac{3}{8}\gamma^*(-\omega;\omega,0,0)\Bigg(2+\exp\left[-4\frac{kT}{k_\theta}\right]+\tfrac{1}{5}\exp\left[-12\frac{kT}{k_\theta}\right]\Bigg)\\
&\qquad+\frac{1}{4kT}\beta^*(-\omega;\omega,0)\mu^*\Bigg(2-\tfrac{8}{3}\exp\left[-3\frac{kT}{k_\theta}\right]+\exp\left[-4\frac{kT}{k_\theta}\right]\\
&\qquad-\tfrac{8}{15}\exp\left[-9\frac{kT}{k_\theta}\right]+\tfrac{1}{5}\exp\left[-12\frac{kT}{k_\theta}\right]\Bigg)\\
&\qquad+\frac{1}{8kT}\alpha^*(-\omega;\omega)\alpha^*(0;0)\Bigg(1+\tfrac{1}{3}\exp\left[-4\frac{kT}{k_\theta}\right]\\
&\qquad-\tfrac{23}{15}\exp\left[-8\frac{kT}{k_\theta}\right]+\tfrac{1}{5}\exp\left[-12\frac{kT}{k_\theta}\right]\Bigg)\\
&\qquad+\frac{1}{16}\left(\frac{1}{kT}\right)^2\mu^{*2}\alpha^*(-\omega;\omega)\Bigg(1-\tfrac{8}{3}\exp\left[-3\frac{kT}{k_\theta}\right]+\tfrac{1}{3}\exp\left[-4\frac{kT}{k_\theta}\right]\\
&\qquad+\tfrac{56}{15}\exp\left[-7\frac{kT}{k_\theta}\right]-\tfrac{23}{15}\exp\left[-8\frac{kT}{k_\theta}\right]-\tfrac{16}{15}\exp\left[-9\frac{kT}{k_\theta}\right]\\
&\qquad+\tfrac{1}{5}\exp\left[-12\frac{kT}{k_\theta}\right]\Bigg)\Bigg]. \qquad (3.1.3)
\end{aligned}$$

It is interesting to evaluate each term in these susceptibilities to lowest order in $1/k_\theta$. Keeping only the lowest-order nonvanishing term in the parentheses in Eqs. (3.1.2) and (3.1.3) yields:

$$\begin{aligned}
\chi^{(3)}_{1133}(-\omega;\omega,0,0)=\frac{N}{15}\Bigg[&\gamma^*(-\omega;\omega,0,0)+\frac{4}{3}\frac{\beta^*(-\omega;\omega,0)\mu^*}{k_\theta}\\
&-\frac{4}{3}\frac{\alpha^*(-\omega;\omega)\alpha^*(0;0)}{k_\theta}-\frac{1}{3}\frac{\mu^{*2}\alpha^*(-\omega,\omega)}{k_\theta^2}\Bigg], \qquad (3.1.4)
\end{aligned}$$

and

$$\begin{aligned}
\chi^{(3)}_{3333}(-\omega;\omega,0,0)=\frac{N}{15}\Bigg[&3\gamma^*(-\omega;\omega,0,0)+4\frac{\beta^*(-\omega;\omega,0)\mu^*}{k_\theta}\\
&+\frac{8}{3}\frac{\alpha^*(-\omega;\omega)\alpha^*(0;0)}{k_\theta}+\frac{2}{3}\frac{\mu^{*2}\alpha^*(-\omega,\omega)}{k_\theta^2}\Bigg]. \qquad (3.1.5)
\end{aligned}$$

We note that these results for the high elasticity limit (Eqs. 3.1.4 and 3.1.5) are in agreement with the results of Kuzyk and coworkers, who used a nonthermodynamic calculation [15]. In the previous work, however, the $[\mu^{*2}\alpha^*(-\omega;\omega)]/k_\theta^2$ term was dropped because it was considered a higher-order term.

3.2. Low Elasticity Limit

In the low elasticity limit ($k_\theta \ll kT$) Eqs. (2.5.6) and (3.0.1) give:

$$\chi^{(3)}_{1133}(-\omega;\omega,0,0) = \frac{N}{6}\left\{\left[6\gamma^*(-\omega;\omega,0,0) + \left(\frac{4}{kT}\right)\beta^*(-\omega;\omega,0)\mu^*\right]\right.$$
$$\times\left[\frac{1}{15} - \frac{11}{3375}\left(\frac{k_\theta}{kT}\right)\right]$$
$$+\frac{2\alpha^*(-\omega;\omega)}{45kT}\left[2\alpha^*(0;0) + \left(\frac{1}{kT}\right)\mu^{*2}\right]$$
$$\left.\times\left[-1 + \frac{46}{225} - \left(\frac{k_\theta}{kT}\right)\right]\right\} \tag{3.2.1}$$

and

$$\chi^{(3)}_{3333}(-\omega;\omega,0,0) = \frac{N}{6}\left\{\left[6\gamma^*(-\omega;\omega,0,0) + \left(\frac{4}{kT}\right)\beta^*(-\omega;\omega,0)\mu^*\right]\right.$$
$$\times\left[\frac{1}{5} - \frac{76}{1125}\left(\frac{k_\theta}{kT}\right)\right]$$
$$+\frac{4\alpha^*(-\omega;\omega)}{45kT}\left[2\alpha^*(0;0) + \left(\frac{1}{kT}\right)\mu^{*2}\right]$$
$$\left.\times\left[1 - \frac{46}{225}\left(\frac{k_\theta}{kT}\right)\right]\right\}. \tag{3.2.2}$$

3.3. Impeded Rotator

In the case of the impeded rotator and for Δ small, we get:

$$\chi^{(3)}_{1133}(-\omega;\omega,0,0) = \frac{N}{6}\left\{6\gamma^*(-\omega;\omega,0,0) \times \left(\tfrac{1}{15} + \tfrac{1}{90}\sin^2\Delta - \tfrac{1}{50}\sin^4\Delta\right)\right.$$
$$+ \left(\frac{4}{kT}\right)\beta^*(-\omega;\omega,0)\mu^* \times \left(\tfrac{2}{45}\sin^2\Delta + \tfrac{1}{75}\sin^4\Delta\right)$$
$$+ \left(\frac{2}{kT}\right)\alpha^*(-\omega;\omega)\alpha^*(0;0)$$
$$\times \left(-\tfrac{4}{45}\sin^2\Delta + \tfrac{44}{675}\sin^4\Delta\right)$$
$$\left. - \frac{16}{675}\left(\frac{1}{kT}\right)^2 \alpha^*(-\omega;\omega)\mu^{*2}\sin^4\Delta\right\} \tag{3.3.1}$$

and

$$\chi^{(3)}_{3333}(-\omega;\omega,0,0) = \frac{N}{6}\left\{6\gamma^*(-\omega;\omega,0,0) \times \left(\tfrac{1}{5} - \tfrac{2}{15}\sin^2\Delta + \tfrac{1}{25}\sin^4\Delta\right)\right.$$
$$+ \left(\frac{4}{kT}\right)\beta^*(-\omega;\omega,0)\mu^* \times \left(\tfrac{2}{15}\sin^2\Delta - \tfrac{2}{75}\sin^4\Delta\right)$$
$$+ \left(\frac{2}{kT}\right)\alpha^*(-\omega;\omega)\alpha^*(0;0)$$
$$\times \left(+\tfrac{8}{45}\sin^2\Delta - \tfrac{88}{675}\sin^4\Delta\right)$$
$$\left. + \frac{32}{675}\left(\frac{1}{kT}\right)^2 \alpha^*(-\omega;\omega)\mu^{*2}\sin^4\Delta\right\} \tag{3.3.2}$$

where $\Delta = \Delta\theta/2$.

4. ELECTRO-OPTIC MEASUREMENTS

While the early electrochromic (quadratic electroabsorption) studies of Havinga and Van Pelt were focused on determining transition dipole moments of guest molecules in polymers [20,21,22], their measurements were—for all practical purposes—third-order susceptibility measurements. Quadratic electro-optic effects have been recently used to measure the third-order nonlinear-optical susceptibility of polymers. The depolarization of light in a doped polymer in response to a static field was reported by

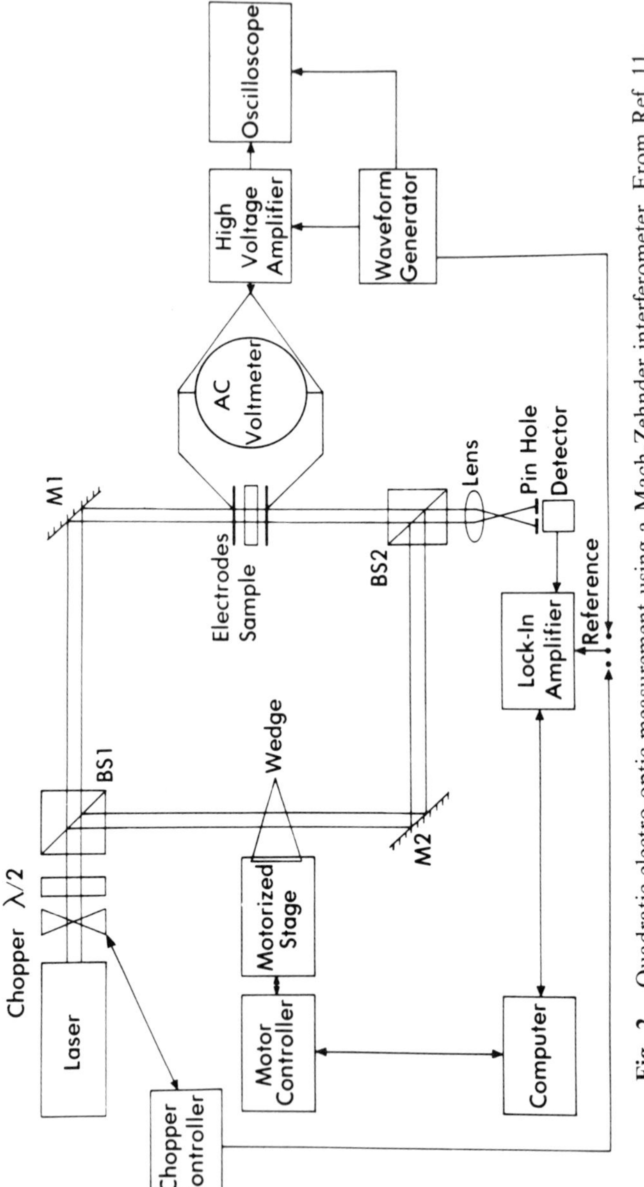

Fig. 2. Quadratic electro-optic measurement using a Mach-Zehnder interferometer. From Ref. 11.

Khanarian and coworkers [34]. Later, Uchichi and Kobayashi [35,36] reported using a Fabry-Perot interferometer to measure the frequency dependence of the third-order susceptibility. Other quadratic electro-optic measurements were reported by Luong and coworkers [37]. Dentan and coworkers used a prism-coupling technique that separated the contributions of the faster electronic and reorientational response from the slower electrostrictive effects [38]. More recently, Kuzyk and coworkers showed that a scanning Mach-Zehnder interferometer provided a quick and simple measure of the third-order nonlinear-optical susceptibility that could be simply related to the fast electronic response as measured with third harmonic generation [14]. These studies were used to identify guest molecules with large electronic nonlinear-optical susceptibilities.

In common to all electro-optic experiments is that an applied static field changes the propagation parameters of an optical field. Figures 2 and 3 show two such experiments. In the Mach-Zehnder interferometer (Fig. 2) the change in refractive index as induced by the static field results in a phase

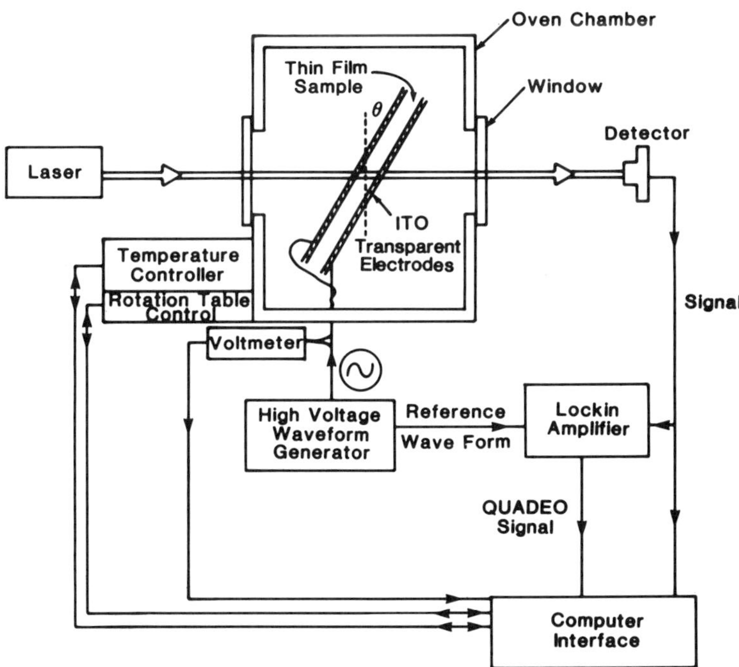

Fig. 3. Quadratic electroabsorption apparatus.

shift of the light in the sample arm of the interferometer. For thin-film samples, the static field is applied with transparent indium tin oxide (ITO) electrodes. Alternatively, the sample can replace one of the mirrors of the interferometer where the transparent electrode defines the front surface while the bottom electrode reflects the light. In both configurations, the induced phase change in the sample arm results in an intensity change at the detector.

The Mach-Zehnder interferometer measures a combination of the real and imaginary part of the third-order susceptibility. The electroabsorption experiment (Fig. 3), on the other hand, measures the imaginary part only. In this experiment, the applied static field results in a change of the material's optical absorption coefficient. Using both the quadratic electroabsorption and quadratic electro-optic measurements, the real and imaginary part of the third order susceptibility can be determined. Note that the electro-absorption experiment is identical to the Mach-Zehnder experiment with the reference arm blocked.

In typical thin-film quadratic electro-optic experiments, the induced intensity changes are small compared with the total light intensity. To increase the sensitivity, the static field is slowly modulated and a lock-in amplifier is synchronized at twice the modulation frequency to pick out the quadratic response from background. The strength of this modulation is then related to the sample's nonlinear response. The derivation of this relationship is as follows.

In the Mach-Zehnder interferometer, the first beam splitter separates the incoming beam into the two arms. Let us define the electric fields at the combining beam splitter in the reference arm as $\vec{E}_1$ and the field in the sample arm as $\vec{E}_2$ with incoming field amplitudes E_{01} and E_{02} respectively. After passing through the sample, the field E_2 is given by

$$E_2 = E_{02} \exp\left[i\left(\phi_2 + \frac{2\pi d(n-1)}{\lambda} \right) \right], \tag{4.0.1}$$

where λ is the wavelength, d the sample thickness, ϕ_2 the phase of the light if the sample were removed, and n the refractive index of the sample. Note that ϕ_2 is real while n is complex. Similarly, the light in the reference arm is given by

$$E_1 = E_{01} \exp[i\phi_1], \tag{4.0.2}$$

where the phase, ϕ_1, is adjusted by a glass slide or wedge that is placed in the reference arm. Note that ϕ_1 is assumed real.

The refractive index of the sample, n, is a function of the applied static field, $\vec{E}^0$ [15]:

$$n_{ij} = (n_0)_{ij} + \frac{(n_0)_{ij}^3 s_{ijkl}}{2} E_k^0 E_l^0, \tag{4.0.3}$$

where the quadratic electro-optic tensor, s, is related to the third-order susceptibility tensor, $\chi^{(3)}$, and where n_0 is the zero-field refractive index. Because the electric field is applied in the laboratory 3-direction, and the optical field polarization is either in the 1-3 or 2-3 plane, Eq. (4.0.3) reduces to:

$$n_{ii} = (n_0)_{ii} + \frac{(n_0)_{ii}^3 s_{ii33}}{2} E_3^0 E_3^0, \tag{4.0.4}$$

where the refractive index tensor and the quadratic electro-optic tensor are both complex, i.e., $n = n_R + in_I$, $n_0 = n_{0R} + in_{0I}$, and $s = s_R + is_I$.

The quasi-static electric field, E, results from a voltage, V, that is slowly modulated at frequency Ω:

$$E = 2^{1/2} \frac{V_{\text{rms}}}{d} \cos(\Omega t), \tag{4.0.5}$$

where d is the sample thickness and V_{rms} the root-mean-square voltage. Because the samples are assumed isotropic, the zero-field refractive index tensor is also isotropic. Defining

$$A_i = \frac{\pi}{\lambda} (n_0^3 s_{ii33})_R \frac{V_{\text{rms}}^2}{d} \tag{4.0.6}$$

and

$$B_i = \frac{\pi}{\lambda} (n_0^3 s_{ii33})_I \frac{V_{\text{rms}}^2}{d}, \tag{4.0.7}$$

and provided that the two light beams remain similarly polarized after being recombined at the second beam splitter, the total light intensity at the detector, I_i, for i-polarized light at the sample is

$$\begin{aligned} I_i &= |E_1 + E_2|^2 \\ &= E_{01}^2 + E_{02}^2 \exp\left[-4\pi n_{0I} \frac{d}{\lambda}\right] \exp[-4B_i \cos^2(\Omega t)] \\ &\quad + 2E_{01}E_{02} \exp\left[-2\pi n_{0I} \frac{d}{\lambda}\right] \exp[-2B_i \cos^2(\Omega t)] \\ &\quad \times \cos(\phi + 2A_i \cos^2(\Omega t)), \end{aligned} \tag{4.0.8}$$

where ϕ is the zero-field phase difference between the two arms of the interferometer.

The electro-optic measurement is performed in two consecutive steps. First, the absolute intensity and contrast of the interferometer is determined from a zero-field phase sweep: The zero-field intensity at the detector varies sinusoidally as a function of the phase difference ϕ. Defining I_{max} as the maximum intensity at the peak and I_{min} the intensity at the troughs, the difference between these two values yields:

$$I_{max} - I_{min} = 4E_{01}E_{02} \exp\left[-2\pi n_{0I} \frac{d}{\lambda}\right]. \tag{4.0.9}$$

To get a clean signal, the light source is modulated with a chopper and the signal voltage from the detector measured with a lock-in amplifier that is synchronized with the chopper frequency. The lock-in amplifier measures the root-mean-square intensity $\langle I \rangle$. Because the peak-to-peak amplitude is I, $\langle I \rangle = I/2^{3/2}$ so that:

$$\begin{aligned} I_{max} - I_{min} &= 2^{3/2}(\langle I_{max} \rangle - \langle I_{min} \rangle) \\ &= 4E_{01}E_{02} \exp\left[-2\pi n_{0I} \frac{d}{\lambda}\right]. \end{aligned} \tag{4.0.10}$$

The upper curve in Fig. 4 shows the theoretical zero-field trace. Note that when I_{min} is small, $\langle I_{max} \rangle - \langle I_{min} \rangle \approx I_0$, the zero-field peak value as shown in Fig. 4.

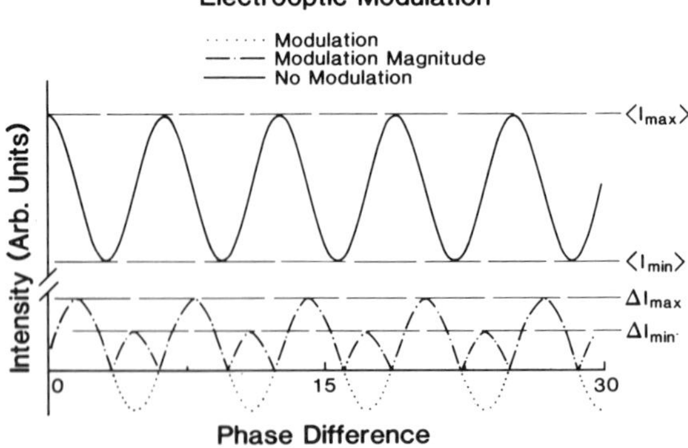

Fig. 4. Theoretical plot of interferometer intensity (top curve) and intensity-modulated amplitude at 2ω (bottom) as a function of phase difference.

Next, a voltage at frequency Ω is applied to the sample, and the lock-in amplifier is synchronized to twice the modulation frequency. We assume that the field-induced refractive index change is small so that the modulated intensity of Eq. (4.0.8) can be expanded in a power series in the applied field. The lowest-order term with Fourier component at 2Ω is proportional to the square of the applied field. Using (4.0.10) and the definition of I_0, the root-mean-square intensity at 2Ω of Eq. (4.0.8) is:

$$\langle I_{2\Omega}^i \rangle = \left| 2^{1/2} B_i E_{02}^2 \exp\left[-4\pi n_{0I} \frac{d}{\lambda} \right] + I_0[(1 - B_i)A_i \sin\phi + B_i \cos\phi - A_i B_i \sin\phi] \right|, \quad (4.0.11)$$

where summation convention does not apply to $A_i B_i$. Note that the absolute value around Eq. (4.0.11) was added to reflect that the lock-in amplifier measures the magnitude of the signal.

The second term in Eq. (4.0.11) varies sinusoidally with phase difference, ϕ, while the first term results in an overall offset of the curve. The theoretical dotted curve in Fig. 4 shows the signal strength while the dotted-dashed curve shows its absolute value as calculated from Eq. (4.0.11). The offset term is responsible, then, for the bimodal peak amplitude. By definition of an extrema, the measured signal at 2Ω is then maximum when the slope of a tangent to the curve vanishes. The phase corresponding to these peaks can thus be calculated from (4.0.9): $\partial\langle I_{2\Omega}^i \rangle/\partial\phi = 0$. The higher peak intensity, ΔI_{max}, and the lower peak intensity, ΔI_{min}, is evaluated by substituting this phase back into Eq. (4.0.11).

If the offset term in Eq. (4.0.11) (see Fig. 4) is smaller than the amplitude, then I_{sig}, the amplitude, and δ, the offset, can be calculated from Eq. (4.0.11) according to:

$$I_{sig} = \frac{\Delta I_{max} + \Delta I_{min}}{2} \quad (4.0.12)$$

and

$$\delta = \frac{\Delta I_{max} - \Delta I_{min}}{2}. \quad (4.0.13)$$

Using Eqs. (4.0.11) through (4.0.13) we get:

$$I_{sig}^i = I_0(A_i^2(2B_i - 1)^2 + B_i^2)^{1/2} \quad (4.0.14)$$

and

$$\delta^i = 2^{1/2} B_i |E_{02}|^2 \exp\left[-4\pi n_{0I} \frac{d}{\lambda} \right]. \quad (4.0.15)$$

Equations (4.0.14) and (4.0.15) are expressed in terms of measured quantities except for the field amplitude, E_{02}, in the sample arm and the parameters A_i and B_i, which depend on the real and imaginary part of the electro-optic coefficient, respectively. If the field amplitude is determined, Eqs. (4.0.14) and (4.0.15) can be solved for A_i and B_i, thus yielding the real and imaginary part of s_{ii33}.

To measure E_{02}, we simply block the reference arm of the interferometer and measure the intensity of light at the detector. This is equivalent to the electroabsorption configuration. With $E_{01} = 0$ in Eq. (4.0.8), the zero-field root-mean-square intensity $\langle I_i^{EA} \rangle$ is:

$$2^{3/2}\langle I_i^{EA} \rangle = E_{02}^2 \exp\left[-4\pi n_{0I} \frac{d}{\lambda}\right]. \tag{4.0.16}$$

Substituting Eq. (4.0.16) into (4.0.15) yields:

$$\delta^i = 4B_i \langle I_i^{EA} \rangle. \tag{4.0.17}$$

Substituting Eq. (4.0.7) into Eq. (4.0.17) and solving for the imaginary part of the quadratic electro-optic coefficient yields:

$$|(n_0^3 s_{ii33})_I| = \frac{\lambda d}{4\pi V_{\text{rms}}^2} \frac{\delta^i}{\langle I_i^{EA} \rangle}. \tag{4.0.18}$$

Equation (4.0.18) represents the electroabsorption measurement and relates the imaginary part of the quadratic electro-optic tensor to experimentally determined parameters. The electro-optic experiment as quantified by Eq. (4.0.14) measures a combination of both the real and imaginary parts of the susceptibility. When far from resonances, the imaginary part is small, and the interferometer measurement can be used to determine the real part of the susceptibility. When close to resonance both the electroabsorption and electro-optic measurements need to be performed to determine the real part of s_{ii33}. Thus, combining Eqs. (4.0.14) and (4.0.17) with the help of (4.0.6) and solving for $(n_0^3 s_{ii33})_R$ yields:

$$|(n_0^3 s_{ii33})_R| = \frac{\lambda d}{4\pi V_{\text{rms}}^2} \left[\left(\frac{4I_{\text{sig}}^i}{I_0} \right)^2 - \left(\frac{\delta^i}{\langle I_i^{EA} \rangle} \right)^2 \right]^{1/2}. \tag{4.0.19}$$

Equations (4.0.18) and (4.0.19) can be solved for both the real and imaginary part of s_{ii33} provided that the complex refractive index is known:

$$\text{Re}[s_{ii33}] = \frac{1}{2(n_R^3 - 3n_R n_I^2)} [(n_0^3 s_{ii33})_R - (n_0^3 s_{ii33})_I] \tag{4.0.20}$$

and $$\mathrm{Im}[s_{ii33}] = \frac{1}{2(n_I^3 - 3n_I n_R^2)}[(n_0^3 s_{ii33})_R + (n_0^3 s_{ii33})_I]. \tag{4.0.21}$$

5. DISCUSSION

It is possible to study a wide variety of material properties and nonlinear-optical processes by relating the generalized response function to data from electro-optic characterization techniques. We briefly summarize several applications below.

As evidenced by the models presented above, many forms of molecular reorientation can contribute to the quadratic electro-optic response. For fast device applications, only the purely electronic response (as given by γ^*) is of interest. In contrast, studies that probe the microscopic rheological properties of a polymer matrix rely on the slower reorientational contributions of probe dye molecules. In both types of measurement, it is important to be able to separate these different terms.

Early quadratic electro-optic measurements of freely rotating gases found that the reorientational contributions dominate, making it difficult to deduce the purely electronic response [39]. Similar orientational contributions are found to dominate in liquid solution measurements, where the solute molecules are nearly freely to rotate [40]. Havinga and Van Pelt, however, recognized that chromophore motion is greatly reduced in solid polymer solution [20,21,22]. While the motion was believed to be frozen in place, recent second harmonic generation measurements by Kuzyk and coworkers showed that the orientational contribution was substantial [16].

Figure 5 shows some common dyes that have been measured with quadratic electro-optic modulation techniques. As found in the literature, Table 1 shows typical susceptibility values and elasticities for the disperse red azo dye DR1 in poly(methyl methacrylate) (PMMA) polymer [14,15,16].

Table 1.

Approximate Susceptibilities and Elasticities of DR1/PMMA

μ^* (esu·cm)	α^* (cm^3)	β^* (cm^5/esu)	γ^* (cm^7/esu^2)	k_θ (esu^2/cm)
14×10^{-18}	9×10^{-23}	5×10^{-28}	1×10^{-32}	6×10^{-13}

DRI

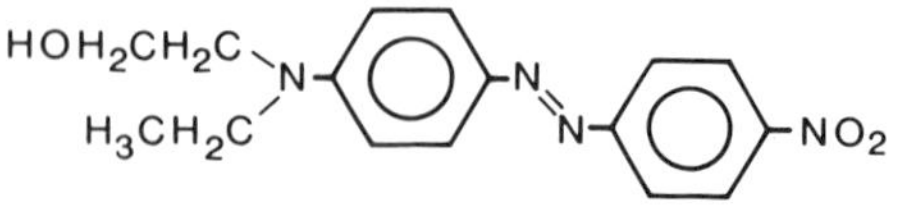

ISQ

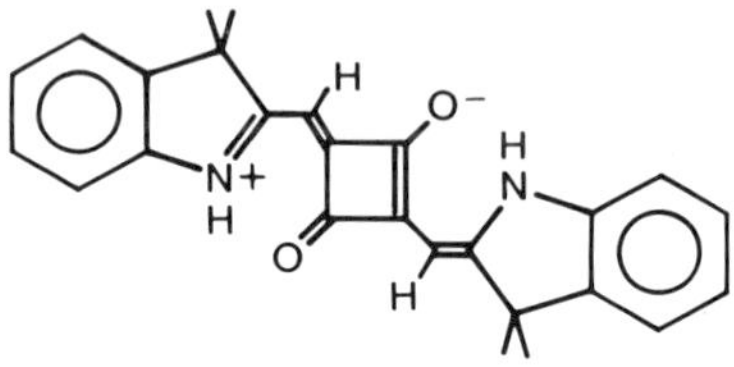

BSQ

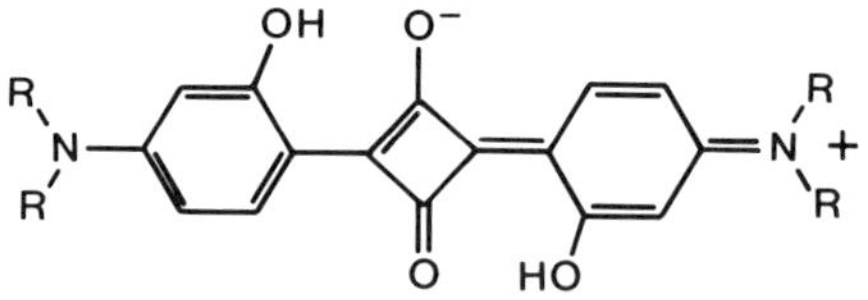

Fig. 5. Some common dyes with large quadratic electro-optic coefficients.

Note that these parameters depend strongly on the wavelength of the probing laser, the modulating frequency, and the host polymer matrix and are only approximate. Table 2 shows the calculated relative contributions to the quadratic electro-optic coefficient in the high elasticity limit (Eqs. 3.1.4 and 3.1.5) and Table 3 the calculated low elasticity limit (Eqs. 3.2.1 and 3.2.2). In the low elasticity limit, the orientational terms dominate while in the high elasticity limit of room temperature PMMA, the reorientational terms are comparable to the electronic ones.

Table 2.

High-Elasticity Terms for DR1/PMMA

$\frac{\beta^*\mu^*}{k_\theta}$ (10^{-32} cm^7/esu)	$\frac{(\alpha^*)^2}{k_\theta}$ (10^{-32} cm^7/esu)	$\frac{(\mu^*)^2}{k_\theta^2}$ (10^{-32} cm^7/esu)	γ^* (10^{-32} cm^7/esu^2)
1.2	1.4	5.2	1.1

Table 3.

Low-Elasticity Terms for Room-Temperature DR1/PMMA

$\frac{\beta^*\mu^*}{kT}$ $(10^{-32}\ cm^7/esu)$	$\frac{(\alpha^*)^2}{kT}$ $(10^{-32}\ cm^7/esu)$	$\frac{(\mu^*)^2}{(kT)^2}$ $(10^{-32}\ cm^7/esu)$	γ^* $(10^{-32}\ cm^7/esu)$
18	22	78	1.1

In an effort to deduce the electronic contribution of DR1 dye in PMMA, Kuzyk and coworkers calculated the reorientational response from the microscopic elasticity (as measured with second harmonic generation) and the previously measured dipole moment, polarizability, and hyperpolarizability [14,15]. The reorientational effects were then subtracted from the measured off-resonant quadratic electro-optic response. The electronic third-order susceptibility so determined with the Mach-Zehnder apparatus was found to be consistent with third-harmonic generation measurements [41]. The quadratic electro-optic measurement technique is a simple alternative to the more difficult third-harmonic-generation measurements and can be used to screen molecules for their third-order nonlinear optical properties [15].

When the optical field is tuned closer to resonance, the imaginary part becomes larger. Figure 6 shows the absorption spectrum of a thin film of DR1 dye in PMMA and DB339 dye in PMMA. Figures 7 and 8 show the phase-scanned intensity for the DR1 and DB339 samples, respectively, as probed with a HeNe laser operating at $\lambda = 633nm$. Note that the offset in the DB339 data is much larger than that of the DR1 owing to the larger contribution of the imaginary part of the third-order susceptibility.

Quadratic electro-optic modulation can also be used as a powerful spectroscopic tool to study the underlying physics of the nonlinear-optical response. As an example, there has been recent interest in developing an understanding that links the nonlinear-optical response of a molecule to features of its structure [1]. The general starting point for developing such an understanding is a quantum perturbative approach, which is used to calculate the polarization of a molecule in the presence of an optical field. The expression for the polarization was first reported in detail by Ward [42] and later corrected for divergent behavior by Orr and Ward [43]. These complicated expressions depend on all the transition dipole moments and energy levels of the molecule and require precise information about the

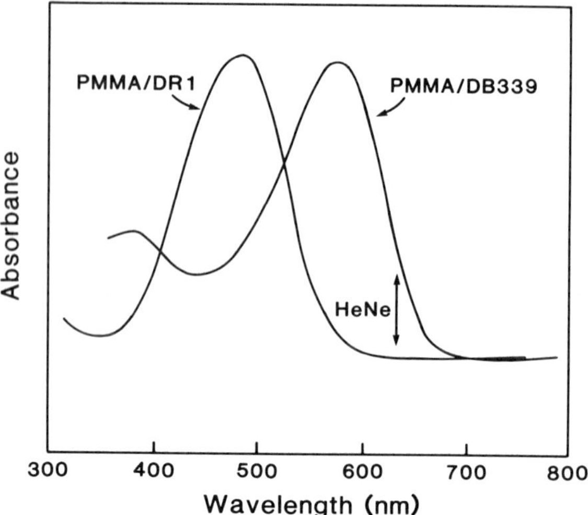

Fig. 6. Absorption spectrum of thin film DR1 dye and DB339 dye in a PMMA polymer.

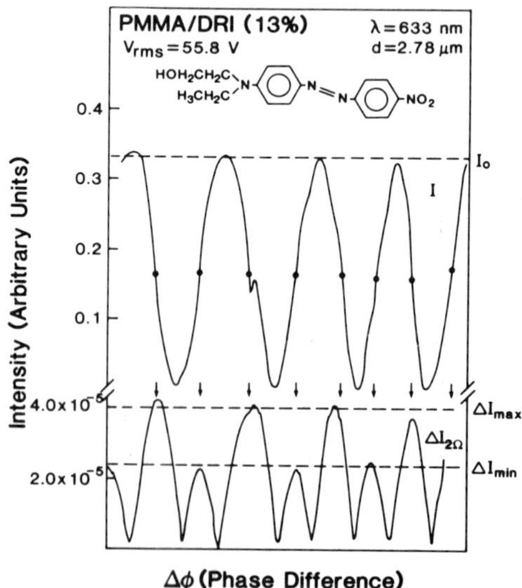

Fig. 7. Phase sweep for DR1/PMMA thin film modulation (bottom) and interferometer intensity (top). Arrows show where modulation peaks are expected. Solid circles are points of maximum slope.

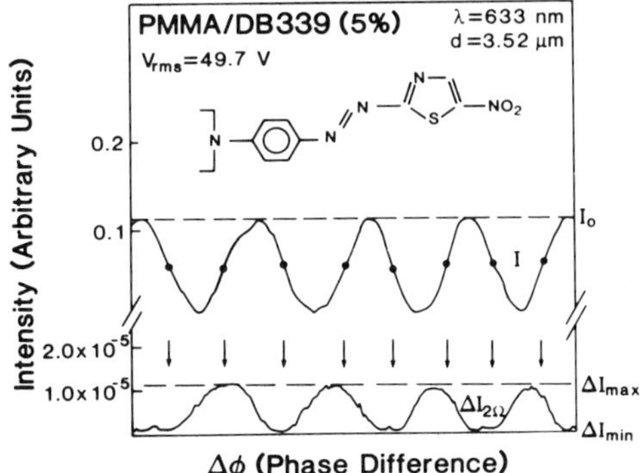

Fig. 8. Phase sweep for DB339/PMMA thin film modulation (bottom) and interferometer intensity (top). Arrows show where modulation peaks are expected. Solid circles are points of maximum slope.

molecular wave functions. Even with exact wave functions in hand, it is not clear from these complicated expressions just what causes a large third-order susceptibility.

In the case of the second-order nonlinear-optical susceptibility, a two-level model was shown by Oudar [44] and later by Lalama and Garito [45] to account for most of the response. Using the two-level model, chemists have designed molecules with enhanced second-order susceptibilities. For the third-order susceptibility, however, the response is more complex. Dirk and Kuzyk showed that many states contribute to the third-order susceptibility [46]. For certain molecules such as octatetraene, Heflin [47] and Soos [48] showed that three levels account for every 80 percent of the total response. With such multilevel systems, it is difficult to know *a priori* what molecules will have the largest third-order susceptibility without either calculating the wave functions or performing measurements.

Recently, Kuzyk and Dirk suggested that a strongly colored two-level centrosymmetric molecule should have the largest third-order response [49]. A highly colored centrosymmetric squarylium dye (ISQ) was identified as having one of the largest third-order susceptibilities [50]. While the linear absorption spectrum shows a two-level behavior (Fig. 9), it is not clear whether one or more two-photon states contributes to the third-order susceptibility while being undetected with linear absorption spectra. The wavelength dependence of the electronic quadratic electro-optic response is calculated and, compared with experiment and the qualitative agreement, confirms that the molecule behaves as a two-level system (see Fig. 9) [15].

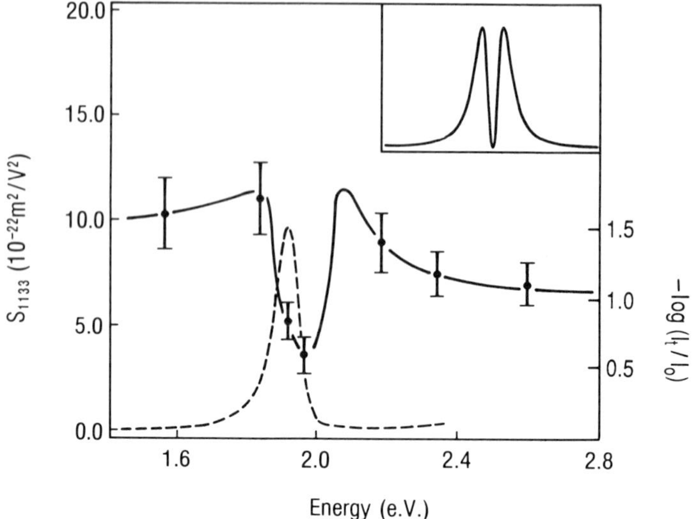

Fig. 9. Linear absorption spectrum and the real part of the third-order nonlinear-optical spectra of the ISQ Squarylium dye. The inset shows the theoretical dispersion of the third-order susceptibility for a two-level model. From Ref. 15.

There are, however, some features in the measured quadratic electro-optic (QEO) spectra that are not accounted for by the two-level model. For example, the QEO susceptibility is asymmetric about the central dip. Furthermore, the dip does not fall off to zero near the molecule's first excited-state transition energy as predicted. These discrepancies were attributed to contributions from higher-lying electronic states and to the orientational contribution [15]. The analysis in Sec. 4 can be used to shed some light on these features as follows.

In the interferometric measurements, the imaginary part of the QEO susceptibility was ignored, and the data were analyzed by setting $\delta_i = 0$ in Eq. (4.0.19). Note that δ_i gets large close to the electronic resonance, resulting in a smaller QEO susceptibility than was calculated in the previous studies. When corrected for the imaginary contribution, the dip becomes deeper. Figure 10 shows the imaginary part of the QEO coefficient as measured with the electroabsorption apparatus (Fig. 3). Note the strong resonance enhancement in the QEO susceptibility at the wavelength of maximum linear absorption. The theoretical curve is derived for a two-level model using the wavelength of maximum absorption and peak width from the linear absorption data. The only adjustable parameter in the model response is the amplitude of the peak. While the shoulder in the linear absorption spectrum

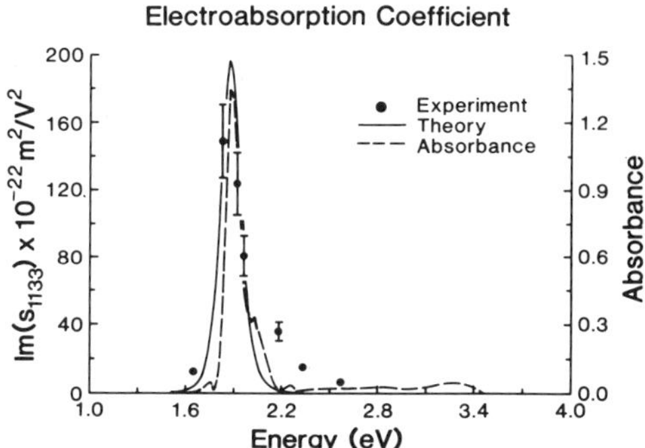

Fig. 10. Linear absorption spectrum and the imaginary part of the third-order nonlinear-optical spectra of the ISQ Squarylium dye. The solid curve shows the theoretical dispersion of the third-order susceptibility for a two-level model.

suggests that another nearby state may contribute, the two-level model is in good agreement except for the shoulder region.

Both the electroabsorption and electro-optic measurements suggest that the response of the squarylium dye (Fig. 5) in PMMA polymer is dominated by the electronic mechanism. Other electro-optic measurements can be used to further test this hypothesis. The even-order susceptibilities vanish ($\mu^* = \beta^* = 0$) for centrosymmetric molecules such as ISQ, and only two terms are nonzero in the high elasticity limit of nonlinear-optical response (Eqs. 3.1.4 and 3.1.5). If the electronic response dominates, the ratio $\chi^{(3)}_{3333}/\chi^{(3)}_{1133} = 3$, while in the limit where the reorientational contribution dominates, $\chi^{(3)}_{3333}/\chi^{(3)}_{1133} = -2$. It is therefore possible to deduce which terms dominate with the *p*- and *s*-polarized angular dependent measurements, where the sample is rotated about an axis in the film plane that is parallel to the beam (see Fig. 3).

In such experiments, the measured electro-optic coefficient for *p*-polarized light, s_P, for a sample whose film normal makes an angle θ with the beam propagation direction is given by

$$s_P(\theta) = s_{1133} \cos^2 \theta + s_{3333} \sin^2 \theta. \tag{5.0.1}$$

The *s*-polarization measurement, on the other hand, is independent of angle and directly probes s_{1133}. The angular dependence of the measured response is thus sensitive to the magnitude and sign of s_{1133} and s_{3333}.

Temperature-dependent measurements can be used to augment these tensor ratio measurements. Above the glass transition temperature of the polymer, where the molecules are nearly free to rotate, Eqs. (3.2.1) and (3.2.2) describe the nonlinear-optical response. At the elevated temperature, the tensor ratio is unchanged at $\chi^{(3)}_{3333}/\chi^{(3)}_{1133} = 3$, if the electronic term dominates and $\chi^{(3)}_{3333}/\chi^{(3)}_{1133} = -2$ when the reorientational term dominates. A temperature/angular dependence of the QEO susceptibility, then, can be used to unambiguously determine the dominant mechanism.

Temperature dependent measurements can also be used to study which models more accurately describe the polymer restraint. When the polymer sample is cooled to low temperatures, the microscopic elasticity decreases. If the cooling rate is fast, the free volume will be frozen in place. The free-volume model, then, predicts that the molecules will reorient more easily at lower temperatures, while the elasticity model predicts that reorientation is strongly inhibited as the polymer stiffens. Such low-temperature studies have not been investigated to date.

The modulation frequency dependence of the QEO susceptibility can also be used to differentiate between these two models. If the polymer elasticity is responsible for the constraint, the orientational motion will be enhanced at the natural resonant frequency of the polymer which, based on static field-induced second harmonic measurements of the polymer elasticity, is in the 30 GHz range [16]. Given the frequency dependence of the elasticity, the resonant frequency may be as far out as 300 GHz. The free volume model predicts no such resonances in the frequency dependence.

The usefulness of quadratic electro-optic spectroscopy can be extended by considering a more reasonable model of the polymer, especially in the high temperature regime (near the glass transition). To more accurately describe the polymer's role in the nonlinear optical response, the response models will need to be generalized to include the distribution of dopant sites as well as the temperature dependence of such a distribution. Studies that measure temperature-dependent second-harmonic generation in doped polymers [51] have applied distribution of site models to characterize the polymer restraint and have successfully predicted the decay of orientational order in a poled polymer below its glass transition temperature [52]. Unfortunately, the observed behavior is nearly model-independent and is not sensitive to large variation in the phenomenological parameters. Nevertheless, a combination of a variety of nonlinear-optical experiments and models will no doubt provide a fruitful path to understanding guest-host polymers.

6. CONCLUSION

It is clear that the quadratic electro-optic effect can be used as a powerful tool for studying the nonlinear-optical and material properties of thin film polymeric materials. By measuring the full parameter space of modeling of modulating frequency, wavelength, temperature, and sample orientation such diverse properties as the electronic nonlinear optical susceptibility, orientational motion, microscopic rheology, and structure-property relationships can be determined. Complemented with other nonlinear optical characterization techniques, such as electric-field-induced second-harmonic generation and third-harmonic generation, measurements of the generalized response function can both help elucidate the microscopic properties of polymers as well as aid in the design of better nonlinear-optical materials.

ACKNOWLEDGMENTS

This work was supported in part by The Donors of the Petroleum Research Fund, administered by the American Chemical Society, and by funds provided by Washington State University. We thank Dr. Carl Dirk for useful discussions and materials, Mike Holstine for the linear absorption measurements, and Fassil Ghebremichael for critical reading of this manuscript.

REFERENCES

1. See, for example, "Materials for Nonlinear Optics," American Chemical Society, ACS Symposium Series, vol. 455, S. R. Marder, J. E. Sohn, and G. D. Stucky, eds. (ACS, Washington, DC, 1991).
2. See, for example, "Nonlinear Optical Properties of Polymers," A. J. Heeger, J. Orenstein, and D. R. Ulrich, eds. (Materials Research Society, Symposium Proceedings, vol. 109, Pittsburgh, 1988).
3. See, for example, "Nonlinear Optical Effects in Organic Polymers," NATO ASI Series, Series E: Applied Sciences, vol. 162, J. Messier, F. Kajzar, P. Prasad, and D. Ulrich, eds. (Kluwer Academic Publishers, Dordrecht, 1989).
4. K. D. Singer, J. E. Sohn, and S. J. Lalama, *Appl. Phys. Lett.* **49**, 248 (1986).
5. M. G. Kuzyk, U. C. Paek, and C. W. Dirk, *Appl. Phys. Lett.* **59**, 902 (1991).
6. T. Kaino, *Japanese J. of Appl. Phys.* **23**, 1661 (1985).
7. T. Kaino, K. Jinguji, and S. Nara, *Appl. Phys. Lett.* **42**, 567 (1983).
8. Y. Koike, N. Tanio, and Y. Ohtsuka, *Macromol.* **22**, 1357 (1989); C. Emslie, *J. Mat. Res. Sci.* **23**, 2281 (1988).
9. K. D. Singer, W. R. Holland, M. G. Kuzyk, G. L. Wolk, and P. A. Cahill, "Guest-Host Polymers for Nonlinear Optics," *Mol. Cryst. Liq. Cryst.* **189**, 123 (1990).

10. J. I. Thackara, G. F. Lipscomb, M. A. Stiller, A. J. Ticknor, and R. Lytel, *Appl. Phys. Lett.* **52**, 1031 (1988).
11. T. E. Van Eck, A. J. Ticknor, R. S. Lytel, and G. F. Lipscomb, *Appl. Phys. Lett.* **58**, 1588 (1991).
12. K. D. Singer, M. G. Kuzyk, R. B. Comizzoli, H. E. Katz, M. L. Schilling, J. E. Sohn, and S. J. Lalama, *Appl. Phys. Lett.* **53**, 1800 (1988).
13. H. L. Hampsch, J. Yang, G. K. Wong, and J. M. Torkelson, *Macromolecules* **21**, 526 (1988).
14. M. G. Kuzyk and C. W. Dirk, *Appl. Phys. Lett.* **54**, 1628 (1989).
15. M. G. Kuzyk, J. E. Sohn, and C. W. Dirk, *J. Opt. Soc. Am. B* **5**, 842 (1990).
16. M. G. Kuzyk, R. C. Moore, and L. A. King, *J. Opt. Soc. Am. B* **7**, 64 (1990).
17. J. Kerr, *Phil. Mag.* **4**, 337 (1875).
18. W. Liptay, *Angew. Chem. Internat. Edit.* **8**, 177 (1969).
19. K. Yamaoka and Elliot Charney, *J. Am. Chem. Soc.* **98**, 8963 (1972).
20. E. E. Havinga and P. van Pelt, *Ber. Bunsenges. Phys. Chem.* **83**, 816 (1979).
21. E. E. Havinga and P. van Pelt, *Mol. Cryst. Liq. Cryst.* **52**, 145 (1979).
22. E. E. Havinga and P. van Pelt, "Electro-optic Dielectric Macromolecular Colloids," B. R. Jennings, ed. (Plenum, New York, 1979).
23. K. D. Singer, M. G. Kuzyk, and J. E. Sohn, *J. Opt. Soc. Am. B* **4**, 968 (1987).
24. M. A. Mortazavi, A. Knoeson, S. T. Kowel, B. G. Higgens, and A. Dienes, *J. Opt. Soc. Am. B* **6**, 733 (1989).
25. R. H. Page, M. C. Jurich, B. Reck, A. Sen, R. J. Twieg, J. D. Swalen, G. C. Bjorklund, and C. G. Willson, *J. Opt. Soc. Am. B* **7**, 1239 (1990).
26. G. T. Boyd, C. V. Francis, J. E. Trend, and D. A. Ender, *J. Opt. Soc. Am. B* **8**, 887 (1991).
27. See, for example, Y. R. Shen, "Principles of Nonlinear Optics," (Wiley Interscience, New York, 1984).
28. P. N. Butcher and D. Cotter, "The Elements of Nonlinear Optics" (Cambridge Studies on Modern Optics 9, Cambridge University Press, Cambridge, 1990).
29. H. Ringsdorf, H. W. Schmidt, G. Baur, R. Kiefer, and W. Windsheid, *Liq. Cryst. (G.B.)* **1**, 4 (1986).
30. C. J. F. Bottcher and P. Bordewijk, *Theory of Electric Polarization*, vol. 2 (Elsevier, New York, 1980), p. 315.
31. I. P. Kaminow, *An Introduction to Electrooptic Devices* (Academic Press, New York, 1974).
32. M. G. Kuzyk, K. D. Singer, H. E. Zahn, and L. A. King, *J. Opt. Soc. Am. B* **6**, 742 (1989).
33. I. S. Gradshteyn and I. M. Ryzhik, *Tables of Integrals, Series, and Products* (Academic Press, New York, 1980), p. 480.
34. G. Khanarian, A. Artigliere, R. Keosian, E. W. Choe, R. DeMartino, D. Stuetz, and C. C. Teng, in "Molecular and Polymeric Optoelectronic Materials," G. Khanarian, ed., *Proc. Soc. Photo-Opt. Instrum. Eng.* **682**, 153 (1986).

35. H. Uchichi and T. Kobayashi, in "Nonlinear Optical Properties of Polymers," A. J. Heeger, J. Orenstein, and D. R. Ulrich, eds. (Materials Research Society, Symposium Proceedings, vol. 109, Pittsburgh, 1988), p. 373.
36. H. Uchichi and T. Kobayashi, *J. Appl. Phys.* **64**, 2625 (1988).
37. J. C. Luong, N. F. Borelli, and A. R. Olszewski, in "Nonlinear Optical Properties of Polymers," A. J. Heeger, J. Orenstein, and D. R. Ulrich, eds. (Materials Research Society, Symposium Proceedings, vol. 109, Pittsburgh, 1988), p. 251.
38. V. Dentan, Y. Levy, M. Dumont, P. Robin, and E. Chasting, *Opt. Commun.* **69**, 379 (1989).
39. N. J. Bridge and A. D. Buckingham, *Proc. Roy. Soc. A* **295**, 334 (1966).
40. J. V. Champion, G. H. Meeten, and C. D. Whittle, *Trans. Faraday Soc.* **66**, 2671 (1970).
41. S. Matsumoto, K. Kubodera, T. Kurihara, and T. Kaino, *Appl. Phys. Lett.* **51**, 1 (1987).
42. J. F. Ward, *Rev. Mod. Phys.* **37**, 1 (1965).
43. B. J. Orr and J. F. Ward, *Mol. Phys.* **20**, 513 (1971).
44. J. L. Oudar, D. S. Chemla, and E. Batifol, *J. Chem. Phys.* **67**, 1626 (1977).
45. S. J. Lalama and A. F. Garito, *Phys. Rev. A* **20**, 1179 (1979).
46. C. W. Dirk and M. G. Kuzyk, *Phys. Rev. A* **39**, 1219 (1989).
47. J. R. Heflin, K. Y. Wong, O. Zamani-Khamiri, and A. F. Garito, *Phys. Rev. B* **38**, 1573 (1988).
48. Z. G. Soos and S. Ramasesha, *J. Chem. Phys.* **90**, 1067 (1989).
49. M. G. Kuzyk and C. W. Dirk, *Phys. Rev. A* **41**, 5098 (1990).
50. C. W. Dirk and M. G. Kuzyk, *Chem. of Materials* **2**, 5 (1990).
51. W. Kohler, D. R. Robello, P. T. Dao, C. S. Willand, and D. J. Williams, *J. Chem. Phys.* **93**, 9157 (1990).
52. K. D. Singer and L. A. King, *J. Appl. Phys.* **70**, 3521 (1991).

Chapter 8

POLYMER PHYSICS OF POLED POLYMERS FOR SECOND-ORDER NONLINEAR OPTICS

Hilary S. Lackritz

School of Chemical Engineering, Purdue University, West Lafayette, Indiana

and

John M. Torkelson

Department of Chemical Engineering, Department of Materials Science and Engineering, Northwestern University, Evanston, Illinois

1. INTRODUCTION

Optical studies have played an important role in the investigation of polymer properties. Organic materials, including polymer solutions, films, single crystals, and liquid crystals, have been studied using birefringence, interferometry, the Kerr effect, and a variety of optical methods to obtain information about the index of refraction, film thickness, and other properties. Polymer films, the object of this study, have also been intensively studied using linear optical properties including absorbance and

ISBN 0-12-784450-3

fluorescence. Nonlinear optical (NLO) techniques also can be used to study physical properties of polymer films [1–4]. Recently, it has been shown that second-order NLO techniques, including second harmonic generation (SHG) and the linear electro-optic effect, are sensitive to local changes in the polymer microenvironment [5–7] surrounding a NLO chromophore. The NLO chromophores are dyes that can display second-order NLO effects when properly oriented in a polymer matrix. This chapter discusses poled polymer films and the polymer physics that dominate the temporal and thermal response of these NLO polymer materials. The ability of nonlinear optical techniques to probe polymer dynamics via the motion of the NLO chromophores is also examined.

Organic nonlinear optical (NLO) systems are of interest in the optics [8–10], semiconductor [11], telecommunications [12], and other industries. Applications include optical switching, frequency doubling, coupling devices [12,13], and storage displays [14,14a]. Polymers have been designed and synthesized for second- and third-order nonlinear optical applications. The second-order NLO materials discussed here are polymers that have been doped or functionalized with efficient NLO chromophores. Third-order NLO polymers, including polydiacetylenes and polysilanes, have been considered in the second volume of this series [12].

Secondary properties, such as physical and optical stability and homogeneity, low absorption and scattering, compatibility with microelectronic processing methods, and mechanical properties, can help determine whether the polymeric materials will be technologically useful [15]. Polymeric NLO systems have been discussed as having many advantages over inorganic materials, including physical strength, chemical stability, low dielectric constants, transparency, low cost, and the ability to cover large surfaces while maintaining properties on a microscopic level [11,12,15]. A wide range of molecular properties may be created by utilizing varied synthetic techniques. Organic NLO materials include organic single crystals, doped glassy polymers, functionalized polymers with chromophores attached to a side chain or incorporated into the polymer backbone [7,16–18], liquid crystalline polymers [19], copolymer composites [20,21], functionalized and doped epoxies [22,23], and Langmuir-Blodgett films [24–26].

For second harmonic generation (SHG), or frequency doubling, to be observed the NLO chromophores in the polymer matrix must be aligned into a noncentrosymmetric orientation. The SHG technique is sensitive to small degrees of chromophore rotational mobility. The initial local free volume surrounding the chromophore and changes with time in the microenvironment thus affect the SHG intensity. Electric field-induced chromophore orientation occurs in regions of sufficient local free volume and segmental mobility. Contact and corona poling techniques, involving large magnitude electric fields applied across the polymer film, are used to orient the chromophores into the noncentrosymmetric structure required to obtain a second harmonic signal. The disorientation of the chromophores is due in part to mobility of the polymer chains and free volume present in the vicinity of the chromophore; the relaxations of the polymer chains and the presence of free volume prevent the "freezing in" of the imposed orientation. Chromophore orientation can be examined over a wide range of time and temperature scales as a function of the local free volume and segmental mobility in the polymer matrix. The temporal and thermal dependence of the space charges generated in the insulator during poling also influence dye-orientation behavior. Thus, a variety of phenomena can be investigated using second-order nonlinear optical techniques.

The purpose of this chapter is to describe the information about polymer physics that can be and has been obtained using second-order nonlinear optical techniques. Following a short description of second harmonic generation and polymer relaxation and mobility phenomena, the experimental techniques involving in poling are outlined. A review of the poled polymer literature will be given, followed by specific examples of the influence of polymer physics on the temporal and thermal stability of the NLO chromophore orientation. It will be seen that there is a significant relationship between the basic polymer physics and the material characteristics and performance for technological applications. This approach involves an understanding of basic polymer physics properties, including local free volume and polymer mobility and relaxation phenomena, which will allow a method of probing how these advanced materials function and can thus be properly modified and improved. The most straightforward approach is to investigate chemical and physical modifications to these systems, based on known polymer physics and examine changes in the polymer behavior as observed through changes in the temporal stability of chromophore orientation.

2. BACKGROUND

2.1. Second Harmonic Generation and the Linear Electro-optic Effect

Second harmonic generation (SHG) is the conversion of light of frequency ω to light of frequency 2ω. NLO effects occur at the molecular level in the

NLO chromophores due to a deviation in the electronic potential energy from a simple harmonic dependence on the field [27]. Typical NLO organic chromophores are polar aromatic molecules that usually have a conjugated π electron system with an electron donor and an electron acceptor on either end, creating a large magnitude charge transfer resonance and a molecular hyperpolarizability usually dominant along a single direction [28,29]. NLO dyes such as 4-dimethylamino-4′-nitrostilbene (DANS) exhibit solvatochromism [30,31] which can be studied using fluorescence techniques [32] including emission anisotropy [33]. The chromophores are also electrochromic, and large spectral changes have been observed upon application of electric fields to solutions [34,35], organic glasses [36], and polymer films [37]. This led to the assumption that the dyes would make good Kerr effect media [38]. Large values for the excited and ground state dipoles were measured using changes in optical density [39,40] and microwave conductivity [41], implying that the dyes would be useful as second-order NLO materials. Dye orientation via electric fields were studied in liquid crystals [40,42] and NLO properties were measured [43–45].

The electromagnetic field of a lightwave propagating through a medium exerts forces on the loosely bound outer or valence electrons that are usually small [27]. In a linear isotropic medium the resulting electric polarization is parallel with and directly proportional to the applied field, and, if the field is harmonic, the polarization can be written as [27]

$$p = \varepsilon_0 \chi E \tag{2.1.1}$$

where χ is the dimensionless constant known as the electric susceptibility. In the extreme case of very high fields, the polarization will become saturated, and a nonlinearity will appear and increase. Polarization can be induced by outside influences, including an electric field. The field-induced microscopic polarization, **p**, is written as a power series expansion in the dipole approximation [11]

$$\mathbf{p} = \alpha \cdot \mathbf{E} + \beta : \mathbf{EE} + \gamma \vdots \mathbf{EEE} + \cdots, \tag{2.1.2}$$

where α, β, and γ are the linear, second- and third-order hyperpolarizabilities and are tensors determining the field dependence in the molecular coordinate system. E represents the local optical field, E(ω). The magnitude of β depends on the charge-transfer induced delocalization giving rise to the NLO response and can be independently measured using DC-field-induced second harmonic generation (EFISH) [12].

Second-order NLO properties arise from the second-order NLO susceptibility $\chi^{(2)}$ tensor found in the expansion of the bulk polarization **P**(E) in powers of the electric field E [11]:

$$\mathbf{P} = \chi^{(1)} \cdot \mathbf{E} + \chi^{(2)} : \mathbf{EE} + \chi^{(3)} \vdots \mathbf{EEE} + \cdots \tag{2.1.3}$$

$\chi^{(1)}$ and $\chi^{(3)}$ are nonzero in all materials. Since the second-order polarization is proportional to the optical field squared, SHG is zero in a centrosymmetric system. In molecular solids the macroscopic second-order coefficient is related to the microscopic coefficient for SHG by [46]

$$\chi^{(2)}(-2\omega; \omega, \omega) = N\beta F(\omega)F(\omega)F(2\omega)\mathrm{O}(\theta, \varphi) \tag{2.1.4}$$

where $\chi^{(2)}(-2\omega; \omega, \omega)$ is the second-order bulk susceptibility for second harmonic generation (hereafter referred to as $\chi^{(2)}$ or $\chi^{(2)}$), N is the number density of molecules, β is the second-order hyperpolarizability tensor (describing the molecular second-order NLO properties of the chromophore), the $F(i\omega)$'s are Onsager local-field factors and $\mathrm{O}(\theta, \varphi)$ is an orientation factor that provides the projection of β onto the macroscopic coordinate [46]. It can thus be seen that the chromophore orientation is related to the experimentally measured values of $\chi^{(2)}$.

The second order susceptibility $\chi^{(2)}_{333}$ to describe the polar ordering of dipoles by a dc electric field E_0 is given by [47]

$$\chi^{(2)}_{333} = NF(\omega)^2F(2\omega)(\beta\langle\cos^3\theta\rangle + \gamma F(0)E_0\langle\cos^4\theta\rangle), \tag{2.1.5}$$

where $\langle h(\theta)\rangle$ represents the orientational averaged value for the distribution function $h(\theta)$. γ is the third-order molecular hyperpolarizability and is related to third-order NLO properties such as third harmonic generation. The subscript 3 refers to the direction of the orienting (poling) field, and z represents the axis of the molecule parallel to the molecular dipole moment (assuming $\beta = \beta_{zzz}$, implying one predominant component of the molecular tensor). θ represents the angle between the z and 3 directions. A Boltzmann distribution is used to calculate the distribution function for an individual chromophore when perturbed by the applied field [46]. Contributions due to the third-order term for doped polymer films were calculated to be between 1 and 10 percent as a function of chromophore size [47]. According to Boyd [47], the term arises from the second-order effects observed from distortions of the electronic cloud caused by the presence of the chromophore in the matrix and is independent of poling. The magnitude and effect of the γ term on the observed $\chi^{(2)}$ behavior for a given film as a function of electric-field processing is insufficiently studied and will be examined in future work.

A schematic of an apparatus used to measure SHG is shown in Fig. 1. Light is generated by a Q-switched Nd:YAG laser with a 1.064 μm

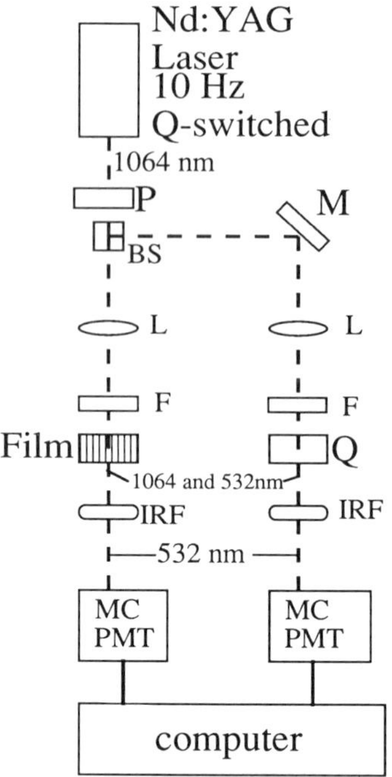

Fig. 1. Schematic of an apparatus used to measure second harmonic generation. P = polarizer, BS = beam splitter, M = mirror, L = lens, F = filter, Q = quartz reference, IRF = infrared filter, MC = monochromator, PMT = photomultiplier tube

fundamental (yielding the SHG signal at a wavelength of 532 nm). The beam is *p*-polarized and split so that a *y*-cut quartz reference and the sample are tested simultaneously. Lenses are used to vary the beam diameter. The sample beam is passed through the vertically mounted polymer film at an angle and then passed through an IR filter so that only the light at 532 nm continues to the detection system. The sample angle affects the effective film thickness. The beam passes through a second *p*-polarizer, and the emergent SHG light continues to a monochromator and photomultiplier. The optical signal is ratioed to a quartz reference (quartz crystallizes in a noncentrosymmetric structure and is capable of SHG) to minimize error due to laser drift or changes in the laboratory environment. The reference beam is similarly polarized, filtered, and analyzed.

Relationships between experimentally measured parameters and relative values for $\chi^{(2)}_{333}$ have been developed [48–50]. After simplification, the effective second-order nonlinear susceptibility $\chi^{(2)}_{\text{film}}$ assuming no dispersion (the index of refraction is independent of frequency) can thus be given by

$$\chi^{(2)}_{\text{film}} \approx (I_{\text{film}}/I_Q)^{1/2}(2/\pi)[(n^3/l^2)^{1/2}]_{\text{film}}[l_c^2/n^3)^{1/2}]_Q\chi^{(2)}_Q, \qquad (2.1.6)$$

where n is the index of refraction of the film, Q represents a quartz reference, I is the SHG intensity, I_{film}/I_Q is the experimentally measured intensity ratio, l is the film thickness, and l_c is the coherence length of the quartz [19]. The square root of the experimentally measured SHG intensity decay ratio is thus directly related to $\chi^{(2)}_{\text{film}}$, which is related to the NLO chromophore orientation.

Under low field conditions, $\chi^{(2)}_{311} \approx (1/3)\chi^{(2)}_{333}$, and

$$\langle\cos^3\theta\rangle \approx (I_{\text{film}}/I_Q)^{1/2} \approx \chi^{(2)}_{\text{film}}/\chi^{(2)}_Q. \qquad (2.1.7)$$

However, at higher fields, this relationship no longer holds, and

$$(I_{\text{film}}/I_Q)^{1/2} \approx (a\langle\cos^3\theta\rangle + b\langle\cos\theta\sin^2\theta\rangle), \qquad (2.1.8)$$

where a and b are constants dependent on the orientation of the film. The important point is that the measured SHG intensity is related to the angle of chromophore orientation. More detailed and quantitative descriptions of these relationships are available [11,12,46,51].

The linear electro-optic (LEO) effect arises from changes in the material index of refraction because of an applied electric field. Electro-optic polymers have been developed recently that have figures of merit and nonlinear coefficients close to that of lithium niobate, a commercially available inorganic crystal used as the industry standard. In the presence of an applied DC or low frequency field $E_0(\Omega \approx 0)$, where Ω is the modulating frequency, the optical dielectric constant $\varepsilon(\omega, E_0)$ of a medium is a function of E_0. For sufficiently small E_0 [27],

$$\varepsilon(\omega, E_0) = \varepsilon^{(1)}(\omega) + \varepsilon^{(2)}(\omega+\Omega)\cdot E_0 + \varepsilon^{(3)}(\omega+2\Omega)\cdot E_0E_0 + \cdots. \qquad (2.1.9)$$

Since $E + 4\pi p = \varepsilon\cdot E$, and using equations (2.1.1) and (2.1.9),

$$\varepsilon^{(2)}(\omega+\Omega) = 4\pi\chi^{(2)}(\omega+\Omega). \qquad (2.1.10)$$

In a medium with no inversion symmetry, the electro-optic effect is dominated by the $\varepsilon^{(2)}$ term linear in E_0, known as the linear electro-optic (LEO) or Pockel's effect. The term proportional to the field squared is related to the Kerr effect. The LEO effect is only observed in materials with no inversion symmetry. The field-induced refractive indices give rise to linear birefringence or double refraction. In the following, all bold type has been dropped so that the tensor nature of the variables is understood, and n^2_{xx} represents the

xx element of $\mathbf{n}\cdot\mathbf{n}$. The EO effect is described using the index ellipsoid

$$x^2/n_{xx}^2 + y^2/n_{yy}^2 + z^2/n_{zz}^2 + 2yz/n_{yz}^2 + 2zx/n_{zx}^2 + 2xy/n_{xy}^2, \qquad (2.1.11)$$

with $n_{ij}^{-1} = (\varepsilon - 1)_{ij}^{(1/2)}$. The power series expansion is carried out for all of the coefficients n_{ij}^{-2} of the index ellipsoid

$$1/n_{ij}^2(E_0) = (1/n_{ij}^2)_0 + \Sigma_k r_{ijk} E_{0k} + \Sigma_{k,l} p_{ijkl} E_{0k} E_{0l} + \cdots, \qquad (2.1.12)$$

where the coefficient r_{ijk} is the linear electro-optic tensor and p_{ijkl} is the quadratic electro-optic tensor. The longitudinal phase shift for a crystal aligned with the beam propagation direction along the optical axis is given by [52]

$$\Delta\phi = 2\pi n_0^3 r V/k_0, \qquad (2.1.13)$$

where r is the electro-optic constant (m/V), n_0 is the ordinary index of refraction, V is the potential difference (V), and k_0 is the vacuum wavelength (μm) [53]. Since the electro-optic coefficient is defined by the electric-field dependence of the optical indicatrix, r is related to the second-order bulk susceptibility through

$$\chi_{ijk}^{(2)}(-\omega;\omega,0) = -\{\varepsilon_{ii}(\omega)\varepsilon_{jj}(\omega)r_{ijk}(-\omega;\omega,0)\}/2 \qquad (2.1.14a)$$

or

$$r_{ijk}(-\omega;\omega,0) = -(2/n^4)\chi_{ijk}^{(2)}(-\omega;\omega,0) \qquad (2.1.14b)$$

and

$$\chi_{ijk}^{(2)}(-2\omega;\omega,\omega) = \chi_{ijk}^{(2)}(-\omega;\omega,0)F(W,\omega), \qquad (2.1.14c)$$

where $F(W,\omega)$ is the dispersion term. Equation (2.1.14c) is only valid away from any absorption bands that would give resonant contributions to $\chi_{ijk}^{(2)}(-2\omega;\omega,\omega)$ [54].

LEO effects result from either ionic and/or molecular movement and distortion of the electronic cloud induced by the applied electric field. Even if the induced refractive index change is 10^{-5} (typical values of r_{ijk} in crystalline materials are around 10^{-10} to 10^{-8} cm/V), a 1 cm-long medium can impose on a visible beam a phase retardation of more than $\pi/2$ [52]. These materials can thus be used as optical modulators, and devices have been made from ammonium dihydrogen phosphate, potassium dihydrogen phosphate, and potassium dideuterium phosphate [52].

2.2. Polymer Relaxation and Mobility

Polymers are long-chain molecules made up of many identical repeat units, or monomers, and are either amorphous or semicrystalline. In structural terms, amorphous materials have no long-range order (but can have order along the polymer backbone). Amorphous polymers, used in many technological applications, include poly(methyl methacrylate), polystyrene, bisphenol-A-polycarbonate, polyimides, and epoxies. By varying the structural unit of a polymer, a wide range of physical properties, including temperature stability, optical scattering, and processability can be obtained. Thus, when a polymeric system is synthesized for NLO applications, a wide variety of desired properties can be "molecularly engineered" into the materials design. In order to create carefully a system with a set of desired properties, the polymer physics underlying the material behavior must be understood and characterized.

Most polymers investigated for NLO applications are amorphous. Large-scale segmental or bulk mobility is observed in amorphous polymers above an experimentally determined glass transition temperature, T_g. Factors such as pressure, temperature, the presence of diluents, molecular weight, and other parameters all affect T_g. The main types of polymeric NLO materials to be discussed in the literature include guest–host (GH), side-chain or main-chain functionalized polymers, and crosslinked networks. In doped polymers, or GH systems, chromophores or dopants with large second-order hyperpolarizabilities are doped into host polymer matrices. The polymer does not show SHG effects, but acts simply as a host material for the oriented chromophores. These composite materials must be designed carefully to avoid dopant aggregation or precipitation that can occur at sufficiently high concentration. Functionalized polymers have chromophores covalently attached either to the polymer side chain or incorporated into the backbone. Higher dopant concentrations can be achieved in these materials. These different systems can be characterized by differences in their mobility. In functionalized systems, it is expected that the temporal stability of the chromophore orientation following poling would be improved due to the decreased mobility of the dye in the matrix. To further fix the chromophore orientation, polymers such as epoxies can be crosslinked to form a network with low local mobility. Local mobility will be defined here as small-scale motions that allow the NLO chromophore to rotate.

In order to understand how the chromophore orientation is related to the characteristics of the local microenvironment surrounding it, the local free volume and mobility of the proximal segments must be defined. The field of glassy polymer physics is very broad, and excellent reviews have been written on this subject [55–57]. In amorphous polymers the transition from liquidlike to glasslike behavior, called the glass transition, is a kinetic phenomenon [57,58]. The glass transition temperature T_g is defined as the temperature at which large-scale cooperative mobility of the polymer matrix occurs, or, alternatively, where the thermal expansion coefficient α changes from the rubbery to the glassy state with its value being dependent on the rate at which data are measured [55]. The secondary or β transition, associated with limited molecular mobility below T_g at a temperature T_β, is a broad relaxational process related to side chain rotations (as in the methacrylate group in PMMA or the wagging of the phenyl unit in PS), the motion of a small number of monomeric units in the main chain, or short chain portions [59]. These motions can occur at temperatures as low as or lower than room temperature, and thus if of sufficient magnitude may affect the temporal stability of dopant orientation. These relaxations are particularly important to consider in guest–host systems where the mobility is high due to the plasticizing effect of the dopants. Plasticization is the increase in mobility and related decrease in the glass transition temperature that occurs because small molecules in the glassy matrix act as defects in the lattice packing. It is important to note that the relaxations characterized by both T_g and T_β represent a distribution of relaxation times, and that T_g and particularly T_β occur over a fairly wide temperature range.

There are two general types of models for the glass transition. The first type assumes that, independent of the time scale, the glass transition is purely thermodynamic in nature. The other type of model for the glass transition involves free volume, first considered by Doolittle [61–63]. An empirical relationship was derived by Williams, Landel, and Ferry (WLF) to describe changes in viscoelastic properties that occur in polymers above T_g [64]. Assuming that the spectrum of relaxation times changes with temperature by a simple horizontal shift along the time axis, it was found that the shift factor a_T for viscosity was described empirically by the WLF equation [64]

$$\log a_T = \log(\eta_0/\eta_{00}) = -C_0^1(T - T_0)/(C_0^2 + T - T_0), \qquad (2.2.1)$$

where η_0 is the viscosity at temperature T, η_{00} is the viscosity at temperature T_0, and C_0^1 and C_0^2 are constants related to a reference temperature T_0 (often taken as T_g). The WLF equation predicts a rapid rise in viscosity or a (slower) decrease in free volume as the temperature approaches the reference temperature, causing glass formation. The WLF equation is valid for describing viscosity data over the temperature range $T_g + 10°C$ to $T_g + 100°C$ [55,56]. The WLF equation is not applicable for describing free volume characteristics at temperatures below T_g.

Other models, both phenomenological and thermodynamic, have been examined when considering glassy phenomena at temperatures near or slightly below T_g. The phenomenological models, including those developed by Kovacs *et al.* [65–67], Chow [68–70], Matsuoka [71, 72], Moynihan and coworkers [73,74], and Ngai and coworkers [75–78] all involve multiordering parameter equations with a variety of distribution functions. These models cannot predict T_g, but can represent volumetric and enthalpy relaxation and recovery processes. The basis for the original multiordering parameter models involves the kinetic aspect of the glass transition temperature. Essential features of these models involve a multiplicity of ordering parameters necessary to characterize fully the kinetic behavior and the dependence of the relaxation time on structure [65–67,73,79]. Molecular arguments to describe the glass transition, based on statistical thermodynamics, have been developed [80–86]. Molecular theories describe the kinetics of structural recovery and can relate the glass transition temperature to molecular and composition parameters but do not predict the time dependence behavior of the fundamental thermodynamic properties.

Phenomenological models were developed in order to describe the kinetic aspects of the glass transition. The above information indicated that the behavior of the polymers was both nonexponential and nonlinear. The original mathematical formalism related to viscoelastic responses was developed by Narayanaswamy [87]. It assumed that structural relaxation was due to a single nonexponential mechanism with constant activation energies (equivalent to several exponential mechanisms with identical activation energies) and that the shape of the response function was fixed. The phenomenological models developed were able to describe the nonexponential response function using multiple retardation mechanisms characterized by a distribution of retardation times [56,67]. The departure from equilibrium was related to time and temperature by a linear superposition principle but with each retardation time dependent upon the temperature and on the instantaneous structure of the glass [65].

Ngai and coworkers [75–77,88] developed a cooperativity model. It involved a primary relaxing species described by a time-independent relaxation rate coupled to low energy excitations considered to be generic to disordered systems. The coupled equations are of the form of the Williams-Watts (WW) [89] stretched exponential equation discussed below, with the characteristic relaxation time related to an effective time that is dependent on a relaxation rate with time. This model attempted to describe phenomena at lower temperatures below the glass transition, and has been shown to describe certain observed relaxations [78]. More importantly, it attempts to give physical significance to the model parameters.

2.2.1. Physical Aging

As the amount of free volume decreases with decreasing temperature, the molecular mobility also decreases until the time scale for molecular rearrangement is longer than the time scale for the experiment [55]. The polymer glass, a supercooled liquid, is in a nonequilibrium state so that the volume, entropy, and other thermodynamic parameters will approach their equilibrium values resulting in changes in the relaxation times [57,90,91]. The polymer mobility is thus nonzero below T_g. The specific volume will spontaneously decrease with time, at a rate characterized by the deviation from the equilibrium line at a given aging temperature, causing a densification. This causes increased chain packing, thus decreasing the segmental mobility (the rate factor for changes in chain configuration) of the polymer chains. This process, illustrated in Fig. 2, is called physical aging [91]. Since the segmental mobility of the system depends on the degree of packing, local barriers to rotation, and intermolecular interactions, and is a function of temperature, aging affects properties through changes in relaxation times [91,93]. The state of zero mobility can only be approached asymptotically, since free volume never vanishes completely [90,83] Physical aging also changes the mechanical properties of the material (increasing stiffness and brittleness) and decreases volume, dielectric constant, and dielectric loss [90,94].

During aging, not only does the amount of local free volume decrease, but the available free volume redistributes as well. Photochromism studies using nonpolar probes dispersed into glassy polymers [95,96] such as polystyrene and poly(methyl methacrylate) have shown that larger regions of local free volume decrease more rapidly than smaller regions during the aging process.

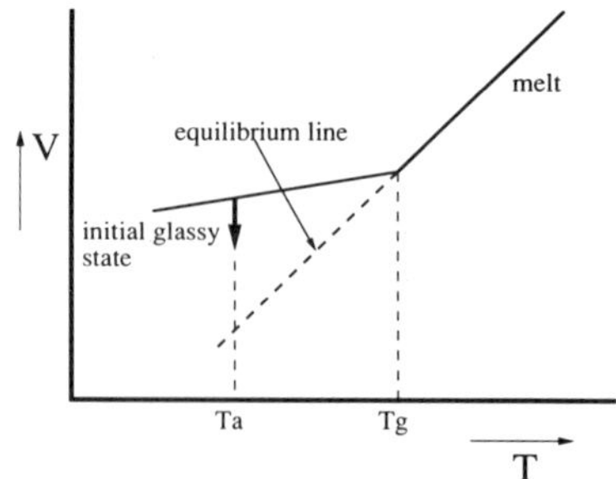

Fig. 2. Specific volume versus temperature for an amorphous polymer (similar to Fig. 20, [92]). (Reprinted with permission from Macromolecules. Copyright 1990 American Chemical Society.)

Photochromism involves the reversible change of the chromophore that can be studied by changes in the absorbance spectra. It should be stressed that free volume should not be considered as "holes" in the material, but as fluctuations of local low density [97,98]. These low density fluctuations occur over a range of size scales. There are not physical voids in the polymer material of the order of hundreds of cubic angstroms. In all future references to free volume holes or regions the implication is that of regions of local low density.

Sub-T_g polymer relaxations in physically aged polymers are often analyzed using the Williams-Watts (WW) stretched exponential function [89]:

$$y = A \exp[-(t/\tau)^{\beta}], \tag{2.2.2}$$

where y is the relaxation function, τ is the characteristic relaxation time, A an arbitrary constant, and β is a constant between zero and one that describes how the relaxation deviates from single exponential behavior or measures the cooperativity of the relaxation process [99–102]. β is thus a measure of how many relaxation processes are contributing to the observed behavior. This empirical relation is usually applied to macroscopic phenomena, such as enthalpy relaxation as measured by DSC [100], which occur during aging. During aging, it is predicted that τ should increase, since the densification should reduce the rate of the relaxation processes that can take place. Similarly, β should increase as the breadth of the deviation decreases since the densification removes some of the relaxation processes present in the quenched system. The densification also decreases the cooperativity and the number of segments participating in the relaxation. The exact physical significance of β relating to the molecular characteristics of certain polymers is uncertain [103]. However, the increases in τ and β upon aging that have been observed [3,100] are consistent qualitatively with a cooperativity argument.

Glass and secondary transition temperatures of amorphous polymers can be studied by a variety of methods [56,59,104–106]. Certain measurement techniques are able to give information on bulk or average polymer properties, and others are more specific to local volumes, for example those surrounding plasticizers or chromophores. The changes in the material properties caused by aging, such as brittleness and increased density, allow the characterization of these changes. It is important to relate changes in observed physical properties to changes in the polymer relaxation behavior, either in the bulk or at the local level. The aging phenomenon is universal in glasses, so the study of this important issue is of general applicability and not specific to the study of the temporal stability of chromophore orientation in glassy nonlinear optical polymers.

Polymer mobility and aging has been examined using a variety of experimental techniques, including differential scanning calorimetry (DSC) [100, 107,108], dilatometry [104,105,109], dielectric relaxation [72], mechanical testing [91,94,110,111], electrical conductivity [112], positron annihilation [113], and thermal [65,67] testing. Molecular sensitivity to physical aging has been investigated using photochromic [96,114–116], photophysical [117,118], x-ray scattering [97], NMR [119], and forced Rayleigh scattering [120,121] techniques. Thus, certain techniques are applicable to obtaining bulk or macroscopic properties, whereas others test local or small-scale environments in the polymer matrix.

These experiments indicate that, when using conventional techniques, it is difficult to measure mobility and local free volume at temperatures well below the glass transition. Understanding the dynamics of the polymer mobility and its relationship to the observed chromophore orientation is important in comprehending and potentially manipulating the temporal stability of chromophore orientation in the glassy polymer host and, more significantly, in understanding properties by observing the effect of changes upon the temporal stability of dopant orientation. The next section will describe the nonlinear optical techniques used to study the local microenvironment surrounding the chromophores dispersed into the amorphous polymer host. By understanding the basic polymer physics outlined in this section, it will be possible to relate changes in the observed nonlinear optical intensity with time and temperature to changes in the local microenvironment surrounding the optical chromophore.

2.3. Poling Techniques

Electric field-induced corona and contact poling are used to orient the NLO chromophores [3,12,49,122–125]. Typical poling devices are illustrated in Fig. 3. Poling involves applying a large electric field across the film normal to the surface so that the chromophores align in the field direction. Poling can be performed at room temperature or at temperatures up to and above T_g, as long as the polymer has enough mobility and/or local free volume to allow the chromophores freedom to rotate along the electric field vector [5,126]. Figure 4 illustrates a schematic of the $\chi^{(2)}$ relative to time that the field is removed as a function of time for corona-poled, doped glassy polymers. As the voltage is applied across the film at time zero, the chromophores in regions of sufficient mobility orient in response to the field gradient and the second harmonic signal increases. The permanent dipoles experience a net force or torque that tends to align them in the direction of

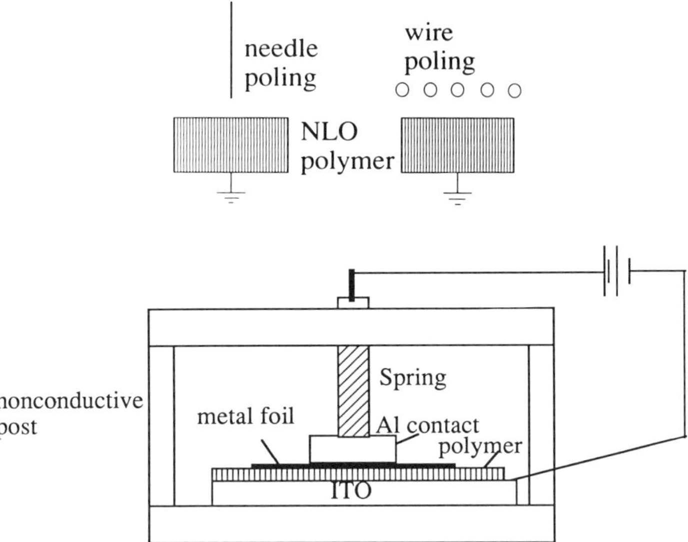

Fig. 3. (a) Schematic diagram of the corona-poling apparatus for both needle and wire poling. (b) Schematic diagram of the contact-poling apparatus. ITO = indium-tin-oxide-coated glass.

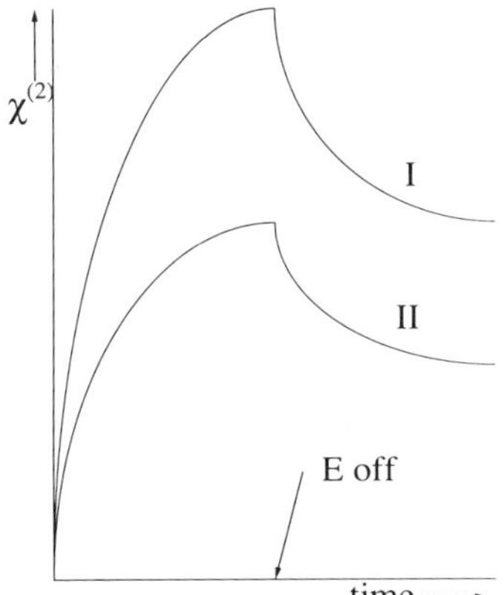

Fig. 4. Schematic diagram of $\chi^{(2)}$ versus time for *in situ* corona poling of doped polymer films. Curves are described in the text.

the field. The net orientation is determined by the potential energy of the molecular dipoles in the applied field relative to the energy associated with thermal fluctuations of the system, as well as the dielectric and electric properties of the medium and interfaces. The poling process induces a polar axis in the film in the applied field direction that is basically an infinite-fold rotational axis with an infinite number of mirror planes. This is designated as ∞mm symmetry [127].

When the electric field is removed, the relaxation and mobility processes of the polymer release the chromophore from its field-induced orientation. Chromophores that are in a local microenvironment with sufficient mobility or local free volume are free to rotate out of their imposed orientation if they can overcome local viscous forces and thus no longer contribute to the SHG intensity (experimentally seen as a decay of the SHG signal with time) [3,122]. The net alignment retained in the film under test conditions depends on the local microenvironment surrounding the dopant and the amount of local mobility as a function of sample history. For a given chromophore, the SHG observed for the film with the polymer host represented in curve I indicates greater dopant mobility than the one represented by curve II under the given experimental conditions. This indicates how qualitative information about the local microenvironment surrounding the chromophore may be obtained using SHG.

Contact poling [3,12,48,49,122,124,125,128] has been employed to induce polarizations in polymers for applications such as electret formation and piezoelectricity. It has been used to orient NLO chromophores such as disperse red 1 (DR1), 2-methyl-4-nitroaniline, and DANS doped in PMMA [3,122,124], PS, and polycarbonate [3,122], and functionalized on PS [16,122,129], and PMMA [125]. Contact or electret poling, illustrated in Fig. 3a, involves contacting metallized electrodes to the film surfaces to pass the electric field gradient across the material. The film is usually heated to temperatures above the glass transition, poled for a certain length of time, and then cooled to room temperature with the field still applied. This prevents the chromophores from rotating out of the poling imposed alignment during cooling. The field is then removed, and the SHG intensity is measured as a function of time. This contact method is susceptible to sample breakdown due to pinhole defects in the films and gives lower SHG intensities due to its inherently low electric field, but has the advantage of a known voltage drop across the sample. Theory for poling of NLO polymers in an electric field has been developed for side chain [125,130] and liquid crystalline [131] polymers in order to quantify the polar order obtained in the materials.

Corona poling [3,18,49,122,132,133] creates an electric field across the polymer film by generating a discharge from a metallic point source that deposits ions across an air gap onto the film surface. This is illustrated in Fig. 3b. A compensation charge is built up at a ground placed behind the film generating the electric field. The field generated at the needle tip is intense, accelerating free electrons near the electrode and creating gas ions with the same polarity as the needle. Corona onset and stability are a function of atmospheric conditions and electrode geometry [3,122,133]. Gas ions are accelerated toward and deposited on the film surface, creating a uniform high-magnitude field across the material capable of efficiently orienting the NLO chromophores [3,49]. A net buildup of opposite polarity charge carriers (compensation charges) at the metal bottom electrode (or ground) ensures uniformity of the field. The polar nature of the chromophores will also affect the poling process because of contributions to the net dipole moment. Corona discharges are difficult to characterize completely [133–136]. Corona poling is more efficient than contact poling and yields greater SHG signal (greater chromophore orientation) in shorter time periods and at lower temperatures due to its greater field. However, the polymer mobility and free-volume behavior and the effect of surface charge decay must be considered simultaneously when examining the overall temporal stability of corona poled materials.

The electric-field gradient crated by the corona discharge is due to the ions deposited on the surface and to a smaller extent the carriers inherent in the material (space charges). These surface and space charges are known as real charges, as opposed to a true dipole polarization [135,135b]. This gradient creates a net polarization across the film via the real charges that generate a net force capable of rotating the NLO chromophores into orientation in regions of sufficient polymer segmental mobility and local free volume. The magnitude of the charge density is a function of the strength and polarity of the applied field, time relative to the application of the field, and temperature [133–135]. Real charge effects have been examined for polymer electrets [135], made with contacting electrodes, as a function of temperature. This information is used to predict some of the field-induced behavior in corona-poled samples when considering the extra contribution due to the ion deposition on the surface. Corona treatments [133,137,138] are also used in other industrial surface treatment processes, such as xerography, adhesion/welding, and printing. Corona poling is an attractive alternative to contact poling because it efficiently orients a large number of chromophores in a shorter time and at lower temperatures than the other technique. The effect of corona discharges has been studied in polymers for piezoelectric [139], pyroelectric [140–142], and nonlinear optical uses.

Ion current carriers form a space charge between the electrodes, since the ambient gases are good insulators. The reactive ions accelerate toward the grounded film and accumulate near the surface region, generating a very high magnitude electric field across the film [133,135,143]. This field orients the dopants, thereby inducing second-order nonlinear activity. Positive polarity coronas deposit positive (+) ions while negative polarity coronas deposit negative (−) ions on the surface. An opposite charge (compensation charge) is built up at the grounded bottom surface plane [135,144]. The corona current from a point to a plane, assuming no surface potential on the plane, is given [133,143]

$$i \propto \gamma V_1(V_1 - V_T), \tag{2.3.1}$$

where V_1 is the gap voltage, V_T the corona-onset threshold voltage, and γ is a constant. V_T is the minimum voltage required to create a corona at a given temperature and pressure and is a function of atmosphere and electrode geometry [133]. The constant γ is determined by the geometry of the setup and the ion mobility. Increasing γ and decreasing V_T produces a larger corona current for a given applied voltage that should orient more dopants during poling and increase the maximum SHG intensity.

If the field E_1 within the air gap is homogeneous, the gap voltage V_1 ($=s_1E_1$; s_1 = gap thickness) can be expressed as a function of surface charge σ_r and the externally applied voltage V_0. Upon substitution and setting $i = d\sigma_r/dt$, Sessler [135] obtained the differential equation

$$d\sigma_r/dt = -\gamma\{V_T(V_T - V_0 s_1/\varepsilon_1 S) + \sigma_r(s s_1/\varepsilon_0\varepsilon_1\varepsilon S)\}, \tag{2.3.2}$$

where s is the film thickness, ε and ε_1 are the dielectric constants of the material and the gap, respectively, and

$$S = (s_1/\varepsilon_1) + (s/\varepsilon). \tag{2.3.3}$$

The solution to Eq. (2.3.2) was found to be

$$\sigma_r(t) = \sigma_m(1 - e^{-t/\tau}) \tag{2.3.4}$$

with the final value $\sigma_m = (\varepsilon_0\varepsilon/s)(V_0 - \varepsilon_1 V_T S/s_1)$ and a time constant $\tau = \varepsilon_0\varepsilon_1\varepsilon S/\gamma s s_1 V_T$. The current has only a weak thermal dependence due to V_T and γ [135].

A very rough estimate of $\sigma_r(t)$ was calculated. For a 2 μm thick PMMA film ($\varepsilon \approx 3$–3.6) on a soda lime glass (($\varepsilon \approx 3.8$–4) slide, the following parameters were used: $\varepsilon_1 = 1$; $\varepsilon = 3.8$; $s = 0.2$ cm; $s_1 = 0.6$ cm; $V_0 = -3000$ V; $V_T = -4400$ V [145]; and γ was assumed to be 1. This leads to a predicted value of $\sigma_r(t = 0) \approx 3.4 \times 10^4$ V/cm. The value of V_T is assumed to be the same as the corona-onset voltage observed in the laboratory. This field strength is about 10 times smaller than that used to contact pole the NLO films (usually about 0.5 to 0.7 MV/cm), but is quite large considering the insulating nature of the material.

3. DEVELOPMENT OF POLYMERIC MATERIALS FOR SECOND-ORDER NONLINEAR OPTICS

Reviews of the optical theory behind SHG [28,146] and the specific role of organic materials in SHG applications are available [11,12,15,25,29,54,127]. The use of polymers as potential NLO devices was considered in the late 1970s and early 1980s by a variety of groups. The pyroelectric and nonlinear optical properties of electric field poled polyvinylidene fluoride (PVDF) were examined [147–149] for contact-poled films. Using fields of 10^6 V/cm at 120°C, pyroelectric coefficients of the same order as those found for $LiNbO_3$ and SHG coefficients similar to those of quartz were observed in rolled, oriented PVDF [147]. No SHG was observed in unpoled films. Swalen [1,2] indicated that the determination of the index of refraction, film thickness, and optical loss could be examined using integrated optical (IO) techniques. Other techniques used to investigate thin polymeric and Langmuir-Blodgett (LB) films included attenuated total reflectance-Fourier transform infrared spectroscopy, ellipsometry, and surface plasmon resonance techniques via prism coupling [2,150]. Second- and third-order EO methods for characterizing polymers and bipolymers [151] were also examined.

Copolymers have been developed as NLO materials. Block copolymers should be reasonably straightforward to study using second-order NLO techniques, with the ability to isolate and probe different microenvironments experimentally useful. Contact poling an amorphous copolymer of vinylidene cyanide and vinyl acetate at T_g led to observable SHG intensities [152], as did corona poled copoly(vinylidene fluoride-trifluoroethylene) doped with a nonlinear guest [20]. Hill *et al.* examined the absorbance of the materials and noted an electrochromic effect following corona poling at room temperature. The temporal stability of the internal field and polarization was discussed, but the temporal stability of chromophore orientation and SHG intensity was not considered [20,153]. The same group created an LEO modulator using functionalized acrylate polymers [153], examined the SHG intensity and briefly discussed the temporal stability of the SHG intensity in functionalized acrylic ester polymers [153]. Optical waveguides have also been fabricated and characterized using polyester and acrylic polyurethane films [154]. Thus, it should be reasonable for future studies to examine local polymer dynamics in copolymer systems with the techniques discussed above.

Polymeric optical devices have also been demonstrated. Polymer waveguides have been prepared from PMMA doped with *m*-nitroaniline, and other chromophores dissolved into the polymer matrix [155], and Mach-Zehnder interferometers were constructed to show device viability. A 20 GHz electro-optic polymer Mach-Zehnder modulator has also been demonstrated. An Akzo EO polymer spun onto a 50 ohm microstrip drive electrode showed a modulation depth of 90 percent and a half-wave voltage of 9 V at 2 kHz [156]. A polymer-based optical switch was made from dye tagged poly(tetrafluoroethylene) latex particles in an aqueous dispersion using electric-field pulses [157] and observed via polarized fluorescence. Waveguides have been fabricated from spin-coated polyester and acrylic polyurethanes, but high losses were obtained [154]. Noncollinear phase-matched SHG has been observed in thin film waveguides of side chain stilbene functionalized polymers spun on glass [158].

Quasi-phase matching has been used to make waveguides in NLO copolymers via periodic poling techniques, the periodicity dependent on the effective coherence length of the waveguide [159]. (The coherence length is the distance required for the harmonic and second harmonic waves to become completely in phase.) A 1 μm-thick film of a 50/50 copolymer of methyl methacrylate and a monomer containing 4-oxy-4′-nitrostilbene was spin coated onto a 1-μm-polysiloxane buffer layer on an indium tin oxide (ITO) slide or a silicon wafer. A layer of polyvinyl alcohol was used as top buffer layer. The ITO slides had aluminum finger electrodes with a 12-μm periodicity deposited via lithography. Films poled at 70 V/μm at high temperature showed an induced Δn of 0.006 at 514 nm, and a quasi-phase matching efficiency of about 1.2×10^{-4} percent was obtained [159]. This method of material preparation shows strong potential for device applications.

Not only can materials be developed to make optical waveguides, but the waveguiding itself can be used to study the physical properties of the polymer. Side-chain functionalized polymers, guest–host systems of Phenol Blue or Disperse Red 1 doped into PMMA, and an epoxy system spin coated onto ITO glass were studied by optical waveguiding [160]. Electrochromism was used to monitor corona poling in real time. By measuring small changes in the refractive index anisotropy via shifts and intensity decreases of the charge transfer bands of the chromophores, a measure of poling induced order and its relaxation was obtained [160]. The time required for maximum poling efficiency was also examined.

The ability of polymer to react with light has also been utilized in NLO device applications. Side-chain NLO polymers have been made into low-loss waveguides by photobleaching of a DANS moiety to cause changes in the index of refraction [161]. Recently, a new fabrication technique for forming channel waveguides in spin-coated electro-optic polymers was demonstrated [162]. They used projection printing to form inverted-ridge channel patterns in a UV curing optical epoxy-cladding spin-coated on top of a Hoechst Celanese NLO polymer, allowing the waveguides to be generated in a noncontacting manner that required no post processing. Photoisomerization has also been used to make polymer waveguides. PMMA doped with azo dyes were made into waveguides by photoisomerizing *N*,*N*-dihexyl-4-amino-4′-nitroazobenzene or methyl red doped in PMMA, and polarization-dependent optically induced refractive index changes were observed [163]. Optical nonlinearities and bistability in organic photochromic thin films were examined by creating a Fabry-Perot etalon from spun films of biazo dyes doped in PMMA [164]. An active complementary optical tap has also been demonstrated [165] involving an NLO polymer sandwiched between two cladding layers with a gold top and aluminum bottom electrode. This device is used in an optical railtap, which is a system capable of providing many optical interconnects with a single laser source. The device was fabricated by selective photobleaching of DANS moieties functionalized in the side chain of an EO polymer. It has thus been shown that devices can be made from these NLO polymers.

Liquid crystals and liquid crystal polymers (LCP) have also been used in NLO studies and to make NLO devices. Nematic LCs have been used to create a spatial light modulator [43] and tunable electro-optic filters [166]. The use of electric fields for orienting LCs in solutions of rigid macromolecules has also been studied [167,168]. Polysiloxane LCP storage displays for thermal laser writing were created [169] and dielectric relaxation was used to characterize the side chain LCPs aligned with an electric field [169,170]. A series of experiments studying the reorientational properties of nematic LCPs was performed using dielectric relaxation, Kerr electro-optic relaxation, and switching studies [131]. The Kerr experiment consisted of applying an external field to the isotropic material and determining the birefringence induced by the reorientation motion of the polymers. They modeled the EO relaxations using a phenomenological approach and found that the dynamical quantities such as the viscosity and response times were consistent for the three measurement techniques examined. They concluded that the basic process controlling the three phenomena was thus the reorientational motion of the mesogenic unit about the short axis and that the rotational process was controlled by fluctuations rather than by small step rotational diffusion [131].

The temporal and thermal stability of the chromophore orientation in the polymer matrix following poling was not considered in the early open literature in NLO polymers. In 1982, the SHG properties of a thermotropic acrylic copolymer LCP doped with 2 wt.% DANS and poled with interdigitated electrodes were examined [19]. This important study found that very high SHG intensities could be obtained upon poling at or above T_g, and considered the effect of LC relaxation on the time-dependent response of the signal. They achieved homeotropic alignment of the nematic host and polar ∞mm symmetry upon poling. There was also an initial attempt to relate the observed SHG to the local environment or polymer "cage" surrounding the NLO chromophore [19]. When discussing guest–host, doped-polymer films as potential materials for optical devices, Zyss [29] notes that the NLO "chromophores should be locked in" to the poling imposed configuration, but does not discuss the issue of the stability of the chromophore orientation following poling any further than this.

Singer, Sohn, and Lalama [48,124] examined the SHG of contact poled PMMA doped 10 wt.% disperse red 1 (DR1). They found that the azo dye molecules behaved as a noninteracting solution under the influence of the poling field and that there was no concentration dependence of the poling efficiency of the films. It was speculated that trapped charge at the electrode interface was lowering the effective poling field and thus the observed second harmonic signal. They considered the effect of local fields and interactions on the measurement of molecular nonlinear susceptibilities, including evidence for solvent–solvent interactions. It was noted in their discussion that "since polymer glasses are not in thermodynamic equilibrium, molecular relaxations tend to reduce the polarization gradually with time" [124], but the relaxation or chromophore-orientation behavior was not characterized. In review articles as recent as August, 1988, it was stated that "if a fluid's microscopic viscosity can be altered to allow poling and then lock in the alignment, a usable $\chi^{(2)}$ results" [171], with no mention of temporal stability of chromophore orientation.

The first paper concerned with the effect of polymer relaxations on the temporal stability of chromophore orientation considered contact poled PMMA and PS doped with DANS and DR1 [122]. It was seen that the SHG intensity of contact-poled, doped films decayed at room temperature, with 50 percent of the intensity a few minutes after the applied field was removed being lost after 4 hours for PMMA films and 8 hours for PS films. Later the same year, a study of the SHG signal magnitude and temporal stability and electrochromic effects in corona poled (+6 kV, 1 cm gap, 1 μA,

125°C) PMMA doped with DR1 was presented [132,172]. They noted a 50 percent decay in the SHG intensity about 10 days following the removal of the applied field and indicated that the "cause of this decay has not been elucidated and may be due to various factors such as chemical reaction and molecular relaxation of the ordered structure" [172]. No evidence was presented in the text for chemical reactions in the film. They noted a large drop and increased maximum wavelength in the absorbance intensity following poling, with the absorbance spectrum returning towards its unpoled shape as the time after poling increased. Films corona poled with a thin tungsten wire instead of a needle showed a significantly more rapid decrease in SHG intensity following poling, but gave a more uniform poling region. No explanation for this change was given by Mortazavi *et al.* [132], but it may relate to the surface voltage persistence and magnitude.

The LEO and SHG coefficients of corona poled dicyanovinyl- (DCV-) and tricyanovinyl-doped and functionalized polymers and PMMA doped with DR1 was examined [49] and gave an indication of the decrease of SHG intensity observed following the removal of the applied field. They found that the functionalized polymers showed a smaller drop in intensity following poling than the doped films, with doped PMMA + DCV losing 70 percent of its initial $\chi^{(2)}$ magnitude after the first six days followed by a relatively stable signal as compared to the functionalized polymer, which lost only 10 percent of its initial magnitude over the first two to three days and then remained stable for the following 30 days.

Doped-epoxy matrices [22] and chromophores functionalized onto polystyrene [16,22] were also examined in terms of signal magnitude and temporal stability, with both systems showing improved temporal stability of chromophore orientation over doped glassy polymers. The temporal stability of DANS in the optical epoxy was less than disperse orange 1 (DO1), a larger optical dye, with the smaller dye losing about 60 percent of its initial intensity and the larger dye losing about 35 percent of its intensity about 1000 hours after contact poling [22]. These results are consistent with results obtained using doped glassy polymers which show that increasing chromophore size increases the temporal stability of chromophore orientation [5]. SHG studies have also been performed on functionalized epoxies [18], with the chromophores attached to a crosslinkable prepolymer that is then cured during poling. This allows a great number of the chromophores to orient and then significantly reduces the polymer matrix mobility via the crosslinking. No detectable decay was found in the SHG intensity of the epoxy films after 500 hours at room temperature and 30 minutes at 85°C following corona poling (140°C, no corona conditions listed). This is another promising method of designing polymeric materials for second-order NLO device applications.

The effect of temperature on the temporal stability of chromophore orientation in NLO polymers has been insufficiently discussed in the open literature. Kanevskii *et al.* [37] poled DANS doped in PS between indium-tin-oxide (ITO) sheets and noted that at high temperatures chromophore orientation was observed via changes in optical density following poling during an electrochromic experiment. Gribanov [173] further noted that slow rotational diffusion of 10 wt.% polar azo dyes could be observed by electrochromism in PS and PMMA. Decay was observed as a function of temperature, with no motion observed in either polymer at temperatures less than or equal to 60°C. Little attention was paid to this information until recently. The effect of temperature on the temporal stability of chromophore orientation in doped glassy polymers has been studied [126]. The effect of temperature on SHG intensity was examined in contact-poled PS-functionalized polymers [129], with the SHG intensity decay much more rapid as the temperature increased. The drop in the SHG intensity was greatest within the first few minutes after removing the applied field. A recent study that examined the thermal stability of guest–host film of polyimide doped with Eriochrome Black T has found that samples poled and cured at 250°C showed an electro-optic response stable up to 150°C for over 300 hours [174]. By utilizing a polymer host known for its high temperature stability, it was thus possible to choose a system that would be likely to have enhanced thermal stability.

Recent in-situ SHG studies have been performed utilizing both contact [16,22,175,176] and corona [18] poling of functionalized NLO polymers. Gerbi *et al.* [176] examined the decrease in SHG intensity with time following contact poling for PMMA fllms doped with 3-chloro-4-nitroaniline. For a film poled at 110°C and quenched to room temperature with the field still applied, the SHG intensity dropped 50 percent within the first 30 seconds after the applied field was removed. Sixty-five hours later, the intensity dropped by only another 10 percent. Thus the SHG intensity of the contact-poled doped films indicates a large, rapid drop in intensity at very short times following the removal of the applied field, followed by a more gradual loss in signal over much longer time scales. It has been suggested that this drop may be caused by the electronic contribution represented by the γ term in Eq. (2.1.3), but recent studies [177] indicate that this phenomenon may be explained largely without this contribution and simply by polymer physics. Similar results involving a large rapid drop in SHG intensity are reported by Eich *et al.* [18], who examined the SHG intensity behavior of a functionalized polymer corona poled for 25 minutes above T_g. There was no intensity maximum observed during poling. The intensity dropped by a factor of five immediately following the field removal and decayed to zero within the following 15 to 20 minutes. The polymer used by Eich *et al.* [18], a polyethylene backbone functionalized with a small NLO chromophore, is different from the majority of the NLO materials studied in previous applications. This system should have a great deal of mobility above T_g, and a rapid decay in intensity would be expected. This behavior is different from that observed in doped PMMA and PS films corona poled at 95°C to 110°C [3]. This may be attributed to the different magnitudes and decay rates of the surface changes created within the various types of materials by the different poling techniques. To examine these characteristics, the charge behavior with time and temperature must be investigated.

4. USING POLYMER PHYSICS TO DESCRIBE SECOND HARMONIC GENERATION IN DOPED GUEST–HOST POLYMERS

Typical SHG results obtained by the authors for doped guest–host glassy polymer systems will be summarized below in order to emphasize the importance of the role of polymer physics in understanding the optical behavior. The systems that were studied included a variety of well characterized NLO chromophores with a size range of about 100–275 Å^3 doped into glassy polystyrene (PS), poly(methyl methacrylate) (PMMA), and bisphenol-A-polycarbonate (PC). The effect of dopant size, dopant/polymer interactions, physical aging, electric field, and thermal effects have been examined [3,4,6].

The effect of dopant size should be sensitive to the amount and size distribution of local free volume surrounding the NLO chromophore. Assuming that two chemically similar dopants are oriented via poling above T_g, when the polymer has a great deal of mobility, a large chromophore should have a more difficult time rotating out of its poling-imposed orientation as a function of time than a smaller chromophore. The smaller the chromophore, the less local free volume/mobility required for rotation, and the more likely the chromophore will lose its orientation in a given time frame at a given temperature. During an SHG experiment, this is observed as a decreased temporal stability of dopant orientation with time following removal of the applied poling field. It was found that as the dopant size increased, the temporal stability of the SHG intensity and $\chi^{(2)}(t)/\chi^{(2)}(t = 0)$

increased following removal of the applied poling field [4,122] for contact-poled, doped films of PC, PS and PMMA. Under the experimental conditions used, the best sensitivity to changes in the polymer matrix were observed with dopants having volumes greater than or equal to about 200 Å^3. Dopant mobility was greater in PMMA and PC than in PS for the same dopants and thermal histories. Dopant size has also been examined [47] with similar dyes doped in PC and PMMA, with dye concentrations up to 30 percent. They found for the two smaller chromophores used ($<$200 Å^3) that greater mobility was observed in PMMA than in PC but for the largest dye (256 Å^3) that the mobility was greater in PC. Both of these studies noted that size alone in an insufficient parameter to characterize the temporal stability of dopant orientation in the glassy polymer films and that dopant/polymer interactions must be considered.

If the NLO dopants can interact in any way with the surrounding polymer matrix, the temporal stability of the dopant orientation will be affected. Even low energy bonds such as hydrogen bonds (5 Kcal/mole) may significantly affect the ability of the dopant to rotate with time and temperature. Dopants and polymers that could interact via hydrogen bonding showed improved temporal stability of dopant orientation following contact poling. The temporal stability of dopant orientation in PMMA doped with disperse orange 3, which can hydrogen bond, and DANS, which cannot, was examined via contact poling and via corona poling as a function of temperature. An activation energy for hydrogen bonding of about 4 to 5 Kcal/mole was determined via an Arrhenius-type calculation, agreeing with values determined in the literature for hydrogen bonding in amine-oxygen systems [145,178]. Enhanced temporal stability of dopant orientation was also observed in a PC system doped with DO3, but was not observed for similarly doped PS films.

Physical aging before and during poling was performed as a function of aging time and temperature, with the SHG technique sensitive to small changes in dopant orientation relating to the changing local microenvironment surrounding the NLO dopants [3,178]. The stability of the dopant orientation in the polymer matrix depends strongly on the microenvironment surrounding the dopant, including segmental mobility and local free volume. Aging the materials before corona poling in the glassy state decreased the amount of free volume in the system and in particular decreased the number of larger regions of local free volume or mobility required for the dopants to rotate. In this case, the polymer matrix was aged or densified around randomly oriented dopants. This is illustrated in Fig. 5 for PMMA and PS doped with DANS. This decreased the number of dopants able to rotate into

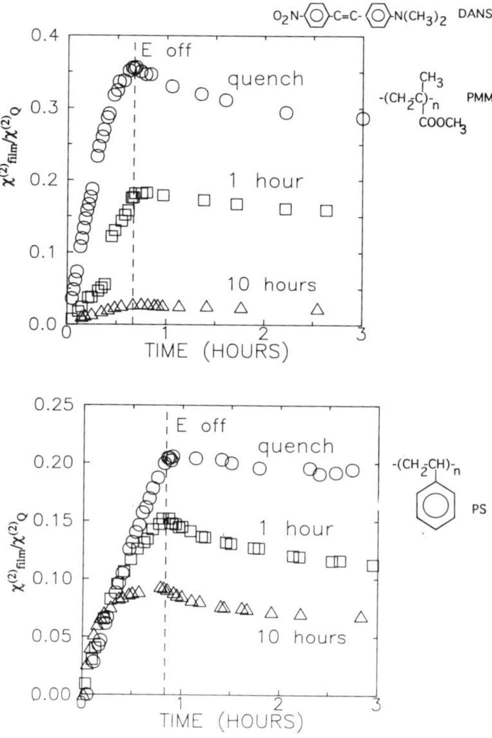

Fig. 5. Effect of physical aging at 25°C before corona poling at 25°C in (a) poly(methyl methacrylate) and (b) polystyrene films doped with 4 wt.% DANS. Films (2 μm) were either quenched (circles), aged one hour before poling (squares), or aged ten hours before poling (triangles). Dashed line indicates the time the corona voltage (−3000 V, 1 cm air gap) was shut off.

alignment in response to the applied field when poling was subsequently performed below T_g where the mobility was relatively low. The greater the amount of densification the more difficult it was for a dopant of a given size to rotate, decreasing the amount of SHG intensity observed. Changes in the observed dopant orientation, related to SHG intensity, due to physical aging effects were observed even at short aging time scales (after 1 hour at room temperature). Similar results were observed in PS films doped with the same chromophore, and for PS and PMMA films poled at 60°C (still below the glass transition temperature at 85°C). As is illustrated in Fig. 6, aging a PMMA +4 wt.% DANS film for 100 hours at 25°C before poling annealed out the mobility sufficient in size to allow dopant disorientation, with no

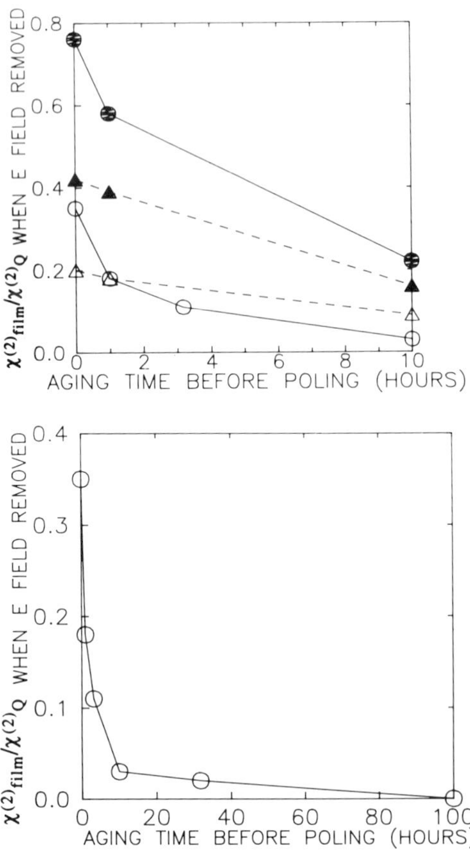

Fig. 6. (a) $\chi^{(2)}_{\text{film}}/\chi^{(2)}_Q$ when the applied field is removed versus aging time before poling for poly(methyl methacrylate) and polystyrene +4 wt.% DANS films aged at 25°C before corona poling at 25°C or 60°C. PMMA films are poled at 60°C (filled circles) or 25°C (open circles) and aged at 25°C. PS films are poled at 60°C (filled triangles) or 25°C (open triangles) and aged at 25°C. (b) $\chi^{(2)}_{\text{film}}/\chi^{(2)}_Q$ when the applied field is removed versus aging time before poling for poly(methyl methacrylate) +4 wt.% DANS films aged at 25°C before corona poling at 25°C over longer time scales.

SHG intensity observed upon corona poling at 25°C [4]. This would be impossible to detect with techniques sensitive to bulk or averaged properties such as dilatometry [91].

Aging the material during poling allowed the matrix to densify around the oriented dopants, decreasing their ability to rotate out of alignment once the applied electric field was removed and improving the temporal stability

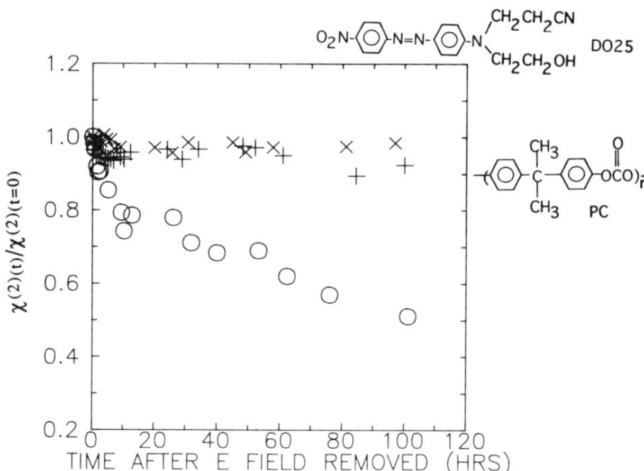

Fig. 7. $\chi^{(2)}(t)/\chi^{(2)}(\mathrm{t}=0)$ versus time after the applied field is removed for quenched and aged contact poled polycarbonate +4 wt.% disperse orange 25. The aged films were annealed at 25°C for 12 hours with the field still applied after contact poling (0.5 MV/cm at 100°C).

of the SHG intensity. Figure 7 illustrates $\chi^{(2)}(t)/\chi^{(2)}(t=0)$ versus time after the applied field is removed for quenched and aged contact poled PC +4 wt.% disperse orange 25. The aged films were annealed at 25°C for 12 hours with the field still applied after contact poling (0.5 MV/cm at 100°C). The temporal stability of dopant orientation was greatly enhanced by simply aging the films at room temperature with the field applied [5,6]. The SHG signal for both the quenched and aged films eventually approach zero, with the quenched films losing their signals more rapidly. The enhancement of the temporal stability of dopant orientation in PS and PMMA films doped with 4 wt.% chromophore and aged for about ten hours at 25°C was greater the larger the chromophore (chromophore sizes ranging from about 120 to 260 Å^3). If the dopant was capable of hydrogen bonding to the polymer matrix, a further improvement was observed. The importance of these results is not that the temporal stability of dopant orientation is improved to the point where these films can be used for device applications, but that it clearly indicates that the temporal stability of dopant orientation can be manipulated so that performance can be improved. Since physical aging is a phenomenon common to all glassy polymers, the augmented temporal stability of dopant orientation obtained by understanding and manipulating the basic polymer physics can be applied to a variety of more commercially promising systems. Thus, it can be seen that the interrelationship between polymer physics and advanced materials development cannot be overstressed.

Experiments were also performed to consider the effect of temperature on the temporal stability of dopant orientation in corona-poled films. Poling the doped films at higher temperatures (but still in the glassy state) increased the SHG intensity magnitude at early poling times with the greater segmental mobility allowing larger numbers of dopants to orient during poling. As the temperature was decreased from a higher poling temperature to a lower one or during long time, higher temperature poling, the SHG intensity decayed even with the electric field still applied. This was caused by space charge effects due to the corona-generated ions deposited on the film surface. The implications of the interaction between the corona-induced electric field and the polymer matrix is particularly significant when examining the kinetics of the dopant orientation processes in amorphous matrices.

The effect of temperature above and below T_g on the rotation of NLO dopants in corona-poled, amorphous polymer matrices was examined using in-situ SHG [6]. The polymer relaxation behavior and corona-induced electric field, both dependent on temperature and time, influence the magnitude and temporal stability of the dopant orientation following removal of the applied field. When the field was applied, the SHG intensity and $\chi^{(2)}_{\text{film}}/\chi^{(2)}_{Q}$ increased as dopants in regions of sufficient mobility and local free volume aligned in the field direction. At poling temperatures above T_g, the SHG intensity and $\chi^{(2)}_{\text{film}}/\chi^{(2)}_{Q}$ increased, reached a maximum and then decayed with the field still applied. At poling temperatures below T_g, the rate of signal increase was slower, and the maximum intensity and the amount of decay during poling was lower. This decay during poling was due to surface/space charge effects induced in the matrix by the applied field. Once the applied field was removed, the SHG signal decayed with time as a function of temperature, with increasing temperature creating greater segmental mobility and local free volume that allowed more facile dopant rotation. Dopant disorientation is related to polymer-relaxation phenomena and surface-charge persistence. The surprisingly low surface voltage observed on these films following corona poling significantly affects the dopant-rotation dynamics. The persistence of the surface voltage increases the temporal stability of the dopant orientation. The nearly instantaneous, large magnitude drop in SHG intensity seen in contact-poled films upon removal of the applied field is not observed in these corona poled films due to the persistence of the surface charge. The surface voltage is more temporally stable at lower temperatures. Since the surface voltage and the polymer relaxations at short times occur simultaneously, both processes affect the SHG measurements over the time frame of these experiments.

Some studies have begun trying to relate the observed optical behavior directly to motion in the polymer chains in functionalized polymers. Recent studies [7,179] attempted to relate the observed thermal dependence of the SHG intensity to information about mobility and relaxations obtained using thermally stimulated current techniques and dielectric relaxation for side chain and main chain functionalized polymers. Simultaneous measurements of SHG and thermally stimulated discharge currents (TSDC) were performed on the same polymers to study the relaxation of the electric field induced orientation following contact poling [7]. They found that the relaxation of a polymer with polar chromophores functionalized to the side chain occurred through local orientation, but if the chromophores were functionalized into the backbone then a local rotation and the reorientation of the end-to-end vector of the chain were detected. The combination of two techniques performed simultaneously was found to generate information about the poling behavior and relaxation dynamics of the polymers with the elimination of sample-to-sample fluctuations, drifts due to sample age, and hysteresis effects [7].

Dielectric relaxation, because it measures directly the motions of the ground-state dipole moments of the NLO chromophores, is a useful technique for studying the dynamic behavior of the polymers [179–181]. Dielectric relaxation studies were performed on functionalized NLO polymers [179] and related guest–host systems. They found that in addition to local relaxation modes of the chromophores found in all systems, an additional relaxation mode attributed to global reorientation of the end-to-end vectors of the chains was found in the main-chain-functionalized systems. In addition, the ac and dc conductivity were examined, and the screening effects at the partly blocking electrodes due to charge accumulation were considered. By examining the same chromophore either doped into or functionalized onto a side group or into the backbone of a polymer at similar loadings, optical and physical effects that vary from chromophore to chromophore are eliminated, and the polymer mobility is the dominant feature considered when comparing experimental trends. Dielectric relaxation was also used to study main-chain functionalized epoxies [180]. The temporal response of the SHG intensity decay following corona poling was studied at ambient temperature. By examining the dielectric relaxation behavior of poled and unpoled films, they were able to find an α and β relaxation in both films, with the α relaxation attributed to the glass transition and the β relaxation associated with local segmental motion of the main chain. The α relaxation, able to be fit with the WLF equation, showed a large decrease with increasing poling field. The relaxation times of the orientation decay, fitted with the Williams-Watts equation, were found to be consistent with the dielectric α relaxation characteristics of the poled samples [180]. These studies illustrate the importance of relating the molecular motions of the polymer chains to the observed optical properties of the materials.

Corona-processing variables have also been examined using SHG in nonlinear optical doped polymer films. Many variables affect corona poling, including sample geometry, humidity, atmosphere, polarity induced current, and others [133]. Positive and negative polarity DC-coronas were used to pole PMMA + DANS films in air, helium, and nitrogen environments at room temperature [126,182]. Changes in intensity decay and magnitude with differences in polarity and atmosphere were examined. For a given atmosphere and poling time, films poled in a positive corona show greater SHG intensities than those poled with negative coronas. The temporal stability of the SHG intensity for films poled with a positive corona increases from air to nitrogen to helium, and for a negative corona is similar in nitrogen and helium atmospheres, and less stable in air [126]. If films are poled in a humid or noninert atmosphere at too large a corona current, the corona discharge will cause oxidation of the polymer, forming various nitrous oxides and carboxylic acids [136]. Preliminary attempts were made to understand the electric field behavior in the corona-poled materials by measuring the temporal dependence of the surface voltage following corona poling as a function of temperature. Ion penetration, space charge effects, and surface charging are areas under current study, with a great deal more needing to be understood before corona discharge becomes a practical method for poling NLO polymer films.

5. SUMMARY

In-situ SHG has been shown to be a sensitive technique for probing amorphous polymer physics at temperatures ranging from well below to well above T_g and has been used to examine the effects of dopant size, dopant/polymer interactions, physical aging, and thermal, electric field, and corona processing in doped glassy polymer films [3,47–49,122,124,172,183], functionalized polymer films [7,17–19,21,46,125,150,153,160], and doped and functionalized epoxies [18,129]. The dynamics of the dopant rotation, related to the local microenvironment surrounding the NLO chromophore, gives information on the local mobility and free volume in the polymer matrix. The relationship between the basic polymer physics and the performance and characterization of these advanced polymer materials has been developed. By observing the rotation of chromophore molecules in the polymer host, information was obtained about the local microenvironment surrounding the NLO dye. This will allow more effective molecular engineering of NLO materials since the information in this chapter concerning the effects of physical modifications to the polymer matrix has more general applications to all polymeric NLO systems. In addition, the thermal and physical aging effects are generally observed in all glassy polymer systems, and SHG has been shown to be a sensitive technique for examining these effects. For the guest–host systems, dopant/polymer interactions and dopant size effects are important for understanding the properties of filled composites and other technologically important materials. Thus, the information obtained in this work has applicability both to the area of NLO polymers and to the more general concern of mobility/property relationships in amorphous polymers.

REFERENCES

1. J. D. Swalen, M. Tacke, R. Santo, and J. Fischer, *Opt. Comm.* **18**, 387 (1976).
2. J. D. Swalen, M. Tacke, R. Santo, K. E. Reickhoff, and J. Fischer, *Helv. Chim. Acta* **61**, 960 (1978).
3. H. L. Hampsch, G. K. Wong, J. M. Torkelson, S. J. Bethke, and S. G. Grubb, *Proc. SPIE* **1104**, 268 (1989).
4. H. L. Hampsch, J. M. Torkelson, S. J. Bethke, and S. G. Grubb, *J. Appl. Phys.* **67**, 1037 (1990).
5. H. L. Hampsch, J. Yang, G. K. Wong, and J. M. Torkelson, *Polym. Commun.* **30**, 40 (1989).
6. H. L. Hampsch, J. Yang, G. K. Wong, and J. M. Torkelson, *Macromolecules* **23**, 3640 (1990).

6a. H. L. Hampsch, J. Yang, G. K. Wong, and J. M. Torkelson, *J. Polym. Sci., Polym. Phys.*, in press.

7. W. Koehler, D. R. Robello, P. T. Dao, C. S. Willand, and D. J. Williams, *J. Chem. Phys.* **93**, 9157 (1990).
8. A. M. Glass, *Science* **235**, 1003 (1987).
9. S. D. Smith, *Nature* **316**, 319 (1985).
10. M. Zahn, M. Hikita, K. A. Wright, C. A. Cooke, and J. Brennan, *IEEE Trans. Elec. Insul.* **EI-22**, 181 (1987).

11. D. J. Williams, ed., "Nonlinear Optical Properties of Organic and Polymeric Materials," ACS Symposium Series #233, American Chemical Society, Washington, DC, 1983.
12. D. S. Chemla and J. Zyss, eds., "Nonlinear Optical Properties of Organic Molecules and Crystals," 2 vols. (Academic Press, New York, 1987).
13. C. Flytzanis and J. L. Oudar, eds., "Nonlinear Optics: Materials and Devices" (Springer-Verlag, Berlin, 1986).
14. R. Simon and H. J. Coles, *Polymer* **27**, 811 (1986).
14a. G. S. Attard and G. Williams, *Polym. Commun.* **27**, 66 (1986).
15. D. J. Williams, *Angew. Chem. Int. Ed. Eng.* **23**, 690 (1984).
16. C. Ye, N. Minami, T. J. Marks, J. Yang, and G. K. Wong, *Macromolecules* **21**, 2901 (1987).
17. M. Eich, A. Sen, H. Looser, D. Y. Yoon, G. C. Bjorklund, R. Tweig, and J. D. Swalen, *Proc. SPIE* **971**, 128 (1988).
18. M. Eich, A. Sen, H. Looser, G. C. Bjorklund, J. D. Swalen, R. Tweig, and D. Y. Yoon, *J. Appl. Phys.* **66**, 2559 (1989).
19. G. R. Meredith, J. G. VanDusen, and D. J. Williams, *Macromolecules* **15**, 1385 (1982).
20. J. R. Hill, P. L. Dunn, G. J. Davies, S. N. Oliver, P. Pantelis, and J. D. Rush, *Elec. Lett.* **23**, 700 (1987).
21. J. R. Hill, P. Pantelis, F. Abbasi, and P. Hodge, *J. Appl. Phys.* **64**, 2749 (1988).
22. M. A. Hubbard, N. Minami, C. Ye, T. J. Marks, J. Yang, and G. K. Wong, *Proc. SPIE* **971**, 136 (1988).
23. M. Eich, B. Reck, D. Y. Yoon, C. G. Wilson, and G. C. Bjorklund, *J. Appl. Phys.* **66**, 3241 (1989).
24. L. M. Hayden, T. S. Kowel, and M. P. Srinivasan, *Opt. Comm.* **61**, 351 (1987).
25. A. J. Heeger, J. Orenstein, and D. R. Ulrich, eds., "Nonlinear Optical Properties of Polymers," Materials Research Society #109, Materials Research Society, Pittsburgh, 1988).
26. I. R. Girling, N. A. Cade, P. V. Kolinsky, J. D. Earls, G. H. Cross, and I. R. Peterson, *Thin Solid Films* **132**, 101 (1985).
27. Y. R. Shen, *Principles of Nonlinear Optics* (John Wiley & Sons, New York, 1984).
28. J. Zyss, *J. Non-Cryst. Solids* **47**, 211 (1982).
29. J. Zyss, *J. Molec. Eng.* **1**, 25 (1985).
30. A. T. Peters and S. K. Cheung, *J. Chem. Tech. Biotechnol.* **35A**, 335 (1985).
31. T. Kobayashi, H. Ohtani, and K. Kurokawa, *Chem. Phys. Lett.* **121**, 356 (1985).
32. H. E. Lessing and M. Reichert, *Chem. Phys. Lett.* **46**, 111 (1977).
33. A. Kawski, J. Kukielski, and J. Kaminski, *Z. Naturforsch.* **34a**, 1066 (1979).
34. H. Bischof and W. Baumann, *Z. Naturforsch.* **40a**, 874 (1985).
35. L. A. Blumenfeld, F. P. Chernyakovskii, V. A. Gribanov, and I. M. Kanevskii, *J. Macromol. Sci.-Chem.* **A6**, 1201 (1972).
36. K. Seibold, H. Navangul, and H. Labhart, *Chem. Phys. Lett.* **3**, 275 (1969).

37. I. M. Kaneviskii, V. A. Gribanov, F. P. Chernyakovskii, A. V. Ryazanova, and L. A. Blumenfeld, *Russ. J. Phys. Chem.* **45**, 232 (1971).
38. A. E. Lutskii, B. A. Veretenchenko, and I. S. Romodanov, *Russ. J. Phys. Chem.* **50**, 576 (1976).
39. A. N. Shchapov, A. I. Kornilov, R. M. Basaev, and F. P. Chernyakovskii, *Russ. J. Phys. Chem.* **53**, 1464 (1979).
40. R. E. Sah, *Mol. Cryst. Liq. Cryst.* **129**, 315 (1985).
41. J. M. Warman, M. P. De Haas, A. Hummel, C. A. G. O. Varma, and P. H. M. VanZeyl, *Chem. Phys. Lett.* **87**, 83 (1982).
42. G. S. Attard and G. Williams, *Polym. Commun.* **27**, 2 (1986).
43. D. Armitage and J. I. Thackara, *Appl. Opt.* **28**, 219 (1989).
44. V. G. Rumyanantsev, V. M. Muratov, V. G. Chigrinov, and T. S. Pluysnina, *Mol. Cryst. Liq. Cryst.* **129**, 209 (1985).
45. H. J. Coles and R. Simon, *Polymer* **26**, 1801 (1985).
46. C. S. Willand and D. J. Williams, *Ber. Bunsenges. Phys. Chem.* **91**, 1304 (1987).
47. G. T. Boyd, C. V. Francis, J. E. Trend, and D. A. Ender, *J. Op. Soc. Am. B* **8**, 867 (1991).
48. K. D. Singer, M. G. Kuzyk, and J. E. Sohn, *J. Opt. Soc. Am. B* **4**, 968 (1987).
49. K. D. Singer, M. G. Kuzyk, W. R. Holland, J. E. Sohn, S. J. Lalama, R. B. Comizzoli, H. E. Katz, and M. L. Schillin, *Appl. Phys. Lett.* **53**, 1800 (1988).
50. S. Singh, in *CRC Handbook of Laser Science and Technology, Volume III: Optical Materials Part 1: Nonlinear Optical Properties/Radiation Damage*, M. J. Weber, ed. (CRC Press, New York, 1986).
51. J. W. Wu, *J. Opt. Soc. Am. B* **8**, 142 (1991).
52. E. Hecht, *Optics*, 2nd ed. (Addison-Wesley, Reading, MA, 1987).
53. N. Bloembergen, *Nonlinear Optics* (W. A. Benjamin, New York, 1965).
54. B. K. Nayar and C. S. Winter, *Opt. Quant. Electron.* **22**, 297 (1990).
55. J. D. Ferry, *Viscoelastic Properties of Polymers* (John Wiley & Sons, New York, 1980).
56. G. B. McKenna, in *Comprehensive Polymer Science, Vol. 2: Polymer Properties*, C. Booth and C. Price, eds. (Benjamin, Oxford University Press, 1989).
57. M. R. Tant and G. L. Wilkes, *Polym. Eng. Sci.* **21**, 874 (1981).
58. C. A. Angell, J. M. Sare, and E. J. Sare, *J. Phys. Chem.* **82**, 2622 (1978).
59. J. Kolarik, *Adv. Polym. Sci.* **46**, 120 (1982).
60. J. H. Gibbs and E. A. DiMarzio, *J. Chem. Phys.* **28**, 373 (1958).
61. A. K. Doolittle, *J. Appl. Phys.* **22**, 1471 (1951).
62. A. K. Doolittle, *J. Appl. Phys.* **23**, 236 (1952).
63. A. K. Doolittle and D. B. Doolittle, *J. Appl. Phys.* **28**, 901 (1957).
64. M. L. Williams, R. F. Landel, and J. D. Ferry, *J. Am. Chem. Soc.* **77**, 370 (1955).
65. A. J. Kovacs, *J. Polym. Sci.* **30**, 131 (1958).
66. A. J. Kovacs, *Fortschr. Hochpolym.-Forsch.* **3**, 394 (1963).
67. A. J. Kovacs, J. J. Aklonis, J. M. Hutchinson, and A. R. Ramos, *J. Polym. Sci., Polym. Phys. Edn.* **17**, 1097 (1979).

68. T. S. Chow, *Macromolecules* **13**, 362 (1980).
69. T. S. Chow, *Macromolecules* **17**, 2336 (1984).
70. T. S. Chow, *Macromolecules* **22**, 698 (1989).
71. S. Matsuoka, H. E. Bair, S. S. Bearder, H. E. Kern, and J. T. Ryan, *Polym. Eng. Sci.* **18**, 1073 (1978).
72. S. Matsuoka, G. Williams, G. E. Johnson, E. W. Anderson, and T. Furukawa, *Macromolecules* **18**, 2652 (1985).
73. M. A. DeBolt, A. J. Easteal, P. B. Macedo, and C. T. Moynihan, *J. Am. Cer. Soc.* **59**, 16 (1976).
74. C. T. Moynihan, A. J. Easteal, M. A. DeBolt, and J. Tucker, *J. Am. Cer. Soc.* **59**, 12 (1976).
75. K. L. Ngai, *Commun. Solid State Phys.* **9**, 127 (1979).
76. K. L. Ngai, A. K. Rajagopal, and S. Tetler, *J. Chem. Phys.* **88**, 5086 (1988).
77. K. L. Ngai and D. J. Plazek, *J. Polym. Sci., Polym. Phys.* **23**, 2159 (1985).
78. K. L. Ngai and G. Fytas, *J. Polym. Sci., Polym. Phys.* **24**, 1683 (1986).
79. C. T. Moynihan, P. B. Macedo, C. J. Montrose, P. K. Gupta, M. A. DeBolt, J. F. Dill, B. E. Dom, P. W. Drake, A. J. Easteal, P. B. Elterman, R. P. Moeller, H. Sasabe, and J. A. Wilder, *Ann. New York Acad. Sci.* **279**, 15 (1976).
80. G. S. Grest and C. H. Morrel, *Phys. Rev. B* **21**, 22 (1980).
81. R. E. Robertson, *J. Polym. Sci., Polym. Symp.* **63**, 173 (1978).
82. R. E. Robertson, R. Simha, and J. G. Curro, *Macromolecules* **17**, 911 (1984).
83. R. E. Robertson, *Macromolecules* **18**, 953 (1985).
84. R. E. Robertson, R. Simha, and J. G. Curro, *Macromolecules* **18**, 2239 (1985).
85. R. E. Robertson, R. Simha, and J. G. Curro, *Macromolecules* **21**, 3216 (1988).
86. R. J. Roe, *J.Appl. Phys.* **48**, 4085 (1977).
87. O. S. Narayanaswamy, *J. Am. Cer. Soc.* **54**, 491 (1971).
88. K. L. Ngai, *Commun. Solid State Phys.* **9**, 141 (1980).
89. G. Williams and D. C. Watts, *Trans. Faraday Soc.* **66**, 80 (1970).
90. L. C. E. Struik, *Polym. Eng. Sci.* **17**, 165 (1977).
91. L. C. E. Struik, *Physical Aging in Amorphous Polymers and Other Materials* (Elsevier, Amsterdam, 1978).
92. G. B. McKenna, in *Comprehensive Polymer Science*, Vol. 2, C. Booth and C. Price, eds. (Pergamon Press, Oxford, 1989).
93. R. E. Robertson, *J. Polym. Sci., Polym. Phys.* **17**, 597 (1979).
94. J. Heijboer, in *Static and Dynamic Properties of the Polymeric Solid State*, R. Penthrick and R. Richards, eds. (D. Reidel, Amsterdam, 1982), p. 197.
95. J. G. Victor and J. M. Torkelson, *Macromolecules* **20**, 2951 (1987).
96. J. G. Victor and J. M. Torkelson, *Macromolecules* **21**, 3490 (1988).
97. J. J. Curro and R. Roe, *Polymer* **25**, 1424 (1984).
98. R. J. Coe and H. H. Song, *Macromolecules* **18**, 1603 (1985).
99. I. M. Hodge and A. R. Berens, *Macromolecules* **15**, 762 (1982).
99a. I. M. Hodge and A. R. Berens, *Macromolecules* **14**, 31 (1981).
99b. I. M. Hodge, *Macromolecules* **20**, 2897 (1987).

100. I. M. Hodge and A. R. Berens, *Macromolecules* **18**, 1980 (1985).
101. I. M. Hodge, *Macromolecules* **19**, 936 (1986).
102. I. M. Hodge, *Macromolecules* **20**, 2897 (1987).
103. I. M. Hodge and G. S. Huvard, *Macromolecules* **16**, 371 (1983).
104. R. Greiner and F. R. Schwarzl, *Rheol. Acta* **23**, 378 (1984).
105. R. Greiner and F. R. Schwarzl, *Coll. Polym. Sci.* **267**, 39 (1989).
106. J. Countandin, D. Ehlich, H. Sillescu, and C. H. Wang, *Macromolecules* **18**, 587 (1985).
107. S. E. B. Petrie, R. S. Moore, and J. R. Flick, *J. Appl. Phys.* **43**, 4318 (1972).
108. J. J. Tribone, J. M. O'Reilly, and J. Greener, *Macromolecules* **19**, 1732 (1986).
109. R. Pixa, B. Grisoni, T. Gay, and D. Froelich, *Polym. Bull.* **16**, 381 (1986).
110. G. B. McKenna and A. J. Kovacs, *Polym. Eng. Sci.* **24**, 1138 (1984).
111. N. G. McCrum, B. Read, and G. Williams, *Anelastic and Dielectric Effects in Polymeric Solids* (John Wiley & Sons, London, 1967).
112. V. Ademac and E. Mateova, *Polymer* **16**, 166 (1975).
113. B. D. Malhurota and R. A. Pethrick, *Eur. Polym. J.* **19**, 457 (1983).
114. J. G. Victor and J. M. Torkelson, *Macromolecules* **20**, 2241 (1987).
115. W.-C. Yu and C. S. P. Sung, *Macromolecules* **21**, 365 (1988).
116. W.-C. Yu, C. S. P. Sung, and R. E. Robertson, *Macromolecules* **21**, 355 (1988).
117. E. F. Meyer, A. M. Jamieson, and R. Simha, *Polymer* **31**, 243 (1990).
118. J. S. Royal and J. M. Torkelson, *Macromolecules* **25**, 4792 (1992).
119. J. F. O'Gara, A. A. Jones, C. C. Hung, and P. T. Inglefield, *Macromolecules* **18**, 1117 (1985).
120. J. Zhang and C. H. Wang, *Macromolecules* **21**, 1811 (1988).
121. C. H. Wang, J. L. Xia, and L. Yu, *Macromolecules* **24**, 3638 (1991).
122. H. L. Hampsch, J. Yang, G. K. Wong, and J. M. Torkelson, *Macromolecules* **21**, 526 (1988).
123. K. D. Singer, S. J. Lalama, and J. E. Sohn, *Proc. SPIE* **578**, 130 (1985).
124. K. D. Singer, J. E. Sohn, and S. J. Lalama, *Appl. Phys. Lett.* **49**, 248 (1986).
125. K. D. Singer, J. E. Sohn, L. A. King, H. M. Gordon, H. E. Katz, and C. W. Dirk, *J. Opt. Soc. Am. B* **4**, 1339 (1989).
126. H. L. Hampsch, J. Yang, G. K. Wong, and J. M. Torkelson, *Macromolecules* **23**, 3648 (1990).
127. P. N. Prasad and D. J. Williams, *Nonlinear Optical Effects in Molecules and Polymers* (Wiley Interscience, New York, 1991).
128. S. R. Marder, J. E. Sohn, and G. D. Stuky, eds., *Materials for Nonlinear Optics: Chemical Perspectives*, ACS Symposium Series #455 (American Chemical Society, Washington, DC. 1991).
129. M. A. Hubbard, T. J. Marks, J. Yang, and G. K. Wong, *Chem. Mater.* **1**, 167 (1989).
130. C. P. J. van der Vorst and S. J. Picken, *J. Opt. Soc. Am. B* **7**, 320 (1990).
131. R. Birenheide, M. Eich, D. A. Jungbauer, O. Herrmann-Schonherr, K. Stoll, and J. H. Wendorff, *Mol. Cryst. Liq. Cryst.* **177**, 13 (1989).

132. M. A. Mortazavi, D. Yankelvich, A. Dienes, A. Knoessen, S. T. Kowel, and S. Dijaili, *Appl. Opt.* **28**, 3278 (1989).
133. E. M. Williams, *The Physics and Technology of Xerographic Processes* (Wiley Interscience, New York, 1984).
134. J. R. Li and H. J. Wintle, *J. Appl. Phys.* **65**, 4617 (1989).
135. G. M. Sessler, ed., *Electrets* (Springer-Verlag, Berlin, 1980).
135a. J. van Turnhout, "Thermally Stimulated Discharge of Polymer Electrets" (Elsevier, Amsterdam, 1975).
136. J. J. O'Dwyer, *The Theory of Electrical Conduction and Breakdown in Solid Dielectrics* (Clarendon Press, Oxford, 1973).
137. T. Hjertberg, B. A. Sultan, and E. M. Sorvik, *J. Appl. Polym. Sci.* **37**, 1183 (1989).
138. R. B. Comizzoli, *J. Electrochem. Soc.* **134**, 424 (1987).
139. T. T. Wang and J. E. West, *J. Appl. Phys.* **53**, 6552 (1982).
140. J. G. Bergman, Jr., J. H. McFee, and G. R. Crane, *Appl. Phys. Lett.* **18**, 203 (1971).
141. M. G. Broadhurst, G. T. Davis, A. S. DeReggi, S. C. Roth, and R. E. Collins, *Polymer* **23**, 22 (1982).
142. P. D. Southgate, *Appl. Phys. Lett.* **28**, 250 (1976).
143. E. W. McDaniel and E. A. Mason, "The Mobility and Diffusion of Ions in Gases" (Wiley Interscience, New York, 1978).
144. D. A. Seanor, ed., *Electrical Properties of Polymers* (Academic Press, New York, 1982).
145. H. L. Hampsch, J. Yang, G. K. Wong, and J. M. Torkelson, in *Advanced Organic Solid State Materials: Materials Research Society Symposium Series Volume 173*, L. Y. Chiang, P. M. Chaikin, and D. O. Cowan, eds. (MRS Symposium Series, Pennsylvania, 1990), p. 625.
146. Y. R. Shen, "Recent Advances in Nonlinear Optics", *Rev. Mod. Phys.* **48**, 1 (1976).
147. J. H. McFee, J. G. Bergman, Jr., and G. R. Crane, *Ferroelectrics* **3**, 305 (1972).
148. F. Kerherve, *Opt. Commun.* **9**, 420 (1973).
149. F. I. Mopsik and A. S. DeReggi, *Appl. Phys. Lett.* **44**, 65 (1984).
150. G. G. Cross, I. R. Peterson, I. R. Girling, N. A. Cade, M. J. Goodwin, N. Carr, R. S. Sethi, R. Marsden, G. W. Gray, D. Lacey, A. M. McRoberts, R. M. Scrowston, and K. J. Toyne, *Thin Solid Films* **156**, 39 (1988).
151. B. R. Jennings, *Pure Appl. Chem.* **54**, 395 (1982).
152. H. Sato, T. Yamamoto, I. Seo, and H. Gamo, *Opt. Lett.* **12**, 579 (1987).
153. J. R. Hill, P. Pantelis, F. Abbasi, and P. Hodge, in *Organic Materials for Nonlinear Optics*, R. A. Hann and D. Bloor, eds. (Royal Soc. Chem. London, London, 1989).
154. S. K. Kapoor, C. D. Pandey, J. C. Joshi, A. L. Dawar, K. N. Tripathy, and V. L. Gupta, *Appl. Opt.* **28**, 37 (1989).
155. M. J. Goodwin, *Proc. SPIE* **836**, 265 (1987).
156. D. G. Girton, S. L. Kwiatkowski, G. F. Lipscomb, and R. S. Lytel, *Appl. Phys. Lett.* **58**, 1730 (1991).

157. B. R. Jennings and P. J. Ridler, *Polymer* **28**, 581 (1987).
158. Y. Shuto, H. Takara, M. Amano, and T. Kaino, *Jap. J. Appl. Phys.* **28**, 2508 (1989).
159. G. Khanarian, R. A. Norwood, D. Haas, B. Feuer, and D. Karim, *Appl. Phys. Lett.* **57**, 977 (1990).
160. R. H. Page, M. C. Jurich, B. Reck, A. Sen, R. J. Tweig, J. D. Swalen, G. C. Bjorkland, and C. G. Wilson, *J. Opt. Soc. Am. B* **7**, 1239 (1990).
161. M. B. J. Diemeer, F. M. M. Suyten, E. S. Trommel, A. McDonach, J. M. Copeland, L. W. Jennesdens, and W. H. G. Horsthuis, *Electron. Lett.* **26**, 379 (1990).
162. P. R. Ashley and T. A. Tumolillo, J., *Appl. Phys. Lett.* **58**, 884 (1991).
163. T. Luckemeyer, H. Franke, and W. F. X. Frank, *Proc. SPIE* **1213**, 126 (1990).
164. C. J. G. Kirkby, R. Cush, and I. Bennion, *Opt. Commun.* **56**, 288 (1985).
165. T. E. van Eck, A. J. Ticknor, R. S. Lytel, and G. F. Lipscomb, *Appl. Phys. Lett.* **58**, 1588 (1991).
166. S. T. Wu, *Appl. Opt.* **28**, 48 (1989).
167. R. Menzel and W. Rapp, *Z. Naturforsch.* **41a**, 983 (1986).
168. A. R. Khokhlov and A. N. Semenov, *Macromolecules* **15**, 1272 (1982).
169. H. J. Coles and R. Simon, *Mol. Cryst. Liq. Cryst. Lett.* **3**, 37 (1986).
170. G. S. Attard, J. J. Moura-Ramos, and G. Williams, *J. Polym. Sci., Polym. Phys.* **25**, 1099 (1987).
171. G. R. Meredith, *MRS Bull.* (August 24, 1988).
172. A. Knoesen, M. A. Mortazavi, S. T. Kowel, A. Dienes, and B. G. Higgins, in *Nonlinear Optical Properties of Materials; Technical Digest Series*, vol. 9 (Optical Society of America, New York, 1988), p. 244.
173. N. H. Gribanov, *Chem. Phys. Lett.* **164**, 253 (1989).
174. J. W. Wu, J. F. Valley, S. Ermer, E. S. Binkley, J. T. Kennedy, G. F. Lipscomb, and R. S. Lytel, *Appl. Phys. Lett.* **58**, 225 (1991).
175. D. Gerbi, G. T. Boyd, D. A. Ender, R. M. Henry, K. K. Kam, P. C. W. Leung, and J. J. Stofko, presentation at the "Topical Workshop on Organic and Polymeric Nonlinear Optical Materials," American Chemical Society, Division of Polymer Chemistry, May 1988.
176. R. N. DeMartino, U.S. Patents 4,717,508 and 4,720,355 (1988).
177. A. Dhinojwala and J. M. Torkelson, *Polym. Preprints* **32, 373** (1991).
178. H. L. Hampsch, J. Yang, G. K. Wong, and J. M. Torkelson, in *New Materials for Nonlinear Optics*, G. Stucky, S. Marder, and J. E. Sohn, eds. (ACS Symposium Series #455, American Chemical Society, Washington, DC, 1991), p. 294.
179. W. Koehler, D. R. Robello, P. T. Dao, C. S. Willand, and D. J. Williams, *Macromolecules* **24**, 4589 (1991).
180. I. Teroka, D. Jungbauer, B. Reck, D. Y. Yoon, R. Tweig, and C. G. Wilson, *J. Appl. Phys.* **69**, 2568 (1991).
181. D. Rei, J. Runt, A. Safari, and R. E. Newnham, *Macromolecules* **20**, 1797 (1987).
182. S. J. Bethke, S. G. Grubb, H. L. Hampsch, and J. M. Torkelson, *Proc. SPIE* **1216**, 260 (1990).
183. G. T. Boyd, *J. Opt. Soc. Am. B* **6**, 685 (1989).

Chapter 9

INTEGRATED OPTICS WITH NONLINEAR POLYMERS

Vincent Lemoine and Jean Paul Pocholle

Laboratoire d'Optique Intégrée et d'Optronique, Thomson-CSF, Orsay, France

and

Pierre le Barny and Philippe Robin

Laboratoire de Chimie pour l'Electronique, Thomson-CSF, Orsay, France

1. INTRODUCTION

Nonlinear optical polymers are of great interest for low-cost devices and nonlinear applications including frequency mixing, switching, and

ISBN 0-12-784450-3

modulators. They have recently been the focus of intensive fundamental and applied studies by government, academic, and industrial laboratories. For many applications, polymers offer advantages over bulk and waveguide materials such as crystals ($LiNbO_3$, $LiTaO_3$, ...) and semiconductors (GaAs/GaAlAs, InGaAs/InP, ...). The advantages of nonlinear polymers are:

- Low-cost technology using the conventional thin film deposition methods from solution over a large area.
- Ease of modification and adjustment of the optical properties (refractive index, transparency, optical nonlinearity) based on the molecular engineering versatility.
- Large substrate choice, leaving hybrid integration with optics and electronics, promising to offer a powerful new device. They can be deposited on almost any substrate using relatively low temperatures and can be spin-coated directly on hybrid circuits.
- High nonlinear coefficients.
- Low dielectric constant (large electrical band width, and high RF coupling for, electro-optic applications).

Intrinsically the optical waveguides are essential structures for practical applications of those materials. Waveguide structures have several important properties as follows:

- Good overlap between the electric field applied and the optical mode propagating in the waveguide structure (electro-optic efficiency).
- Confinement of optical power, transforming low powers like those emitted by diode lasers into high intensities.
- Selection of guided modes by controlling the layer's thickness to adjust the phase-matching conditions in the nonlinear interactions.

Also, in the guided wave configuration, very active nonlinear interactions are possible.

The first part of this chapter is devoted to the progress in the field of nonlinear polymeric materials for integrated optical devices such as electro-optic modulators and second harmonic generators.

After an introduction discussing the specific requirements concerning these polymers to be included in a nonlinear optical device, the problem of the stability with temperature of these materials is detailed. The different possibilities to increase this stability (doped polymers, side-chain polymers, main-chain polymers, and crosslinked polymers) and the main achievements in each case are reviewed.

The second part of this chapter deals with two measurement methods of the nonlinear coefficients. The first one, based on the Maker's fringes experiment, gives a direct estimation of the second harmonic coefficients. The second one is a simple method to measure the electro-optic coefficients.

The last part is devoted to the devices realized with these nonlinear polymers. The key concepts of nonlinear integrated optics are presented. Some of the technological solutions to obtain two-dimensional optical structures are reviewed. Different options concerning the realization of electro-optic modulators and second harmonic generators are discussed. The main achievements already published are reviewed.

2. MATERIALS FOR WAVEGUIDED QUADRATIC NONLINEAR OPTICS

2.1. Introductory Remarks

Specific requirements have to be fulfilled simultaneously to obtain efficient organic materials for waveguided quadratic nonlinear optics, namely:

- Enhanced quadratic nonlinear optical properties compatible with the transparency required by the type of application considered (electro-optic modulation or frequency doubling).
- Chemical and photochemical stability.
- No decay of nonlinear optical properties with time and, to some extent, with temperature.
- As low optical losses as possible.
- Processability.

Strategy for getting materials with large second-order nonlinear susceptibility tensor $\chi^{(2)}$ consist in organizing molecules with high microscopic second-order nonlinear hyperpolarizability tensor β in such a way that the resulting system has no center of symmetry and an optimized constructive additivity of the molecular hyperpolarizabilities. If we except few molecules such as TED and NOTED [1], the nonlinear optical (NLO) active moieties that have been synthesized so far derived from the donor-π system–acceptor molecular concept.

Extending the dimension of the π electron conjugated system and/or using strong electron-donor and electron-acceptor groups is the classical strategy to increase the hyperpolarizabilities of NLO moieties [2]. But high β values are strongly connected with the existence of low-energy excited states in molecules and, hence, are observed with chromophores having their charge-transfer band in the visible or in the near infrared. The efficiency transparency tradeoff problem is particularly important in integrated optics applications, since the input laser beam is propagated into a thin NLO material over a one centimeter-long distance. However, this problem is less accurate for the electro-optic modulation where the useful laser wavelength for telecommunication will be very likely in the 1.3 μm or 1.5 μm ranges, than for coherent blue-light emission by frequency doubling of 0.8 to 0.9 μm semiconductor laser diodes, that needs fully transparent organic materials with no residual absorption in the visible region. The efficiency–transparency tradeoff problem has already been discussed by J. F. Nicoud in volume I of this series.

Macroscopic organization of NLO moieties, suitable for quadratic applications are currently obtained from single crystals, Langmuir-Blodgett (LB) multilayers, and polymer films.

Advantages of single-crystal organic materials lie in the possibility of obtaining very large stable nonlinear coefficients, but noncentrosymmetric single crystals are not common by nature and their growth in optical thin films of well-defined thickness is very difficult and expensive. Although LB technique seems very attractive to build up multilayer structures with a high degree of order in a direction normal to the substrate, its use in guided-wave devices seems questionable at the moment, since a partial regular structure within the plane of the layer induces important optical losses by scattering. On the other hand, obtention of a thick film of the order of one μm thickness requires more than 300 monolayers to be deposited.

Organic polymers seem to be the best candidates to meet the needs of integrated nonlinear optics since they are compatible with semiconductor technology; they are particularly attractive for getting waveguides owing to the spin coating or the dipping techniques that lead to thin low-cost films. Moreover, they have the advantages of organic materials, namely, high nonlinear coefficients, fast response time, high optical damage threshold, and the tremendous potential for molecular engineering.

Three classes of quadratic NLO polymers have been investigated so far: side-chain liquid crystal polymers (SCLCP), ferroelectric polymers, and amorphous polymers. Currently, amorphous polymers seem to be the most promising class of polymeric materials for NLO applications.

Whatever the type of polymer used, a film of sufficient thickness (about one μm) has to be deposited onto a substrate via spin coating or dipping techniques. The obtained thickness depends on the viscosity of the solution, on the spinning or dipping speed and on the type of substrate. Of course, just after the deposition process the obtained film is centrosymmetric. In order to get a noncentrosymmetric system, a DC external electric field must be applied at a temperature that allows free orientation of the NLO groups (generally above the T_g of the material). The field acts on the ground-state dipole moment of the NLO moieties, causing them to align preferentially with the field, hence destroying the centrosymmetry of the film. Finally the obtained alignment is frozen by cooling the film at room temperature before removing the electric field (Fig. 1). The poling provides the alignment predicted by the Boltzmann's distribution law and imparts an ∞mm symmetry to the film.

The poling process is one of the key points to achieve efficient NLO polymeric materials for quadratic applications. Two main poling techniques are currently used, the parallel-plate poling and the corona poling [3,4].

2.2. Doped Amorphous Polymers

During the last few years, the conception of amorphous polymeric materials has moved from molecularly doped polymers to covalently functionalized polymers and very recently to crosslinked polymers.

The use of ordered guest-host polymers as an alternative to bulk crystals for quadratic NLO applications was first proposed by Williams [5] and since have been extensively studied by Singer *et al.* [6–10].

They have developed a statistical model for one-dimensional molecules in an isotropic host material that relates the bulk largest second-harmonic coefficient component d_{33}, to the molecular hyperpolarizability component along the molecular z axis, β_{zzz}, the ground-state dipole moment of the NLO molecules μ_0, the poling field Ep, the number of NLO molecules per volume unit N, and the local factors $f(\omega)$ and $f(2\omega)$, by:

$$d_{33}(-2\omega, \omega, \omega) = Nf(2\omega)f^2(\omega)\frac{\varepsilon_r(n^2 + 2)}{(n^2 + 2\varepsilon_r)}\frac{\mu_0 Ep}{5kT}\beta_{zzz}, \qquad (2.2.1)$$

where $$f(\omega) = \frac{n^2(\omega) + 2}{3}, \qquad f(2\omega) = \frac{n^2(2\omega) + 2}{3},$$

ε_r is the static dielectric constant; n is the optical index of refraction.

Singer *et al.* have checked the validity of the model by studying the linear dependence of d_{33} versus poling field Ep and number density N in polymethyl-

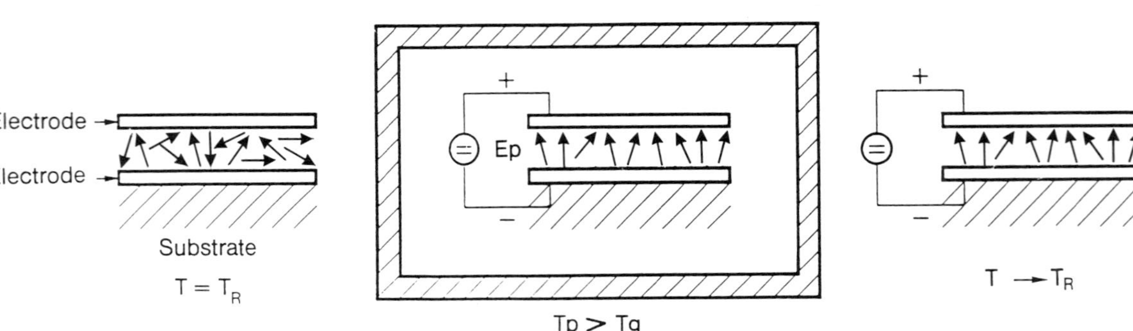

Fig. 1. Electric-field-induced orientation of NLO moieties in a polymer; T_p = poling temperature, E_p = poling field, T_R = room temperature.

methacrylate (PMMA) doped with 4-[ethyl(2-hydroxyethyl)amino]-4′-nitro-azobenzene, an azo dye also called Disperse Red 1.

By using dicyanovinylazo (DCV) guest which exhibits a high $\beta_{zzz}\mu_0$, dissolved in PMMA ($N = 2.3 \times 10^{20}/\text{cm}^3$), a d_{33} value as high as 31 pmV^{-1} has been obtained just after poling at a incident wavelength of 1.58 μm [11]. But, doped polymers suffer from the relatively low solubility of the dye in the host and from the relaxation of the orientational ordering obtained by poling, which occurs at room temperature (Fig. 2). These relaxations may result from a dipole–dipole induced reorientation and a diffusion of the dye in the polymer.

Mechanisms of the relaxation processes are not well understood. Local free volume and/or local mobility in the polymer matrix seem to play an important role [12].

Hampsch *et al.* [12] have tried to fit their experimental results to the Williams-Watts stretched exponential

$$y = e^{-\left(\frac{t}{\tau}\right)^b}, \tag{2.2.2}$$

where y represents the normalized SHG intensity, τ is a characteristic relaxation time, and b reflects the breadth of distribution of relaxation $(0 < b < 1)$.

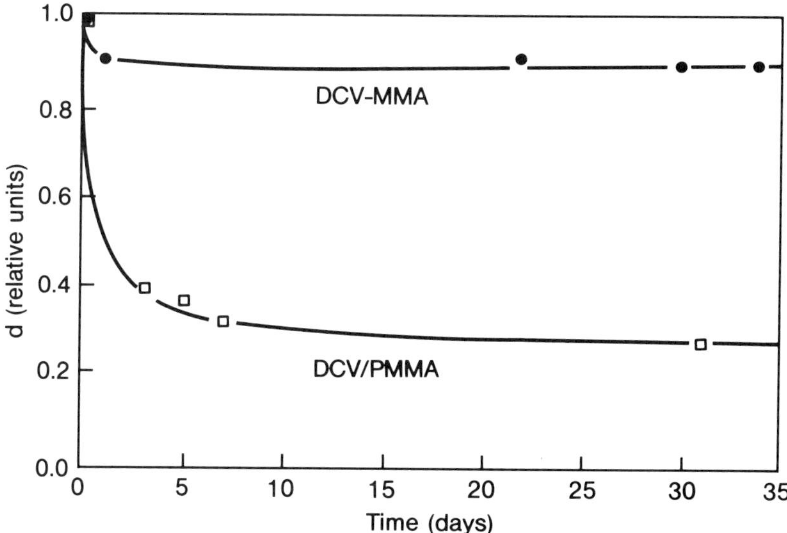

Fig. 2. Decay of second-harmonic coefficient of DCV-MMA and DCV/PMMA corona-poled films. The chemical structure of DCV-MMA copolymer is depicted in Fig. 5.

processes [14]. It appears that the model is insufficient to describe the behavior of all the guest-host systems studied.

A new concept of guest-host systems has recently been published in which a centrosymmetric crystal such as *p*-nitroaniline (*p*-NA) is changed into a noncentrosymmetric crystal when it is allowed to crystallize with an electric field in a poly(oxyethylene) (POE) matrix. Highly efficient SHG activity is obtained with such a system ($d_{33} = 16.3\ \text{pm}\cdot\text{V}^{-1}$ at 1.064 μm), but again it suffers from thermal and temporal instabilities [15,16].

A natural step toward highly efficient and stable NLO polymers was the synthesis of covalently functionalized amorphous polymers. Depending on whether the NLO unit is in the side chain or in the main chain, two classes of such polymers can be considered.

2.3. Side-Chain NLO Amorphous Polymers

By directly linking the chromophore to the polymer backbone, it is possible to achieve high NLO unit densities without phase separation and to increase the temporal stability of the poled material. Covalently functionalized amorphous polymers can be obtained either by copolymerization of a monomer bearing the chromophore unit via a flexible spacer group with a comonomer contributing to the amorphous character of the final copolymer

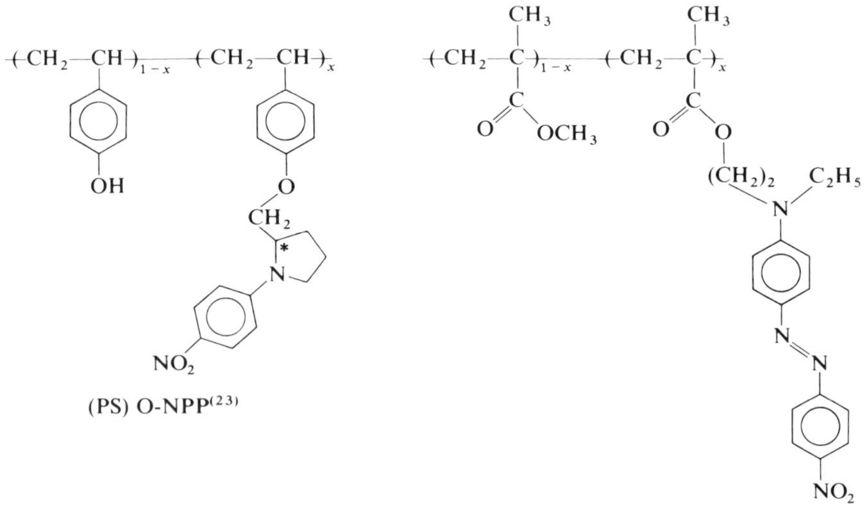

(PS) O-NPP(23)

DRI-MMA(17)

Fig. 3. Chemical structure of the two families of copolymers used to study the behavior of d_{33} as a function of x

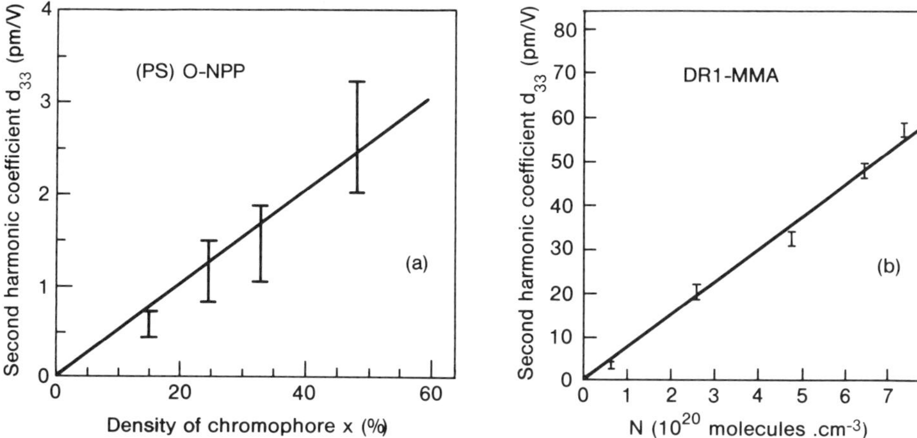

Fig. 4. Second harmonic coefficients d_{33} of (PS) 0-NPP films poled at 30 V μm^{-1} as a function of chromophore functionalization level for 1.064 μm incident radiation (a) and d_{33} of DR1-MMA films poled at 85 V μm^{-1} as a function of x for 1.064 μm incident radiation (b).

[17–21] or by modifying a reactive polymer such as poly(*p*-hydroxy-styrene) [22–24], poly(methylmethacrylate-co-(*N*-ethyl-anilino)ethyl methacrylate) [11,25], poly(hydrogen methyl siloxane) [26] or poly(allylamine hydrochloride) [27]. Chromophore functionalization levels from 0 to 100 percent can thus be achieved at will.

Studies of the nonlinear coefficient as a function of the dye content in two families of copolymers [17,23] (Fig. 3) have shown a fairly linear dependence (Fig. 4).

This result indicates that the noninteracting oriented molecular gas model also works for copolymers having at least 48 percent chromophore functionalization level.

As expected, an improvement of the stability of NLO properties of covalently functionalized polymer (DCV-MMA for example), Fig. 5, compared to guest-host polymer systems (DCV/PMMA), is observed [11] in Fig. 2.

The temporal stability of the orientational ordering obtained by poling covalently functionalized amorphous polymers has been investigated mainly by using (PS) O-NPP copolymers (Fig. 3). Again, the relaxation processes are not well identified. The d_{33} decay can be fitted to a two-exponential model [23,24] written as:

$$d_{33} = \underbrace{Ae^{-t/\tau_1}}_{\text{fast component}} + \underbrace{Be^{-t/\tau_2}}_{\text{slow component}} \qquad (2.3.1)$$

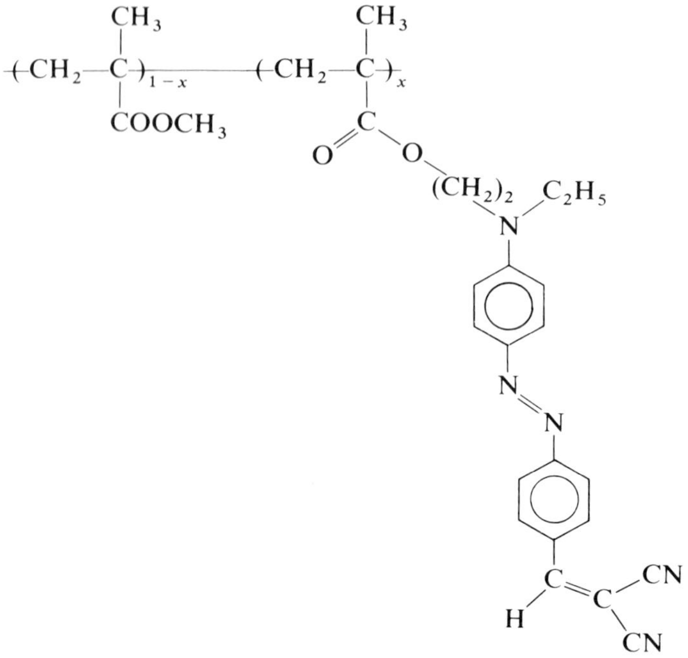

Fig. 5. Chemical formula of DCV-MMA copolymer.

τ_1 and τ_2 are respectively the short-term and long-term SHG decay lifetimes. The maximum derived τ_2 value is as large as 313 days (15 percent functionalization level), but τ_2 is dependent on the functionalization level and decreases as the chromophore content increases (Fig. 6). The behavior could be explained by the decrease in the hydroxy phenyl–hydroxy phenyl hydrogen bonding as the functionalization level increases. On the other hand, it has been found that the amplitude of the short-term process is connected to the presence of local free volume allowing the local movement of chromophores [28] (Fig. 7). τ_2 can be regarded as originated from the reorganization of entire polymer chains, which takes a much longer time than does the local movement.

The mobility of the NLO unit in the glassy state, which is responsible for the relaxation phenomenon, is strongly connected to the T_g of the polymer as shown in Fig. 8 [29]. A higher T_g provides a better stability.

Quadratic nonlinear devices that have been demonstrated so far were made with side-chain NLO polymers. Many examples of such copolymers

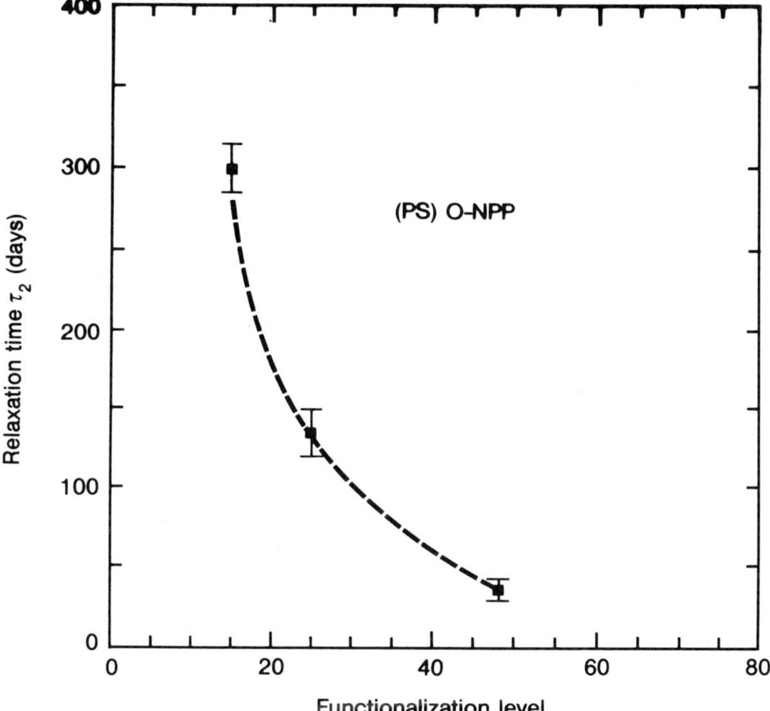

Fig. 6. Dependence of the second harmonic coefficient exponential lifetime τ_2 on functionalization level for (PS)O-NPP.

will be presented in Sec. 4 of this chapter. Most of the synthetic efforts have been devoted to the EOM application and, until now, very few organic materials that meet the requirements for efficient second harmonic generation in the visible short wavelength region have been published. The main property to be fulfilled for SHG is the material transparency in the visible region. This leads to the following limitations on the maximum absorption of the NLO unit:

green laser source application: $\lambda max < 350$ nm

blue laser source application: $\lambda max < 300$ nm

Table 1 summarizes the main properties of three families of SHG copolymers developed at Thomson-CSF.

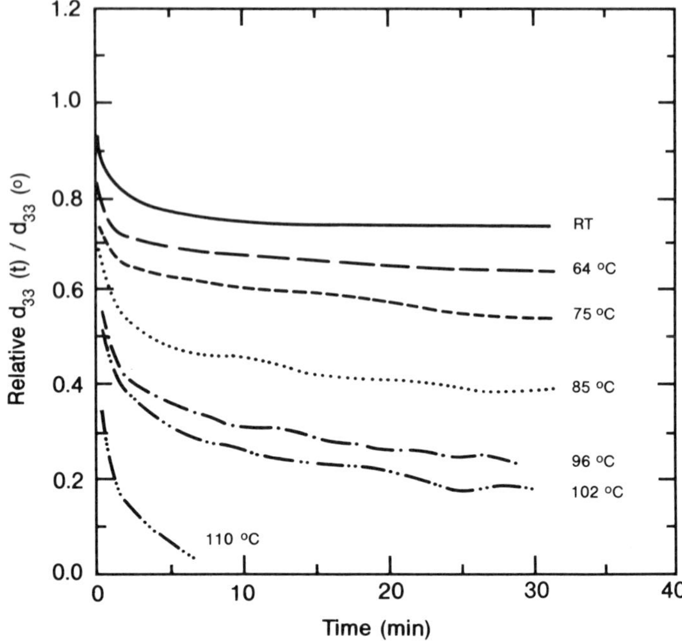

Fig. 7. Decay curves of (PS) O-NPP (48 percent functionalization level) at various aging temperatures.

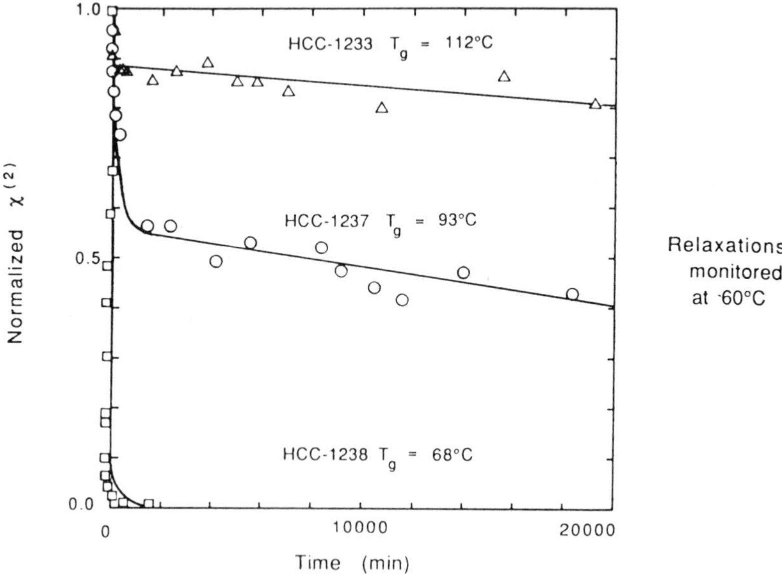

Fig. 8. Effect of T_g on poling relaxation.

Table 1.

Chemical Structure and Properties of SHG Copolymers Developed at Thomson-CSF

$$-\!\!\left(CH_2-\overset{CH_3}{\underset{C(=O)OCH_3}{C}}\right)_{1-x}\!\!-\!\!\left(CH_2-\overset{CH_3}{\underset{C(=O)O-(CH_2)_y-\boxed{\text{NLO Unit}}}{C}}\right)_x$$

NLO Unit	Index	x	y	T_g (°C)	λmax (nm)	d_{33} at 1.06 μm (pm V^{-1})	Applications
$-N(CH_3)-C_6H_4-C_6H_4-CN$	PLBP 80	0.45	3	112	348	12	Green laser source
$-O-C_6H_4-C_6H_4-CN$	PLBP 78	0.50	6	79	297	3.4	Blue laser source
$-N(CH_3)-C_6H_4-CN$	PLBP 85	0.49	3	89	293	3	Blue laser source

SHG has been demonstrated in guided wave structure with PLBP 80.

2.4. Main-Chain NLO Polymers

Study of nonlinear optical polymers with a chromophoric main-chain started a few years ago [30,31]. Figure 9 shows some examples of main-chain NLO polymers.

Main-chain NLO polymers are synthesized by step polymerization which yields low molecular weight materials (Mw < 10 000) with rather low T_g (in most cases $T_g < 90°C$). The magnitude of the nonlinearities exhibited by these polymers is comparable with that of side-chain NLO polymers having nearly identical NLO moieties and comparable number density. In addition, they show an improvement in stability over the corresponding side-chain NLO polymers, with a decay rate slowed by a factor of 4 to 5 [33].

2.5. Crosslinked Polymers

Development of NLO polymers is strongly connected to the possibility of obtaining a stable material. All the solutions that have been suggested so

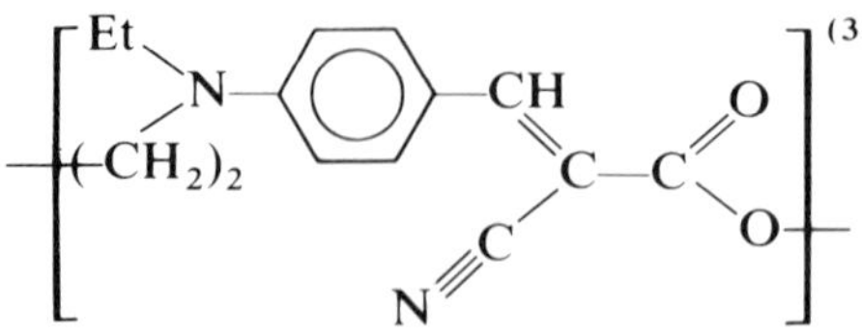

(32)

BISA-NA

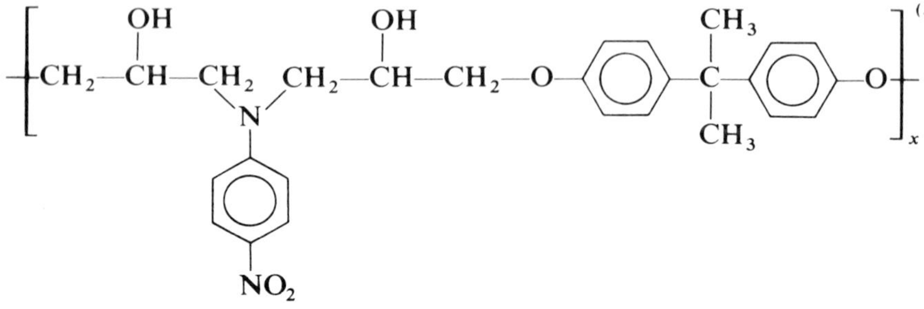

(33)

BISA-DMNPDA

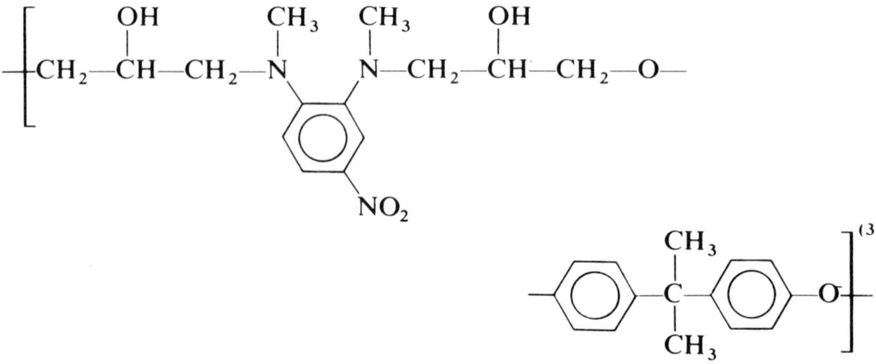

(33)

Fig. 9. Chemical formula of some main-chain NLO polymers.

far are based on the realization of a polymeric network. The NLO unit itself can be either embedded in the pores of the crosslinked polymer [34] or incorporated in the rigid three-dimensional polymeric network with one [35,36] or more than one chemical linkage between each NLO moiety and the network [37–39]. Whatever the approach is, crosslinking has to be achieved under an electric field in order to obtain the desired alignment of the NLO units.

This system, proposed by IBM (Fig. 10), requires the preparation of a soluble prepolymer which is then spin coated onto indium-tin-oxide coated

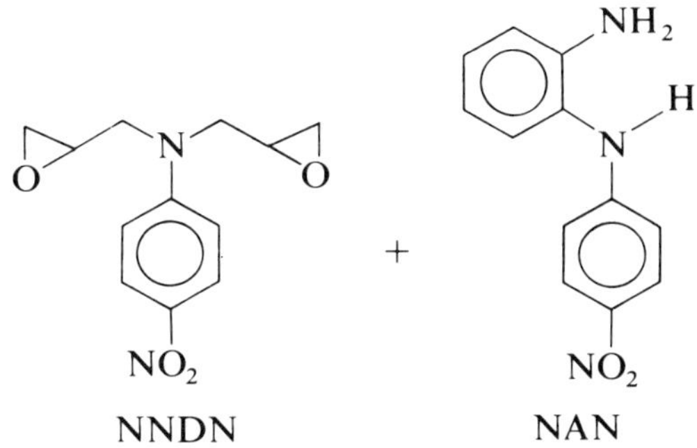

NH2
H
N
N
O
O
+
NO2
NO2
NNDN
NAN

Fig. 10. Chemical formula of the NLO-active monomers used by IBM [40].

wafer. The film thus obtained is fully cured under corona poling at 120°C for 16 hours. Although the curing process is quite tedious, stable nonlinearities with d_{33} as large as ~42 pm/V, even after extended annealing at ~80°C, have been achieved successfully.

Curing time can be reduced by using aliphatic amine instead of aromatic ones (Fig. 11). Nevertheless, several hours are still needed to obtain the final network [39].

An alternative route to get a stable NLO network is the use of photo-crosslinking reactions between the photosensitive chromophores functionalized into NLO molecules and the same or related chromophores appended into a linear polymer acting as a matrix [41] (Fig. 12).

Preliminary results showed that stable d_{33} values in the range of 15 to 30 pm V^{-1} over a long period of time at temperatures ranging from 60 to 85°C were obtained for modest concentration of the NLO molecules.

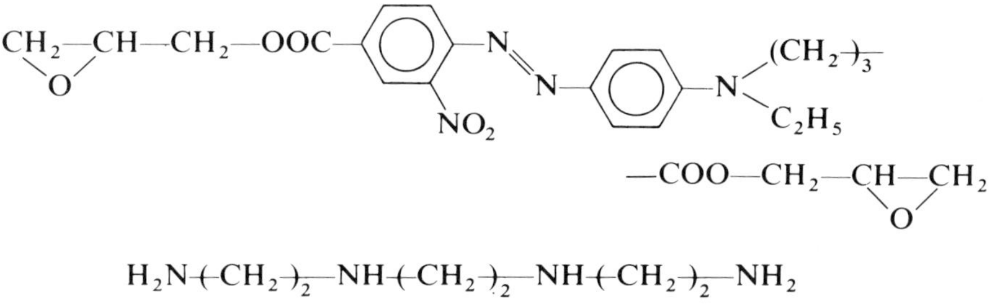

CH2—CH—CH2—OOC
O
N
N
NO2
N
(CH2)3
C2H5
—COO—CH2—CH—CH2
O
H2N(CH2)2—NH(CH2)2—NH(CH2)2—NH2

Fig. 11. Chemical formula of the epoxy-amine system developed at Thomson-CSF [39].

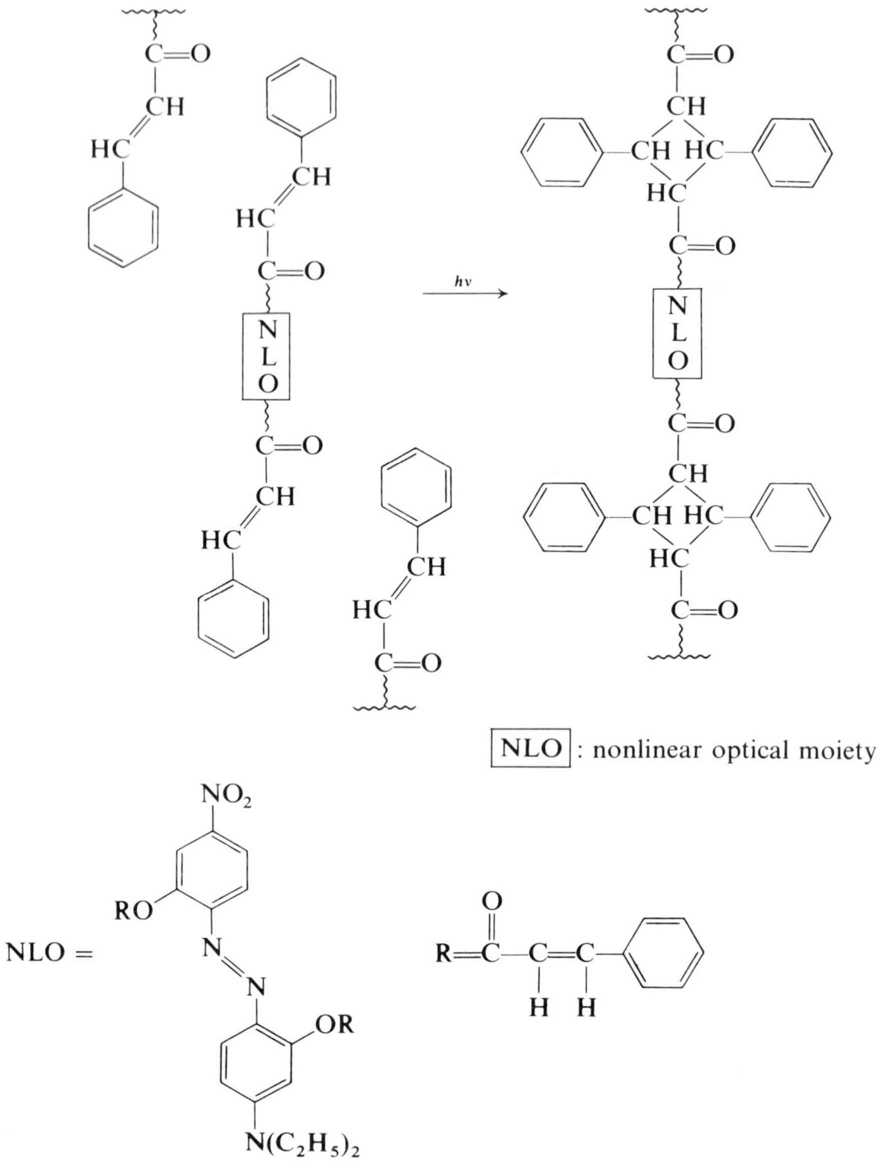

$\xrightarrow{h\nu}$

NLO : nonlinear optical moiety

NLO =

Fig. 12. Example of photo-crosslinkable systems [41].

To our knowledge, no NLO device has been achieved so far with crosslinked NLO materials. Nevertheless crosslinking seems to be the best way to obtain a stable NLO polymeric material. However some improvements have to be done to obtain an easy-to-use crosslinkable system leading to a stable material with all the desired properties (efficiency, transparency, mechanical properties . . .).

3. MEASUREMENTS OF THE NONLINEAR OPTICAL PROPERTIES

Once the polymer material has been synthesized, the electro-optic coefficients and the second harmonic coefficients can be measured. We will describe here the simplest technique to evaluate the nonlinear optical properties of the polymer.

First we will describe the common relations that are used in a poled polymer and that simplify the general calculations.

In a poled polymer we generally assume that $d_{33} = 3d_{31}$. This relation is due to the poling process: dipolar orientation, by the electric field, of the active moieties in an isotropic medium (the host polymer) [11]. It has been shown that other relations can be obtained if the host polymer has not an isotropic macroscopic structure (nematic structure of the active moieties [42] or alignment of the main chains inside the plane substrate [43]). This fact has to be kept in mind because in most of the measurements the hypothesis $d_{33} = 3d_{31}$ is done to calculate the d_{33} coefficient or the r_{33} coefficient.

One can notice that these coefficients are strongly dependent of the wavelength. Using the two-level model [44] we can obtained the following relationships:

$$d_{33}^{\omega} = d_{33}^{0} F(\omega),$$

where d_{33}^{0} is the static component and $F(\omega)$ a dispersion factor given by:

$$F(\omega) = \frac{1}{\left\{1 - \left(\frac{2\omega}{\omega_0}\right)^2\right\}\left\{1 - \left(\frac{\omega}{\omega_0}\right)^2\right\}},$$

where ω_0 is the frequency corresponding to the excited state.

The electro-optic and second-harmonic-generation coefficients are, of course, not independent [11]:

$$r_{ij,k}(-\omega; \omega, 0) = -\frac{4d_{k,ij}(2\omega'; \omega', \omega')}{n_i^2(\omega)n_j^2(\omega)} \frac{f_{ii}^{\omega} f_{jj}^{\omega} f_{kk}^{0}}{f_{kk}^{2\omega'} f_{ii}^{\omega'} f_{jj}^{\omega'}} \times \frac{(3\omega_0^2 - \omega^2)(\omega_0^2 - \omega'^2)(\omega_0^2 - 4\omega'^2)}{3\omega_0^2(\omega_0^2 - \omega^2)^2},$$

taking into account the local field corrections (f_{ii}^{ω}) and dispersion effects.

3.1. Second-Harmonic-Generation Coefficients

The second-harmonic-generation coefficients are measured by using a Q-switched YAG laser at 1.06 μm or 1.32 μm. Beam power of 10 kW is often used with a rate of 1 kHz and a pulse duration of 250 ns. It must be pointed out that the wavelength of measurement has to be outside the absorption band of the polymer. Otherwise the calculations have to be corrected by an absorption factor, and the results are not very accurate.

The polymer film is deposited onto an ITO-coated glass substrate. The thickness of the film has to be very low to avoid any phase-matching effect. This effect can be calculated, but the results are very dependent on the precision of refractive index measurements. In a doped polymer the coherence length is usually longer than a few microns, so a thickness of one micron for the polymer film is a good range of magnitude for the second-harmonic measurements. The ITO (indium tin oxide) layer is a transparent conducting electrode. The polymer film is then poled by the corona or parallel-plate method described above. In the case of the parallel-plate method, the counter electrode has to be removed before the second harmonic experiment. NaOH solutions can be used to remove aluminum electrode. The experiment is described on the following scheme (Fig. 13).

The experimental setup and the calculations have been described in detail by Jerphagnon and Kurtz [45].

3.2. Electro-optic Coefficients

Several methods can be employed to measure the electro-optic coefficients. We shall describe here the simplest one, based on the birefringence of the polymer film. The phase shift between TE and TM propagation is measured upon the applied electric field. As previously published [46], the polymer film to be measured is sandwiched between two electrodes. On an ITO-coated glass plate, a polymer film is spin coated. Then a second electrode in aluminum is evaporated onto the surface. The film is poled, as explained above. The sample is then set up on a θ motor of a goniometer while the photodiode is driven by a 2θ motor. The wavelengths that can be

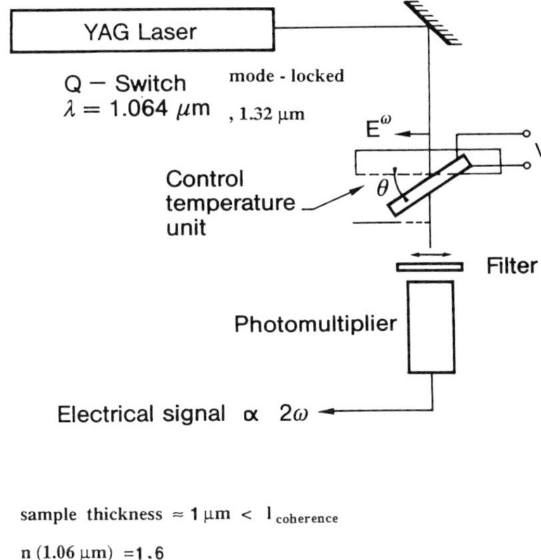

Fig. 13. Second-harmonic-generation coefficients measurement; experimental setup.

used are 632.8 nm of an HeNe laser, 830 nm of a laser diode, 1.06 μm and 1.32 μm from YAG lasers, or others from other laser beams. The static-phase shift between TE and TM propagation is modified by a babinet compensator positioned between the optical beam and the sample.

For the measurements, an electric voltage (10 Volts at 1 kHz) is applied on the sample, and the resulting modulation of the beam Im is detected with a lock-in amplifier. The general setup is represented in Fig. 14a.

The intensity of the reflected optical beam I varies when changing the phase shift $\Delta\phi$ between TM and TE components. The maximum value Ic of I occcurs when $\Delta\phi = \pi$. According to [46] an electro-optic signal Im at point A (Fig. 14b) must be of the opposite sign to the one at point B, and the signal at point O is equal to zero. It has been shown [47] that this is a rather suitable simplification when no absorption occurs nor multiple reflections. In this case the calculations will lead to

$$r_{33} = (3\lambda/4\pi)(Im/Ic)(\cos r_0/\sin^2 r_0)(1/n_0^3)(1/V)$$

for a poled polymer in which $r_{33} = 3r_{13}$ [11] and where r_0 is the incident angle in the polymer film, λ the incident wavelength, n_0 the refractive index

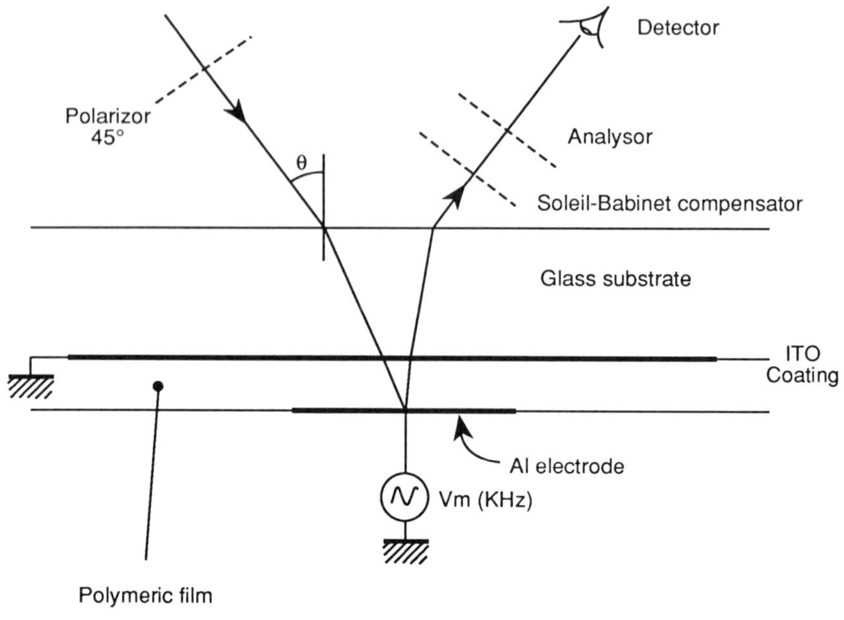

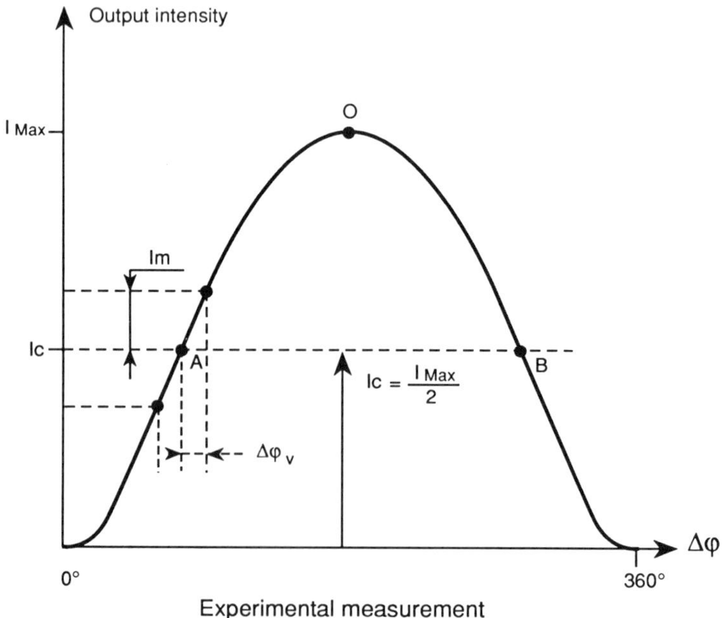

Fig. 14. Electro-optic coefficient measurement; experimental setup and experimental measurement.

of the polymer (the low birefringence does not affect this calculation). In this notation axis 3 is the poling axis, perpendicular to the plane of the substrate, and axis 1 is parallel to the plane of the substrate.

One can notice that when the simplifications are valid (no absorption and no multiple reflections), the thickness of the film is not needed. Moreover that means that the thickness of the polymer does not have to be homogeneous over the whole surface of measurement, even under the laser beam spot. This fact gives a great importance to this method to test rapidly the electro-optic coefficients of a newly synthesized polymer.

Of course if a whole electro-optic modulator has been set up, the measurement of the $V\pi$ gives the electro-optic coefficients (see Sec. 4.2).

4. REALIZATIONS

4.1. Passive Waveguides in Polymer Layers

The efficiency of nonlinear optic interactions depends directly on the optical intensity in the active material. As a consequence, guided wave optics plays an important role in that field as it allows high-power density to be obtained over long interaction lengths. In addition, the use of guided waves for electro-optic devices is also very attractive as it could lead to compact structures with increased performances (in particular low driving power and broadband characteristics).

The one-dimensional or bidimensional confinement of an optical beam in an active material implies the realization of a complex structure, defined by at least two media presenting different refractive indices. The active polymer and all the other materials involved in the structure must be transparent at the working wavelength with low scattering losses and processable in high-quality thin layers.

The ultimate aim being the realization of active devices with these guiding structures, a poling step is necessary leading to some limitations. For example it requires the realization of at least one electrode, and a buffer layer has to be processed between the active layer and the metallic layers to avoid coupling and absorption of the guided light in the electrode. In addition the poling field depends on the relative resistivity of the materials involved in the structure and a highly resistive buffer layer will significantly reduce the effective poling field applied on the active film.

In spite of these limitations, numerous guiding structures have been realized with polymers. Different criteria could be defined to evaluate the relative advantages of these structures including: the propagation losses intrinsic to the structure, the technological feasibility of the structure, the versatility of the structure, and the possibility to achieve a good polarization and to adapt the structure to the ultimate use.

In the following, after a theoretical approach of the dielectric optical waveguides and a brief description concerning planar waveguides, we will describe some of the structures used to obtain two-dimensional optical waveguides.

4.1.1. Theoretical Approach of Dielectric Optical Waveguides

In this chapter, we study a number of devices that involve propagation of electromagnetic waves in dielectric films [48] with thickness or geometries comparable to the optical wavelength.

4.1.1.1. Basic Planar Waveguide. The propagation of electromagnetic plane waves in homogeneous and symmetric dielectric slab waveguide obey Maxwell's equations. In isotropic dielectric medium with dielectric constant $[\varepsilon]$ and magnetic permeability μ_0, the wave equation in rectangular coordinates can be written symbolically as:

$$\nabla^2\psi_i - \mu_0\varepsilon\frac{\partial^2\psi_i}{\partial t^2} = 0, \tag{4.1.1}$$

where $i = x, y, z$; ψ is E_i or H_i and $\nabla^2 = \partial^2/\partial x^2 + \partial^2/\partial y^2 + \partial^2/\partial z^2$.

A mode of a waveguiding structure is an allowable field configuration that satisfies Maxwell's equations and all of the boundary conditions. We will assume that a dielectric waveguide propagates electromagnetic energy in a given direction z (longitudinal axis of the waveguide) with a dependence in the form $\exp(-j\beta z)$. β is the longitudinal component of the propagation vector. We will also assume that the permittivity $\varepsilon(x, y)$ does not depend on z but can vary with x and y. In Eq. (4.1.1) the vectorial form of the field is taken as:

$$\mathbf{E}(\mathbf{r}, t) = \mathbf{E}(x, y)\exp j(\omega t - \beta z). \tag{4.1.2}$$

If we would like to express the transverse components in terms of longitudinal components, from Eqs. (4.1.1) and (4.1.2) we obtain:

$$
\left.\begin{aligned}
E_x &= \frac{-j}{\kappa^2}\left(\omega\mu\frac{\partial H_z}{\partial y} + \beta\frac{\partial E_z}{\partial x}\right)\\
E_y &= \frac{-j}{\kappa^2}\left(\beta\frac{\partial E_z}{\partial y} - \omega\mu\frac{\partial H_z}{\partial x}\right)\\
H_x &= \frac{-j}{\kappa^2}\left(\beta\frac{\partial H_z}{\partial x} - \omega\varepsilon\frac{\partial E_z}{\partial y}\right)\\
H_y &= \frac{-j}{\kappa^2}\left(\beta\frac{\partial H_z}{\partial y} + \omega\varepsilon\frac{\partial E_z}{\partial x}\right)
\end{aligned}\right\} \tag{4.1.3}
$$

with $\kappa^2 = \omega^2\mu\varepsilon - \beta^2$.

From equation (4.1.3) we obtain a partial differential equation in E_z and H_z only.

$$\nabla_T^2\psi_z + \kappa^2\psi_z = 0 \tag{4.1.4}$$

where $\nabla_T^2 = \partial^2/\partial x^2 + \partial^2/\partial y^2 = \Delta_T$ and $\psi_z = E_z, H_z$.

These equations are modified forms of a wave equation. Equations (4.1.4) in E_z and H_z are uncoupled equations in the longitudinal component of the fields only. The coupling of the two longitudinal fields are generally governed by the boundary conditions of the problem. The various types of modal solutions that can exist in a dielectric waveguide are listed below.

TEM (transverse electromagnetic)	$E_z = 0,\quad H_z = 0$
TE (transverse electric)	$E_z = 0,\quad H_z \neq 0$
TM (transverse magnetic)	$H_z = 0,\quad E_z \neq 0$
HE or EH (hybrid)	$H_z \neq 0,\quad E_z \neq 0$

4.1.1.2. Symmetric Dielectric Slab Waveguide. The waves in a symmetric waveguide with $\varepsilon_1 > \varepsilon_2$ propagating in an infinite slab geometry in the y direction (Fig. 15a) can be expressed as

$$\mathbf{E} = \mathbf{E}_0(x)\exp j(\omega t - \beta z),$$

with no variation in the field distribution in the y direction ($\partial/\partial y = 0$).

For TE modes ($E_z = 0$, $H_z \neq 0$), the field components E_y, H_x, H_y exist; and for TM modes ($H_z = 0$, $E_z \neq 0$), the field components are E_x, E_y, H_y.

In TE mode case, only the E_y component of the electric field exists. The functional form of E_y is

$$E_y(x, z) = E_y(x)\exp j(\omega t - \beta z), \tag{4.1.5}$$

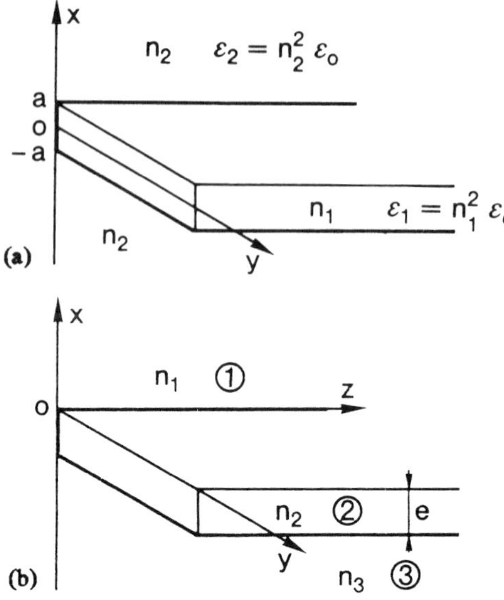

Fig. 15. Basic symmetric (a) and asymmetric (b) dielectric slab waveguide.

and the wave Eq. (4.1.4) is of the form

$$\frac{d^2E_y}{dx^2} + \kappa^2 E_y = 0, \tag{4.1.6}$$

with $\kappa^2 = \omega^2 \varepsilon \mu_0 - \beta^2$.

The solution for Eq. (4.1.6) will be different inside the slab and in the surrounding medium. Solutions for E_y inside the slab can be written as:

$$E_{y1} = (A \cos ux + B \sin ux) \exp(-j\beta z); \qquad |x| \leq a,$$

where $u^2 = n_1^2 k_0^2 - \beta^2$, and $k_0 = 2\pi/\lambda$.

This solution can be decomposed into even and odd modes:

$A \cos ux$	even TE mode
$B \sin ux$	odd TE mode

In the cladding, because the field approaches zero as x approaches infinity, the solution is

$$E_{y2} = C \exp(-w(|x| - a) - j\beta z); \qquad |x| \geq a,$$

with: $w^2 = \beta^2 - n_2^2 k_0^2$. u and w are positive real constants to be determined.

$$H_{z1} = juA/\omega\mu \sin ux \exp(-j\beta z); \qquad |x| \le a,$$
$$H_{z2} = \mp jwC/\omega\mu \exp(-w(|x| - a) - j\beta z); \qquad |x| \ge a.$$

The $(-)$ sign is used with $x \ge a$ and $(+)$ for $x \le -a$.

From the boundary conditions applied at the core cladding interface for the tangential field components E_y and H_z at $x = \mp a$: $H_{z1} = H_{z2}$, $E_{y1} = E_{y2}$ it follows that:

$$\text{tg}(ua) = w/u. \tag{4.1.7}$$

This equation is the so-called characteristic equation for the even TE mode.

For the odd TE modes of the dielectric slab waveguide, the characteristic equation is of the form:

$$\text{tg}(ua) = -w/u. \tag{4.1.8}$$

A simple graphical form can be used to find a solution for these characteristic equations. The u and w relations can be combined to give the relation:

$$V^2 = a^2(u^2 + w^2) = k^2a^2(n_1^2 - n_2^2). \tag{4.1.9}$$

This equation can be plotted graphically in the form of a circle

$$V^2 = U^2 + W^2, \tag{4.1.10}$$

where $W = aw$ and $U = au$.

With this representation, the effect of increasing the frequency is to increase the radius of the circle, and for the even TE modes, the characteristic equation takes the form:

$$W = U\,\text{tg}(U), \tag{4.1.11}$$

rewriting in terms of U and W, for the odd TE modes Eq. (4.1.8) becomes:

$$W = -U\,\text{cotg}(U). \tag{4.1.12}$$

Equations (4.1.10), (4.1.11), and (4.1.12) can be plotted on an U, W diagram to obtain the propagation constants of the guided TE modes of the waveguide.

The intersections of the circle in Eq. (4.1.10) and the curve of Eq. (4.1.12) define the propagation conditions for the modes in the waveguide structure. Each intersection with a $W > 0$ corresponds to a confined mode, and their number is proportional to V.

The general features of TM modes are similar to those of TE modes. The characteristic equations for the TM even and odd modes are respectively:

$$\mathrm{tg}(ua) = n_1^2/n_2^2(w/u) \qquad \text{(TM even modes)} \tag{4.1.13}$$

$$\mathrm{tg}(ua) = -n_2^2/n_1^2(u/w) \qquad \text{(TM odd modes)} \tag{4.1.14}$$

The corresponding values of W are smaller for TE models compared to a TM mode of the same order, indicating a lesser degree of confinement for the TM modes and a larger fraction of the total TM mode power propagates in the cladding.

4.1.1.3. Asymmetric Dielectric Slab Waveguide. Basic asymmetric waveguide structure is shown in Fig. 15b. By considering that there is no variation in the y direction (infinite slab geometry approximation $\partial/\partial y = 0$), the transverse electric field distribution is given by

$$E_y = A\exp(-w_1 x); \qquad 0 \le x < \infty$$

$$E_y = B\exp j(ux) + C\exp -j(ux); \qquad -e \le x \le 0$$

$$E_y = D\exp(w_2(x+e)); \qquad -\infty \le x \le -e.$$

where $w_1^2 = \beta^2 - k_0^2 n_1^2$ and $w_2^2 = \beta^2 - k_0^2 n_3^2$. Applying the continuity requirement at both interfaces leads to the eigenvalue equation:

$$u_m e = \mathrm{Arctg}\left(\frac{\dfrac{w_{1m}}{u_m} + \dfrac{w_{2m}}{u_m}}{1 - \dfrac{w_{1m}w_{2m}}{u_m^2}}\right) + m\pi,$$

where the symbol m denotes the mth confined TE mode characterized by β_m.

For the fundamental TE mode, this equation becomes:

$$\tan(ue) = \frac{w_1 + w_2}{u\left(1 - \dfrac{w_1 w_2}{u^2}\right)}.$$

In the same manner, for the TM mode the eigenvalue equation is given by:

$$u_m e = \mathrm{Arctg}\left(\frac{\dfrac{\tilde{w}_{1m}}{u_m} + \dfrac{\tilde{w}_{2m}}{u_m}}{1 - \dfrac{\tilde{w}_{1m}\tilde{w}_{2m}}{u_m^2}}\right) + m\pi,$$

where $\tilde{w}_1 = \left(\dfrac{n_1}{n_2}\right)^2 w_1$ and $\tilde{w}_2 = \left(\dfrac{n_2}{n_3}\right)^2 w_2.$

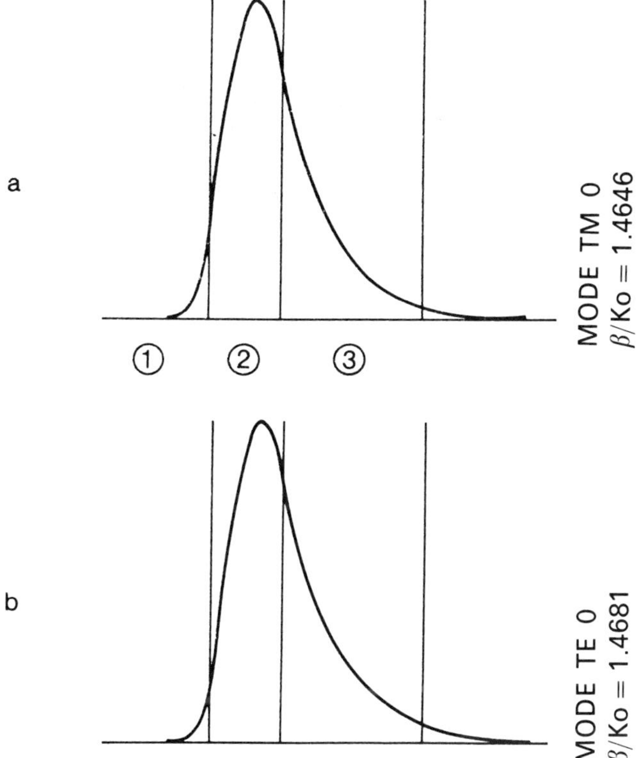

Fig. 16. Field distribution of TE_0 and TE_1 modes.

The field distributions of TE modes are plotted in Fig. 16. Notice the different structure of the fields in the guide for the even and odd modes and the presence of the evanescent fields in the various cladding.

4.1.1.4. Two-Dimensional Waveguide. A rectangular dielectric waveguide of dimensions 2a, 2d, as shown in Fig. 17 whose refractive index n_1 is surrounded on four sides by materials of indices n_2, n_3, n_4, n_5 is simi!ar to a two-dimensional waveguide.

The modes supported by such waveguide structure are of two types. The first type, $E_{x\mu}$, is polarized in the x direction; the other type, denoted $E_{y\mu}$, is polarized principally in the y direction.

If the assumption is made that the cladding indices are slightly smaller than the core index n_1, then the modal eigenvalue problem is similar to that describing the propagation in two slab waveguides at right angles.

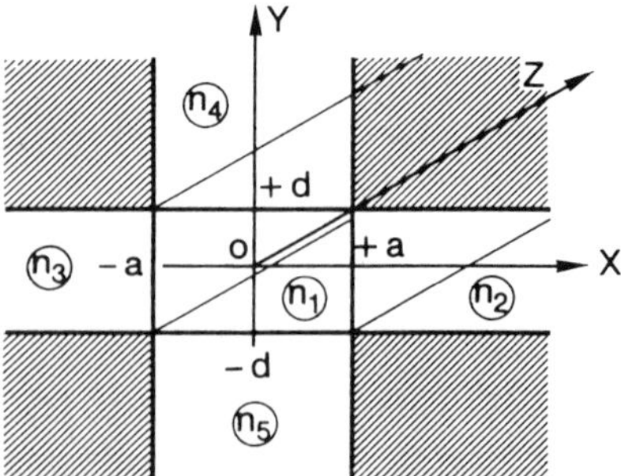

Fig. 17. Two-dimensional waveguide.

4.1.2. Slab Waveguides

The simplest way to realize a waveguide with polymers is to spin coat a solution composed of the polymer dissolved in a solvent on a low index substrate (e.g., silica or quartz) and to cure this sample around the glass temperature of the polymer to evaporate the solvent.

This simple structure, easy to modelize, is used to characterize the mechanical quality (such as adhesion properties of the film or roughness of the upper face) and optical properties (such as optical losses or effective refractive index) of the polymeric layer.

Coupling light into and out of a waveguide is a problem, especially to achieve high coupling efficiencies. A good overlap between the different waves involved is essential. In this chapter different techniques will be presented. Prism couplers and grating couplers are well adapted to excite guided modes in slab waveguides with a laser beam.

In a multimode waveguide, from the measured effective index values of at least two modes, it is possible to know precisely the thickness and the refractive index of the guiding layer.

The same kind of results are possible with grating couplers. This technique supposes the realization of a grating, which represents an additional technological step.

4.1.3. Bidimensional Waveguides

Bidimensional confinement structures in polymers were first studied because of their potential as passive waveguides for integrated optic applications. The possibility of realizing low-cost optical waveguides on nearly any kind of substrate gave rise to a real interest in this field. Another fundamental point is relative to the coupling possibilities offered by the structure: the refractive indices of the polymers are small and close to the refractive index of optical fibers (1.45), thus structures with input and output ends perpendicular to the propagation direction can be easily coupled into, either with an optical fiber put in close proximity to the waveguide side or simply by focusing a laser beam onto the side. To achieve high coupling efficiency, the waveguide mode size must be comparable with the focused laser beam or the fiber's mode size. One of the main criteria to evaluate the potential interest of the waveguide is the propagation losses brought by the structure. There are two types of contributions: scattering losses due to the opto-geometrical definition of the guiding zone and absorption of the materials involved. To measure the losses of bidimensional waveguides different methods are used. The measurement of the light scattered under the waveguide is one of them. A cutback method is also frequently used; it consists of measuring the total losses (input and output coupling losses and propagation losses) and separating the different contributions by cutting the sample to reduce its length (assuming that the coupling losses do not change in each experiment). A method based on the Fabry-Perot resonator can also be used [49]. The mirrors of the resonator are the waveguides' end faces. This method is well adapted to low-loss monomode waveguides. It is limited by the small reflection coefficient, typically 5 percent for polymer waveguides, and always gives pessimistic values (end-face parallelism and quality of the mirrors).

Two types of bidimensional waveguide fabrication technique in polymer films were studied: the first results from a local modification of the guiding-layer refractive index, either by a physical or chemical process; the second one consists of spin coating the guiding layer on a substrate previously processed to obtain the lateral confinement. The first one involves mainly UV-sensitive polymers. It is an attractive method because it needs only the realization of a classical UV mask. Several UV modifications can increase the refractive index.

Up to now only two second-order nonlinear optical devices using bidimensional waveguides based on a first type method have been reported. Under the RACE programme of the European Community, project 1019 "polymeric optical switches," M. B. J. Diemeer (PTT Research Neher Laboratories) has reported a photobleaching method to realize bidimensional waveguides in a side-chain polymer with the nonlinear molecule 4-dimethylamino-4′-nitrostilbene (DANS) [50]. In that case the bleaching effect goes with a refractive index decrease (0.03 at 1335 nm for both polarization in a completely bleached film). It is due to a conformational change (*trans-cis*) in the active molecule. Monomode waveguides were produced after exposure to an unfiltered 200 W Hg lamp for one to three hours through a mask.

Average losses of 0.9 dB/cm for the polymer deposited on a microscope slide and 1.2 dB/cm in a multilayer structure compatible with the realization of a phase modulator were measured at 1335 nm. In spite of the long exposure time, this method, technologically one of the simplest, is well adapted to the realization of electro-optic modulators. It would be interesting to know the stability to temperature and light exposure of the waveguides' characteristics.

The other method of the first type was proposed by a Lockheed R&D Division team. It is based on the birefringence induced by the orientation of the active molecules. The poling process induces an extraordinary axis along the poling-field direction. In most cases, n_e is greater than the refractive index of the unpolled polymer. This implies that the refractive index is increased for a light polarization corresponding to the highest element of the nonlinear susceptibility tensor: χ_{333}. J. I. Thackara *et al.* [51,52] have reported the realization of bidimensional waveguides based on the structure described in Fig. 18. Two types of active polymers were studied, PC6S and C22 from Hoechst Celanese Research Company. The refractive index increase for a light polarization parallel to the poling orientation was large for PC6S (0.06 at 830 nm) and small for C22 (0.005 at 830 nm). For electro-optic applications this method has a great drawback: If the poling electrode is also used to carry the electric modulation signal, the overlap between the modulation field and the electromagnetic field of the guided mode will not be optimized. For a rib electrode of 8.5 μm the measured mode spot size at 632.8 nm was 25 μm. It seems significant that in recent publications the photobleaching method presented above was used by this team to realize the bidimensional confinement in optical waveguide structures.

With this type of technique, the waveguide boundaries are defined by a gradual variation of the refractive index. The UV mask or the electrode edge roughness are significantly reduced on the waveguides' side boundaries; this leads to low diffraction losses brought by the structure.

The etching of the active layer to realize either a rib or a ridge is an extreme case of the first type. In that case the diffraction losses brought by the edge

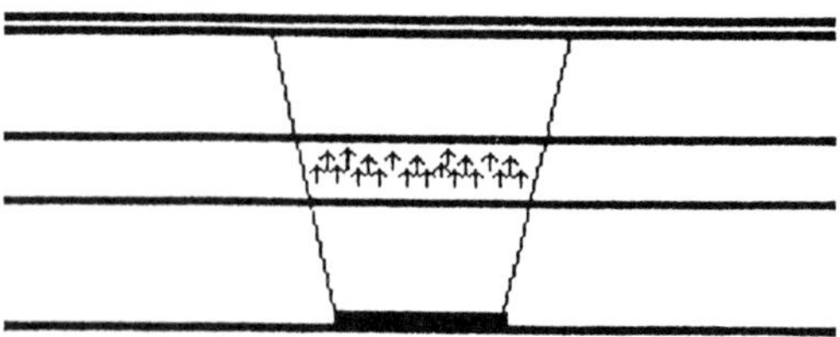

Fig. 18. Two-dimensional confinement obtained by birefringence induced by poling.

roughness are critical. D. R. Haas *et al.* [53] from Hoechst Celanese have obtained propagation losses of 2 dB/cm at 1.3 μm with this type of structure, the active material was referred to as HCC 1232.

The second type of fabrication technique, based on the active polymer spin coating on a substrate previously processed to assure the lateral confinement, is more versatile than the first one. The principle is no longer to increase the refractive index of the active material but to increase the effective index of the structure. Two solutions have been studied: either a local increase of the active layer thickness with an inverted ridge preformed in the substrate, or a high index strip defined on or in the substrate.

P. R. Ashley *et al.* [54] have realized inverted ridge in a UV-curing optical epoxy layer. To complete the waveguide, the active polymer (C6ST provided by Hoechst Celanese) is spin coated on this structure. Propagation losses of 3 dB/cm at 1.3 μm were measured in single-mode waveguides.

V. Lemoine *et al.* [55] from Thomson-CSF realized grooves on a silicon substrate; a low-refractive-index polymer was spin coated as a buffer layer. In order to achieve an optical waveguide the active polymer (copolymer of MMA in DR1) was spin coated (Fig. 19). Propagation losses of nearly 1 dB/cm at 1.3 μm were measured in monomode waveguides. An inverted ridge structure was also realized. A groove was etched directly in the first buffer layer (Fig. 20). In that case, the losses were 1.6 dB/cm at 1.3 μm in monomode waveguides.

Strip-loaded waveguide consists of the patterning of a dielectric strip on top of or below the active layer. The dielectric refractive index has to be greater than the refractive index of the material in which it is patterned. This method was used by P. D. Townsend (Bellcore) [56] to realize waveguides in a polymer presenting high third-order effects: poly (4-BCMU). The

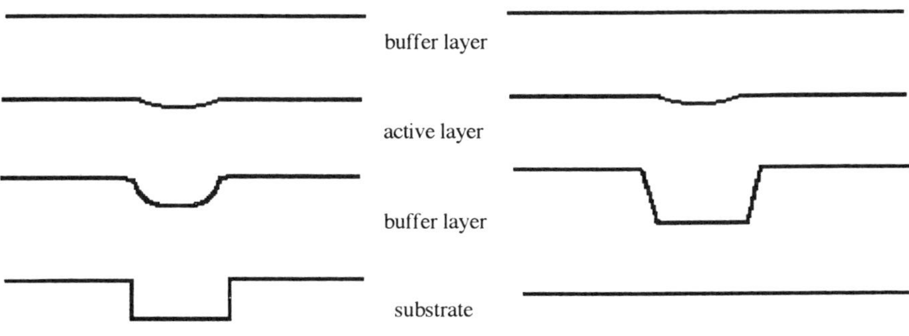

Fig. 19. Grooves in Si.

Fig. 20. Inverted ridge.

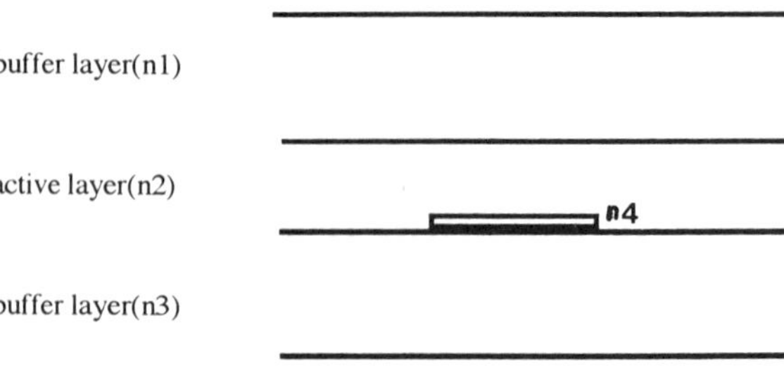

n4 > n2

Fig. 21. High-index buried-rib structure.

substrate is patterned with high-index ion-exchanged channels prior to spin coating of the active polymer.

A similar technique was developed at the Thomson-CSF central research laboratory [54]: the bidimensional confinement is obtained by burying a very thin rib of high-refractive-index material between the polymeric active layer and the substrate (as shown in Fig. 21). In that case, the refractive index of the rib material has to be greater than the active layer refractive index. The advantage of this fabrication technique over the previous one is that it is easier to control the opto-geometrical definition of rib than, for example, those of a diffused strip. On the other hand, the losses due to the roughness of the rib walls are more important. Different ribs were realized. Monomode waveguides were fabricated with 2 μm width 800 Å thick SiNH ($n = 1.8$ at 1.3 μm) ribs.

4.2. Electro-optic Modulation

We have seen that a wide variety of solutions was possible to achieve bidimensional confinement in polymer layers. Most of these studies deal with the realization of electro-optic modulators. Electro-optic polymers offer potential advantages over conventional materials for integrated electro-optic modulators:

- The material properties: Polymers with nonlinear coefficients comparable to those of $LiNbO_3$ have been synthetized and larger coefficients are potentially achievable. Polymers have low dielectric constant (typically $\varepsilon_r = 3$, $\varepsilon_r = 30$ for $LiNbO_3$), implying low capacitance and wide frequency bandwidth. On the other hand, the refractive index of the polymers is small, and it appears as the third power in the expression of the induced phase shift in a Pockels' effect. The dispersion of the dielectric constant with frequency is weak (no ionic contribution), implying small velocity mismatch between electrical and optical signals.

- The fabrication technology: Polymers are fully compatible with thin-film technologies. This advantage has already been underlined in the previous paragraph for the realization of confined structures. It also allows the fabrication of microstrip lines, easier to realize and more efficient than coplanar lines realized on crystals. In the case of microstrip lines, if the strip electrode is larger than the mode size, the overlap integral between the electromagnetic wave and the modulation electric field can be optimized because the electric field lines are straight. With coplanar lines the useful electric lines of the modulation field are bent, which can reduce this overlap integral and consequently the modulation efficiency. Confined structures can be processed on nearly any kind of substrate and then electro-optic modulators could be easily integrated in more complex hybrid devices. The processing technology requires standard semiconductor fabrication equipment and is compatible with mass production.

4.2.1. The Pockels' Electro-optic Effect

This chapter deals with the electric modulation of an optic signal using the linear Pockels' effect.

To combine the integrated optic and the Pockels' effect equations, we will consider the effect as a two-wave mixing problem, one of the two wave frequencies being zero. The nonlinear polarization induced by an external electric field ε in a medium characterized by its nonlinear second-order susceptibility $\chi^{(2)}$ seen by an electromagnetic wave $\mathbf{E}(\omega)$ is defined by:

$$P_j(\omega = \omega + 0) = \varepsilon_0 \chi_{jik}^{(2)}(\omega = \omega + 0) E_i(\omega)/\varepsilon_k(0).$$

From Maxwell's equations we know:

$$\Delta E_j - \mu_0 \varepsilon_0 \varepsilon_j \frac{\partial^2 E_j}{\partial t^2} = \mu_0 \frac{\partial^2 P_j}{\partial t^2}. \tag{4.2.1}$$

Using a perturbation method [57] we consider that the external electric-field influence on the electromagnetic wave is simply of the form:

$$E_j(x, y, z, t; \varepsilon) = \tilde{E}_j(x, y, z, t; 0) T_j(z, \varepsilon)$$

where $\tilde{E}_j$ is the electromagnetic field when the external electric field is not applied. Consequently we have:

$$\Delta\tilde{E}_j - \mu_0\varepsilon_0\varepsilon_j \frac{\partial^2 \tilde{E}_j}{\partial t^2} = 0,$$

and so:

$$\tilde{E}_j\left(\frac{\partial^2 T_j(z, \varepsilon)}{\partial z^2} - 2j\beta_j \frac{\partial T_j(z, \varepsilon)}{\partial z}\right) = \mu_0 \frac{\partial^2 P_j}{\partial t^2}. \tag{4.2.2}$$

Assuming "slow variations"

$$\left|\frac{\partial^2 T_j(z, \varepsilon)}{\partial z^2}\right| \ll \left|2\beta_j \frac{\partial T_j(z, \varepsilon)}{\partial z}\right|,$$

and by taking the product of Eq. (4.2.2) by $\tilde{E}_j^*$ and after integration over the whole xy plan, we obtain:

$$\frac{\partial T_j(z, \varepsilon)}{\partial z} = j \frac{\mu_0 \omega^2 \varepsilon_0}{2\beta_j} T_j(z, \varepsilon) \frac{\int\int_{-\infty}^{+\infty} \chi_{jjk}\varepsilon_k \tilde{E}_j \tilde{E}_j^* \, dx\, dy}{\int\int_{-\infty}^{+\infty} \tilde{E}_j \tilde{E}_j^* \, dx\, dy}. \tag{4.2.3}$$

We have assume that no coupling between TE and TM modes occurs when the electric field is applied (we have equalized the two first indices of the nonlinear tensor). In the case of modulators realized with polymers there is generally no nonlinear coefficient corresponding to this coupling, the applied electric field is parallel to the optical extraordinary axis (structure and material axis).

T_j can be written as: $e^{j(2\pi/\lambda)\overline{\Delta\beta}_j z}$ with:

$$\overline{\Delta\beta}_j = \frac{1}{2\bar{\beta}_j} \frac{\int\int_{-\infty}^{+\infty} \chi_{jjk}\varepsilon_k \tilde{E}_j \tilde{E}_j^* \, dx\, dy}{\int\int_{-\infty}^{+\infty} \tilde{E}_j \tilde{E}_j^* \, dx\, dy}. \tag{4.2.4}$$

An important remark concerning this result is that the $\Delta\beta$ depends on an overlap integral between the applied electric field and the electromagnetic field. For most of the electro-optic modulators realized with polymers, the active layer is sandwiched between two passive buffer layers and two electrodes whose size is bigger than the electromagnetic wave extension. Consequently we will generally assume that:

$$\Delta\beta_j = \frac{1}{2\beta_j} \sum_k \chi_{jjk}\varepsilon_k \theta, \tag{4.2.5}$$

where

$$\theta = \frac{\int\int_{AL} \tilde{E}_j \tilde{E}_j^* \, dx\, dy}{\int\int_{-\infty}^{+\infty} \tilde{E}_j \tilde{E}_j^* \, dx\, dy}, \tag{4.2.6}$$

with AL representing the "Active Layer." Knowing that:

$$\chi_{jjk}(\omega = \omega + 0) = -n_j^4 r_{jjk}(\omega = \omega + 0) \qquad \left(\Delta\left(\frac{1}{n^2}\right) \approx -2\frac{\Delta n}{n^3}\right),$$

we deduce:

$$\left.\begin{aligned} \Delta\phi_{TE} &= -\frac{2\pi}{\lambda} L \frac{n_0^4 r_{13}\varepsilon}{2\bar{\beta}_{TE}} \theta, \\ \Delta\phi_{TM} &= -\frac{2\pi}{\lambda} L \frac{n_e^4 r_{33}\varepsilon}{2\bar{\beta}_{TM}} \theta. \end{aligned}\right\} \tag{4.2.7}$$

Most of the time, the difference between the two refractive indices and also the two effective indices is neglected in this expression: $n_0 \approx n_e \approx \bar{\beta}_{TM} \approx \bar{\beta}_{TE}$.

4.2.2. First Demonstrators

Demonstration of many devices is possible using the Pockels' effect in nonlinear optic polymers. It goes from phase modulators to directional couplers and for intensity modulators from modulators between cross polarizers to Mach Zehnder interferometers. Most of the time these modulators are fabricated to work at 1.3 or 1.5 μm, the appropriate spectral range for optical telecommunication applications.

The simplest modulator consists of a slab waveguide sandwiched between two electrodes (as shown in Fig. 22). The phase velocity of light propagating in the waveguide depends on the applied electric field. Mounted between cross polarizers whose axes form an angle of 45° with the axes of the structure, this phase modulator becomes an intensity modulator.

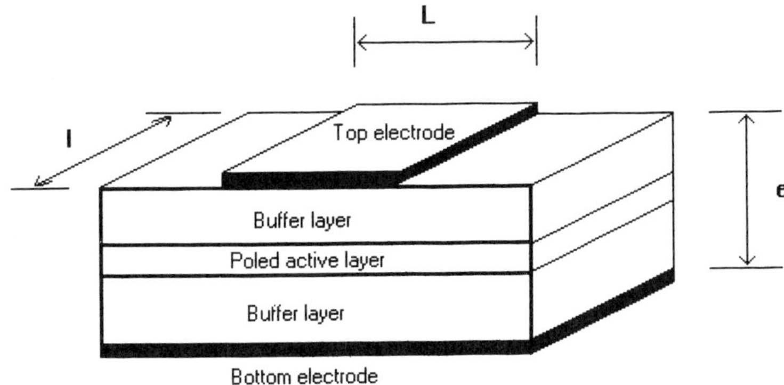

Fig. 22. Slab modulator.

The difference in phase delay between the TE and TM modes is:

$$\Delta\phi = -\frac{2\pi}{\lambda} L \frac{n^3 r_{\text{eff}} \varepsilon}{2} \theta, \tag{4.2.8}$$

with $r_{\text{eff}} = r_{33} - r_{13} = 2/3 r_{33}$.

The output intensity is modulated with this effective electro-optic coefficient:

$$I(E) = I_0 \left[1 - \cos\left(\frac{2\pi}{\lambda} L \frac{n^3 r_{\text{eff}} \varepsilon}{2} \theta \right) \right]. \tag{4.2.9}$$

An important characteristic parameter of an electro-optic modulator is the $V\pi$: the applied voltage yielding $\Delta\Phi = \pi$. It defines the electric signal useful to achieve the most efficient modulation. It is proportional to the inverse of the nonlinear coefficient multiplied by the useful length. It also takes into account structural considerations, such as the distance between the electrodes or the resistivity of the different layers. To achieve a linear modulation of the intensity with the electric signal it is necessary to work around $\Delta\Phi = \pi/2$, with an amplitude of the electric signal lower than the $V\pi$. This working point is fixed either electrically (with a bias voltage $= V\pi/2$) or optically (with a Babinet compensator). Generally the $V\pi$ does not depend on the modulation frequency except at very low frequency, where thermal effects or very slow electrical effects can contribute to the optical modulation.

The Mach-Zehnder interferometer is an advantageous solution to minimize the $V\pi$ by structural considerations. First the effective electro-optic coefficient is directly the r_{33} coefficient. The working point is fixed by changing the length or the shape of one arm (to create a phase delay of $\pi/2$ between the two waves). The realization of the Y junctions with low losses and exhibiting a good symmetry to equalize the power in each arm is the main technological difficulty of the Mach-Zehnder interferometer.

E. Van Tomme *et al.* have published [58] results concerning a Mach-Zehnder interferometer working at 1.32 μm. The electrodes were 1.4 cm long, the $V\pi$ was measured to be 4.5 V at dc operation and 11 V/cm at 200 Hz. The on-off ratio was in excess of 10 dB. The active material is a side-chain homopolymer with DANS as side group (provided by Akzo Research Laboratories) exhibiting an r_{33} coefficient of 34 pm/V at 1.32 μm; waveguides are realized by photobleaching.

In the same paper the fabrication of a directional coupler with the same polymer and technology is also reported. The coupling constant between two waveguides depends on the distance between them, on the refractive index of the different media involved in the guiding structure and on the propagation constant of each mode (the directional coupler can be modelized either considering the even and odd supermode of the complete structure or the coupling between the modes of each waveguide). Using the electro-optic effect to modify this coupling constant efficiently both waveguides must be driven independently. An integrated $2 * 2$ switch was fabricated with a switching voltage of 7.5 V for a 14-mm-long electrode and with a modulation depth of 17 dB.

The same polymer was used by T. E. Van Eck *et al.* from Lockheed [59]. to realize an active complementary optical tap. This system distributes an electrically controlled optical energy to different receivers with a single laser source. The realized device consists of a main waveguide that carries the optical power, on both sides of which single mode waveguides are disposed symmetrically. The light coupled in the side waveguides is controlled with an electric signal. The system is symmetric to maintain optical power in the main waveguide independent of the power coupled in the side complementary waveguides. The realized coupler is far from being optimized, leading to high drive voltage (200 V) and small modulation depth.

The other important characteristic parameter of an electro-optic modulator is its frequency bandwidth, especially in the case of polymers that are particularly adapted to high-frequency modulation. The capacitance of the electric circuit determines an upper frequency cutoff: $Fc = 1/(2\pi RC)$, R being the output impedance of the generator; and $C = \varepsilon_0\varepsilon_r\left(\frac{L \cdot l}{e}\right)$ (Fig. 22), e is fixed by the structure (to insulate the propagating wave from the electrodes, typically $e = 5$ μm); the length of the electrode is generally about 1 cm to work with reasonably low $V\pi$.

Under the ESPRIT program of the European Community, V. Lemoine *et al.* [60] (Thomson-CSF) have realized an electro-optic modulator with confined waveguides mounted between cross polarizers. The width of the upper electrode was $1 = 250$ μm, implying an upper frequency cutoff $Fc = 245$ MHz, probably limited by the contact circuit.

The transit time of the optical beam between the electrodes defines another upper frequency cutoff. The electric field seen by the optical beam must stay the same during this transit time. If we suppose that the electrodes are equipotentials, a phase delay of $\pi/2$ between the electric and optic signals is obtained for a frequency: $F_{\pi/2} = c/(4Ln)$, c/n being the speed of light in the waveguide and L the length of the sample. For example, in a 1 cm interaction length the calculated RF frequency cutoff is 4.7 GHz.

To achieve very high frequency bandwidth modulators with a significant overlap integral between electrical modulation and optical modulated signals, it is necessary to achieve electro-optic modulation by using a travelling wave configuration in confined structures.

4.2.3. Traveling Waves

The electrostatic hypothesis (equipotential electrodes) is no more valid when the wavelength associated to the applied electric signal is of the same order of magnitude than the sample length $\lambda = c/(Fn)$.

The electrode system forms a microwave line. The propagation characteristics of this line depend on the geometry and electrical properties of the constitutive materials. The propagation of the electric signal and the geometry of the line have to be studied with Maxwell's electromagnetic equations. The design of a line requires computation of its impedance. The microstrip line is a basic line often used for microwave hybrid integrated circuits because of its simplicity and planar structure. Numerous papers deal with numerical methods of microstrip impedance calculation [61,62,63]. Figure 23 shows the characteristic parameters of a microstrip line. The impedance of the line (Zm) depends on the microstrip width (W) and thickness (t) and on the dielectric thickness (h) and constant (ε_r).

To allow a large bandwidth, the line is adapted to the load, which is connected at its end. Generally this load is 50 Ω, implying that the characteristic impedance of the line must also be 50 Ω. With $h = 5$ μm, $\varepsilon = 3$; $t = 4$ μm, and $W = 10$ μm, the characteristic impedance of the line is calculated to be 50.6 Ω. These values are compatible with classical processing techniques.

The limitations in terms of frequency of such a modulator should be of different types: the velocities of the optic wave and microwave are generally not matched with each other, and the bandwidth is limited by this velocity

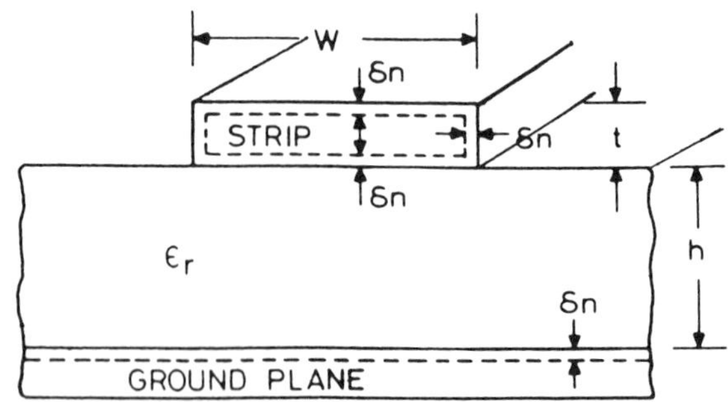

Fig. 23. Microstrip line.

mismatch, due to the dispersion of the dielectric constant with frequency $n = \sqrt{\varepsilon_{eff}}$. Organic materials show lower intrinsic velocity mismatch than classical inorganic materials ($LiNbO_3$) and lead to a limitation that is estimated to be greater than 100 GHz for 1-cm-long organic waveguides.

For a gold strip line with the geometrical parameters described above the attenuation due to the strip resistivity is equal to $\alpha_c = 1.3$ dB/cm for a 40 GHz electric wave and increases with the square root of the frequency.

The dielectric losses can be neglected as they are generally lower than the conductor losses. Finally, one of the main contributions to the frequency limitation of the electro-optic modulator is generally related to the impedance mismatch at the electric input and output.

D. G. Girton *et al.* [64] from the Lockheed R&D division have reported the fabrication of a 8 GHz electro-optic Mach-Zehnder modulator with microstrip electrodes. A 9-V half-wave voltage was measured (the electrodes are 1.7 cm long). An optical modulation response from 20 to 40 dB above the noise floor was observed over the 2 to 8 GHz bandwidth (the rolloff was due to the combination of rolloffs in the detector amplifier cables and electrodes). In [65] the same team has reported a measured frequency response at 20 GHz with the same device structure. More recently, C. C. Teng from Hoechst Celanese reported the fabrication of an intensity modulator with a similar structure and material, presenting a −3 dB electrical bandwidth of more than 40 GHz [66].

4.3. Second Harmonic Generation

The research on compact laser sources emitting in the blue region shows real industrial interests, mainly because of its potential use in optical disc memory to achieve a four-time increase in recording density (the minimum size of a bit of information being proportional to the square of the wavelength used to write and read it). High-definition television (large screen projection) and medical applications are also asking for this range of wavelength. The obtention of a blue or at least a green compact laser source by second harmonic generation (SHG) seems to be a very promising solution to this need.

Efficient SHG has already been demonstrated in bulk crystals (the phase matching is achieved using the birefringence properties of the crystal) and nonlinear phase-matchable crystals are commercially available. Guided optics offers the advantage of high-power density and so more efficient SHG. For the past three years, an increasing amount of research results has been published in this field with a large effort on inorganic crystals such as $LiNbO_3$ or KTP.

Up to now all the work published on second harmonic generation in polymer films deals with slab waveguides, the simplest guiding structure. The phase-matching problem supposes an accurate control of the opto-geometrical parameters of the structure over long interaction length (typically 1 cm). It will often be useful to look for noncritical phase-matching conditions.

4.3.1. *Second Harmonic Generation in Optical Waveguide* [67]

Consider a basic slab waveguide consisting of an organic film (active layer) of thickness e_{AL} and index of refraction n_2 sandwiched between two media with indices n_1 for the superstrate and n_3 for the substrate. A finite number of confined TE and TM modes are supported by the structure. The wave equation for the magnetic component H_y of a TM mode excitation of the waveguide at the pump frequency ω is given by (z: poling axis, x: propagation direction):

$$[\partial_y^2 + \partial_z^2 - (1/\varepsilon_x^\omega)\partial_z\varepsilon_x^\omega + \omega^2\mu_0\varepsilon_x^\omega - (\beta^\omega)^2\varepsilon_x^\omega/\varepsilon_z^\omega]H_y = 0, \tag{4.3.1}$$

where $\varepsilon_x = \varepsilon_0\bar{\varepsilon}_x$, $\varepsilon_z = \varepsilon_0\bar{\varepsilon}_z$, and $\partial_x, \partial_z = \partial/\partial x, \partial/\partial z$. β^ω is the propagation constant at the fixed frequency of the guided mode.

The components of the nonlinear polarization are:

$$\left.\begin{aligned} P_x &= 2d_{31}E_xE_z, \\ P_y &= 2d_{31}E_yE_z, \\ P_z &= d_{31}E_x^2 + d_{31}E_y^2 + d_{33}E_z^2. \end{aligned}\right\} \tag{4.3.2}$$

The electric components attached to TM mode at fundamental frequency are:

$$E^\omega \quad \left.\begin{matrix} j/\omega\varepsilon_x^\omega\,\partial_z Hy \\ 0 \\ -\beta^\omega/\omega\varepsilon_z^\omega Hy. \end{matrix}\right\} \tag{4.3.3}$$

It follows that the z-axis nonlinear polarization may be written as:

$$P_{NL^z}^{2\omega} = U(y, z)e^{2j(\omega t - \beta_x^\omega)}, \tag{4.3.4}$$

with $U(y, z)$ defined as:

$$U(y, z) = \frac{1}{\omega^2}\left(\frac{d_{33}}{(\varepsilon_z^\omega)^2}(\beta^\omega)^2 H(y, z)^2 - \frac{d_{31}}{(\varepsilon_x^\omega)^2}(\partial_z H(y, z))^2\right).$$

The electric-field solution of the wave equation at the harmonic frequency polarized along the 0_z direction can be written as a superposition of confined modes:

$$\mathscr{E}^{2\omega} = \sum_{m \text{ modes}} \mathscr{E}_m^{2\omega}(x, y, z) = \sum_m A_m(x)\mathscr{E}_m(y, z)e^{j(2\omega t - \beta_m^{2\omega} - x)}. \quad (4.3.5)$$

$A_m(x)$ and $\mathscr{E}_m(y, z)$ are respectively the amplitude and transverse modal distribution of the mth mode order.

We assume "slow variation," so that

$$\partial_x^2 \mathscr{E} \ll 2j\beta^{2\omega}\partial_x \mathscr{E};\ \partial_y^2 \mathscr{E} \approx \partial_z^2 \mathscr{E} \ll 2j\beta^{2\omega}\partial_x \mathscr{E};\ d_{31}(\partial_x H(x, y))^2 \ll d_{33}(H(x, y))^2.$$

The wave equation of 2ω can be written as:

$$\sum_m e^{j(2\omega t - \beta_m^{2\omega} x)}(4\omega^2 \mu_0 \varepsilon_z^{2\omega} - (\beta_m^{2\omega})^2 - 2j\beta_m^{2\omega}\partial_x)A_m(x)\mathscr{E}_m(y, z) = -4\omega^2 \mu_0 P_{NL}^{2\omega} z \quad (4.3.6)$$

By taking the product of (4.3.6) with $\mathscr{E}^*$ and after integration from $-\infty$ to ∞ over $dy\,dz$ we obtain:

$$\frac{dA_n(x)}{dx} = -\frac{2j\omega^2\mu_0}{\beta_n^{2\omega}} e^{j\Delta\beta_{m,n}x} \frac{\int\int U_m(y, z)\mathscr{E}_n^*(y, z)\,dy\,dz}{\int\int |\mathscr{E}_n(y, z)|^2\,dy\,dz} \quad (4.3.7)$$

with the normalization condition:

$$\int\int \mathscr{E}_m \mathscr{E}_n\,dy\,dz = \delta_{m,n} \int\int \mathscr{E}_n^2\,dy\,dz. \quad (4.3.8)$$

The phase mismatch is given by

$$\Delta\beta_{m,n} = \beta_m^{2\omega} - 2\beta_n^{\omega}. \quad (4.3.9)$$

By writing

$$\mathscr{R} = \frac{\int\int U_m(y, z)\mathscr{E}_n^*(y, z)\,dy\,dz}{\int\int |\mathscr{E}_n(y, z)|^2\,dy\,dz}, \quad (4.3.10)$$

after integration of Eq. (4.3.7), the length dependence of the field amplitude is given by:

$$A_n(x) = \frac{-j\omega(\mu_0)^{1/2}}{(\varepsilon_{\text{eff},n}^{2\omega})^{1/2}}\, x\, \frac{\sin\left(\Delta\beta_{m,n}\dfrac{x}{2}\right)}{\Delta\beta_{m,n}\dfrac{x}{2}}\, \mathscr{R}\, e^{j\left(\Delta\beta_{m,n}\frac{x}{2}\right)}. \quad (4.3.11)$$

The harmonic power generated into the waveguide structure is calculated by taking the average power per unit area carried in the direction of propagation:

$$\langle P_n^{2\omega}(x)\rangle = \frac{\mu_0^{5/2}\omega^2 d_{33}^2}{2(\varepsilon_{\mathrm{eff},n}^{2\omega})^{1/2}} \operatorname{sinc}^2\left(\Delta\beta_{m,n}\frac{x}{2}\right)x^2\mathscr{I}^2 \iint |\mathscr{E}_n(y,z)|^2\,dy\,dz. \quad (4.3.12)$$

The efficiency conversion from ω to 2ω can be written as:

$$\frac{\langle P_n^{2\omega}(x)\rangle}{\langle P_m^{\omega}\rangle} = \frac{2\mu_0^{3/2}\omega^2 d_{33}^2}{(\varepsilon_{\mathrm{eff},n}^{2\omega})^{1/2}}\frac{1}{(\varepsilon_{\mathrm{eff},m}^{\omega})}\operatorname{sinc}^2\left(\Delta\beta_{m,n}\frac{x}{2}\right)x^2\mathscr{A}^2\langle P_m^{\omega}\rangle. \quad (4.3.13)$$

We notice that the conversion efficiency is proportional to the overlap integral $\mathscr{A}^2$ between the TM modes at pump and harmonic frequencies. The factor $\mathscr{A}^2$ is given by:

$$\mathscr{A}^2 = \frac{\{\int_{\text{active layer}}\int H_m^2(y,z)\mathscr{E}_n^*(y,z)\,dy\,dz\}^2}{\int\int_{-\infty}^{+\infty}|\mathscr{E}_n(y,z)|^2\,dy\,dz\{\int\int_{-\infty}^{+\infty}|H_{ym}^{\omega}|^2\,dy\,dz\}^2}. \quad (4.3.14)$$

The second harmonic signal reaches a maximum value when the phase-matching conditions are respected: $\Delta\beta_{mn} = 0$. The coherence length defined by

$$L_c = \frac{\lambda^{\omega}}{4|n_{\mathrm{eff}}^{2\omega} - n_{\mathrm{eff}}^{\omega}|} \quad (4.3.15)$$

is determined from the waveguide structure characterized by effective indices at the fundamental and second harmonic frequencies.

In numerical applications, the asymmetric slab waveguide was considered. The substrate is a PMMA based polymer, and the planar waveguide is the nonlinear polymer. At 1.064 μm the refractive index distribution is ($n_1 = 1$, $n_2 = 1.610$, $n_3 = 1.50$), and at second harmonic wavelength ($\lambda^{2\omega} = 0.53$ μm), the refractive indices are ($n_1 = 1$, $n_2 = 1.666$, $n_3 = 1.51$).

4.3.2. *Quasi-Phase-Matched Second Harmonic Generation*

The mismatch in the k-vector between the fundamental and second harmonic waves can be compensated by introducing a periodic structue [68]

$$\Delta k = k^{2\omega} - 2k^{\omega} = 2\pi/\Lambda = \Delta\beta_{m,n}, \quad (4.3.16)$$

where Λ is the period of the perturbation. Quasi-phase-matching technique based on periodic modulation of the nonlinearity is a versatile method to phase match a structure. A periodic reversal of the sign of the nonlinear coefficient of the active medium at odd integer multiples of the coherence length prevents the accumulation of phase mismatch between the interacting waves. The second harmonic field is $2/\pi$ lower than in the case of real phase matching.

The nonlinear polarization can be described by the function: $P_{NLz}(x) = U(x)P_{NLz}$ where P_{NLz} is the nonperturbed nonlinear polarization of the material and $U(x)$ is the periodic modulation function of $\chi^{(2)}$.

From Eq. (4.3.7) the distance dependence of the second harmonic field is given by:

$$A_n(L) = -A_n \int_0^L e^{j\Delta\beta_{m,n}x} U(x)\,dx, \tag{4.3.17}$$

where
$$A_n = \frac{2j\omega^2\mu_0\mathscr{R}}{\beta_n^{2\omega}}.$$

For $U(x) = 1$, we find again the expression previously obtained in (4.3.11) with $x = L$. For a really phase-matched structure ($\Delta\beta_{mn} = 0$): $A_n(L) = AL$.

Now if the nonlinear polarization is periodically modulated such as:

$$\frac{2n\Lambda}{2} < x < \frac{2(n+1)\Lambda}{2}; \quad U(x) = 1,$$

$$\frac{2(n+1)\Lambda}{2} < x < \frac{2(n+2)\Lambda}{2}; \quad U(x) = -1,$$

where Λ is defined in (4.3.16), from (4.3.17) integrated over a propagation distance integer multiple of the half modulation period $L = b\Lambda$ (b: integer), the second harmonic intensity produced in a waveguide is given by:

$$A_n(L) = \frac{j2AL}{\pi}.$$

In Figs. 24a, b the harmonic-signal dependences on the propagation distance are plotted. Figure 24a corresponds to $U(x) = + - + - + -$, describing a periodic reversal of the $\chi^{(2)}$ coefficient; Fig. 24b corresponding to $U(x) = +0 + 0 + 0$ is less efficient.

In such a configuration, the overlap integral and consequently the frequency conversion efficiency can be optimized by using the nonlinear coupling between TM_0 (at λ^{ω}) and TM_0 (at $\lambda^{2\omega}$). The potential of this method was demonstrated in periodically poled lithium niobate channel waveguides [69].

The coherence length of nonlinear polymers is generally greater (about 5 μm at 1.06 μm) than those of inorganic crystals (at 1.06 μm, $Lc \approx 1.5$ μm for $LiNbO_3$); consequently it seems technologically possible to work with a first-order grating. Different solutions to achieve a periodic modulation of

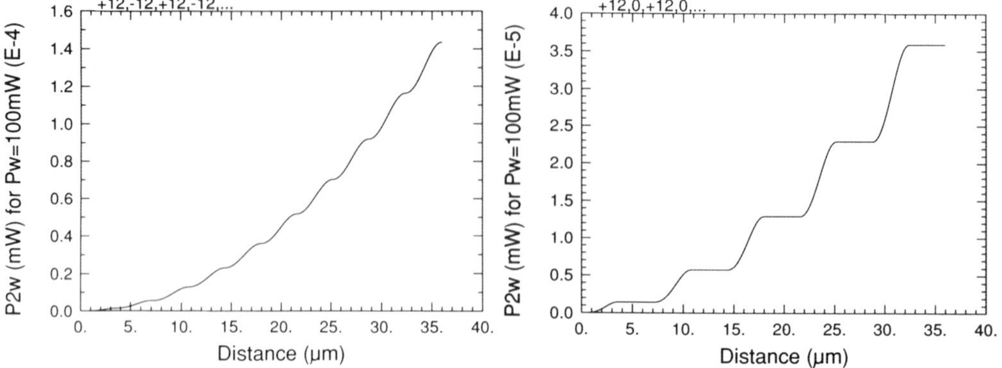

Fig. 24. Second harmonic signal for a quasi-phase-matched structure with various periodic modulation functions. a: $+-+-+-$; b: $+0+0+0$.

the nonlinear coefficient have been studied for cross-linkable polymers, but no results have been published up to now. With non-cross-linkable polymers the two poling steps (up and down) have to be done simultaneously; otherwise the orientation achieved during the first poling step is lost during the second one. A poling process with a temperature profile presenting a period of about 10 µm and a contrast of nearly 100°C is rather difficult to achieve. A preferred solution seems to be the realization of two combs of alternated electrodes with a period of 10 µm over a reasonable surface; it is a very critical technological step to avoid short circuits between the two electrodes. A second difficulty is the very high electric field applied between the two electrodes during the poling process (two times the usual poling field: 100 V/µm on a very small distance is often greater than the electric breakdown threshold). This is probably the reason why no results involving a periodic reversal of the nonlinear susceptibility of polymers have been reported so far.

Another theoretical solution to achieve quasi-phase matching in guided structure is to alternate poled and unpoled zones. The second harmonic generation efficiency is four times lower than in the case of periodic reversal. Up to now two poling processes have been published.

The first one was proposed by G. Khanarian and coworkers from Hoechst Celanese [70]. As in the case of periodic reversal, it is based on the periodic poling with a special electrode structure. A single-comb electrode is processed on one side of the structure and a planar ground electrode on the other side. With this kind of electrode geometry, G. Khanarian [70] has demonstrated that the electric field seen by the active layer is composed of a continuous and a modulated part. The relative amplitude of these two parts depends

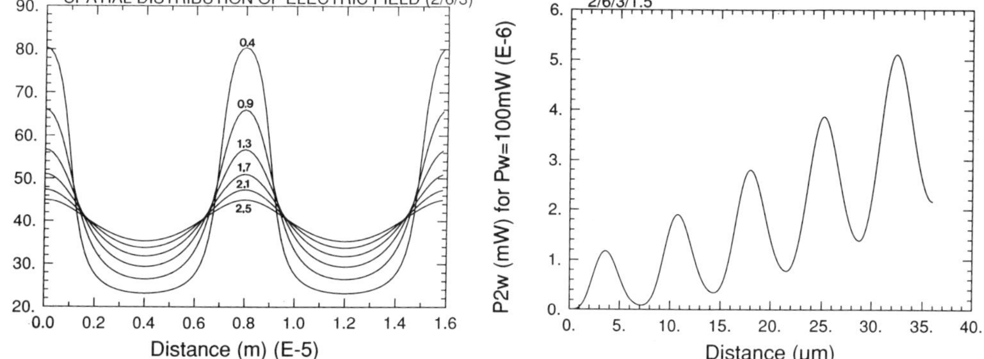

Fig. 25. a: Vertical electric-field shape at different depth (the figures on the drawing correspond to the depth: distance is given in microns between the comb electrode and the plane where the field is calculated). b: Second harmonic generation with this type of periodic modulation.

on the ratio: electrode width/period, the distance between the comb electrode and the ground electrode, and of course the distance between the active layer and the comb electrode. Figure 25a represents the vertical electric field calculated at different depths for an electrode width of 2 μm, a space between electrodes of 6 μm, and a total thickness between the ground and comb electrodes of 3 μm. Figure 25b represents the second harmonic signal in that case.

G. Khanarian *et al.* [71,72] have reported second harmonic generation from 1.34 to 0.67 μm in polymer quasi-phase-matched slab waveguides. The polymer used was a copolymer of MMA and a monomer containing 4-oxy, 4′-nitrostilbene, the d_{eff} was 2 pm/V, and the coherence length between the two fundamental modes (TM_0^{ω}, $TM_0^{2\omega}$) was measured to be close to 6 μm in the slab waveguide. The period of the realized comb electrode was 12 μm (3 μm width, 9 μm spacing between electrodes). By tuning the wavelength, the matching between the period of the grating and two times the coherence length was optimized. From the dependence of the SHG power with the temperature, the wavelength, and the angle formed by the incident beam and the grating, they deduced an effective quasi-phase-matched interacting length of 5 mm with an SHG conversion efficiency of 0.01%/W.

A second solution was proposed by G. L. J. A. Rikken and coworkers from Philips Research Laboratories [73]. They have noticed that the UV bleaching effect results in a very strong loss of nonlinearity (the polymer was an MMA copolymer with 4-alkoxy-4′-alkylsulfone stilbene side chain). The spatial modulation of the nonlinear coefficient was realized with this

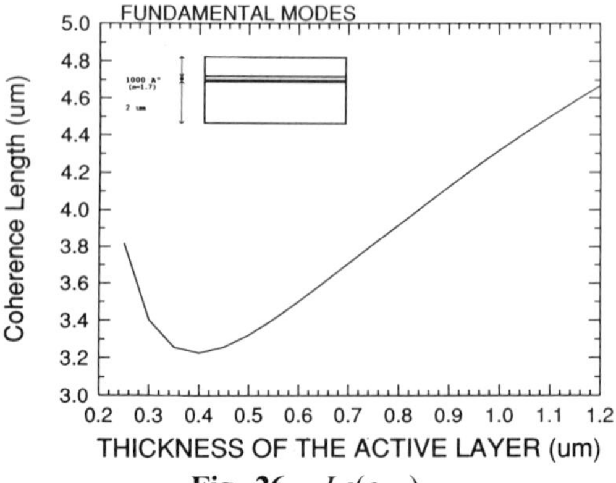

Fig. 26. $Lc(e_{AL})$.

bleaching effect on a poled polymer layer. With a tunable dye laser (780 to 840 nm) second harmonic generation was demonstrated. No quantitative figures were published.

The tolerance of the phase-matching conditions to small fluctuations of the active layer thickness (e_{AL}) can be estimated from the dependence of the coherence length (Lc) on e_{AL}. The function $Lc(e_{AL})$ can be determined using the effective refractive index method. Figure 26 presents such a function. A small change $d(e_{AL})$ on the active layer thickness at the quasi-phase-matching conditions induces a change on the effective index dispersion approximated by:

$$d(\Delta\beta) = \pi/Lc^2(\delta Lc/\delta e_{AL})\, d(e_{AL}).$$

A new effective coeherence length can be defined:

$$Lnc = Lc^2/\{(\partial Lc/\partial e_{AL}) * d(e_{AL})\}.$$

In the case of the structure described in Fig. 26, for $e_{AL} > 0.5\ \mu m$ the slope of the curve $Lc(e_{AL})$ is nearly constant and equal to $(\delta Lc/\delta e_{AL}) = 2$.

For example, $Lnc(e_{AL} = 0.9\ \mu m) = 8/d(e_{AL})\ 10^{-12}$ m; an interactive length of at least $Lnc = 1$ cm implies $d(e_{AL}) \leq 8$ Å. For $e_{AL} = 0.4\ \mu m$ there is a noncritical phase-matching condition; second-order derivatives have to be taken into account. In that case, a change of nearly 10 nm on the optimal active-layer thickness can be tolerated for having $Lnc \geq 1$ cm.

The typical amplitude of the surface roughness of a spin-coated layer of good quality is about 5 to 10 nm. It seems interesting to analyze the influence of this surface roughness on the second harmonic generation. Figure 27 results from a numerical calculation of the second harmonic power along

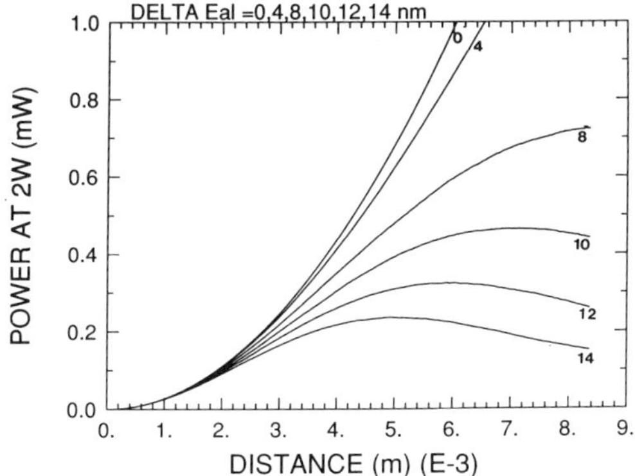

Fig. 27. Influence of the surface roughness on the second harmonic generation.

the propagation direction in the case of a critical quasi-phase-matched structure (the ideal thickness is e_{AL}); the surface roughness of the active layer was modelized by considering the active layer thickness as a random function along the propagation direction varying between $e_{AL} \pm \partial e_{AL}/2$.

A fundamental limitation of this type of phase-matching option is the losses due to the refractive index grating, defined together with the periodic modulation of the nonlinear coefficient. The orientation of the active molecules significantly increases the refractive index of the polymer for an optical polarization parallel to the poling axis (TM polarization). In the case of G. Khanarian's polymer the change in "average" index was measured to be $\Delta n = 0.006$ at 514 nm. The bleaching effect goes with a decrease in refractive effect. In the case of G. L. J. A. Rikken's polymer it is ranging from 0.003 at 820 nm to 0.009 at 410 nm.

4.3.3. Phase Matching by Modal Dispersion

By controlling the thickness of the active film, we can adjust the phase-matching condition between the TM modes at the pump and harmonic frequencies.

Figure 28 shows the dependence of the different modes effective index of a classical structure on the active-layer thickness, for both the fundamental and the second-harmonic fields. The crossing points exhibit the possibility to phase match a guided structure by modal dispersion. In Fig. 29 are shown the dependence of the coherence length on the film thickness by considering $TM_0^{\omega} \Rightarrow TM_1^{2\omega}$ interaction.

TM_0 and TM_1 fields are respectively even and odd functions, and, in the case of a symmetric structure, the overlap integral between these two modes

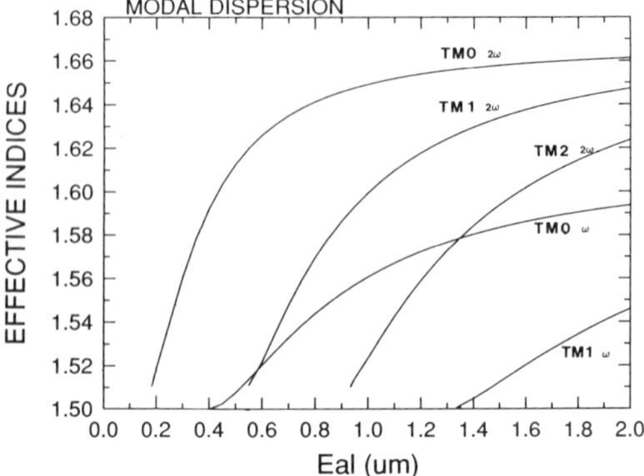

Fig. 28. Effective index. Pomp modes: TM_0, TM_1; harmonic modes: TM_0, TM_1, TM_2.

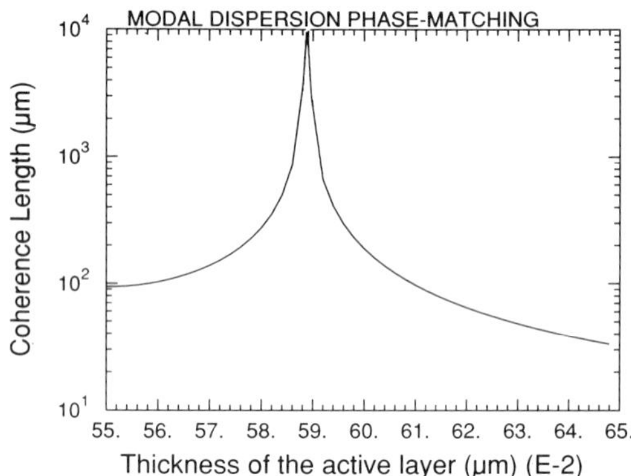

Fig. 29. Coherence length. Pomp modes: TM_0; harmonic modes: TM_1.

is equal to zero; thus in spite of a phase-matched structure no second harmonic power will be generated with these two modes.

This integral is related to the overlap of two different modes on the active layer and at two different wavelengths. With special waveguide geometry it is possible to adjust the field distribution on the different layers in a way to achieve reasonably high overlap integral (nearly half of the overlap integral between two fundamental modes).

As in the previous case, it is interesting to study the accuracy needed on the active layer thickness to achieve a 1-cm-long useful interaction length. The dependence of this coherence length on the active layer thickness is shown in Fig. 29 (for TM_0^{ω} and a $TM_1^{2\omega}$ interacting modes). Using the effective index method $\partial(1/Lc)/\partial e_{AL}$ is calculated. A small change $d(e_{AL})$ on the active layer thickness from nominal phase-matching conditions induces a change on the effective index dispersion. The coherence length is approximated by:

$$Lc = 1\Bigg/\left\{\left(\partial\left(\frac{1}{Lc}\right)\Bigg/\partial e_{AL}\right)d(e_{AL})\right\}.$$

In the case of the structure shown in Fig. 29, $Lc(e_{AL} = 0.738\ \mu m) = 1/\{0.18 \times 10^{12}\ d(e_{AL})\}$. For an interacting length of 1 cm, $d(e_{AL})$has to be smaller than 5.6 Å.

Noncritical phase-matching conditions can be obtained, but adjusting the three requirements simultaneously—to define (1) a phase-matched structure with (2) noncritical conditions and (3) a maximized overlap integral—is not an easy task.

With this phase-matching technique V. Lemoine *et al.* [60] from Thomson CSF have reported an enhancement of the second harmonic generation in slab waveguides (for TM_0^{ω} and $TM_2^{2\omega}$ interacting modes). Second harmonic generation efficiency of 2.10^{-3} %/W/cm^2 was obtained from a 1.06 μm cw YAG laser beam; the active polymer was a PMMA-based side-chain polymer with dialkyl-amino-cyano-biphenyl dye ($d_{33} = 12$ pm/V).

O. Sugihara and coworkers from the Keio University [74,75] have succeeded in realizing a phase-matched structure at 1.06 μm between the first mode at ω and the TM_2 mode at 2ω. The polymer was an MNA doped PMMA ($d_{33} = 0.71$ pm/V). The key point of the process was the realization of a graded active-layer thickness to adjust it to the phase-matching conditions. The experiment was performed with two prism couplers 1 mm distant; for a 43.4 W fundamental wave power measured after the decoupling prism, a 3.2 mW second harmonic power was measured.

5. CONCLUSION

From the first steps of molecular engineering toward the realization of integrated optic devices with nonlinear polymers, the gap was quickly jumped over. Very important points have been demonstrated. The first one was the fabrication of low-loss bidimensional waveguides with these materials (losses lower than 1 dB/cm at 1.32 μm have been measured). Various solutions to achieve bidimensional confinement were tested giving good results. If the technological process is not always easy to define (temperature process, solvent compatibility, ...) the equipment used is compatible with standard silicon technology equipment, and, consequently, the fabrication of these structures represents a low-cost money and time investment. Besides, most of these structures could be processed on nearly any kind of substrate, probably the main advantage of the nonlinear polymers. This results in potentially low-cost devices: standard technology and low-cost material (the polymer would be produced in large quantity).

The electro-optic modulation is probably one of the most promising applications for polymers. In nonlinear polymers the electro-optical effect arises from electrons' motions, allowing very fast switching properties. Devices presenting wide frequency bandwidth (up to 20 GHz) with low driving voltage (a few volts) were realized. The possibility of fabricating these devices on any kind of substrate creates new fields of applications impossible with inorganic materials. Diodes, waveguides, electro-optic functions, detectors, among others, could be integrated on a single chip; this is of great importance knowing the difficulties related to the coupling of optical or microwave signals between devices.

The nonlinear polymers also show true promise for second harmonic generation in the blue-green spectral range. The results are not as advanced as in the case of the electro-optic modulator, mostly because the focus was put only recently on the phase-matching problems. For example the precision and the small roughness needed on the active layer thickness to keep the phase-matching condition are new kinds of problems, and the solutions are still to be found. Poled polymer area (spatially distributed) remains a new investigation field in integrated optics. From this point of view, the method of quasi-phase matching in polymeric waveguides using periodic modulation of the material nonlinear susceptibility, already demonstrated, is a good example of the versatility of these materials.

The devices presented in this chapter illustrate the tremendous potential of the nonlinear polymers. Some problems still remain, slowing the evolution of these materials from research laboratories to industrial development units, but all of them, after a few years of research, seem possible to solve. For example the stability with temperature and environment of the optical properties is often considered as a fundamental limitation of the polymers for active devices. In the first part of this chapter the huge progress made to increase this stability has been presented. It clearly shows that the classical aim: total stability at 100°C will soon be achieved.

The number of useful nonlinear organic materials is very large, and various organic forms can be synthesized. Some possibilities of these materials are still to be studied. Our discussion has been limited to second-order effects and to the devices already realized, but a great deal of research has also been done on third-order effects and, concerning second-order effects, other properties already studied could be used in devices (for example, parametric effects).

REFERENCES

1. P. Mihailovic, P. Bassoul, and J. Simon, *Chem. Phys. Letters* **141**(5), 462 (1989).
2. K. D. Singer, J. E. Sohn, L. A. King, H. M. Gordon, H. E. Katz, and C. W. Dirk, *J. Opt. Soc. Am. B* **7**, 1339 (1989).
3. H. L. Hampsch, J. M. Torkelson, S. J. Bethke, and S. G. Grubb, *J. Appl. Phys.* **67**(2), 1037 (1990).
4. M. A. Mortazavi, A. Knoesen, S. T. Kowel, B. G. Higgins, and A. Dienes, *J. Opt. Soc. Am B* **6**(4), 733 (1989).
5. D. J. Williams, *Angew. chem. Intl. ed. Eng.* **23**, 690 (1984).
6. K. D. Singer, S. J. Lalame, and J. E. Sohn, *Proceedings of SPIE* **578**, 168 (1985).
7. K. D. Singer, J. E. Sohn, and S. J. Lalama, *Appl. Phys. Lett.* **49**, 5, 248 (1986).
8. R. D. Small, K. D. Singer, J. E. Sohn, M. G. Kuzyk, and S. J. Lalama, *Proceedings of SPIE* **682**, 160 (1986).
9. K. D. Singer, S. J. Lalama, J. E. Sohn, and R. D. Small, *Nonlinear Properties of Organic Molecules and Crystals*, D. S. Chemla and J. Zyss, eds. (Academic Press, Orlando, 1987).
10. K. D. Singer, M. G. Kuzyk, and J. E. Sohn, *J. Opt. Soc. Am. B* **4**(6), 968 (1987).
11. K. D. Singer, M. G. Kuzyk, W. R. Holland, J. E. Sohn, S. J. Lalama, R. B. Comizzoli, H. E. Katz, and M. L. Schilling, *Appl. Phys. Lett.* **53**(19), 1800 (1988).
12. H. L. Hamsch, J. Yang, G. R. Wong, and J. M. Torkelson, *Polymer Commun.* **30**, 40 (1989).
13. H. L. Hampsch, J. Yang, G. K. Wong, and J. M. Torkelson, *Macromolecules* **21**, 526 (1988).
14. J. M. Hodge, *Macromolecules* **16**, 898 (1983).
15. T. Watanabe and S. Miyata, *J. Chem. Soc. Chem. Commun.* **250** (1988).
16. T. Watanabe and S. Miyata, *Proceedings of SPIE* **1147**, 101 (1989).
17. S. Esselin, P. LeBarny, P. Robin, D. Broussoux, J. C. Dubois, J. Raffy, and J. P. Pocholle, *Proceedings of SPIE* **971**, 120 (1988).
18. R. N. DeMartino and H. N. Yoon, U.S. Patent 4,801,670.
19. R. N. DeMartino, European Patent 0 316 662.
20. C. E. Won, European Patent 0 306 893.
21. R. N. DeMartino, European Patent 0 312 856.
22. C. Ye, T. J. Marks, J. Yang, and G. K. Wong, *Macromolecules* **20**, 2322 (1987).
23. C. Ye, N. Minami, T. J. Marks, J. Yang, and G. K. Wong, *Mat. Res. Soc. Symp. Proc.* **109**, 263 (1988).
24. D. Li, N. Minami, M. A. Ratner, C. Ye, and T. J. Marks, *Synthetic Metals* **28**, D585 (1988).

25. M. L. Schilling, H. E. Katz, and D. I. Cox, *J. Org. Chem.* **53**, 5538 (1988).
26. T. M. Leslie, U.S. Patent 4,801,659.
27. M. Eich, A. Sen, H. Looser, G. C. Bjorklund, J. D. Swalen, R. Twieg, and D. Y. Yoon, *J. Appl. Phys.* **55**, 6, 2259 (1989).
28. N. Minami, C. Ye, T. J. Marks, and G. H. Wong, *Polymer Preprints, Japan, SPSJ, 38th Symposium on Macromolecules* **38**, 3KO2 (1989).
29. H. T. Man, K. Chiang, D. Haas, C. C. Teng, and H. N. Yoon, *Proceedings of SPIE* **1213**, 7 (1990).
30. G. D. Green, J. I. Wenschenk, J. E. Mulvaney, and H. K. Hall, Jr., *Macromolecules* **20**, 722 (1987).
31. C. S. Willand and D. J. Wiliams, *Ber. Bunsenges. Phys. Chem.* **91**, 1304 (1987).
32. J. D. Strenger-Smith, J. H. Fischer, L. M. Hayden, R. A. Henry, J. M. Hoover, and G. A. Lindsay, *Polym. Prep.* (*Am. Chem. Soc. Div. Polym. Chem.*) **31**, 1, 375 (1990).
33. I. Teraoka, D. Jungbauer, B. Reck, D. Y. Yoon, R. Twieg, and C. G. Willson, *J. Appl. Phys.* **69**(4), 2568 (1991).
34. M. A. Hubbard, J. J. Marks, J. Yang, and G. K. Wong, *Chemistry of Materials* **1**, 2, 167 (1989).
35. P. LeBarny, D. Broussoux, S. Esselin, J. P. Pocholle, and J. Raffy, French Patent 88 05790.
36. J. Park, T. J. Marks, J. Yang, and G. K. Wong, *Chem. Mater.* **2**, 229 (1990).
37. M. Eich, B. Reck, D. Y. Yoon, C. Willson, and G. C. Bjorklund, *J. Appl. Phys.* **66**, 7, 3241 (1989).
38. B. Reck, M. Eich, D. Jungbauer, R. J. Twieg, C. G. Willson, D. Y. Yoon, G. C. Bjorklund, *Proceedings of SPIE* **1147**, 74 (1989).
39. P. LeBarny, E. Chastaing, J. C. Dubois, S. Muller, and F. Soyer, French Patent 91 08233.
40. D. Jungbauer, B. Reck, R. Twieg, D. Y. Yoon, C. G. Willson, and J. D. Swalen, **56**, 26, 2610 (1990).
41. B. K. Mandal, J. Kumar, J. C. Huang, and S. Tripathy, *Makromol. Chem. Rapid Commun.* **12**, 63 (1991).
42. D. J. Williams, in *Nonlinear Optical Properties of Organic Molecules and Crystals*, vol. 1, D. S. Chemla and J. Zyss, eds. (Academic Press, Orlando, 1987).
43. P. Robin, P. LeBarny, D. Broussoux, J. P. Pocholle, and V. Lemoine, "Organic molecules for nonlinear optics and photonics," J. Messier *et al.* (eds) (Kluwer Academic Publishers, Dordrecht, 1991), pp. 481, 488.
44. M. Barzoukas, D. Josse, P. Fremaux, J. Zyss, J. F. Nicoud, and J. O. Morley, *J. Opt. Soc. Am. B* **4**, 6, 977, 986 (1987).
45. J. Jerphanion and D. K. Kurtz, *J. Appl. Phys.* **41**, 4, 1667, 1681 (1970).
46. C. C. Teng and H. T. Man, *Appl. Phys. Lett.* **56**, 1734 (1990).
47. Y. Levy, M. Dumont, E. Chastaing, P. Robin, P. A. Chollet, G. Gadret, and F. Kajzar, *Nonlinear Optics* **4**, 1–19 (1993).

48. D. Marcuse, *Theory of Dielectric Optical Waveguides* (Academic Press, New York, 1974).
49. R. G. Walker, *Electron. Lett.* **21**, 581 (1985).
50. M. B. J. Diemeer, F. M. M. Suyten, E. S. Trommel, A. McDonach, J. M. Copeland, L. W. Jenneskens, and W. H. G. Horsthuis, *Electron. Lett.* **26**, 6, 379 (1990).
51. J. I. Thackara, G. F. Lipscomb, M. A. Stiller, A. J. Ticknor, and R. Lytel, *Appl. Phys. Lett.* **52**, 13, 1031 (1988).
52. R. Lytel, G. F. Lipscomb, M. A. Stiller, J. I. Thackara, and A. J. Ticknor, *Proc. SPIE* **971**, 218 (1988).
53. D. R. Haas and H. T. Man, conference paper, "Integrated Photonic Research," Monterey (1991).
54. P. R. Ashley and T. A. Tumolillo, Jr., *Appl. Phys. Lett.* **58**, 9, 884 (1991).
55. V. Lemoine, M. Papuchon, J. P. Pocholle, P. Robin, and P. Le Barny, private communication.
56. P. D. Townsend, J. L. Jackel, G. L. Baker, J. A. Shelburne III, and S. Etemad, *Appl. Phys. Lett.* **55**, 18, 1829 (1989).
57. Y. Bourbin, M. Papuchon, S. Vatoux, A. Enard, M. Werner, and C. Puech, *Revue Technique Thomson-CSF* **15**, 3 (1983).
58. E. Van Tomme, P. Van Daele, R. Baets, G. R. Mohlmann, and M. B. J. Diemeer, *J. Appl. Phys.* **69**, 9, 6273 (1991).
59. T. E. Van Eck, A. J. Ticknor, R. S. Lytel, and G. F. Lipscomb, *Appl. Phys. Lett.* **58**, 15, 1588 (1991).
60. V. Lemoine, M. Papuchon, J. P. Pocholle, P. Robin, and P. LeBarny, *Synthetic Metals* **54**, 147–153 (1993).
61. K. C. Gupta, R. Garg, and I. J. Bahl, *Microstrip Lines and Slotlines* (Artech, Dedham, MA, 1979).
62. M. V. Schneider, "Microstrip lines for micro-wave integrated circuits," *BSTJ* **48**, pp. 1421–1444.
63. E. O. Hammerstad, *Proc. of the 5th European Micro-wave Conference* (Hamburg, Sept. 1975).
64. D. G. Girton, S. L. Kwiatkowski, G. F. Lipscomb, and R. S. Lytel, Conf. Integrated Photonic Research (1991).
65. D. G. Girton, S. L. Kwiatkowski, G. F. Lipscomb, and R. S. Lytel, *Appl. Phys. Lett.* **58**, 16, 1730 (1991).
66. C. C. Teng, *Appl. Phys. Lett.* **60**, 13, 1528 (1992).
67. G. I. Stegemann and C. T. Seaton, *J. Appl. Phys.* **58**, 12, R57, R78 (1985).
68. N. Bloembergen and A. J. Sievers, *Appl. Phys. Lett.* **17**, 483 (1970).
69. E. J. Lim, M. M. Fejer, R. L. Byer, and W. J. Kozlovsky, *Electronics Letters* **25**, 11, 731 (1989).
70. G. Khanarian, R. A. Norwood, and P. Landi, *Proc. SPIE* **1147**, 129 (1989).
71. G. Khanarian, R. A. Norwood, D. Haas, B. Feuer, and D. Karim, *Appl. Phys. Lett.* **57**, 10, 977 (1990).

72. R. A. Norwood and G. Khanarian, *Electron. Lett.* **26**, 25, 2105 (1990).
73. G. L. J. A. Rikken, C. J. E. Seppen, S. Nijhuis, and E. Staring, *Proc. SPIE* **1337**, 35 (1990).
74. O. Sugihara, T. Kinoshita, M. Okabe, S. Kunioka, Y. Nonaka, and K. Sasaki, *Appl. Optics* **30**, 21, 2957 (1991).
75. Y. Nonaka, S. Kunioka, R. Aizawa, O. Sugihara, T. Kinoshita, Y. Koike, and K. Sasaki, paper presented CLEO 91.

Chapter 10

THIRD-ORDER NLO PROCESSES IN POLYDIACETYLENES: PHYSICS, MATERIALS, AND DEVICES

Shahab Etemad

Bell Communications Research, Red Bank, NJ

Gregory L. Baker

Department of Chemistry, Michigan State University, East Lansing, NJ

and

Zoltan G. Soos

Department of Chemistry, Princeton University, Princeton, NJ

1. INTRODUCTION

Conjugated polymers are organic semiconductors with electronic structures closer to those of wide-band semiconductors, like GaAs and Si, than narrow-band molecular semiconductors, such as perylene. The electronic and optical properties of conjugated polymers are dominated by delocalized π-electrons distributed one-dimensionally along the conjugated backbone. In the widely

ISBN 0-12-784450-3

studied polyacetylene and polydiacetylenes, these electrons form valence and conduction bands separated by an optical gap, E_g, of about 2 eV. In σ-conjugated polymers such as polysilanes, the optical gap is larger ($E_g > 3$ eV), and thus they have received less attention for optoelectronic applications.

Polydiacetylenes (PDAs) all share the same conjugated backbone of delocalized electrons in π_z orbitals and localized electrons in π_x orbitals. Because many side groups can be attached to the polydiacetylene backbone without significantly perturbing the electronic structure of the polymer, polydiacetylenes are found in morphologies ranging from single crystals [1] to amorphous films to isolated chains in solution. A particularly useful series of polydiacetylenes is the *n*BCMU family, where the side groups are *n*-butoxycarbonylmethylurethane linked as a chain of *n* CH_2 units attached to the polydiacetylene backbone. Because the side groups impart solubility, polymers such as poly(4BCMU) yield thin films of optical quality suitable for making optical devices [2,3]. These PDAs were used in the construction of nonlinear polymeric waveguides [4,5], the building block of an optical switching device.

In this contribution, we review the progress toward construction of all-optical switching devices from poly(4BCMU). The switching operation is based on the intensity-dependent change in the index of refraction, n_2, in a polymer waveguide. Because n_2 is large and electronic in origin, the switching time scale roughly corresponds to the period of an optical oscillation. To understand the operation of such a device, it is important to characterize the linear and nonlinear optical properties of polydiacetylenes. We begin with a general discussion of the nonlinear optical processes in polydiacetylenes, then focus on the material properties important for applications in all-optical switching devices. Based on these data, we describe our work on the construction and characterization of a nonlinear directional coupler, a device formed from a pair of coupled waveguides made of poly(4BCMU). Finally, we speculate on the usefulness of such devices in all-optical switching schemes.

2. ONE-DIMENSIONAL EXCITONS IN POLYDIACETYLENES

The optical response of PDAs is dominated by a one-dimensional exciton that contains much of the oscillator strength associated with the π–π^* transition. Figure 1 shows the complex dielectric response function obtained directly by an ellipsometric technique using a freshly cleaved thick single

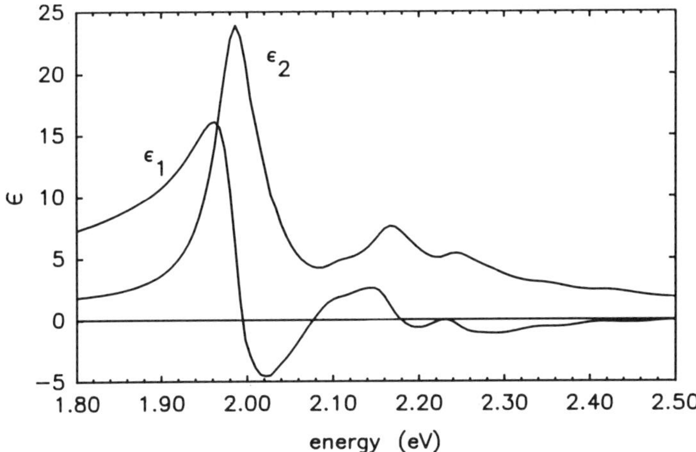

Fig. 1. The real (ε_1) and imaginary (ε_2) parts of the dielectric constant along the backbone direction in PTS. From Ref. [6].

crystal of PTS [6] (a PDA where the side groups are CH_2O-*p*-toluenesulfonate). The singlet exciton and its two strong phonon side bands are typical of crystalline PDAs and reflect the presence of a strong electron-phonon coupling constant [7]. The interband transition, corresponding to the transition to delocalized electronic states, is seen in electromodulation experiments [8] or by the onset of photoconductivity [9] at a higher energy, ~2.4 eV. This difference yields a large binding energy of ~0.4 eV, a direct result of the confinement of the exciton to one-dimensional electronic states and the resultant large overlap of the electron-hole wave function. This large overlap also shifts the oscillator strength from the interband into the exciton and its phonon side bands, since the total π–π^* transition oscillator strength must be conserved.

The width of the excitonic transition in PTS is ~60 meV, among the sharpest excitonic transitions for any conjugated polymer. However, it is broader than expected based on the ~2 psec lifetime measured by resonant bleaching of the exciton [10]. This decay is much faster than the ~1 nsec expected for a radiative transition, and nonradiative decay channel(s) common to all PDAs apparently preempt the radiative decay of excitons.

Using a broad band probe, the spectral dependence of pump-probe experiments show a uniform suppression of the excitonic absorption band [11], as schematically drawn in Fig. 2. Using a phase filling argument, the

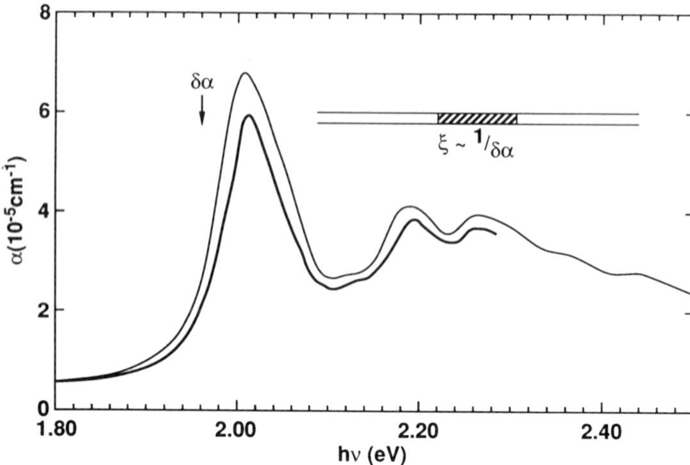

Fig. 2. Schematic description of resonant bleaching of the 1D exciton in PDAs. The inset sketches the relationship between the bleaching signal and the length of an exciton for one absorbed photon.

magnitude of the change in absorption $\int \delta\alpha$, is inversely proportional to the size of the exciton, as depicted in the inset. The basis for the phase-space-filling argument is the exclusion principle, since excitons are composed of an electron and a hole (Fermions). From the absorption coefficient sum rule, the absorption of a photon results in appearance of an exciton of length ξ, making that volume of the chain unavailable to another exciton. This formalism was successfully applied to 2D excitons in quantum wells [12]. In 1D, the exciton length is inversely proportional to the transient change in the absorption; i.e., $\xi \sim (\delta\alpha)^{-1}$. For 1D excitons in LB films of PDAs where $\delta\alpha$ was measured quantitatively, the exciton size was $\xi \sim 50 \pm 10$ Å [13]. Using ultrathin single crystals of PTS, the exciton size was found to be ~40 Å [14].[1]

Linear optical spectroscopy and resonant bleaching experiments yield information on the position, confinement, strength, and the size of 1D excitons in PDAs. Figure 3 schematically highlights the interaction of PDA excitons with off-resonant electromagnetic radiation. As in any two-level system, the presence of an intense optical field is expected to cause a Stark shift of the energy levels and a renormalization of the exciton energy. Such

[1] Earlier attempts to obtain the size from the reflection signal did not measure the exciton correctly, since $\delta R/R$ contains the change in both the real and imaginary parts of the dielectric constant. (See [14] and also Greene *et al. Phys. Rev. Lett.* **58**, 2750 [1987].)

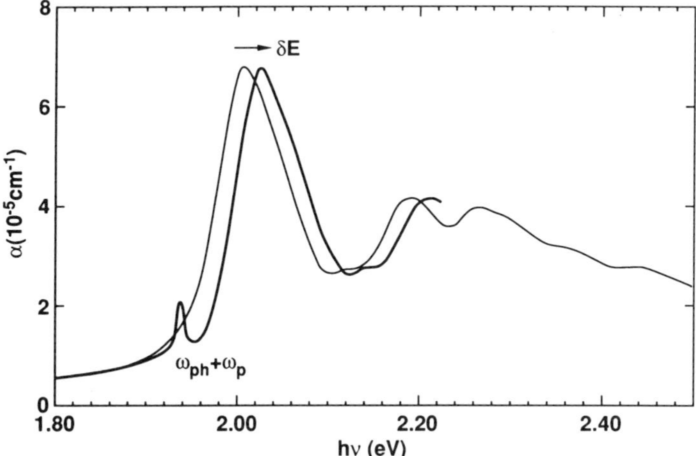

Fig. 3. Schematic description of the optical Stark shift of the 1D exciton in PDAs. The structure at ($\omega_{\mathrm{ph}} + \omega_{\mathrm{p}}$) is caused by the phonon-mediated optical Stark effect.

effects were recently observed [6]. The schematic in Fig. 3 shows that the exciton peak shifts *instantaneously* to higher energies in response to a laser pulse with an energy less than the exciton energy. The resulting optical nonlinearity is electronic in origin and large because of the large oscillator strength of the exciton. Similar results were seen in 2D excitons in AlGaAs-based quantum wells [15]. In PDAs, however, the optical nonlinearity for pumping below the optical gap is dominated by phonon-mediated processes because of the large electron-phonon coupling [6,11,16]. As sketched in Fig. 3, when the energy of a pump photon, ω_p, plus a phonon, ω_{ph}, matches the exciton absorption, the induced transient change in the dielectric response of the medium is large. Scanning ($\omega_p + \omega_{ph}$) through the exciton position from low to higher energies results in a $\delta\alpha$ that is first dispersive (see Fig. 3), then appears as a hole in the exciton absorption spectrum, and finally becomes antidispersive [6,11].

Figure 4 shows the data documenting the optical Stark effect and the phonon-mediated optical Stark effects in PTS. The dashed line is the energy of the exciton absorption peak. The zero-crossing of the data at this point corresponds to the Stark shift of the exciton absorption caused by the intense pump well below the exciton energy. As the pump photon energy is increased, the dispersive line shape of $\delta\alpha$ in Fig. 4a transforms to a hole in the absorption spectrum (see Fig. 4b). Thus, the excitonic transition in PDAs is best described by a three-level system [17].

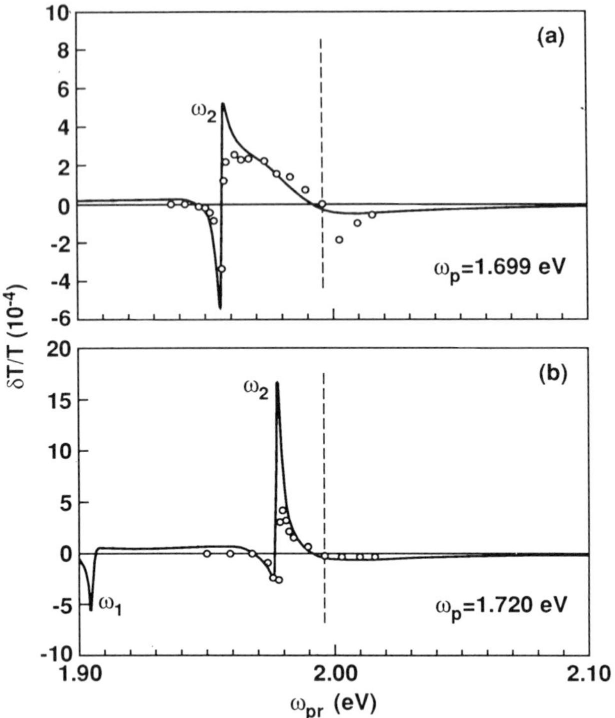

Fig. 4. The change in the absorption spectrum of the 1D exciton in an ultrathin PTS single crystal in response to the pump at ω_p. The points are the data, and the line is the calculation. The optical Stark effect, signified by the zero crossing at the exciton peak (dashed line), is dominated by the phonon-mediated nonlinearity at $\omega_2 = \omega_p + \omega_{ph}$. From Ref. [6].

3. CORRELATION EFFECTS AND ELECTRONIC STRUCTURE

That Coulomb correlations are important for determining the electronic structure of one-dimensional semiconductors has long been recognized. In artificially structured 1D covalent semiconductors based on AlGaAs, correlation effects are small because of the large diameter of the "quantum wire" compared to the size of 1D excitons. In conjugated polymers 1D excitons are confined to carbon backbones on an atomic scale and are ideal quantum wires. Because the large side groups of polydiacetylenes separate the electronically active carbon backbones, correlation effects should play a dominant role. However, the role of correlations in PDAs has been controversial for a variety of reasons including interchain screening effects [18].

To quantify the role of correlations in 1D semiconductors, we studied the linear and nonlinear optical (NLO) spectroscopy of excited states in such polymers. NLO spectroscopy and Coulomb correlations are connected since the spectra of NLO coefficients are sensitive to all excited states irrespective of their symmetry, and the interplay of the odd and even parity optical transitions are a sensitive measure of correlation effects [19].

The lowest energy one-photon accessible excited state corresponds to the optical gap. As discussed in Sec. 2, this energy in PDAs is separated by a large binding energy from the optical transitions to delocalized band states. Extreme 1D confinement and nonvanishing Coulomb correlations (for simplicity represented by the on-site repulsion U), combine to make the binding energy large. We note that estimating U from the binding energy is not simple, since it is difficult to obtain the extent of 1D confinement; however, the lowest-energy one-photon accessible state is generally not the lowest singlet excited state. For example, in finite polyenes the lowest excited state is the $2A_g$, a two-photon accessible state. Experimental observations of large Coulomb correlations were made on finite polyenes, but what is the lowest energy excited state in a conjugated polymer? As shown later, there is no unique answer to this important question.

We consider two limits, $U \gg 2t$ and $U \ll 2t$, where t is the electronic transfer integral. In the small U limit, the excitation gap is equivalent to the optical gap and is largely determined by structural effects. This is schematically shown in Fig. 5a for an idealized polydiacetylene chain with identical bond lengths for the triple and double bonds. The gap (Peierls) is $\sim 2\alpha\langle u\rangle$, where α is the electron-phonon coupling constant and $\langle u\rangle$ is an average Peierls distortion. In this limit, the odd and even parity states alternate, and the gap is a one-photon gap measured from the $1A_g$ ground state to the $1B_u$ excited state. In the large U limit, the states are correlated and cannot be described as single particle states. However, we can still characterize them in terms of one- and two-photon transitions based on their parity. The optical gap is large, $E_g \sim U$, and corresponds to a charge-transfer gap. In this limit, electrons are localized and a real space picture, rather than single-particle energy states, is more appropriate. The gap is now determined by spin excitations that for uniform spacing are spin waves of a Heisenberg 1D antiferromagnetic chain. The excitation gap of conjugated polymers appears to be a balanced competition between the electron-phonon and electron-electron interactions.

Figure 6 shows the matrix elements for the two-photon transition to discrete correlated states in three oligomers ($N = 8$ atoms) that model polyacetylene, polydiacetylene, and polysilane, respectively. These are results

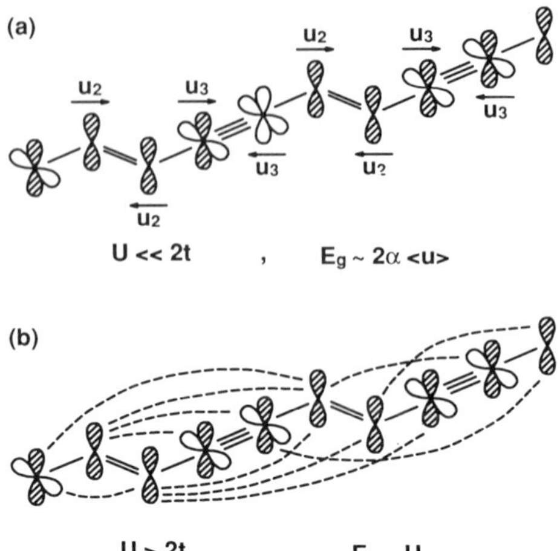

Fig. 5. Schematic drawing of π-electrons in a polydiacetylene chain: (a) the noninteracting electron model where E_g is the Peierls gap; (b) the correlated electron model where E_g is the correlation gap.

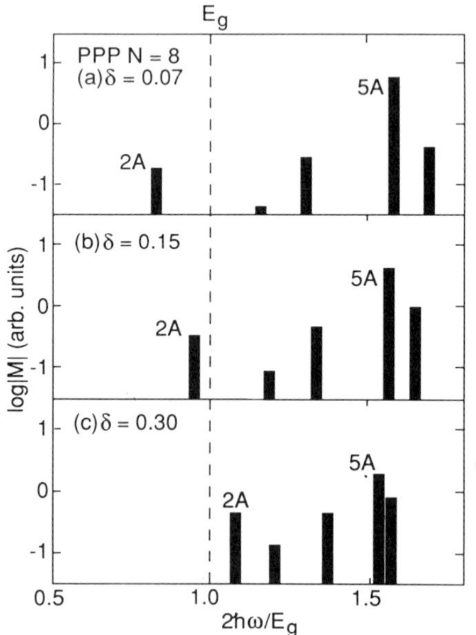

Fig. 6. Exact two-photon transition moments for all the A_g states below $1.8E_g$ in an eight-site PPP model with bond length alteration $\delta = 0.07$, 0.15, and 0.30, and otherwise unchanged polyene parameters and geometry. These δ values are appropriate for PA, PDA, and PS, respectively [20].

from an exact PPP calculation with realistic parameters for the molecular Coulomb interaction and the dimerization distortion, but because of their length, should only be considered representative of oligomers. However, the comparison of the position and strength of the associated TPA in three conjugated polymers in Fig. 7 is instructive. As the relative strength of the dimerization gap increases with respect to the correlation gap in going from polyacetylene to polydiacetylene to polysilane, the position of the $2A_g$ state rises compared to the $1B_u$ state, the optical gap. This is in stark

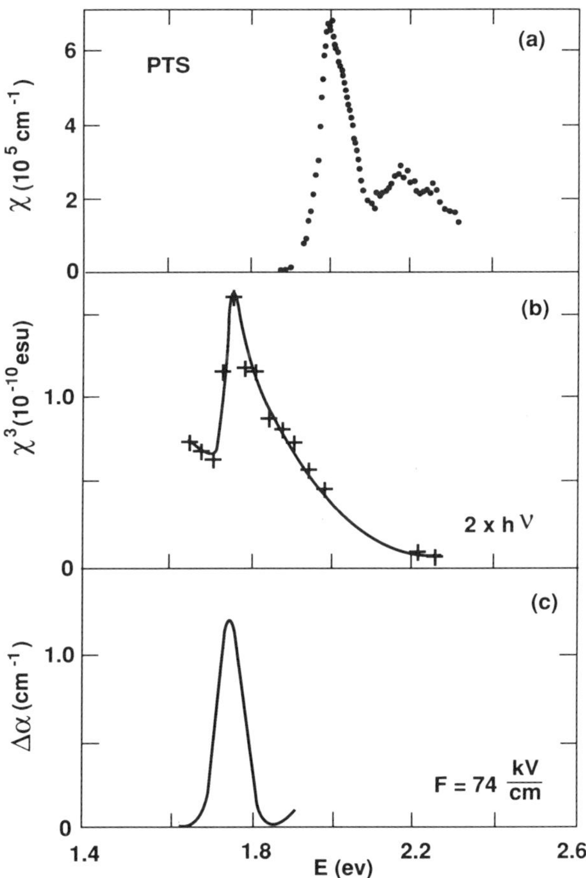

Fig. 7. (a) Linear absorption of ultra-thin PTS crystal (from Ref. [21] with permission from the publisher, Elsevier Science Publishers BV); (b) two-photon resonance enhancement in the THG spectrum of a PTS crystal; (c) electroabsorption signal below the optical gap of a PTS single crystal.

contrast to GaAs quantum wires where the free electron model is a good approximation. Thus, "geometric" knowledge of the material defined by the boundary conditions allows predictions of the position of the states using a free-electron model.

In PDAs the $2A_g$ state was observed to lie below the $1B_u$ state in at least two representative PDAs. Figure 7a shows the linear absorption obtained from a thin PTS crystal [21], and in Figure 7b we show the position of a two-photon accessible state responsible for resonance enhancement of the third harmonic generation efficiency, $\chi^{(3)}$. The assignment is confirmed by an electroabsorption peak at the same energy and with the same width. These results suggest that the lowest-energy two-photon state lies below the lowest-energy one-photon state. We come to a similar conclusion by directly comparing the one- and two-photon absorption spectra [20]. In Fig. 8 we show the spectrum of two-photon absorption with the linear absorption plotted as a function of $h\nu/2$ to compare the position of one- and two-photon accessible states. (The TPA spectrum was obtained using polymer waveguides as described in Sec. 7.) There are two significant results. First, the

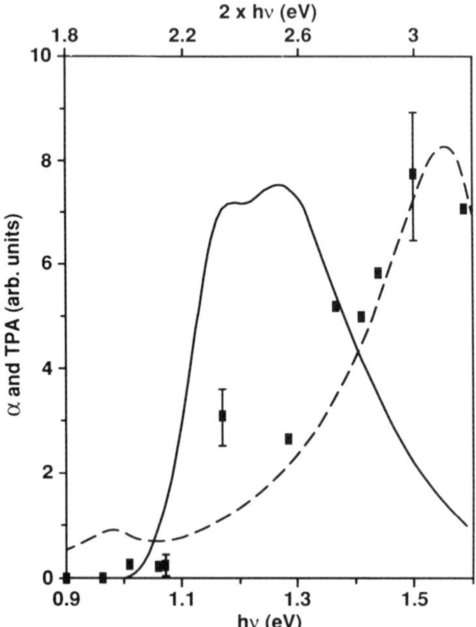

Fig. 8. The two-photon absorption spectrum (solid squares with error bars) compared with the one-photon absorption spectrum (solid line).

onset for two-photon absorption is below the one-photon absorption. At $h\nu \sim 1.0$ eV, there is a measurable TPA that exceeds the signal-to-noise ratio with no detectable linear absorption. Assuming that the broadening of the absorption edge is from disorder, the results of Fig. 8 place the covalent gap $E_c \sim 0.2$ eV below the optical gap $E_g \sim 2.2$ eV, the point of inflection in the absorption spectrum. Second, because the TPA rises for all energies above E_g, the two-photon accessible state responsible for most of the TPA oscillator strength lies at energies well above E_g.[2] Both results are inconsistent with the predictions of a free-electron model [22,23], but are well-explained by a correlated electron theory.

In the free-electron model [22,23], the TPA spectrum has an onset at $E_g/2$ and peaks sharply just above $E_g/2$. Like the one-photon absorption, the two-photon absorption is overshadowed by states near the band edge. In contrast, the TPA spectrum from a correlated electron model has an onset below $E_g/2$ corresponding to the two-photon resonance associated with the $2A_g$ state, and is in qualitative agreement with the experimental spectrum of Fig. 8. However, its oscillator strength is overshadowed by a two-photon resonance enhancement associated with a dominant A_g state well above E_g. The importance of a dominant A_g state has been noted for octatetraene, a polyene with eight carbon atoms [24,25]. We call this excited state "nA_g" since its order in the A_g states manifold changes with the total number of sites in the calculation; however, its relative position to E_g is fixed. In our calculations the nA_g state is at $\sim 1.5E_g$ for polydiacetylenes [18]. The nA_g state, which can be viewed as a biexciton on the same chain, is a correlated version of two-photon–two-electron transition that in the absence of Coulomb interactions occurs at $2E_g$. As the chain length increases, there are additional two-photon one-electron excitations in the TPA spectrum below nA_g. To fit the experimental spectrum a large life-time broadening has been applied to the TPA associated with the dominant nA_g state. It is clear from the data of Figs. 7 and 8 that the overall behavior of the experimental TPA spectrum for poly(4BCMU) is in accord with the prediction of the correlated electron model and differs from that of the free-electron model.

[2] Strong TPA above E_g was inferred from an analysis of the coherent Raman spectra in 4BCMU gels by R. R. Chance, M. K. Shand, C. Hogg, and R. Silby (*Phys. Rev.* **B22**, 3540 [1980]). Figure 8 is the spectrum of this TPA that extends to below E_g. The origin of the TPA above E_g was identified by Chance *et al.* as the $2A_g$ state, whose position in their free-electron model of finite chains would lie well above E_g. This interpretation is inconsistent with the recent report of B. E. Kohler and B. E. Schilke (*J. Chem. Phys.* **86**, 5214 [1987]) who find the $2A_g$ state below E_g in polydiacetylene oligomers.

4. THIRD-ORDER OPTICAL NONLINEARITIES IN POLYDIACETYLENES

Interest in potential device applications for the large off-resonance third-order optical susceptibilities, $\chi^{(3)}$, in conjugated polymers and a desire to obtain a consistent theoretical understanding of the optical properties of one-dimensional systems have stimulated many recent theoretical and experimental investigations. The source of the large $\chi^{(3)}$ for these semiconductors can be traced to the highly polarizable electrons, large bandwidth W, and the small optical gap, E_g, of conjugated polymers. In principle, there is no sum rule over the spectrum of the $\chi^{(3)}$ response. Furthermore, theoretical treatments predict that the off-resonance $\chi^{(3)}$ is inversely proportional to E_g to a high power and increases dramatically as E_g approaches zero. For the case of a simple but realistic Hückel Hamiltonian, it was shown that $\chi^{(3)} \approx [W/E_g]^6$ [26]. Thus, recent assertions that the ratio of $\chi^{(3)}_{\text{res}}/\alpha$ is the same for all materials may be true only for resonance conditions, and its generalization to the off-resonance condition is misleading [17].

The third-harmonic-generation efficiency has usually been the first choice for characterizing the nonlinear response of a given material because the measurement is easier than those for other NLO coefficients. In Fig. 9 we show the composite THG spectrum for a thin poly(4BCMU) film spun from

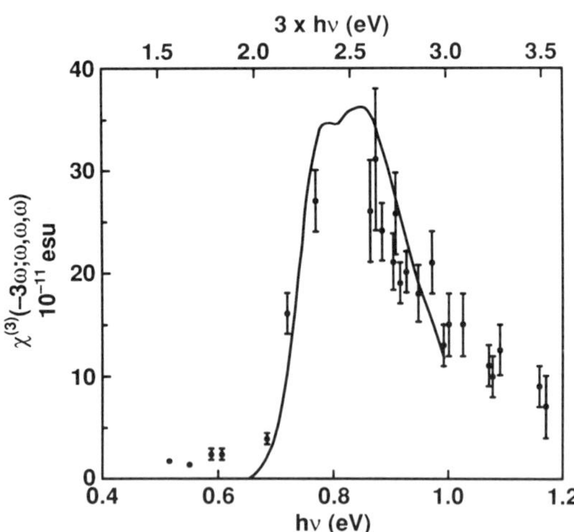

Fig. 9. The THG spectrum as a function of pump energy, $h\nu$, for poly(4BCMU) (filled circles) compared with the linear absorption spectrum plotted as a function of $h\nu/3$ (solid curve in arbitrary units). (From Ref. [27] with permission from the publisher, Elsevier Science Publishers BV.)

solution (individual data points) and compare it to the linear absorption spectrum, α, plotted as a function of $h\nu/3$ (solid line) [27]. The value of α at its peak is $\sim 10^5\ \text{cm}^{-1}$. As we noted in our previous work on polyacetylene, a detailed interpretation of THG spectra can be ambiguous, since the three-photon resonance largely dominates the spectrum [28]. For poly(4BCMU), disorder further complicates its interpretation.

The most significant data of Fig. 9 is the appearance of a true off-resonance regime at $\omega < E_g/3$ where $3 \times h\nu$ is not in resonance with any band state. Here there is no resonance enhancement of $\chi^{(3)}(-3\omega; \omega, \omega, \omega)$, and thus we can validly apply the relation $|\chi^{(3)}(-3\omega; \omega, \omega, \omega)| \sim |\chi^{(3)}(-\omega; \omega, -\omega, \omega)|$. Using the four data points that lie in the off-resonant regime, we find a value of $\chi^{(3)} = 2 \times 10^{-11}$ esu for the optical nonlinearity along the polymer backbone. These data also provide an estimate of the optical Kerr effect or n_2 for poly(4BCMU). Assuming the polymer chains lie isotropically in the plane of the substrate, we find $n_2 = (3.8 \pm 0.6) \times 10^{-14}\ \text{cm}^2/W$ for amorphous thin films [27]. This is in good agreement with the recent measurement at $h\nu = 0.9$ eV using strip-loaded channel waveguides of poly(4BCMU). The samples had a similar morphology and gave $n_2 = (4.8 \pm 2.7) \times 10^{-14}\ \text{cm}^2/W$ [29].

5. OPTICAL LOSSES AND THERMAL STABILITY IN POLYDIACETYLENES

For some time, researchers have recognized that the magnitude of n_2 is only one of many parameters that will determine the usefulness of conjugated polymers for all-optical switching applications. Just as important is a collection of physical properties that must be compatible with the high-intensity optical fields that are expected in NLO devices. Among these are optical clarity, high optical damage thresholds, and thermal stability. Measurements of the absorption losses in the near-IR are particularly important since absorbed photons cause heating and a refractive index change, an effect that competes with the electronic nonlinearity. Because the magnitude of the absorption in the near-IR is small and comparable to scattering losses for most polymers, calorimetric techniques such as Photothermal Deflection Spectroscopy (PDS) are used to measure directly the absorptive losses. For polyacetylene [30], PDS measurements reveal a shallow edge that extends far into the near-IR. Only for wavelengths less than 1.6 μm does the absorption fall below $10\ \text{cm}^{-1}$, probably the practical limit for absorptive losses in a waveguide device. Thus, scattering losses [31] and near-IR absorption render polyacetylene impractical for waveguide devices despite its having the largest optical nonlinearity of any semiconductor.

Because they scatter little light, isotropic, amorphous films of conjugated polymers are attractive nonlinear materials for waveguide applications. We found that low loss films of poly(4BCMU) can readily be spun from solutions of cyclopentanone or chlorobenzene. In contrast, poly(3BCMU) films spun from various solvents have well-defined domains that scatter light. They are too "lossy" for most NLO experiments, particularly those with pathlengths greater than a few mm. Sinclair *et al.* used PDS measurements to characterize the near-IR absorptive losses for several amorphous conjugated polymers [32]. For poly(phenylene vinylene) (PPV) and a 2,6-dialkoxy substituted PPV, the electronic absorption edge was broader and reached limiting absorptions in the near-IR of 6 and 1 cm^{-1}, respectively. For spin-cast poly(4BCMU) (Fig. 10), the electronic absorption edge is significantly sharper, and the minimum absorption is roughly constant at 1 cm^{-1} from 1 eV to 1.5 eV. The PDS spectra of the three polymers are similar below 1 eV where vibrational overtones from CH and NH stretching modes become more prominent and eventually dominate the spectrum [33]. For comparison, the crystalline polydiacetylene PTS has an absorption of 1 cm^{-1} at 1.06 μm, declining to 0.5 cm^{-1} at 1.2 μm [34]. Because of their sharp absorption edges and low near-IR losses, polydiacetylenes compare favorably to other conjugated polymers for waveguide device applications.

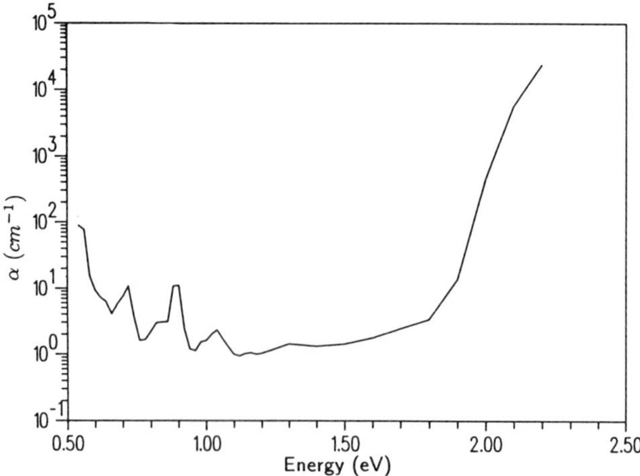

Fig. 10. PDS spectrum of spin-cast poly(4BCMU). Peaks at low energies correspond to IR overtones.

Since polymer films have small optical absorptions, it is important to understand the effects of heating on their physical and optical properties. As-spun films of poly(4BCMU) are kinetic products of the spinning process, and thus are unstable at elevated temperatures toward the formation of increased order. We found, however, that the instability can be overcome and exploited to form films with improved stability and sharper absorption edges [35]. By heating as-spun films under vacuum for an extended period, some of the disorder is lost, and the characteristic leading-edge exciton in the poly(4BCMU) absorption spectrum sharpens. The effect shown in Fig. 11 increases for annealing at higher-temperatures and then is lost for annealing temperatures above the disordering temperature for poly(4BCMU). An examination of concurrent scanning calorimetry traces shown in Fig. 12 for the same series of films shows that a broad featureless thermal transition in the as-spun film is lost while a transition near 110°C, the disordering transition, sharpens. From the combined optical and calorimetric results, we surmise that for the as-spun film, the first thermal transition is broadened because of the random alignment of chains, heating allows the chains to order, and the manifestation of the increased order is the sharpened absorption edge in the optical spectrum. Consistent with this view is the much sharper absorption edge for poly(4BCMU) gels [36]. Since gels are formed from solutions of polymer where the polymer chains have more mobility than films, it is natural to expect that gels should have much sharper edges than films.

Because of the thermal instability of unannealed films, optical experiments can show artifacts corresponding to the reorientation of polymer chains

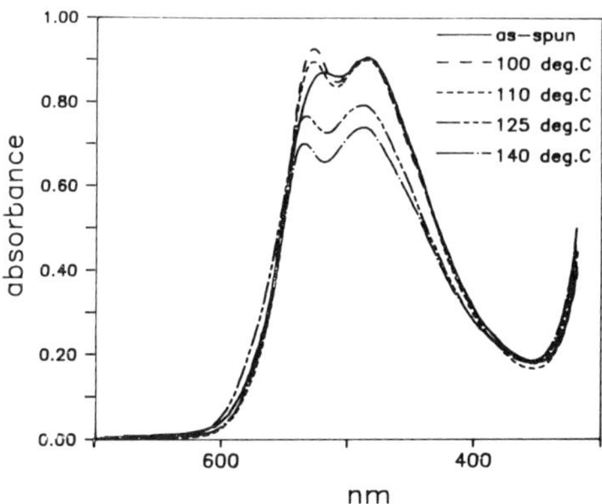

Fig. 11. Changes in the linear absorption spectrum for poly(4BCMU) films annealed for 24 hours at the indicated temperatures.

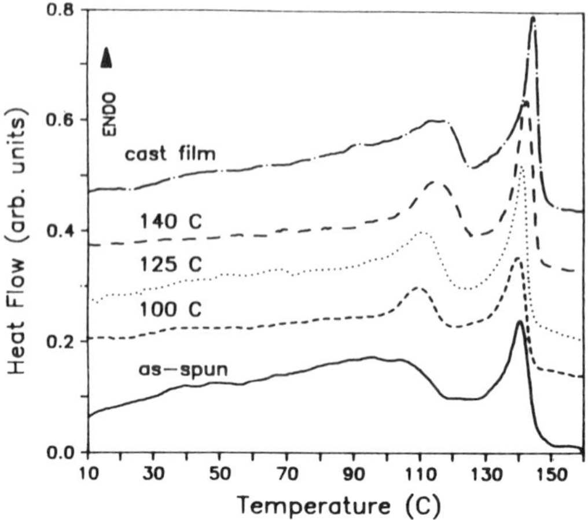

Fig. 12. Scanning calorimetry traces for the films of Fig. 7. The first transition corresponds to melting of the hydrogen-bond lattice; the second is a disordering transition.

during the experiment. For example, our early attempts to measure the nonlinear refractive index in a prism-coupled waveguiding setup (see below) yielded inconsistent results for intensity-dependent coupling angle. Later we realized that at high laser powers, the sample absorbed enough photons to heat to a few degrees above ambient temperature, allowing the chains to relax and cause a local refractive index change unrelated to the nonlinear optical properties of the material. Annealed films show no such effects.

6. NLO ACTIVE POLYMER WAVEGUIDE

Guiding measurements in a slab waveguide configuration are the most direct test for the transmission and nonlinear optical properties of the spun films. We first showed that the losses in poly(4BCMU) films were small enough to allow planar waveguiding [2], where light pulses are confined to propagate in the plane of the film but can disperse within the plane. Figure 13 shows planar guided mode propagation for a train of 1.06 μm pulses in a ~1 μm thick film of poly(4BCMU) over a distance of about 3 cm. The picture is taken with an infrared camera with light coupled in and out of the film using the prism-coupling technique. In these experiments, the coupling angle is varied, and the ratio of the output to input intensity is monitored using silicon diode photodetectors. The coupling angle provides an accurate

Fig. 13. Planar waveguiding of 1.06 μm light in poly(4BCMU) thin film. The distance between the two prisms used to couple the light in and out of the guided mode is ~3 cm [2].

measure of the dielectric constant of the thin film. For single-mode guiding, the refractive index of the film can be obtained from the coupling angle and the known film thickness. When there are two or more guided modes, both the film thickness and refractive index can be found. The TM (electric field normal to the film) and TE (electric field in the plane of the film) guided modes show different propagation properties in poly(4BCMU) spun films. We find $n_{TE} = 1.60$ and $n_{TM} = 1.53$. The propagation losses have also been measured by monitoring the decrease in the intensity as a function of distance. For the two modes we find a transmission loss of ~5 dB/cm and ~1 dB/cm for the TE and TM modes, respectively. The difference between the propagation parameters for the two modes is related to the spin-casting process. During spinning, a solvent-swollen film containing an isotropic ensemble of polymer chains dries to form a film. Because the film shrinks in thickness but not in lateral dimensions, the polymer backbones are constrained to lie primarily in the plane of the substrate. As a result, light propagating as a TE mode sees a larger index of refraction, optical nonlinearity, and transmission losses compared to TM mode. Similar orientation effects have been observed in spun films of nonconjugated polymers, although the anisotropies were smaller [37].

Prism coupling is an inconvenient technique for assessing a device in integrated optics. The preferred approach is end-fire coupling, where a microscope objective or fiber optics can be used to couple light into the polymer waveguides. To have low coupling losses by either approach, it is essential that the ends of the waveguide be optically flat, a condition usually achieved by polishing techniques. Having shown that the losses of poly(4BCMU) films were acceptable for NLO measurements in a waveguide format, we next attacked the problem of defining channel waveguide structures in polydiacetylene films by methods that would also be compatible with end-fire coupling. Several patterning techniques based on methods used in semiconductor fabrication were readily adaptable to patterning conjugated polymers. In the first technique, bilayer lithography (Fig. 14) [38], an etch match is defined in a polymer layer deposited on top of the polydiacetylene substrate. The layer, usually a silicon-containing polymer, is sensitive to light, *e*-beam, or other radiation, and, when exposed, degrades to products that can be removed selectively by an organic solvent (developer). When placed in an oxygen-reactive ion-etching environment, the patterned polymer develops a skin of SiO_x and resists further etching. Unprotected portions of the diacetylene layer have minimal etch resistance and are readily removed by the plasma. Because the oxygen ions are accelerated normal to the surface, steep vertical wall profiles are formed during etching. Thus the pattern is faithfully transferred from the top imaging layer through the polydiacetylene layer to the substrate. The etched structure can then be back-filled with polymers having appropriate refractive indices. Bilayer lithography is attractive because the process is independent of the chemical reactivity of the polydiacetylene and fails only for those polymers that form a refractory oxide like silicon-containing polymers.

Nearly all conjugated polymers are subject to photo-oxidation in varying degrees. For example, PTS surfaces degrade and become dull, but ellipsometric measurements of the surfaces of freshly cleaved single crystals are

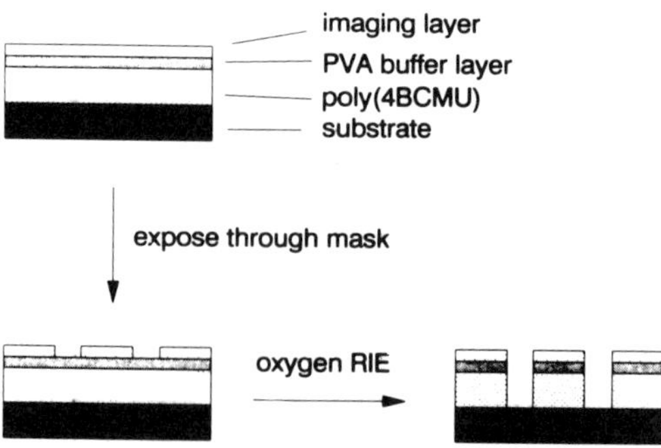

Fig. 14. The bilayer lithographic route to channel waveguides.

consistent with pristine material [6]. In contrast, amorphous films of poly(4BCMU) degrade slowly and uniformly under room lights in an ambient atmosphere, becoming increasingly more soluble in polar solvents. When significantly degraded, the polymer dissolves in acetone to give a yellow solution; pristine poly(4BCMU) is insoluble in acetone. This annoying property can be exploited if properly controlled. Deep-UV ($\lambda < 300$ nm) photolithography makes use of the sensitivity of most conjugated polymers to photo-oxidation. We [39] and others [40,41] showed that selective photo-oxidation using visible and deep-UV radiation allows the formation of gratings, waveguides, and other optical structures (Fig. 15). As implied above, exposure to deep-UV in oxygen degrades the molecular weight and makes the polymer more polar. Using *n*-butanol as a developer, the degraded polymer can be removed, leaving behind the unchanged polydiacetylene. Unfortunately, the degradation chemistry is limited to an optical absorption depth, and thus it is difficult to form structures thicker than ≈ 0.5 μm. A typical patterning step requires about 5 J/cm^2 of UV light, about a minute for conventional UV sources. Because of the lower energy per photon and the strong absorption of visible light by poly(4BCMU), exposure with visible may require as much as 100 times as much energy to define patterns. There is a large refractive-index change associated with the loss of the conjugated bonds during photo-oxidation; the refractive index of exposed regions decreases, and the pattern can be seen immediately following exposure.

A third technique for forming waveguide structures is epitaxial alignment of polymer films [42]. In this approach, a thin (<200 nm) layer of polyester or nylon is deposited on a substrate by spin casting, and then the dried polymer is rubbed with a polyester cloth. The rubbing induces alignment in

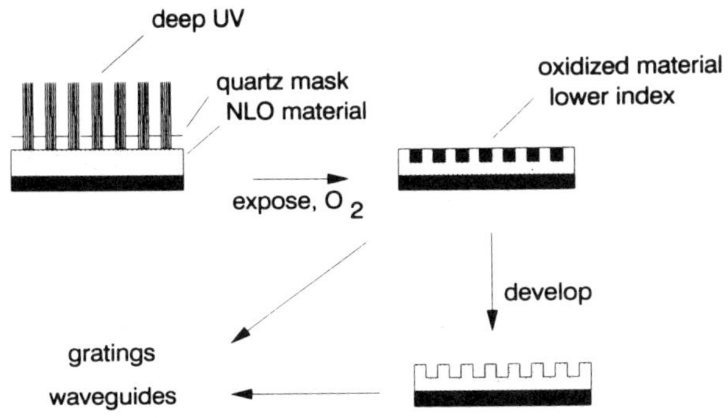

Fig. 15. The photo-oxidative route to waveguide structures.

Fig. 16. Poly(4BCMU) patterned by epitaxial alignment.

the surface of buffed films, and small molecules like liquid crystals align along the buffing direction. We found that the method could be extended to polymers. Thus, thin films of poly(4BCMU) orient along the buffing direction when heated. (See the discussion of thermal annealing, described above.) The birefringence in the near-IR caused by the alignment can be as high as 0.14. By patterning the buffed layer, the technique can be used as a route to waveguides. The alignment layer can be patterned by standard lithographic techniques. Areas where the alignment polymer was removed remain isotropic during the thermal treatment, while the remainder aligns along the buffing direction. An example is shown in Fig. 16.

The fourth, and our preferred, method for forming waveguides is through the formation of composite channel waveguides. These guides are composed of a patterned glass substrate top-coated with a layer of the NLO polymer. The polymer layer is intentionally kept too thin to support guided TM modes, so that any mode that propagates through the structure must have part of its intensity in the polymer and the remainder in the glass substrate. The high-index channels are formed by an ion-exchange process [44]. The fabrication sequence (steps I through IV) is summarized in Fig. 17. The substrates are soda-lime glass microscope slides or the optically superior B-270 Schott glass. The waveguide pattern is first defined in an aluminum mask covering the glass substrate; then the glass is immersed in a melt of

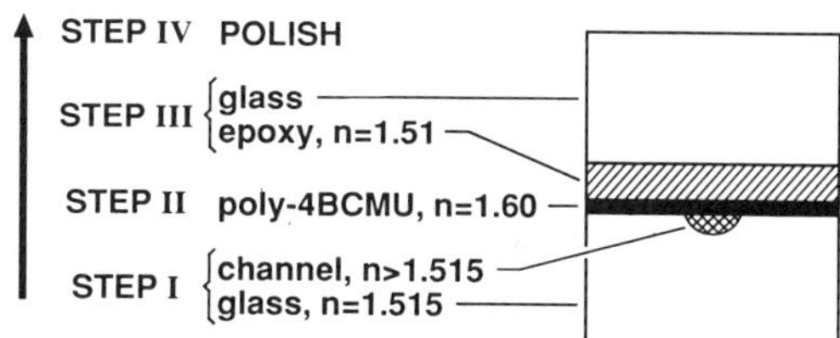

Fig. 17. The fabrication of composite NLO active polymer waveguide. (Reprinted with permission from Ref. [45].)

potassium nitrate at 375°C for one hour. Potassium ions exchange for sodium ions in the unmasked regions of the substrate, and the refractive index in the exchanged region is raised. Stripping the aluminum completes the fabrication of the patterned substrate. Alone, the channels defined by ion exchange guide light at $\lambda = 0.633$ μm, but not at wavelengths larger than 1 μm. A 0.8 to 1.0-μm-thick layer of the polymer is then spun on the substrate and baked to remove residual solvent. To increase robustness, a second glass slide is attached to the first using an optical epoxy with a refractive index less than the polymer. Finally, the sandwich structure is polished to allow end-fire coupling into the guides. The combination of a low-index soda-lime substrate ($n = 1.515$) with an epoxy of slightly lower ($n = 1.51$) index [45] prevents planar guiding in the epoxy itself. TE and TM light at $\lambda = 0.633$, 1.15, 1.32, and 1.55 μm were coupled into the guide sandwich using a single mode fiber [45]. The output face was imaged onto an infrared video camera using a microscope objective. Absorption is high for $\lambda = 0.633$ μm, and no guiding was seen for either polarization. At $\lambda = 1.15$ and 1.32 μm, both the TE and TM modes guided. We find that the TM mode is close to cutoff at 1.32 μm, and much of the light leaks into the substrate. At $\lambda = 1.55$ μm, only the TE mode guides. These observations are consistent with the prism-coupling measurements, since we expect the lower-index TM mode to be cut off at shorter wavelengths than the higher-index TE mode.

These guides function like normal rib guides [46] but are more difficult to model because the "rib" is a channel with an index gradient, rather than a fixed index region. How the composite wave guides confine light in two dimensions can be understood qualitatively by replacing the gradient index channel with a rectangular cross-section of constant index having the same "effective" index. Such guides have been modeled extensively in the literature, using numerical techniques, the simplest of which is the effective index method [47]. Depending on the relative indices of refraction, there are three possibilities for the position of the guided mode with respect to the polymer

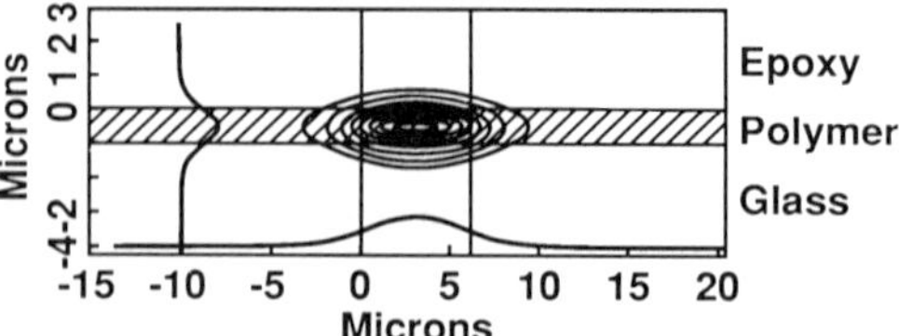

Fig. 18. Numerical simulation of the mode cross-sectional intensity for the waveguide in Fig. 7; the large polymer index pulls in the intensity of the guided mode.

and the channel. If the channel is too deep and/or the polymer has a low refractive index, the guided mode lies in the channel. Thus only the evanescent field reaches the polymer and the nonlinear effects are small. When the nonlinear polymer has an index and thickness similar to that of the channel, the power of the guided mode is equally distributed between the NLO medium and the glass, and the nonlinearity of the polymer is used more efficiently. The most favorable case occurs when the polymer refractive index is larger than the channel, and light is confined to the polymer layer. This case is relevant to our composite wave guides [45]. Figure 18 shows the results of a numerical calculation of the light intensity for a TE guided mode in a composite structure with a 6 μm wide channel and a 1 μm thick poly(4BCMU) film. Because the refractive index of the nonlinear polymer is large compared to the channel, the TE mode is primarily confined to the thin polymer film. Confinement is simply the consequence of Snell's law of refraction.

The flat aspect ratio of the mode cross-section has been verified experimentally. Using a single mode glass fiber, we coupled light into the composite waveguide and imaged the output onto an infrared camera with a microscope objective. The near field image was clearly pancake-shaped [45]. For $\lambda = 1.32$ μm, the TM mode is less well confined than the TE mode, because of the film birefringence. Accordingly, we found that the near-field TM mode output image was rounder than that of the TE mode. If light was propagating in the ion-exchanged channel, both the shape and the position of the guided mode and its polarization characteristics would be different from that observed. Thus the data analysis confirms the modeling results.

Since the light is confined almost entirely in the polymer layer, we expect propagation losses for the channels to be similar to those for planar guides; i.e., $\sim$1 dB/cm for the TM and 6 to 7 dB/cm for the TE modes. For guides a few mm long, we observed no difference between the losses for TE and TM modes. A consequence of the film birefringence, however, is excessive insertion losses. The total insertion loss in our measurements is dominated by the fiber/guide mismatch, because the highly elliptical (see (Fig. 18)) waveguide mode is a poor match to the circular fiber mode. For a typical guide 1 cm in length, we find coupling losses to be more important than propagation losses.

7. ALL-OPTICAL SWITCHING IN POLY(4BCMU)-BASED DIRECTIONAL COUPLERS

Most all-optical signal processing schemes are based on nonlinear optical phenomena. These rely on the observation that the refractive indices of materials are intensity dependent as described in Eq. (7.0.1)

$$n = n_0 + n_2 I, \tag{7.0.1}$$

where n, n_0, and n_2 are the refractive index at intensity I, the zero-intensity refractive index, and the intensity dependent refractive index coefficient, respectively. The building blocks of all-optical switching devices are waveguides with cross-sections of a few square microns suitable for confining intense electromagnetic radiation at optical frequencies over distances of about a cm. (See the previous section.)

The operation of an all-optical switching device can be understood by first considering the directional coupler shown schematically in Fig. 19 [48]. The device consists of a pair of identical waveguides coupled through the overlap of the evanescent tails of their respective guided modes. When light couples into one guide, power is transferred periodically between the two guides as the wave propagates along the device. This is analogous to the problem of two coupled pendula, where the oscillation of one sets the other in motion, and energy is repeatedly transferred between them. The period for the energy

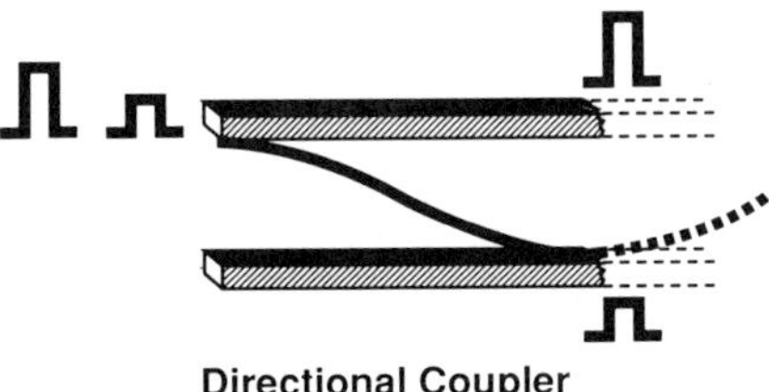

Fig. 19. Operation of a directional coupler: Pulses injected at the upper guide at low (high) intensities appear at the output of the lower (upper) guide. The switching of the outputs can be controlled over the time scale needed to change the index of refraction; i.e., inverse of the optical frequencies.

transfer depends on the strength of the coupling and how nearly identical the pendula are. Similarly in the two optical waveguides, the period of oscillation of the power between the two guides is determined by geometry (mode structure and separation of the two guides) and by an effective index of refraction (strength of coupling).

For a "tuned" device, the length of the device is adjusted so that at low light intensities, full power transfer occurs in that length. The basic requirement for operation of an all-optical device is a light-induced change in phase of $-\pi$ in the optical field. Such a change can be induced by a small change in the index of refraction through n_2. Thus the condition to be satisfied for switching is

$$n_2 \times I \times L/\lambda \approx \pi, \tag{7.0.2}$$

where L is the length of the waveguide. Note that a small change in the index of refraction, $\delta n = n_2 \times I$, is amplified through the interaction length by L/λ. At high intensities, the effective index of refraction of the first guide changes according to $\delta n = n_2 I$, and the two waveguides are no longer identical; periodic power transfer is frustrated. As a result, the output switches between the two guides, depending on the peak power of the input pulse. Since what determines switching is an *electronic* n_2, the switching operation can be as fast as the inverse optical transition rate; i.e., faster than 10^{-14} seconds. The operation of an all-optical switching device has been modeled and demonstrated recently using glass dual-cored optical fiber [49]. This first demonstration of all-optical switching owes its success to the small transmission losses associated with fiber optics [49]. The losses in conjugated polymers are much larger, and only recently have low-loss waveguide structures been developed in polymers [4]. Since the optical nonlinearity in conjugated polymers is orders of magnitude larger than that of glass (Sec. 3), a conjugated-polymer-based device should operate at substantially lower power levels, comparable to those obtainable from diode lasers (see Eq. 7.0.2).

Figure 20 shows a series of directional couplers formed by the composite channel waveguide approach. Each substrate has more than 100 directional couplers with different coupling strengths. The waveguide pairs are coupled by being close ($\sim\lambda$) to each other, with the coupling changing with changes in the separation of the guides. Altering the underlay channel width also affects the coupling and guided mode configuration, but changing the separation between the guides has the most dramatic effect. Since the directional couplers are identical in length, L, "tuning" is simply the selection of a set of couplers with complete power transfer from one to the other guide at low power. Figure 21 shows the output of individual guides for a "tuned"

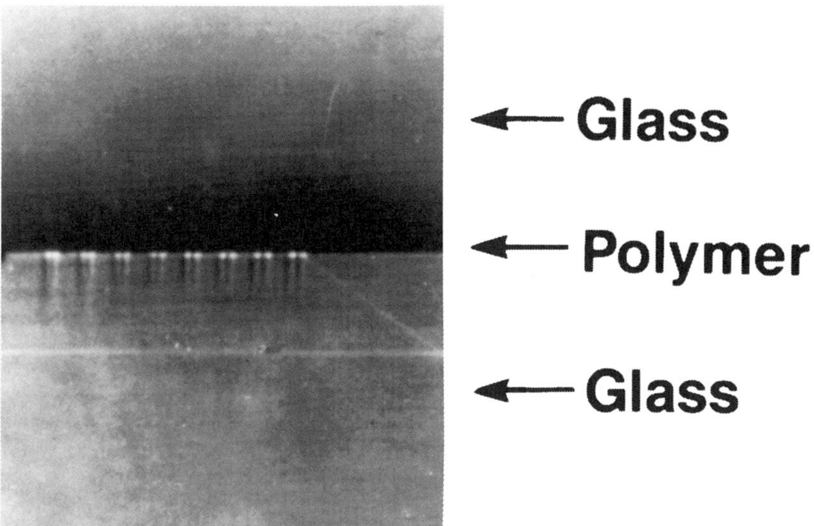

Fig. 20. An optical microscope picture of a series of directional couplers. The different spacings between the waveguides allows variations in the coupling of the two guided modes.

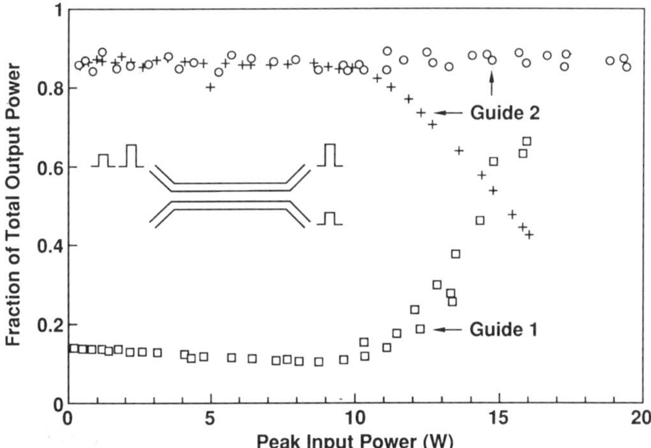

Fig. 21. Intensity-dependent switching caused by thermally induced changes in the refractive index. Note the absence of switching when the repetition rate of the input pulses is changed from 82 MHz (+'s and □'s) to 500 Hz (○'s).

directional coupler as a function of peak input power. The experiment used 5 psec pulses at 1.06 μm with a repetition rate of 82 MHz. The inset outlines the ideal switching behavior expected for a directional coupler. The data (pluses) show a well-defined switching effect, but unfortunately it is caused by a slow thermally induced n_2. That the switching was thermal in nature can be seen from the data (open circles) obtained at the *same peak power* but at a repetition rate of 500 Hz. The slower repetition rate decreases the accumulated thermal load by $\sim 10^5$, and the switching effect is lost. Thus, thermally induced changes dominate the electronic n_2 for experiments under these conditions. Switching behavior caused by the electronic nonlinearity can be seen at much higher peak powers. Figure 22 shows data from two guides obtained by using a low repetition rate (500 Hz) Q-switched laser system operating at $\lambda = 1.06$ μm. The data is plotted as the fraction of the total output power as a function of input power, and its shape suggests switching. Unfortunately, the electronic nonlinearity is dominated by two-photon absorption (TPA) at this wavelength. The characteristics of two-photon absorption can be seen as a decrease in total transmission with increasing input peak power (Fig. 23), and the linear dependence of T^{-1} with input power depicted in Fig. 22. An analysis of the data yields a TPA

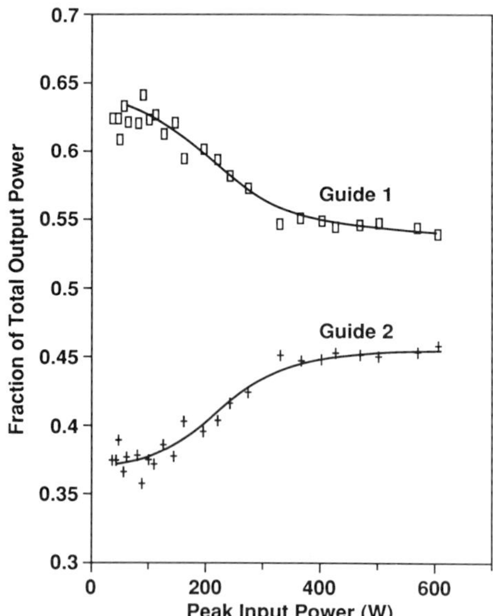

Fig. 22. Intensity-dependent switching resulting from the imaginary part of the ultrafast *electronic* nonlinearity.

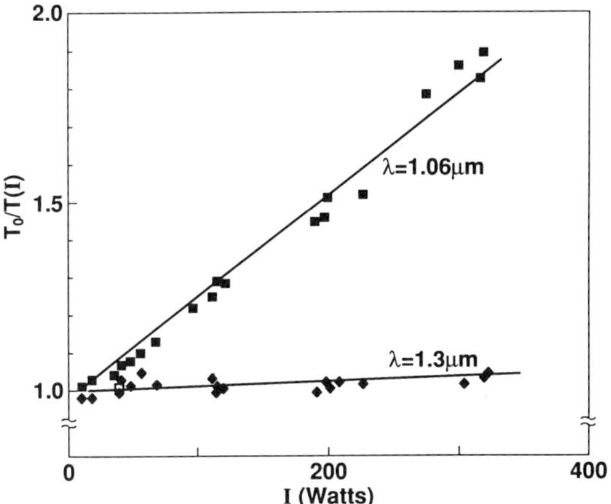

Fig. 23. Normalized total transmission plotted as T^{-1} versus peak input power for two wavelengths. Note the absence of two-photon absorption for $\lambda = 1.3$ μm.

absorption coefficient of $\alpha_2 = 1.3 \times 10^{-9}$ cm/W at 1.06 μm. For wavelengths longer than 1.15 μm, we find [27] that the two-photon absorption cross-section is less than our detection limit of 10^{-10} cm/W.

Further confirmation of two-photon absorption at 1.06 μm as the cause of the apparent switching was obtained from a pump-probe technique designed to monitor the speed of the nonlinearity at high peak powers. When the pump and probe pulses are overlapped [4], there is an "instantaneous" drop in the transmission within the 30 psec cross-correlation time of the laser pulse. TPA at $\lambda = 1.06$ μm ($h\nu = 1.17$ eV) corresponds to a parity conserving transition to an even parity state(s) at $h\nu = 2.34$ eV. This state is apparently degenerate in energy with odd parity state(s) responsible for the first peak observed in the linear absorption spectrum (see Fig. 9). The intensity of this low-energy peak differs in poly(4BCMU) samples with different morphologies and is sensitive to disorder [35]. Thus, the magnitude of TPA at 1.06 μm in solution-cast films is larger than that of poly(4BCMU) spun films [50]. This apparent discrepancy can be traced to the different degrees of disorder in poly(4BCMU) films prepared by different methods. As noted in the discussion of thermal annealing effects, the most disordered films (spin-cast from solution) have the weakest leading-edge exciton in the linear-absorption spectrum. Those films prepared from gels or by simple casting from solution have a more intense exciton absorption. Similarly, the TPA spectrum from the spun films [27] is broad and shifted slightly to higher energies compared with that from solution cast films [50], a result consistent with disorder.

8. PROSPECTS FOR TECHNOLOGY

The TPA spectrum in poly(4BCMU) waveguides helps define the operating window for an all-optical switch based on this material. While TPA dominates at 1.06 μm, data show that TPA is less than 10^{-10} cm/W for $\lambda > 1.15$ μm irrespective of the polydiacetylene used to form the waveguides [4,5,29]. For the telecommunications industry, 1.3 μm is an important wavelength since it is a low-loss window for glass optical fibers. Can *useful, ultrafast, all-optical switching* be attained in a PDA-based device such as our directional coupler?

To answer such a question we need to analyze both the linear and nonlinear transmission losses of the switch. If we assume a not-yet achieved insertion loss as low as that of glass fiber waveguides, the linear transmission loss for such a switch is small. This has been confirmed through loss measurements by us [2] and by others [3,5]. The analysis of the nonlinear loss is more complicated, but Eq. (8.0.1),

$$(\alpha_2/n_2) \times \lambda < 1, \tag{8.0.1}$$

where α_2 is the TPA cross section must be satisfied. In Fig. 24 we plot data from the TPA spectrum (solid squares) together with the effective n_2 for

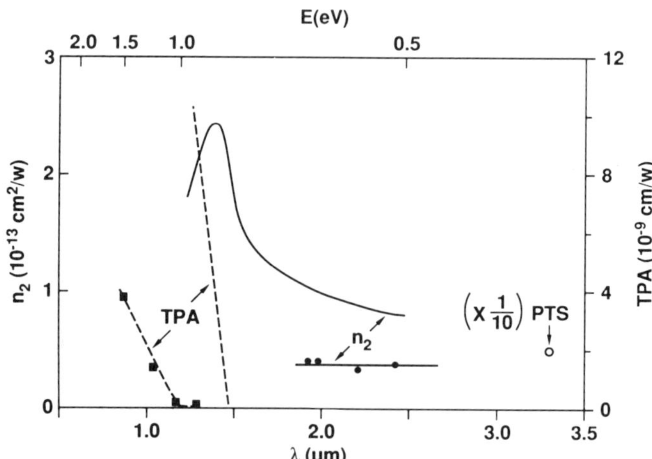

Fig. 24. Comparison of n_2 (solid curves) and TPA (dashed curves) of polydiacetylenes and bandgap engineered GaAl:As-based materials. The plots with no data points are for $Al_{0.18}Ga_{0.82}$:As [51], and ● and ■ are the n_2 and TPA for 4BCMU [27]. The off-resonance n_2 of PTS-PDA crystal is shown multiplied by 1/10 for comparison [52].

poly(4BCMU) waveguides (solid circles). For comparison, we include data for a AlGaAs based material. Using $n_2 \sim 4 \times 10^{-14}$ (cm^2/W) and $\alpha_2 < 10^{-10}$ (cm/W) from Fig. 24, Eq. (8.0.1) is satisfied for $\lambda > 1.15$ μm.

A useful all-optical switch must also be capable of operation at a high repetition rate. Thus the device design must ensure that accumulated heat does not generate a thermal n_2 larger than the electronic n_2. This challenging task has been addressed recently [53]. Since the thermal n_2 scales with τ/τ_{th}, where τ is the pulse duration and τ_{th} is the thermal relaxation time constant, successful ultrafast all-optical switching requires operation at the same peak power but with short pulses. Based on a thermal $n_2 \sim$ of 8×10^{-17} cm^2/W for strip-loaded poly(4BCMU)-based waveguides [24], only repetition rates less than 10^5 Hz are possible using 60 psec pulses. With $\sim$200 fsec pulses from a state-of-the-art mode-locked diode laser,[3] the repetition rate will be limited by that of the operating laser.

Finally, we compare the NLO properties of PDAs with competing materials for all-optical switching devices. First we note that the demonstration of all-optical switching in dual-core glass fiber optics [49] relied not only on a large n_2/α ratio, but also took advantage of the high peak power from a large laser system and a long ($\sim$200 cm) length of fiber. With the technology described here, poly(4BCMU)-based directional couplers of reasonable length ($<$1 cm) are capable of ultrafast all-optical switching in the laboratory environment using mode-locked diode lasers. The apparent competition to conjugated polymers is the bandgap-engineered GaAlAs based materials. Recent demonstrations of all-optical switching effects in $Al_{0.18}Ga_{0.82}As$-based directional couplers prompts us to compare these two classes of materials [54]. The main advantage of bandgap engineering is the ability to increase the optical gap and shift the onset for TPA. Enhancing n_2 by tuning $E_g/2$ to just above $h\nu$, the available pump photon energy, does not address problems of TPA that face both semiconductors and conjugated polymers.

In Fig. 24 we plot the theoretical n_2 and TPA spectra for the $Al_{0.18}Ga_{0.82}As$-based waveguides obtained by modeling AlGa:As materials [55]. A value of $n_2 \sim 0.9 \times 10^{-13}$ cm^2/W, but with a finite TPA, has also been measured at a photon energy 60 meV below the $E_g/2$ of 2D multiple quantum wells of similar composition [56]. Recent measurements of n_2 using bulk AlGaAs waveguides are in reasonable agreement with the plots shown

[3] P. Delfyett *et al.* (preprint, 1991), has recently reported construction of solid state diode lasers with 400 MHz, $\sim$200 fsec pulse duration and 10 mW average power ($>10^0$ W peak power).

in Fig. 24 [55]. In the off-resonance regime, n_2 in poly(4BCMU) is comparable to that obtained from the AlGaAs modeling results. Remarkably, while poly(4BCMU) films have small third-order optical nonlinearities compared to many other π-conjugated polymers, they have a comparable off-resonance n_2 to AlGa:As-based systems, with either 3D or 2D excitons.

Further improvements in performance are expected from switching from the disordered poly(4BCMU) to PDA single crystals. For example, off-resonance measurements of the $\chi^{(3)}$ of PTS crystals at $\lambda \sim 3.3$ µm using an infrared-free electron laser give $\chi^{(3)} = (15 \pm 2)10^{11}$ esu. This leads to $n_2 \sim 5 \times 10^{-13}$ for polarization along the polymer backbone in PTS. This off-resonance value of n_2 is also shown in Fig. 24 for comparison. The maximum nonlinearity possible with the polydiacetylene backbone is considerably larger because the large side groups of PTS translate into a low concentration of the π-conjugated backbone in the unit cell.

All-optical signal processing offers several advantages for computing and communications. For applications that require the inherent parallel processing capabilities of optics, the superiority of optics is clear. Because of the large bandwidth associated with optical frequencies ($\sim 10^{14}$ Hz), most information transmission will be carried out by optics, and all-optical switching would allow transmission and some processing of information without needing to convert from optical information to electrical pulses. We note, however, that the nonlinear direction coupler described in this contribution processes information *serially*, but at a speed that cannot be matched by electronics. To exploit fully the large bandwidth provided by optical fiber requires a marriage such ultrafast switches and wavelength division multiplexing (WDM). The prospect of more than 100 channels each operating at a THz would radically change our ways of transmitting, processing, and using information.

ACKNOWLEDGMENTS

The work described was carried out in collaboration with G. Blanchard, W. S. Fann, J. P. Heritage, J. Jackel, and P. D. Townsend.

REFERENCES

1. M. Thakur and S. Meyler, *Macromol.* **18**, 2341 (1985).
2. P. D. Townsend, G. L. Baker, N. E. Schlotter, C. F. Klausner, and S. Etemad, *App. Phys. Lett.* **53**, 1782 (1988).

3. W. Krug, E. Miao, M. Derstine, and J. Valera, *J. Opt. Soc. Am.* **B6**, 726 (1989).
4. P. D. Townsend, J. Jackel, G. L. Baker, J. Shelburne III, and S. Etemad, *App. Phys. Lett.* **55**, 1829 (1989).
5. M. Thakur and D. M. Krol, *Appl. Phys. Lett.* **56**, 1213 (1989).
6. G. L. Blanchard, J. P. Heritage, A. C. Von Lehmen, M. K. Kelly, G. L. Baker, and S. Etemad, *Phys. Rev. Lett.* **63**, 890 (1989).
7. D. N. Batchelder and D. Bloor, *J. Phys.* **C15**, 3005 (1982).
8. L. Sebastian and G. Weiser, *Phys. Rev. Lett.* **46**, 1156 (1981).
9. K. Lochner, B. Reimer, and H. Bassler, *Phys. Stat. Sol.* **b76**, 533 (1976); R. Chance and R. H. Baughman, *J. Chem. Phys.* **64**, 3889 (1976).
10. G. M. Carter, J. V. Hryniewicz, M. K. Thakur, Y. J. Chen, and S. E. Meyler, *App. Phys. Lett.* **49**, 998 (1986).
11. B. I. Greene, J. F. Mueller, J. Orenstein, D. H. Rapkine, S. Schmitt-Rink, and M. Thakur, *Phys. Rev. Lett.* **61**, 325 (1988).
12. S. Schmitt-Rink, D. S. Chemla, and D. A. B. Miller, *Phys. Rev.* **B32**, 6601 (1985).
13. F. Kajzar, L. Rothberg, S. Etemad, P. A. Chollet, D. Grec, A. Boudet, and T. Jedju, *Thin Solid Films* **160**, 1 (1988).
14. B. I. Greene, J. Orenstein, M. Thakur, and D. Rapkine, *MRS Symp. Proc.* **109**, 159 (1988).
15. A. Von Lehmen, D. S. Chemla, J. E. Zucker, and J. P. Heritage, *Opt. Lett.* **11**, 609 (1986).
16. M. Yoshizawa, M. Taiji, and T. Kobayashi, *IEEE J. Quant. Electron.* **25**, 2532 (1989).
17. B. I. Greene, J. Orenstein, and S. Schmitt-Rink, *Science* **247**, 679 (1990).
18. Z. G. Soos, G. W. Hayden, P. C. M. McWilliams, and S. Etemad, *J. Chem. Phys.* (Nov. 15, 1990).
19. S. Etemad and Z. G. Soos, in *Spectroscopy of Advanced Materials*, R. G. H. Clark and R. E. Hester, eds. (J. Wiley, New York, 1991).
20. P. C. M. McWilliams, G. W. Hayden, and Z. G. Soos, *Phys. Rev.* **B43**, 9777 (1991).
21. G. J. Blanchard, J. P. Heritage, G. L. Baker, and S. Etemad, *Chem. Phys. Lett.* **158**, 329 (1989).
22. G. P. Agrawal, C. Cojan, and C. Flytzanis, *Phys. Rev.* **B17**, 776 (1985).
23. Weikang Wu, *Phys. Rev. Lett.* **61**, 1119 (1988).
24. J. R. Heflin, K. Y. Wong, O. Zamani-Khamiri, and A. F. Garito, *Phys. Rev.* **B38**, 1573 (1988).
25. S. N. Dixit, B. Guo, and S. Mazumdar (unpublished).
26. G. P. Agrawal, C. Cojan, and C. Flytzanis, *Phys. Rev.* **17**, 776 (1985).
27. P. D. Townsend, W.-S. Fann, S. Etemad, G. L. Baker, Z. G. Soos, and P. C. M. McWilliams, *Chem. Phys. Lett.* **180**, 485 (1991).
28. W.-S. Fan, S. Benson, J. M. J. Madey, S. Etemad, G. L. Baker, and F. Kajzar, *Phys. Rev. Lett.* **62**, 1492 (1989).
29. K. Rochford, R. Zanoni, G. I. Stegeman, W. Krug, E. Miao, and M. W. Breanek, *Appl. Phys. Lett.* **58**, 13 (1991).

30. B. R. Weinberger, C. B. Roxlo, S. Etemad, G. L. Baker, and J. Orenstein, *Phys. Rev. Lett.* **53**, 86 (1984).
31. S. Etemad, G. L. Baker, D. Jaye, F. Kajzar, and J. Messier, *Proc. SPIE* **682**, 44 (1986).
32. M. Sinclair, C. H. Seager, D. McBranch, A. J. Heeger, and G. L. Baker, *Proc. Mat. Res. Soc.* (in press); C. H. Seager, M. Sinclair, D. McBranch, A. J. Heeger, and G. L. Baker, *Synth. Met.* (in press).
33. T. Kaino, K. Jinguji, and S. Nara, *Appl. Phys. Lett.* **41**, 802 (1982).
34. M. Thakur, R. C. Frye, and B. I. Greene, *Appl. Phys. Lett.* **56**, 1187 (1990).
35. G. L. Baker, J. A. Shelburne III, and P. D. Townsend, *Material for Nonlinear and Electrooptics, Inst. of Phys. Conf. Proc.* **103**, 227 (1989).
36. J. M. Nunzi and D. Grec, *J. Appl. Phys.* **62**, 2198 (1987).
37. J. D. Swalen, M. Tacke, R. Santo, and J. Fisher, *Optics Commun.* **18**, 387 (1976).
38. G. L. Baker, C. F. Klausner, J. A. Shelburne III, N. E. Schlotter, and J. L. Jackel, *Synth. Met.* **28**, D639 (1989).
39. P. D. Townsend, G. L. Baker, J. L. Jackel, J. A. Shelburne III, and S. Etemad, *Proc. SPIE* **1147**, 256 (1990).
40. K. B. Rochford, R. Zanoni, Q. Gong, and G. I. Stegeman, *Appl. Phys. Lett.* **55**, 1161 (1989).
41. G. Wegner, R. J. Leyrer, and M. A. Muller, German Patent DE-OS 3346716.
42. J. S. Patel, S.-D. Lee, G. L. Baker, and J. A. Shelburne III, *Appl. Phys. Lett.* **56**, 131 (1990).
43. N. E. Schlotter, J. L. Jackel, P. D. Townsend, and G. L. Baker, *Appl. Phys. Lett.* **56**, 13 (1990); G. L. Baker, J. L. Jackel, and N. E. Schlotter, US Patent 4,834,480 (1989).
44. G. L. Yip and J. Albert, *Optics Lett.* **10**, 151 (1985).
45. J. L. Jackel, N. E. Schlotter, P. D. Townsend, G. L. Baker, and S. Etemad, *Proc. SPIE* **971**, 239 (1988).
46. J. E. Goell, *Appl. Opt.* **12**, 2797 (1973).
47. V. Ramaswamy, *Bell Syst. Tech. J.* **53**, 697 (1974).
48. S. M. Jensen, *IEEE J. Quantum Electron.* **QE-18**, 1580 (1982).
49. S. R. Friberg, Y. Silberberg, M. K. Oliver, M. J. Andrejco, M. A. Saifi, and P. W. Smith, *Appl. Phys. Lett.* **51**, 15 (1987).
50. W. E. Torruelas, K. B. Rochford, R. Zanoni, S. Arkakaki, and G. Stegeman, *Opt. Commun.* **82**, 94 (1991).
51. G. L. Baker, S. Etemad, and F. Kajzar, *Proc. SPIE* **824**, 102 (1987).
52. L. Sebastian and G. Wieser, *Chem. Phys.* **62**, 447 (1981).
53. G. I. Stegeman, R. Zanoni, and C. T. Seaton, *MRS Symp. Proc.* **109**, 53 (1988).
54. J. S. Aitchinson, A. H. Kean, C. N. Ironside, A. Villeneuve, and G. I. Stegeman, *Electronics Letters* **27** (Sept. 12, 1991).
55. M. Sheik-Bahae, D. C. Hutchings, D. J. Hagan, and E. W. Van Stryland, *IEEE J.* **QE-27**, 1296 (1991).

56. M. N. Islam *et al.*, post-deadline paper in *NLO Guided Wave Phenomena*, Cambridge, UK, Sept. 2–4, 1991. Similar results are also reported by H. K. Tsang *et al.*; *ibid.*
57. J. S. Aitchinson, A. H. Kean, C. N. Ironside, A. Villeneuve, and G. I. Stegeman, post deadline paper in *NLO Guided Wave Phenomena*, Cambridge, UK, Sept. 2–4, 1991.
58. W.-S. Fann, Ph.D. Thesis, Stanford University (1990).

INDEX

N

Z

Quantum Electronics—Principles and Applications

Edited by Paul F. Liao, *Bell Communications Research Inc., Red Bank, New Jersey*
Paul L. Kelley, *Electro-Optics Technology Center, Tufts University, Medford, Massachusetts*

N. S. Kapany and J. J. Burke, *Optical Waveguides*
Dietrich Marcuse, *Theory of Dielectric Optical Waveguides*
Benjamin Chu, *Laser Light Scattering*
Bruno Crosignani, Paolo DiPorto and Mano Bertoltti, *Statistical Properties of Scattered Light*
John D. Anderson, Jr., *Gasdynamic Lasers, An Introduction*
W.W. Duly, *CO_2 Lasers: Effects and Applications*
Henry Kressel and J.K. Butler, *Semiconductor Lasers and Heterojunction LEDs*
H.C. Casey and M.B. Panish, *Heterostructure Lasers: Part A. Fundamental Principles, Part B. Materials and Operating Characteristics*
Robert K. Erf. editor, *Speckle Metrology*
Marc D. Levenson, *Introduction to Nonlinear Laser Spectroscopy*
David S. Kliger, editor, *Ultrasensitive Laser Spectroscopy*
Robert A. Fisher, editor, *Optical Phase Conjugation*
John F. Reintjes, *Nonlinear Optical Parametric Processes in Liquids and Gases*
S.H. Lin, Y. Fujimura, H.J. Neusser and E. W. Schlag, *Multiphoton Spectroscopy of Molecules*
Hyatt M. Gibbs, *Optical Bistability: Controllong Light with Light*
D.S. Chemla and J. Zyss, editors, *Nonlinear Optical Properties of Organic Molecules and Crystals, Volume 1, Volume 2*
Marc D. Levenson and Saturo Kano, *Introduction to Nonlinear Laser Spectroscopy, Revised Edition*
Govind P. Agrawal, *Nonlinear Fiber Optics*
F.J. Duarte and Lloyd W. Hillman, editors, *Dye Laser Principles: With Applications*
Dietrich Marcuse, *Theory of Dielectric Optical Waveguides, 2nd Edition*
Govind P. Agrawal and Robert W. Boyd, editors, *Contemporary Nonlinear Optics*
Peter S. Zory, Jr., editor, *Quantum Well Lasers*
Gary A. Evans and Jacob M. Hammer, editors, *Surface Emitting Semiconductor Lasers and Arrays*
John E. Midwinter, editor, *Photonics in Switching, Volume I, Background and Components*
John E. Midwinter, editor, *Photonics in Switching, Volume II, Systems*
William K. Burns, editor, *Optical Fiber Rotation Sensing*
Joseph Zyss, editor, *Molecular Nonlinear Optics*

Yoh-Han Pao, *Case Western Reserve University, Cleveland, Ohio,* Founding Editor
1972–1979